Probability and Mathematical Statistics (Continued)

LEHMANN • Testing Statistical Hypotheses

LEHMANN • Theory of Point Estimation

MATTHES, KERSTAN, and MECKE • Infinitely Divisible Point Processes

MUIRHEAD • Aspects of Multivariate Statistical Theory

PRESS • Bayesian Statistics: Principles, Models, and Applications

PURI and SEN • Nonparametric Methods in General Linear Models

PURI and SEN • Nonparametric Methods in Multivariate Analysis

PURI, VILAPLANA, and WERTZ • New Perspectives in Theoretical and Applied Statistics

RANDLES and WOLFE • Introduction to the Theory of Nonparametric Statistics

RAO • Linear Statistical Inference and Its Applications, Second Edition

RAO • Real and Stochastic Analysis

RAO and SEDRANSK • W.G. Cochran's Impact on Statistics

RAO • Asymptotic Theory of Statistical Inference

ROBERTSON, WRIGHT and DYKSTRA • Order Restricted Statistical Inference

ROGERS and WILLIAMS • Diffusions, Markov Processes, and Martingales, Volume II: Îto Calculus

ROHATGI • An Introduction to Probability Theory and Mathematical Statistics

ROHATGI • Statistical Inference

ROSS • Stochastic Processes

RUBINSTEIN • Simulation and The Monte Carlo Method

RUZSA and SZEKELY • Algebraic Probability Theory

SCHEFFE • The Analysis of Variance

SEBER • Linear Regression Analysis

SEBER • Multivariate Observations

SEBER and WILD • Nonlinear Regression

SEN • Sequential Nonparametrics: Invariance Principles and Statistical Inference

SERFLING • Approximation Theorems of Mathematical Statistics

SHORACK and WELLNER • Empirical Processes with Applications to Statistics

STOYANOV • Counterexamples in Probability

Applied Probability and Statistics

ABRAHAM and LEDOLTER • Statistical Methods for Forecasting

AGRESTI • Analysis of Ordinal Categorical Data

AICKIN • Linear Statistical Analysis of Discrete Data

ANDERSON and LOYNES • The Teaching of Practical Statistics

ANDERSON, AUQUIER, HAUCK, OAKES, VANDAELE, and WEISBERG • Statistical Methods for Comparative Studies

ARTHANARI and DODGE • Mathematical Programming in Statistics

ASMUSSEN • Applied Probability and Queues

BAILEY • The Elements of Stochastic Processes with Applications to the Natural Sciences

BARNETT • Interpreting Multivariate Data

BARNETT and LEWIS • Outliers in Statistical Data, *Second Edition*

BARTHOLOMEW • Stochastic Models for Social Processes, *Third Edition*

BARTHOLOMEW and FORBES • Statistical Techniques for Manpower Planning

BATES and WATTS • Nonlinear Regression Analysis and Its Applications

BECK and ARNOLD • Parameter Estimation in Engineering and Science

BELSLEY, KUH, and WELSCH • Regression Diagnostics: Identifying Influential Data and Sources of Collinearity

BHAT • Elements of Applied Stochastic Processes

BLOOMFIELD • Fourier Analysis of Time Series

BOLLEN • Structural Equations with Latent Variables

BOX • R. A. Fisher, The Life of a Scientist

BOX and DRAPER • Empirical Model-Building

BOX and DRAPER • Evolutionary Operation: Process Improvement

BOX, HUNTER, and HUNTER • Statistics for Experimenters: An Introduction to Design, Data Analysis, and Model Building
BROWN and HOLLANDER • Statistics: A Biomedical Introduction
BUNKE and BUNKE • Statistical Inference in Linear Models, Volume I
BUNKE and BUNKE • Nonlinear Regression, Functional Relations and Robust Methods: Statistical Methods of Model Building
CHAMBERS • Computational Methods for Data Analysis
CHATTERJEE and HADI • Sensitivity Analysis in Linear Regression
CHATTERJEE and PRICE • Regression Analysis by Example
CHOW • Econometric Analysis by Control Methods
CLARKE and DISNEY • Probability and Random Processes: A First Course with Applications, *Second Edition*
COCHRAN • Sampling Techniques, *Third Edition*
COCHRAN and COX • Experimental Designs, *Second Edition*
CONOVER • Practical Nonparametric Statistics, *Second Edition*
CONOVER and IMAN • Introduction to Modern Business Statistics
CORNELL • Experiments with Mixtures: Designs, Models and The Analysis of Mixture Data
COX • Planning of Experiments
COX • A Handbook of Introductory Statistical Methods
DANIEL • Biostatistics: A Foundation for Analysis in the Health Sciences, *Fourth Edition*
DANIEL • Applications of Statistics to Industrial Experimentation
DANIEL and WOOD • Fitting Equations to Data: Computer Analysis of Multifactor Data, *Second Edition*
DAVID • Order Statistics, *Second Edition*
DAVISON • Multidimensional Scaling
DEGROOT, FIENBERG and KADANE • Statistics and the Law
DEMING • Sample Design in Business Research
DILLON and GOLDSTEIN • Multivariate Analysis: Methods and Applications
DODGE • Analysis of Experiments with Missing Data
DODGE and ROMIG • Sampling Inspection Tables, *Second Edition*
DOWDY and WEARDEN • Statistics for Research
DRAPER and SMITH • Applied Regression Analysis, *Second Edition*
DUNN • Basic Statistics: A Primer for the Biomedical Sciences, *Second Edition*
DUNN and CLARK • Applied Statistics: Analysis of Variance and Regression, *Second Edition*
ELANDT-JOHNSON and JOHNSON • Survival Models and Data Analysis
FLEISS • Statistical Methods for Rates and Proportions, *Second Edition*
FLEISS • The Design and Analysis of Clinical Experiments
FLURY • Common Principal Components and Related Multivariate Models
FOX • Linear Statistical Models and Related Methods
FRANKEN, KÖNIG, ARNDT, and SCHMIDT • Queues and Point Processes
GALLANT • Nonlinear Statistical Models
GIBBONS, OLKIN, and SOBEL • Selecting and Ordering Populations: A New Statistical Methodology
GNANADESIKAN • Methods for Statistical Data Analysis of Multivariate Observations
GREENBERG and WEBSTER • Advanced Econometrics: A Bridge to the Literature
GROSS and HARRIS • Fundamentals of Queueing Theory, *Second Edition*
GROVES, BIEMER, LYBERG, MASSEY, NICHOLLS, and WAKSBERG • Telephone Survey Methodology
GUPTA and PANCHAPAKESAN • Multiple Decision Procedures: Theory and Methodology of Selecting and Ranking Populations
GUTTMAN, WILKS, and HUNTER • Introductory Engineering Statistics, *Third Edition*
HAHN and SHAPIRO • Statistical Models in Engineering

Influence Diagrams, Belief Nets and Decision Analysis

Proceedings of the Conference entitled
'Influence Diagrams for Decision Analysis,
Inference and Prediction', held at the Engineering
Systems Research Center, University of California
at Berkeley, USA, 9–11 May 1988.

Influence Diagrams, Belief Nets and Decision Analysis

Edited by
R. M. OLIVER
University of California at Berkeley, USA
and
J. Q. SMITH
University of Warwick, UK

JOHN WILEY & SONS
Chichester · New York · Brisbane · Toronto · Singapore

Wiley Editorial Offices

John Wiley & Sons Ltd, Baffins Lane, Chichester,
West Sussex PO19 1UD, England

John Wiley & Sons, Inc., 605 Third Avenue,
New York, NY 10158-0012, USA

Jacaranda Wiley Ltd, G.P.O. Box 859, Brisbane,
Queensland 4001, Australia

John Wiley & Sons (Canada) Ltd, 22 Worcester Road,
Rexdale, Ontario M9W 1L1, Canada

John Wiley & Sons (SEA) Pte Ltd, 37 Jalan Pemimpin #05-04,
Block B, Union Industrial Building, Singapore 2057

Library of Congress Cataloging-in-Publication Data:

Influence diagrams, belief nets, and decision analysis / edited by
R. M. Oliver and J. Q. Smith.
p. cm.—(Wiley series in probability and mathematical
statistics. Applied probability and statistics section)
"Proceedings of the conference entitled 'Influence diagrams for
decision analysis, inference, and prediction,' held at the
Engineering Systems Research Center, University of California at
Berkeley, USA, 9–11 May 1988"—P.
Includes bibliographical references.
ISBN 0 471 92381 8
1. Statistical decision—Congresses. I. Oliver, Robert M.
II. Smith, J. Q., 1953– . III. Series.
QA279.4.153 1989 89-38933
003′.56—dc20 CIP

British Library Cataloguing in Publication Data:

Influence Diagrams for Decision Analysis, Inference and
Prediction.
Influence diagrams, belief nets and decision analysis.
1. Decision theory
I. Title II. Oliver, R. M. III. Smith, J. Q. (James
Quartermaine, *1953–*)
519.5′42

ISBN 0 471 92381 8

Typesetting by Thomson Press (India) Ltd, New Delhi
Printed in Great Britain by Courier International, Tiptree, Essex

Contents

PART V EFFICIENCY AND COMPUTATIONAL ISSUES

Preface

In a series of fortuitous meetings in the summer of 1986 and the spring of 1987 at the University of Warwick, Stanford University and University of California at Berkeley a small group of people recognized a growing interest in the use of influence diagrams, belief nets and graph-related models for prediction and decision analysis in Engineering, Statistics, Operations Research, Management Science, Medicine and Artificial Intelligence. They discussed the possibility of holding a small workshop or conference to which they could invite a number of researchers to informally present theoretical and applied problems of mutual interest. The organizing committee consisted of Professor R. E. Barlow, University of California at Berkeley, Professor R. A. Howard, Stanford University, Professor R. M. Oliver, University of California at Berkeley and Dr. J. Q. Smith, University of Warwick. The end result of these informal discussions was a series of proposals to The National Science Foundation, The United States Nuclear Regulatory Commision and the Lawrence Livermore National Laboratory to help support such a conference. The Conference was held at the University of California at Berkeley on 9–11 May, 1988. Twenty invited speakers presented papers and seventeen discussants presented oral and written critiques of these papers in six sessions over three days.

The conference brought together decision theory and operations research analysts, statisticians, computer scientists, engineers and experts from a variety of disciplines to discuss the theory and applications of Influence Diagrams (IDs). Over the three days the exchange of information was substantive and lively. In this Preface we attempt to summarize the more important issues raised and discussed during the conference and briefly describe the major points made by different speakers as well as provide a brief historical perspective of the development of IDs.

The organization of this Proceedings of the conference mimics the program sessions. The original papers distributed at the conference were drafts that had not been refereed or reviewed; following the conference, author papers and discussant comments were reviewed and final versions were resubmitted by authors based on discussant and referee remarks.

Obviously, papers presented at the conference represent only a small portion of the work now available on the subject of Influence Diagrams; while it was not possible to present all the speakers that have made contributions to this exciting field, we sought to include a diversity of research results and applications. Thus, we hope this collection is only viewed as a sample of the many new developments, models and applications where influence diagrams are being used and where they appear to hold promise for the future. By distributing these proceedings we hope that researchers in different fields will have an opportunity to see how their own work might benefit from an understanding and use of influence diagrams. We also hope that this exposure will continue to enrich collaboration with others in different scientific fields of interest.

There appeared to be common agreement that the study of IDs was moving rapidly and in directions which are difficult to predict at this time. They were originally developed for use in the practice of decision analysis. Although there are many experienced users of IDs working on the resolution of important and complex problems, it has been difficult to find scientific publications in the open literature. We were fortunate to have had at the conference practitioners who had developed many models using IDs as an architectural tool; they shared new and helpful insights into the use of this graphical technique.

When modelling practical problems it is necessary to combine different components of the problem into a coherent and mutually acceptable description. Over the period of the conference it became apparent that IDs were a powerful tool to facilitate communication between groups of clients and the decision analyst or modeller, and to represent changing assumptions in a graphical manner that quickly revealed difficult independence assumptions. They helped focus on internal dependencies as a whole rather than in disjointed sections. Several convincing practical examples were used to provide insight into the structure of a problem which would otherwise have been obscured. There was a consensus that, properly done, this provided a strong argument for making use of IDs in the process of building models.

HISTORICAL BACKGROUND

Many problems in decision-making are characterized by a large number of interrelated alternatives and uncertain quantities. Influence diagrams were designed to bridge the gap between the formulation of a probabilistic model for inference and decision-making and the formal mathematical and statistical analyses of the problem. The central idea behind the development of influence diagrams is that directed graphs can be used as a framework in which to formulate and elicit a probability model for a decision problem and then calculate various quantities of interest. As Shachter (1986a) has recently pointed out, they were originally perceived as the 'front-end' for a decision analysis.

Why are influence diagrams graphs useful? First of all, it may be much easier to trace by eye the conditional dependence and independence of random quantities in a probability model than it is to perform the algebraic analysis for equivalent relationships derived from the original mathematical problem statement. Second, a given factorization of the joint probability of random variables in the original formulation formally corresponds to an ordering of the random variables in the influence diagram and to the resulting inferences which can be derived from observing new data on a set of the original random variables (nodes). Conditional independence statements embedded in a joint probability mass function $p(\mathbf{x})$ over a vector of n variables can be expressed on a graph and useful implicit conditional independence statements can then be deduced directly from the graph without recourse to extensive algebraic manipulations. More importantly, these same graphs are useful in providing a modelling framework in which in-depth aspects of inference, prediction and decision can be discussed at a non-technical level.

In two early papers Wright (1921, 1934) made use of directed graphs to illustrate 'influences' and 'causation' in regression models. The linear models assumed Gaussian statistics, and although arc reversal was not discussed, the notion of directed paths was made explicit. The directed path was useful in that one could actually trace through the specific 'influences' that led to a given result.

In a series of two insightful papers Good (1961a, b) used the notion of 'causal chains' in a directed graph to help explain the tendency of F (an 'indirect predecessor node') to cause E (the node of interest) through a common set of 'direct predecessor nodes', $G=\{G_1, G_2, \ldots, G_n\}$. In his example he explicitly made use of the conditional independence of F and E, given values on the direct predecessor G nodes; he also developed analogies with electrical resistance networks and event trees.

The modern development of infuence diagrams is largely due to the work of Miller *et al.* (1976), Olmsted (1983) and Howard and Matheson (1984) in responding to the important practical need to discuss, formulate and compute decision problems for clients. Howard and Matheson (1984) took the important step of showing the connection between decision trees and influence diagrams and the notion of arc-reversal. A second contribution was the introduction of information arcs to explicitly recognize the timing and influence of new knowledge gathered from observations that could be used to make decisions. There was no explicit inclusion of parameters as unobservable quantities although their models were rich enough to include the notion of unobservability that has been developed more fully since that time. Even though a formal algebra for manipulating such graphs has not become widely available until quite recently, it soon became apparent that the graphical representation of influence diagrams was not only useful in the formulation of the dependence and independence structure among random quantities and decision variables but also in the elicitation of the probabiities themselves.

One of the most important steps in any decision analysis is that of making inferences from observable data; in a Bayesian setting this had been shown to be equivalent to arc-reversal in the directed graph of an influence diagram by Olmsted (1983) and Howard and Matheson (1984). Recently Shachter (1986a, 1988) and Smith (1989a) offered different formal proofs of the arc-reversal theorem.

Shachter (1986a) isolated and identified the important notion of barren nodes and how general results for arc-reversal, combined with barren nodes, lead to elegant simplifications of the influence diagram and the associated decision problem. A later paper (Shachter 1988) yielded a procedure for clarifying what is and is not important in evaluating utilities and making decisions.

Graphical methods for the analysis of statistical problems where no decision variables or utilities are explicitly represented were meanwhile well-developed by the early 1980s. Two questions were being addressed. The first was how can graphs such as influence diagrams help the reading and understanding of implied probabilities and independencies arising from a given factorization of a probability density? The second was: how can graphs be used to guide the efficient computation of required probabilities when the number of variables in the system is large, as is often the case in probabilistic expert systems?

The first question was addressed for probability distributions that were positive everywhere in a series of papers by Darroch *et al.* (1980), Wermuth and Lauritzen (1983), Lauritzen *et al.* (1984), Kiiveri *et al.* (1984), and Lauritzen and Wermuth (1987). The graphs used in these papers were undirected, directed or mixed, with the directed graphs defined equivalently to influence diagrams or knowledge maps. Smith (1989a) generalized these results to arbitrary distributions. Darroch *et al.* (1980) introduced recursive causal models, which identified the important role of conditional independence in statistical models but did not concern themselves with decision problems or the explicit role of arc-reversal or Bayesian inference in these graphs. Wermuth and Lauritzen (1983), Lauritzen *et al.* (1984), and Kiiveri *et al.* (1984) identified classes of directed graphs whose conditional independence statements implied or were implied by a directed graph of interest. It is not surprising that these equivalences are intimately tied in with the notion of arc-reversal.

In the absence of decision or value nodes Verma and Pearl (1988) identified an algorithm for determining whether or not any particular conditional independence statement follows from a given probability factorization or directed graph. Their algorithm has subsequently been simplified using a derived undirected graph by Lauritzen *et al.* (1988).

The second question raised a topic of great interest to computer science, artificial intelligence and inference in large databases: can influence diagrams help in describing and revealing efficient storage and computational schemes? Using a completely different notation the group at UCLA headed by Pearl had also been working on efficient computational techniques in probabilistic expert systems. Kim and Pearl (1983) and Pearl (1986a) developed fast updating algorithms on a restricted class of problems in simply connected networks. Pearl (1986b), again using simply connected networks and a complete Bayesian analysis, dealt with computational algorithms for real-time inference given new evidence on a subset of the nodes in a directed graph. Lauritzen and Spiegelhalter (1988) use the theory of triangulated graphs to construct an undirected graph representation which is particularly efficient for propagation of new evidence and for making sensitivity analyses. Their results made use of linear-time algorithms proposed by Tarjan and Yannakakis (1984) to selectively reduce and simplify graphs of special structure. The current 'state of the art' answer to the second question, together with an excellent review of previous material, is also given in Lauritzen and Spiegelhalter (1988).

Interestingly, the algorithms which have been referred to do not just apply to probabilistic systems but any inferential scheme in which 'relevances' can be defined to satisfy certain axioms by Dawid (1979). Independently, a group of researchers, led by Dempster and Shafer, have shown that analogous graphical methods for tracing relevances, and the consequent updating of beliefs, are applicable to inferences based on belief functions. (See Kong, 1986, Dempster and Kong, 1988, Shenoy and Shafer, 1986, 1988; for linear inference see Smith, 1989a, and for Bayesian decision theory see Smith, 1989b.)

SUMMARY OF PAPERS PRESENTED

A large proportion of papers at the conference involved the analysis of practical problems, emphasised the power of the technique as an aid to Bayesian modelling and its growing development as a useful tool for analysts. Papers were divided into the following sub-headings: Recent Developments using Influence Diagrams in Decision Analysis, Model Formulation and Analysis, Theoretical Foundations, and the Study of Computational Efficiency and Numerical Algorithms.

Model formulation and analysis

Professor R. Howard introduced the idea of a 'clarity test' to check that the description of uncertain events are well-formulated. He also suggested that certain conditioning events, though important to the user, should not be assessed directly but used 'evocatively', that is as part of the assessment of probabilities on well-formed events without themselves being assigned probabilities. He proposed using the term 'relevance' rather than 'influence' for the probabilistic dependence of one uncertain event on another. Another contribution was the discussion and use of knowledge maps in the context of IDs.

Dr. J. Matheson spoke about the value of perfect information, as one might obtain from a 'clairvoyant'. This can be found from an ID by first representing it in a canonical form which can then be used by the client. It is then possible to see whether it is worth obtaining more information about the decision before action is taken. In an analogous way, he introduced the idea of the value of perfect control, as one might obtain from a 'wizard', and argued that when this quantity was elicited it could provoke the client to contemplate other, previously unconsidered, actions.

Professor T. P. Speed addressed the important issues of the complexity and calibration of influence diagrams, the context in which these statistical models are discussed, designed and used, he made a plea for designing better measures of predictive performance to 'honestly' compare different models.

Dr. R. Korsan discussed some of the problems of failing to allow for the existence of correlation in a sensitivity analysis and suggested that, by eliciting conditional moments directly, these problems can be sidestepped. This was referred to in a later talk by Dr. M. Goldstein.

One point continually emphasised throughout the conference was the need to distinguish causal connections from the information dependencies used in IDs that represent relevances or influences between variables. The majority of attendees at the conference believed that causality, although an important concept, acted at too subtle a level of our reasoning processes to be conveniently represented by a directed graph. However the importance of continuing to think about graphical ways of representing causality came up many times during the conference.

Theoretical foundations

In parallel with the use of IDs in practical applications, there have been many developments on the theoretical front. It was argued by some participants that, as a consequence of results obtained over the last several years, strong qualitative consequences could be deduced about a problem directly from its ID without explicit reference to the probability or utility specifications. These results showed how to deduce from a given ID the extent that one variable in a system might be informative about another. Dr. J. Q. Smith suggested that this could provide valuable information about which information variables were needed to make which decisions. Professor Speed argued that such deductions should be treated with skepticism since many conditional independence statements were made by the client for expedience, to keep the model less complex, rather than to reflect true beliefs. Dr. J. Q. Smith argued that feeding back to clients the technical consequences of their statements often highlighted inadequacies of an elicited model, so the results were useful for this purpose. Both produced examples where IDs highlight or disguised certain concepts and principles central to statistical modelling.

Professor J. Pearl discussed ways of characterizing independencies from the graph structure and reported recent results proved by Verma on the use of *d*-separation concepts to verify independencies in directed graphs, based on Dawid's axioms (Dawid 1979). He gave an example which showed sufficient conditions that guarantee the existence of independence.

Two variants and generalizations of probabilistic IDs were discussed: one, based on belief functions, by Professor A. Dempster, and the other, based on the specification of expectations, by Dr. M. Goldstein. Each contribution illustrated new inferential schemes which are less restrictive than a strictly probabilistic formulation. In these cases one can utilise graphs that satisfy the same manipulative rules as probabilistic IDs to track dependencies within a given structure. Both of these presentations provoked heated and thoughtful debate. There was little consensus on whether the two methods represented a valuable extension of the Bayesian theory or new directions for practical modeling. Perhaps by the next conference these controversies will have been better understood and resolved.

Problems and applications

Dr. L. Merkhofer discussed the use of IDs in multi-attribute decision making. Discussing a case study on the evaluation of alternative sites for a high-level nuclear waste repository, he illustrated how IDs provided a valuable aid at many stages of the analysis. They help to identify decision objectives, to establish attribute scales, to facilitate the quantification of uncertainty over attributes, and to compute expected utilities and perform sensitivity analyses.

Two further contributions described analyses of problems faced by the nuclear power industry. Dr. L. Phillips looked at the problem of assessing human reliability in detecting events concerning pressurised thermal shock. By using IDs instead of event or fault trees he combined ideas of decision analysis with methods of small-group processes. He described in some detail how the method worked in a real-life situation.

Professor R. M. Oliver and Mr. H. J. Yang used IDs to build a model to predict high risk accidents based on the idea that high-risk accidents escalate from low-risk incidents and that the knowledge of sub-system failures used in event tree analyses of accident sequences can be sharpened considerably by proper use of shared information. Valuable information about the probability of large scale accidents can therefore be gained by making use of data on low severity incidents.

In a medical application Professor R. Shachter showed how an ID provided the framework to assess mortality risk arising from different treatments following heart attacks. He emphasised how different formulations of the problem could be critically appraised using the ID and gave the results of an analysis based on a linear approximation method, a Monte Carlo run and posterior mode analyses to evaluate non-linear continuous models.

Using a simple medical example Professor C. Pereira showed how IDs help to construct a diagnostic model that is both plausible and easy to analyse. A major contribution of the ID formulation was in the development of the model that correctly represented the updating process and showed how unobservable and observable random quantities interact with one another.

Professor R. E. Barlow and Ms. T. Irony analysed the records of frequency and duration of power station outages due to unforeseen failures. In a Bayesian analysis directed by an ID they showed that a model that takes account of the flaws in the sampling method provided considerably different conclusions to a simple one that did not.

Professor A. Agogino suggested how IDs could be used in real time automated manufacturing applications by reviewing a milling machine example. This talk illustrated how the technique might be used to combine physical sensor readings, statistical data, and subjective experience.

Dr. R. Kenley described a model for trackling large numbers of clustered ballistic objects, designing a Kalman filter that explicitly took into account the independencies formulated in the ID model. He provided a model that did not make independence hypotheses and that claimed to provide a structure to speed up the required calculations.

Computational efficiency and numerical algorithms

Given a model structure represented in the form of an ID, it is used to calculate marginal and conditional probabilities of the events. Professors Howard and Speed, and several other participants, had misgivings about the faithfulness to reality of large networks of knowledge. They argued that in large systems where conditional independence statements were often made for reasons of expediency there was little reason to rely on the preciseness of the probability statements they produced. They felt that simpler but faithful methods would do better in decision problems. Dr. J. Q. Smith also worried that often decisions were not included in the analysis and that without them the models were cumbersome and not helpful.

This argument was countered by the claim by Professor Henrion that the Artificial Intelligence approach was to store large amounts of information to be used for many types of decision making. To structure knowledge around a particular decision problem would only make the system inflexible, and flexibility was its most important property. He showed how medical expert systems could be made probabilistic and that, even when the structure of relationships embedded in the influence diagram of the system was complex, bounds for probabilities and optimal diagnoses given sets of indicants could be formally calculated.

Dr. D. J. Speigelhalter discussed the relationship between graphical representations in genetics and chance-based influence diagrams, showing that the results about the manipulation of probabilities in one system corresponded to analogous and important results in the other.

Professor R. Dechter showed how graphical results for efficient propagation had an exact analogy in graphical methods for finding the maximum of expected utility when one decision variable existed in a system.

CONCLUSIONS

The conference effectively brought together workers from diverse communities that use probabilistic models, mathematical models and structures for the prpagation of evidence, statistical inference, and prediction. Many papers emphasized model building and formulation, including the problems of elicitation of problem structure, practical implications and insights derived from the developed model. Approximately 120 participants came from Australia, Brazil, Canada, Denmark, New Zealand, Norway, France and the United Kingdom, as well as the United States.

The modest size of the conference, combined with the facilities provided at the Bechtel Center, proved particularly conducive to stimulating open discussion of scientific and practical issues. There were no parallel sessions. Attendance was high in every session, resulting in audiences composed of a mixture of disciplines with a range of views. The role of well-prepared formal discussants encouraged lively discussion of each paper. These comments, together with those submitted later by other participants, appear in these proceedings.

The conference brought together a group of people from diverse disciplines who were all vitally interested in Influence Diagrams, who shared their skepticism on many topics but who, in the end, believed that Influence Diagrams yielded yet another important technique of potential value for solving difficult problems associated with probabilistic models, representation of large amounts of information, prediction, inference and decision-making.

ACKNOWLEDGEMENTS

We are grateful to the authors for their papers, the discussants for providing insights and critical discussion during and after the conference, to numerous referees of papers who helped in the review of papers, to Professor Ross Shachter who helped prepare mailing lists and has provided us with helpful suggestions and comments before, during and after the Conference and to Professor G. Anthony Vignaux who kept detailed notes of proceedings as well as helping draft the early summaries and reviews of the conference. We are most appreciative of the careful planning and organization of the conference and to scheduling and travel arrangements by Hilary Mine. Also, we want to thank the graduate students who helped with registration and advice to visitors throughout the conference: Koushik Datta, Telba Irony, Sungchul Kim, Nathan Teske and Heejong Yang.

We are fortunate to have had the financial support of The National Science Foundation, The United States Nuclear Regulatory Commission, Lawrence Livermore National Laboratory and the American Association for Artificial Intelligence. We thank them for their generosity, their willingness to accept risk and their influence on our conference!

Robert M. Oliver
James Q. Smith

BIBLIOGRAPHY

Darroch, J. N., Lauritzen, S. L. and Speed, T. P. (1980) Markov fields and loglinear interaction models for contingency tables. *Annals of Statistics*, **8**, 522–539.

Dawid, A. P. (1979) Conditional independence in statistical theory. *J. R. Statist. Soc.*, **B41**, 1–31.

Dempster, A. P. and Kong, A. (1988) Uncertain evidence and artificial analysis. *J. of Statistical Planning and Inference*, (to appear).

Good, I. J. (1961a) A causal calculus—I. *Brit. J. Phil. of Sci.*, **11**, (44), 305–318.

Good, I. J. (1961b) A causal calculus—II. *Brit. J. Phil. of Sci.*, **12**, (45), 43–51.

Howard, R. A. and Matheson, J. E. (1984) Influence diagrams. In R. A. Howard and J. E. Matheson (eds) *Readings on The Principles and Applications of Decisions Analysis*, vol. II, Strategic Decision Group, Menlo Park, Calif., pp 719–762.

Kim, J. and Pearl, J. (1983) A computational model for combined causal and diagnostic reasoning in inference systems. *Proc. 8th International Conference on Artificial Intelligence*, pp. 190–193.

Kiiveri, H., Speed, T. P. and Carlin, J. B. (1984) Recursive causal models. *J. Austral. Math. Soc. (Ser. A)*, **36**, 30–51.

Kong, A. (1986) Multivariate belief functions and graphical models. Ph.D. Thesis, Department of Statistics, Harvard University.

Lauritzen, S. L., Dawid, A. P., Larsen, B. N. and Leimer, H. G. (1988) Independence properties of directed Markov fields. Research Report R88-32, Institute for Elektroniske Systemer, Aalborg Universitetscenter, Denmark.

Lauritzen, S. L., Speed, T. P. and Vijayan, K. (1984) Decomposable graphs and hypergraphs. *J. Austral. Math. Soc. (Ser. A)*, **36**, 12–29.

Lauritzen, S. L. and Spiegelhalter, D. J. (1988) Local computations with probabilities on graphical structures and their application to expert systems (with discussion), *J. R. Statist. Soc.*, **B50**, 157–224.

Lauritzen, S. L. and Wermuth, N. (1987) Mixed interaction models. Research Report R84-8, Institute for Elektroniske Systemer, Aalborg Universitetscenter, Denmark.

Miller, A. C., Merkhofer, M. W., Howard, R. A. Matheson, J. E. and Rice, T. R. (1976) *Development of automated aids for decision analysis*, Stanford Research Institute, Menlo Park, Calif.

Olmsted, S. M. (1983) On representing and solving decision problems. Ph.D. Thesis, Engineering-Economic Systems Dept., Stanford University, Stanford, Calif.

Pearl, J. (1986a) Markov and Bayes networks: a comparison of two graphical representations of probabilistic knowledge. Technical Rep. CSD 860024 Cognitive Systems Lab., Comp. Sci. Dept., University of California, Los Angeles.

Pearl, J. (1986b) Fusion, propagation and structuring in belief networks. *AI Journal*, **29**, (3), 241–288.

Shachter, R. D. (1986a) Evaluating influence diagrams. *Oper. Res.*, **34**, (6), 871–882.

Shachter, R. D. (1986b) Intelligent probabilistic inference. In L. N. Kanal and J. Lemmer (eds) *Uncertainty and Artificial Intelligence*, North-Holland, Amsterdam, pp 371–382.

Shachter, R. D. (1988) Probabilistic inference and influence diagrams. *Oper. Res.*, **36**, (4), 589–604.

Shenoy, P. P. and Shafer, G. (1986) Propagating belief functions with local computations. *IEEE Expert*, **I**, 43–52

Shenoy, P. P. and Shafer, G. (1988) An axiomatic framework for Bayesian and belief function propagation. *Proc. of 4th Workshop on Uncertainty in Artificial Intelligence*, pp. 307–314.

Smith, J. Q. (1987) *Decision Analysis: A Bayesian Approach*, Chapman and Hall, London.

Smith, J. Q. (1989a) Influence diagrams for statistical modelling. *Annals of Statistics*, vol. 8 (to appear).

Smith, J. Q. (1989b) Influence diagrams for Bayesian decision analysis. *European Journal of Operations Research* (to appear).

Tarjan, R.E. and Yannakakis, M. (1984) Simple linear-time algorithms to test chordality of graphs, text acyclicity of hypergraphs, and selectively reduce acyclic hypergraphs. *SIAM J. Comput.*, **13**, 566–579.

Verma, T. and Pearl, J. (1988) Causal networks: semantics and expressiveness. *Proc. of 4th Workshop on Uncertainty in Artificial Intelligence*, pp. 352–358.

Wermuth, N. and Lauritzen, S. L. (1983) Graphical and recursive models for contingency tables. *Biometrika*, **70**, 537–557.

Wright, Sewell (1921) Correlation and causation. *Journal of Agricultural Research*, **20**, 557–585.

Wright, Sewell (1934) The method of path coefficients. *Annals of Math. Statistics*, **5**, 161–215.

List of Contributors

Ilan Adler	Department of Industrial Engineering and Operations Research, University of California, Berkeley, CA 94720, USA
Alice M. Agogino	Department of Mechanical Engineering, University of California, Berkeley, CA 94720, USA
Richard E. Barlow	Department of Industrial Engineering and Operations Research, University of California, Berkeley, CA 94720, USA
Jack Breese	Rockwell International Scientific Center, Palo Alto Laboratories, 444 High Street, Suite 400, Palo Alto, CA 94301, USA
Thomas R. Casaletto	SSTS Data Systems Engineering, Lockheed Missiles & Space Engineering Company Inc., Sunnyvale, CA 94088–3504, USA
Avi Dechter	Cognitive Systems Laboratory, Computer Science Department, University of California, Los Angeles, CA 90024, USA
Rina Dechter	Cognitive Systems Laboratory, Computer Science Department, University of California, Los Angeles, CA 90024, USA
Arthur P. Dempster	Department of Statistics, Harvard University, Cambridge, MA 02138, USA
Kjell Doksum	Department of Statistics, 310 Evans Hall, University of California, Berkeley, CA 94720, USA
David M. Eddy	Center for Health Policy Research and Education, Duke University, Durham, NC 27706, USA
Ward Edwards	Social Science Research Institute, University of Southern California, Los Angeles, CA 90089, USA
David Embrey	Human Reliability Associates, 1 School House, Higher Lane, Dalton, Parbold, Lancs WN 7RP, UK

Dan Geiger Cognitive Systems Laboratory, Computer Science Department, University of California, Los Angeles, CA 90024, USA

Michael Goldstein Department of Mathematics, University of Hull, Cottingham Road, Hull HU6 7RX, UK

Victor Hasselblad Center for Health Policy Research and Education, Duke University, Durham, NC 27706, USA

David Heckerman Room 215, MSOB, Stanford Medical Center, Stanford University, Stanford, CA 94305, USA

Max Henrion Department of Social and Decision Science, Carnegie–Mellon University, Pittsburgh, PA 15213, USA

Ronald A. Howard 324 Terman Engineering Center, Stanford University, Stanford, CA 94305, USA

Patrick Humphreys Decision Analysis Unit, London School of Economics and Political Science, Houghton Street, London WC2A 2AE, UK

Telba Zalkind Irony Department of Industrial Engineering and Operations Research, University of California, Berkeley, CA 94720, USA

C. Robert Kenley SSTS Data Systems Engineering, Lockheed Missiles & Space Engineering Company Inc., Sunnyvale, CA 94088–3504, USA

Robert J. Korsan Decisions, Decisions!, 2136 Lyon Avenue, Belmont, CA 94002–1639, USA

James E. Matheson Director, Strategic Decisions Group, 2440 Sand Hill Road, Menlo Park, CA 94025–6900, USA

Miley W. Merkhofer Applied Decision Analysis Inc., 3000 Sand Hill Road, Huite 4–255, Menlo Park, CA 94025, USA

Allen C. Miller 925 Rennel Drive, Cincinnati, OH 45226, USA

Robert M. Oliver Department of Industrial Engineering and Operations Research, University of California, Berkeley, CA 94720, USA

Judea Pearl Cognitive Systems Laboratory, Computer Science Department, University of California, Los Angeles, CA 90024, USA

Carlos A. de B. Pereira Universidade de São Paulo, Instituto de Matematica e Estatistica, C. Postal 20570, 10498 São Paulo, Brazil

Lawrence D. Phillips Decision Analysis Unit, London School of Economics and Political Science, Houghton Street, London WC2A 2AE, UK

K. Ramamurthi Department of Mechanical Engineering, University of California, Berkeley, CA 94720, USA

Herman Rubin Department of Statistics, Mathematical Sciences Building, Purdue University, West Lafayette, IN 47907, USA

Douglas L. Selby Engineering Physics Division, Oak Ridge National Laboratory, Union Carbide Corporation, PO Box X, Oak Ridge, TN 37830, USA

Ross D. Shachter Department of Engineering—Economic Systems, Stanford University, Stanford, CA, 94305–4025, USA

S. W. W. Shor Bechtel Power Corporation, San Francisco, California, USA

Nora Smiriga Mathematics and Science Division, L-316, Lawrence Livermore National Laboratory, Livermore, CA 94550, USA

James Q. Smith Department of Statistics, University of Warwick, Coventry CV4 7A1, UK

Jim E. Smith Strategic Decisions Group, 3000 Sand Hill Road, Menlo Park, CA 94025, USA

Terence P. Speed Department of Statistics, University of California, Berkeley, CA 94720, USA

David J. Spiegelhalter MRC Biostatistics Unit, 5 Shaftesbury Road, Cambridge CB2 2BW, UK

Thomas Verma Cognitive Systems Laboratory, Computer Science Department, University of California, Los Angeles, CA 90024, USA

G. Anthony Vignaux Institute for Statistics and Operations Research, Victoria University, Wellington, New Zealand

H. J. Yang Department of Industrial Engineering and Operations Research, University of California, CA 94720, USA

Lotfi Zadeh Department of EECS, Evans Hall, University of California, Berkeley, CA 94720, USA

PART I

Model Formulation and Analysis

CHAPTER 1

From Influence to Relevance to Knowledge

Ronald A. Howard, *Stanford University, USA*

ABSTRACT

Over the years, experience with the influence diagram has shown that it is an effective means for communicating with both decision-makers and computers. The influence diagram has proved to be a new 'tool of thought' that can facilitate the formulation, assessment, and evaluation of decision problems. Practical use has provided several refinements and extensions of the concept that increase effectiveness.

Refinements include the following developments. Since the arrows between uncertainties in the diagram represent an assessment order rather than a chain of physical effects, we prefer the term 'relevance' to describe the relationship. Then it is clear that if A is relevant to B, B is relevant to A. The term 'influence' often carries a causal connotation that is not appropriate. We have also learned the importance of the clarity test to assure that the quantities assessed in the diagram are clearly defined. If we desire to be able to calculate the value of clairvoyance on any uncertain quantity in the diagram, then we must draw it in canonical form. In canonical form, there must be no arrow from a decision node to a chance node other than the value node.

Some extensions of the influence diagram idea arise when we focus on diagrams that contain only chance nodes, which we call 'relevance diagrams'. When we use a relevance diagram to assess the information that an individual or a group has about a set of uncertain quantities, we speak of the result as a 'knowledge map'. Redundant knowledge maps allow us to assess information in ways that do not correspond to any assessment order. Disjoint knowledge maps permit us to reduce the amount of assessment to that required for the decision problem by forgoing the opportunity to ask about other probabilistic results. The distinction between assessed and evocative knowledge maps allows us to think explicitly of the many factors that could affect an assessment without becoming committed to numerical specification.

Influence Diagrams, Belief Nets and Decision Analysis
Edited by R. M. Oliver and J. Q. Smith

These new developments in influence diagrams offer the promise of an even wider range of practical use by both analysts and decision-makers.

INTRODUCTION

As is becoming increasingly well known, an influence diagram was defined in the 1970s as a graphical representation of the relationship of the decisions and uncertainties in a decision problem (Howard and Matheson, 1983). It has since become the most effective tool available for the representation and evaluation of decision problems. Researchers are extending the capability of the tool while practitioners are expanding its use in aiding decision-makers.

In this chapter we will suggest a more precise language for the diagrams that should clarify our thought and eliminate possible sources of confusion. This effort is prompted by the observation that both experienced professionals and well-trained students still create diagrams that break the rules. With the new terminology we will be able to present visually how to resolve certain problems in value of clairvoyance and explain extensions in representing knowledge and preference.

1.1 THE COMPONENTS

The diagram we create to represent a decision or inference problem, which we have traditionally and generically called an influence diagram, consists of nodes and arrows.

1.1.1 Decision and chance nodes

Decisions are represented by square decision nodes that may have an underlying structure of a continuous decision variable or a discrete set of alternatives. Uncertain events or variables are depicted by circular or oval chance nodes. A chance node can represent a continuous or discrete random variable or a set of events. In general we think of a chance node as characterizing a kind of distinction within which there are several degrees. The diagram represents the particular state of knowledge of its author ('author' is used in the sense of 'the beginner, originator, or creator of anything').

1.1.2 Arrows

The knowledge of the author is indicated by connecting the nodes by arrows. An arrow entering a chance node means that the author's probability assignment represented by the chance node is conditional on the node at the other end of the arrow, its input. As we shall see, logic requires that there be no possible cycles of arrows connecting chance nodes.

An arrow entering a decision node means that the author's decision is made with knowledge of the decision made or the outcome of the uncertain quantity at the other end of the arrow. We call this an information arrow. We require that all decisions be thought of as ordered in time; there will be no possibility of a cycle. Since the author is assumed not to forget the previous decisions he has made, every decision node has an information arrow from the previously made decisions.

Relationships in diagrams without cycles like these are easily described. If we think of an arrow as meaning 'parent of' in a possibly multisexual world, we can use everyday relationship terminology, like 'ancestor', to describe the connectivity of the diagrams.

1.1.3 Deterministic nodes

As a special case, nodes may depend deterministically on other nodes. In this case we call them deterministic nodes and depict them by a double-bordered circle or oval. Deterministic nodes will usually represent mathematical relationships like the volume of a cylinder as a function of its height and radius. We can think of the value of a deterministic node as a function of its inputs (or parents). The deterministic relationship means that the author will not have to assess a probability distribution on the node given its inputs and that the computational procedure can be considerably simplified. If the deterministic node has uncertain inputs, then the quantity it represents is uncertain.

Once we have designated a node as deterministic, we lose the possibility of changing our minds and assigning a probability different from one on its value given its inputs. If we wish to represent the possibility of uncertain values for the node given its inputs, then we must declare the node as a chance node whose conditional probabilities happen to be set to one by the author for the moment. This distinction is of primary importance to the programmer who wishes to take advantage of the computational simplifications afforded by functional relationships. In conceptual discussions we can think of the deterministic node as a special case of the chance node where we have asserted graphically that the conditional probability of its value has been set to one.

1.1.4 Value node

A value node is a node designated by the author to be the quantity whose certain equivalent is to be optimized by the decisions. Only one such node can be designated. The value node is a deterministic node; it is usually drawn as a hexagon or octagon ('stop sign').

1.2 THE DECISION DIAGRAM

A decision diagram may be composed of decision nodes, chance nodes, deterministic nodes, and a value node. A decision diagram must contain a value node and at least one decision node that is an ancestor of the value node. The structure becomes significantly more useful, however, when at least one chance node is present. We will now describe the relations and special cases of the decision diagram in more detail.

1.3 RELEVANCE

Consider a diagram that contains only chance nodes—it is clearly not a decision diagram. We observe that the fundamental probabilistic description of the system is the joint probability distribution on the kinds of distinctions represented by the diagram. By the chain rule of probability, a joint distribution on n uncertain quantities can be represented by $n!$ different products of conditional probabilities, each of which we call an assessment order. For example, Figure 1.1 shows the six possible assessment orders for the three distinctions x, y, and z. Every assessment order corresponds to a set of arrows

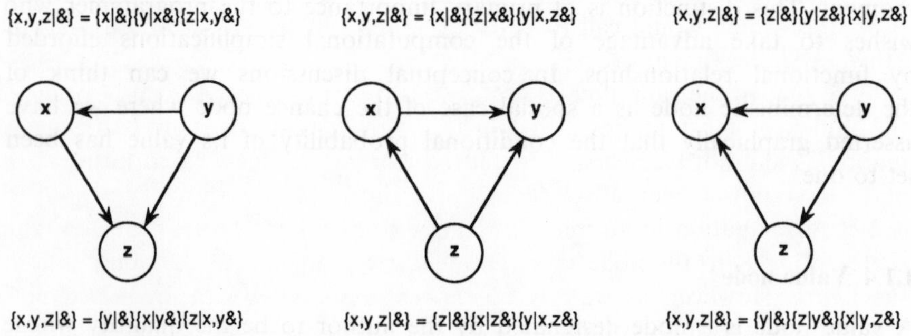

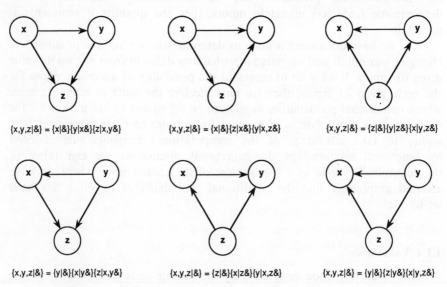

Figure 1.1 Six possible assessment orders of three-distinction diagrams.

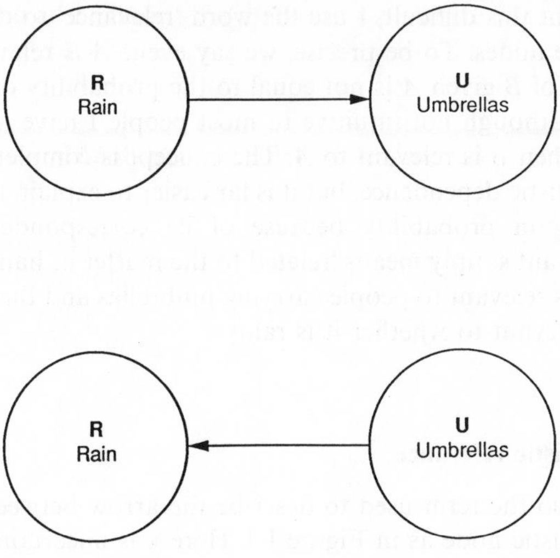

Figure 1.2 Causality or relevance?

in the diagram. Since there are no possible cycles in the expansion, there can be no cycles in the diagram. If an arrow is missing, it means that the possible probabilistic dependence represented by the arrow is asserted not to be present by the author of the diagram.

To provide a specific example for discussion, Figure 1.2 describes two uncertain events, *R*, that it is raining outside, and *U*, that people are carrying umbrellas. An arrow from *R* to *U*, as in the upper part of the figure, means that the author wishes to assess the probability of *R* and hence of its complement R' unconditionally and then to assess the probability of people carrying umbrellas conditional on *R* and on R'. An arrow from *U* to *R* as in the lower part requires first assigning probabilities to *U* and then to *R* given *U*. No arrow would mean that *R* and *U* were independent. Of course it is true that once an assessment is made in one form, it is simply a mathematical or computational procedure (some would call it Bayes's theorem) to convert it into any of the other valid assessment orders.

Note that in the rain–umbrellas example, there is a natural causal order in the minds of most authors. People think that rain causes the carrying of umbrellas and not vice versa. Thus if the word 'influence' is used in the sense of 'to have power over, affect', people will think that the only proper way to depict this situation in an influence diagram is with an arrow from the node representing *R* to the node representing *U*, the diagram in the upper part of the figure. However, as we have seen, representations with arrows in both directions are equally valid. Speaking of arrows between chance nodes as 'influences' can be misleading.

To circumvent this difficult, I use the word 'relevance' to describe an arrow between chance nodes. To be precise, we say event A is relevant to event B if the probability of B given A is not equal to the probability of B given A'. It is easy to show, although not intuitive to most people I have asked, that if A is relevant to B, then B is relevant to A. The concept is completely equivalent to that of probabilistic dependence, but it is far easier to explain to people without formal training in probability because of its correspondence to everyday language. Relevant simply means 'related to the matter at hand'. It is clear that rainy weather is relevant to people carrying umbrellas and that people carrying umbrellas is relevant to whether it is rainy.

1.3.1 Deterministic relevance

Relevance is also the term used to describe the arrow between a chance node and a deterministic node as in Figure 1.3. Here x is uncertain and $y = x^2$. The quantity x is relevant to y because a different probability distribution will be assigned to y when a different value of x is known; in this case the distribution will assign a probability of one to x^2. Here it is clear that the arrow can be reversed, that y is relevant to x. If y is known, then x will be described by a probability distribution that assigns discrete probabilities to the positive and negative square roots of y proportional to the height of the distribution on x at those points. After the reversal, neither node would be in general deterministic.

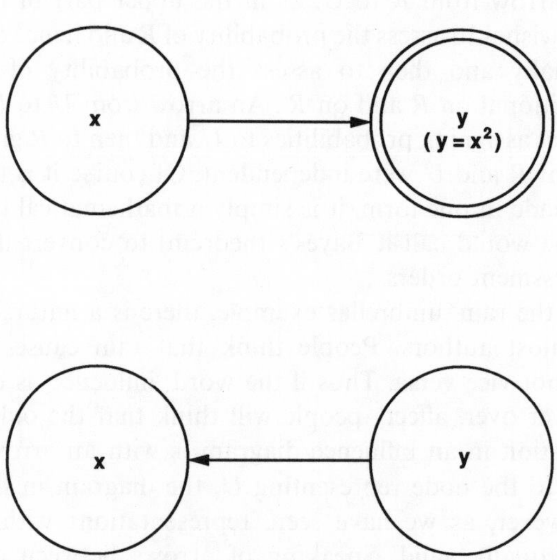

Figure 1.3 Deterministic relevance.

1.4 INFLUENCE

If we are now clear that 'relevance' describes the relation between chance nodes, what are we to make of an arrow from a decision node to a chance node? The arrow says that the probability assignment to the chance node is conditional on the decision taken. This must mean that the decision 'has power over, affects' the chance node. The author believes that the world will be different depending on the decision he makes. I therefore find it appropriate to call an arrow from a decision node to a chance node an 'influence', and this is the only arrow I would give this name. While influences are common in decision diagrams and they pose no logical difficulty in making decisions, we will see in our later discussion of the value of clairvoyance that we may prefer to avoid the use of influences in modeling decisions.

1.4.1 Information/influence confusion

There is a good pedagogical reason to avoid the word 'influence' as a general term to describe arrows. Students learning influence diagrams often reason this way: 'Uncertain quantity x, not known at the time of the decision, is an important consideration in making the decision. It therefore influences the decision—there should be an arrow from the chance node representing x to the decision node.' Their mistake, of course, is that such an arrow means that the quantity x is known at the time of the decision, which is not true in this case. The decision is made looking ahead to the possible values that x may assume with different probabilities and therefore with full consideration of x; but it is not made knowing the value of x.

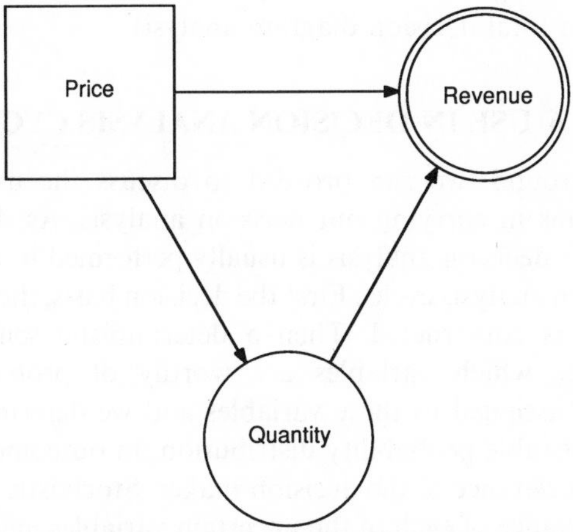

Figure 1.4 Deterministic influence.

1.4.2 Deterministic influence

Deterministic nodes may also have decision nodes as inputs, and hence be thought of as being subject to influences. For example, as shown in Figure 1.4, the deterministic node revenue may be the product of the decision node price and the chance node quantity, which is influenced by price. There is, of course, no problem in computing revenue given price and quantity. For simplicity we describe the arrow from price to revenue as an influence and the arrow from quantity to revenue as a relevance.

1.5 SPECIAL CASES

1.5.1 Relevance diagrams

We call diagrams composed of only chance and deterministic nodes 'relevance diagrams'. They can be used to characterize and compute any problem of uncertainty and inference. We will later discuss in more detail the use of relevance diagrams to represent knowledge.

1.5.2 Deterministic diagrams

We call diagrams composed of only deterministic nodes 'deterministic diagrams'. Now every arrow into a node can be interpreted as an input to a function that produces the node value as its output. Such a deterministic diagram would be an alternative representation to a spreadsheet model. While we would still prefer to avoid cycles, if they should appear, they would be interpreted like circular references in a spreadsheet that would have to be resolved by iteration. This conception allows us to extend considerably the use of influence diagrams in deterministic modeling. As we shall see, such modeling can be a precursor or a component of a total decision diagram analysis.

1.6 USE IN DECISION ANALYSIS CYCLE

With this background, we can proceed to discuss the use of decision and relevance diagrams in carrying out decision analysis. As discussed elsewhere (Howard, 1983), a decision analysis is usually performed in a sequence of steps called the 'decision analysis cycle'. First the decision basis, the formal description of the problem, is constructed. Then a deterministic sensitivity analysis is performed to see which variables are worthy of probabilistic treatment. Probabilities are assigned to these variables and we determine the alternative with the most desirable probability distribution on outcomes according to the values and risk preference of the decision-maker. Stochastic sensitivity analysis reveals the importance of each of the uncertain variables and indicates whether further care in their modeling or assessment is desirable. A value of clairvoyance analysis determines the economic value of resolving any uncertainties in the problem. As a result of these appraisal activities, a new basis may be created and the process repeated, or we may decide there is sufficient clarity of action in view of the total problem setting. The decision diagram is of major assistance in all of these activities.

1.6.1 Conversational use

Let us begin with the process of formulation. We conduct our discussion of the problem with the decision-maker, or the group representing the decision-maker, by drawing a decision diagram. We use decision and chance nodes, but in constructing the diagram we are purposely not concerned about being precise in the definitions of the chance quantities discussed. The reason is that the crisp definitions and logic of decision analysis that are its great strengths at later stages can impede the flow of communication with the decision-maker if imposed too early. Thus at this stage we may tolerate a chance node labeled 'cost of raw materials' without yet inquiring about the exact nature of the raw materials.

1.6.2 Defining the kinds of distinction—the clarity test

Now we proceed to define each of the unanswered questions that we have talked about. We do this by using the 'clarity test'. The question we ask is whether a clairvoyant who knows all physically determinable past and future facts would understand the definition sufficiently to say whether the event had occurred or to state the value of the variable. For example, if we ask the clairvoyant to tell us the price of oil next year, he will not understand. If, however, we say that we are interested in the price for a particular type of oil contract at a named exchange on a certain day as reported in a specified publication, then we would presumably have met the clarity test. In meeting the test we assume that the clairvoyant has unlimited ability to perform any computations on what he knows and to report the results. Thus he would be able to tell us the average closing price of oil as specified above over a given time period. We refine the definition successively until the test can be passed.

1.6.3 Defining relationships

Next we must specify the relationship between the entities in the representation. We must state the desirability or value to the decision-maker of all outcomes that are preference-sensitive. We shall return to this topic later. We must also specify the functional form of the deterministic models that are present.

1.6.4 Deterministic sensitivity

We now usually have a decision diagram composed of deterministic models, uncertain variables, and uncertain events. Before proceeding to probability assignment, we perform a deterministic sensitivity analysis to determine which of the uncertain quantities have the largest effects on the value to the decision-maker and on the choice he must make. We can do this within the decision diagram computational structure if there is the provision for representing complex deterministic models that we discussed earlier. Ideally there should be no need to leave the decision diagram computational environment. Of course some problems are so inherently probabilistic that the deterministic analysis will provide little insight and may be abbreviated.

On the basis of deterministic sensitivity, we often create discrete representations for the continuous variables in the model. We can think of this as establishing for each kind of distinction the degree of distinction required to provide important insight without unnecessary detail.

1.6.5 Probability assessment and stochastic sensitivity

The next step is to assign all the marginal and conditional probabilistic assignments required by the diagram. We shall discuss later how to use relevance diagrams to aid in this assessment. When the assessment is done the standard decision diagram computational procedures will allow us to derive, with the proper specification of risk preference, the best set of decisions and the corresponding certain equivalent or expected utility. By setting the probability of uncertain events or variable values to one and adjusting the probability of other variable values to be conditional on this setting, we can find the stochastic sensitivity of events and variables.

1.7 THE EFFECT OF INFLUENCE ON VALUE OF CLAIRVOYANCE

Let us return to the question of decision diagrams that contain influences, that is, arrows from decision nodes to chance nodes. If we consider the decision tree implied by such a structure, we see there is no difficulty in determining the basic decision; we just insert for the probabilities of the uncertain quantity in the tree the proper ones for the alternative that is taken. Suppose, however, that we wish to calculate the value of clairvoyance on this uncertain quantity. This means that we assume a cost of clairvoyance and find the expected utility of the problem that results when each possible value of the uncertain variable is reported by the clairvoyant. Next we multiply that expected utility by the probability that the clairvoyant will report this value to determine the expected utility of clairvoyance at the assumed cost. If this expected utility is higher than the expected utility without clairvoyance, we increase the cost iteratively until both utilities are equal. At this point we have found the value of clairvoyance. (This type of iteration can and should be supported by the computational environment.)

This is a standard procedure often envisioned as moving the set of branches representing the uncertain quantity to the front of the tree. In decision diagram terms it means assuming that the quantity is known for all decisions. But suppose there is an influence arrow from a decision node to this chance node. Then drawing an arrow from the chance node to the decision node will create a loop in contradiction with the definition of a decision diagram.

Having influence in decision diagrams poses no problems in finding the optimal decision, but poses a major difficulty in computing the value of clairvoyance. The reason is that we have been talking out of both sides of our mouths. On one side we have said that the uncertain quantity could be revealed by a clairvoyant, and on the other that the clairvoyant must know what decision we have made before he can say whether the event occurred. If he could answer without knowing what we did, there would be no free will because our actions would be predetermined. If we were told by the clairvoyant what we were going to do, could we then not do it?

1.7.1 Canonical form

To resolve this dilemma, imagine the clairvoyant filling out a form that will specify for each action we take whether or not the event will occur. Thus if there are m possible actions and n outcomes for the event, there are n^m possible ways to fill out the form. Furthermore, we must anticipate that the answers he gives when we take one action may be relevant to the answers he will give if we take another action. This means that to be completely clear we must replace the original event by m new events, one corresponding to each action we might take, and represent any mutual relevance that exists between them. Once the

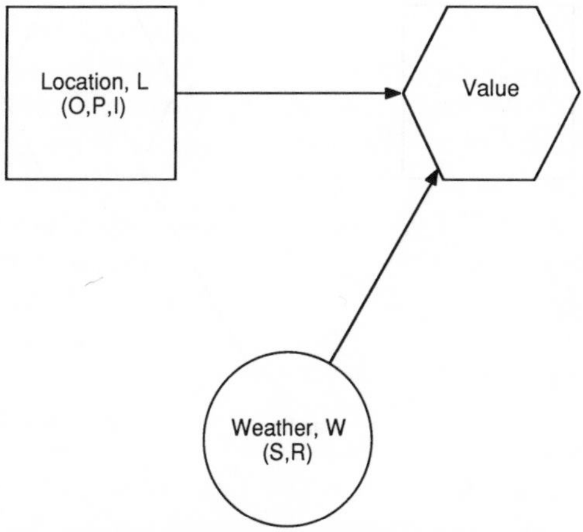

Figure 1.5 The party problem.

associated probabilities have been assigned, we can think of the value of clairvoyance lucidly once again because these new events are not influenced by the decision.

Thus we can replace an influence on a chance node by a constellation of chance nodes without influences, one for each possible alternative in the decision node. In this way, we change a decision diagram with influences into one that isolates its chance nodes in one or more relevance diagrams providing only inputs to other nodes. Then the value of clairvoyance on any uncertainty is determined just by connecting it with an arrow to the decision node.

We call a decision diagram without influences other than those on the value node a decision diagram in canonical form. We define a determinate value of clairvoyance computation as one that can be performed without additional probability assessments. Diagrams in canonical form permit determinate value of clairvoyance computations on any set of uncertain quantities.

1.7.2 Example–Shirley's party

We can illustrate the issues we have been discussing with a simple example based on a party problem I have used in class for over 25 years. A girl planning a party is uncertain about whether the weather W will be sunny S or rainy R; she assigns probabilities to these possibilities. She can have the party location L outdoors O, on the porch (open on the sides to the weather) P, or indoors I. She assigns values to various combinations of location and weather, and specifies a risk preference. The corresponding decision diagram is shown in Figure 1.5. Solution is straightforward. Since the diagram is in canonical

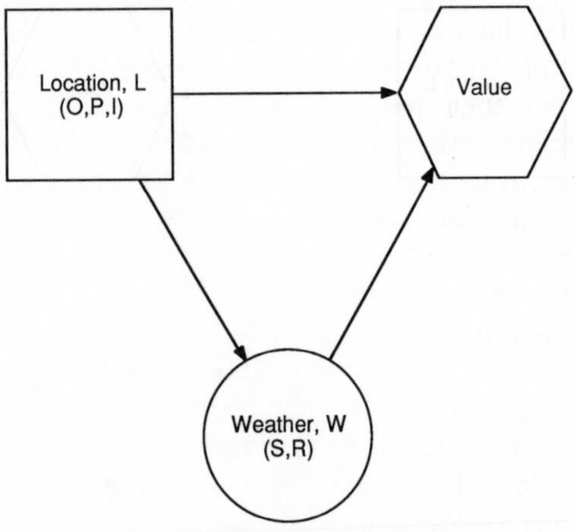

Figure 1.6 Shirley's party problem.

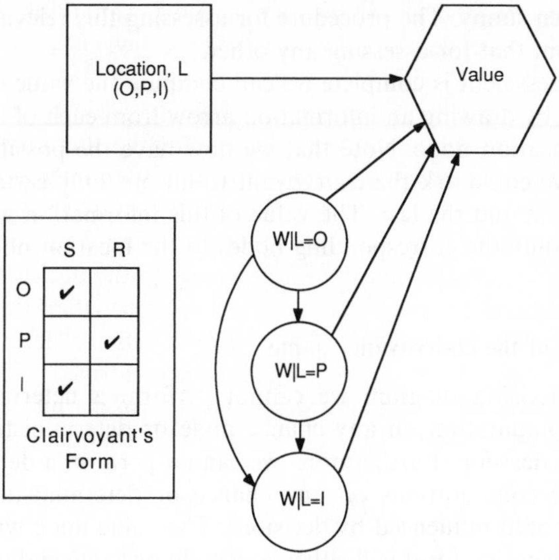

Figure 1.7 The canonical party problem with clairvoyant's form.

form, there is no difficulty in finding the value of clairvoyance on the weather: an information arrow could be drawn from weather to location.

Now consider a variant. The girl, by the name of Shirley, thinks that the weather depends at least partially on where she holds her party; her action will influence the weather. The corresponding decision diagram appears as Figure 1.6. She may think for, example, that if she trusts in fate and holds an outdoor party, the chance of sunshine will increase.

Shirley's decision diagram is not in canonical form. We cannot perform a determinate value of clairvoyance computation on the weather. We know that if we wish to compute the value of clairvoyance on the weather, we shall require additional probability assignments from Shirley. To proceed we replace the original weather event by three new events, each conditioned on her action, as shown in Figure 1.7. The clairvoyant's form, filled out in one of the eight possible ways, also appears in the figure. In this case he is saying that if she holds the party outdoors or indoors, the weather will be sunny; but if she holds the party on the porch, the weather will be rainy.

The three new chance nodes are interconnected in one of the six possible assessment orders—we would choose the one that is most comfortable for Shirley. The order shown indicates that she would first assess the chances on the weather given that she held the party outdoors. Then she would assess, for example, the chances on the weather given that she held the party on the porch, and that the clairvoyant said that if she had held the party outdoors it would have been sunny. The procedure for assessing this relevance diagram is no different from that for assessing any other.

When the assessment is complete we can compute the value of clairvoyance on the weather by drawing an information arrow from each of the new chance nodes to the location node. Note that we now have the possibility of partial clairvoyance: we could ask the clairvoyant to fill out only certain rows of his form, say the first and the last. The value of this information would be found by connecting only the corresponding nodes to the location node.

1.7.3 The rules of the clairvoyance game

In a general decision diagram we cannot perform a determinate value of clairvoyance computation on any chance node or deterministic node that is influenced by a decision. Furthermore, we cannot perform a determinate value of clairvoyance computations on any chance or deterministic nodes whose ancestors have been influenced by decisions. The value node will always be in one of these categories, for it will either be a child or a descendant of a decision in any meaningful decision problem.

1.7.4 Inheritance of influence

Figure 1.8 illustrates the situation using a research and development problem. The technical difficulty of the problem is relevant to the achievement of technical success, which is also influenced by the research effort decision. (While we may have disagreed with Shirley's views on affecting the weather, they are no different in principle from those displayed by the author of this diagram.) Technical success is relevant to market success, which is relevant to value; value is influenced by the research effort. Even though there is no influence arrow from the decision to market success, we cannot request clairvoyance on market success because receiving this information would tell us something about technical success, which would tell us something about the R & D action taken, and thereby create a violation of the free will principle. We can say that an influence on any chance or deterministic node is inherited by all its chance and deterministic node descendants, so that they can no longer be considered to be uncertainties revealable by a clairvoyant.

Note that even when a decision diagram is not in canonical form, it still may be possible to perform a determinate value of clairvoyance computation on some of the chance or deterministic nodes in the diagram. In Figure 1.8, for example, technical difficulty has no direct or inherited influences acting upon it. Finding the value of clairvoyance on technical difficulty would require no additional probability assessments.

When a decision diagram is in canonical form, all chance nodes will constitute

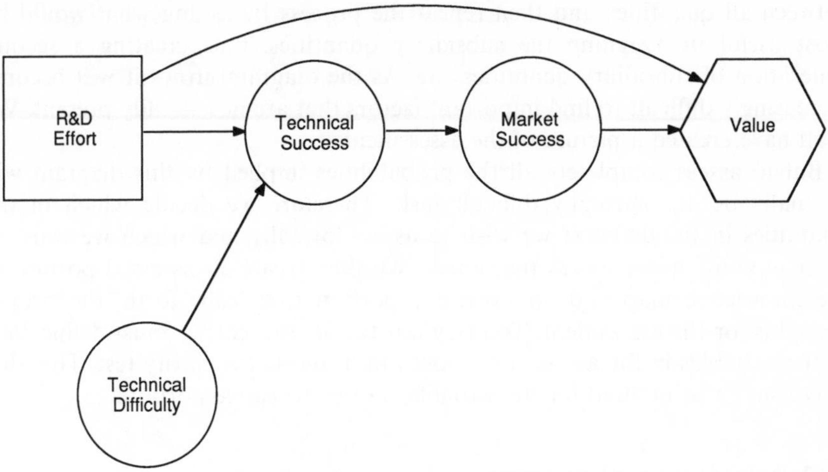

Figure 1.8 An R & D decision.

one or more relevance diagrams that are free from any influences and hence these nodes permit determinate value of clairvoyance computations. There may exist deterministic nodes within these relevance diagrams, and other deterministic nodes that have only nodes in the relevance diagrams as ancestors. These deterministic nodes also permit determinate value of clairvoyance computations.

1.8 KNOWLEDGE MAPS

As we mentioned above, relevance diagrams have an important use in representing carefully the often highly diffuse knowledge that a person or a group of people have on a particular quantity of interest. We call the relevance diagrams representations created for this purpose 'knowledge maps' to emphasize our intent to the participants in the process. The procedures and theory of creating knowledge maps are described elsewhere (Howard, 1989). Here we briefly summarize the concepts and results.

1.8.1 Evocative and assessed knowledge maps

The development of a knowledge map for a probability assignment starts with the same rules as the conversational use of decision diagrams. Only the events or variables whose assessments we seek, the target quantities, must meet the clarity test at the beginning of the process. We construct the relevance diagram by asking, 'What would you like to know to assess the target quantity', and create a first generation of subsidiary quantities. We check for relevances between all quantities and then repeat the process by asking what would be most useful in assigning the subsidiary quantities, thus creating a second generation of subsidiary quantities, etc. As the diagram grows it will become increasingly difficult to find important factors that are not already present. We shall have created a picture of the assessment.

But to assess completely all the probabilities implied by this diagram will normally be an unusually difficult task. Therefore we decide which of the quantities in the diagram we wish to assess formally and which we want to 'bear in mind' as we assess the others. We thus create an assessed portion of the knowledge map and an evocative portion that 'calls forth' the proper thoughts for the assessment. To carry out the assessment we must define each of the variables in the assessed portion until it meets the clarity test. This step may usually be omitted for the variables in the evocative portion.

1.8.2 Redundant knowledge maps

As we have said, the assessed knowledge map represents a possible assessment order for a joint probability distribution, with possible additional assertions of irrelevance within the order. By using the conversational procedure we strive to create from among the many possible orders an assessment order that will be natural and comfortable for the author. Suppose, however, that the assessments that the author is comfortable in making do not correspond to any of the possible assessment orders. How can we reconcile his comfort with the requirements of logic? After all, the whole purpose of creating a knowledge map is to capture in logically correct form the information that is in a person's head.

To illustrate, in the rain–umbrellas example of Figure 1.2, people will usually feel most comfortable making assignments in the 'causal' direction. They will want the arrows to go from R to U and will assign first the probability of rain and then the probability of carrying an umbrella given rain and given no rain. But suppose someone felt most comfortable assigning the chance of rain and the chance of rain given that people were either carrying umbrellas or not carrying umbrellas. If we think of extending our notation for diagrams by indicating a marginal probability assessment of a quantity by an arrow from the general background of the figure to the quantity, then every probabilistic assessment is represented by an arrow in the diagram. In this case we have an arrow directly into R and an arrow from U to R, not a possible assessment order. If we add an arrow into U, implying that we will somehow assess U directly, then we would have two possible, and therefore redundant, assessments on R, one direct, and the other from U and then R given U. The issue is to find a distribution that will, if possible, be consistent with both the direct R assessment and the indirect R given U assessment. We can imagine the person trying different probabilities for U iteratively until he feels comfortable with all the possible distributions in the problem. In more complex cases, it would be possible for the author to specify certain of the assessments about which he is relatively confident so that the 'slack' would be taken up in other assessments.

We can envision the general process as follows. First we create the assessed knowledge map without any consideration of difficulty of assessment. Then we find if there are any required assessments with which the author does not feel comfortable. Usually, he will be comfortable with most, if not all of the assessments, especially if he thinks of them as 'causal'. If there are some with respect to which he is uncomfortable, then we ask him what assessments on the same quantities would be comfortable; these will often be assessments in the other direction, as in the rain–umbrellas example. Under certain conditions (Howard, 1989), we are able to create separate knowledge maps on the same quantities based mostly on comfortable assessments. He will still have to assign tentatively some distributions of an uncomfortable nature, like the one on *U*, but only with the desire to make consistent the two distributions on the quantity he does feel comfortable about, like *R*. This is the source of the name 'redundant'. The tentative assessment plays a role similar to the initial value of a numerical iteration.

Note that we are not primarily concerned with situations where the author feels uncomfortable with whole sections of the knowledge map. In such cases he should seek more information, delegate the assessment to an expert, or treat these sections as purely evocative.

The ability to help people make assessments when they are not comfortable with assessing any possible assessment order will considerably extend the usefulness of the influence diagram. We can think of redundancy as analogous to iteration in spreadsheets, although one must be much more participative in the process.

1.8.3 Disjoint knowledge maps

In the interest of making assessment easy for people, we note that assigning probabilities conditional on more than one variable is quite difficult for virtually everyone. We may ask just what is it that makes the assessment on the target quantities produced by a knowledge map more credible than a direct assessment. We see that it is a matter of faith, that there is no way to prove that a more detailed assessment is superior to a direct one. When we face a knowledge map with multiple conditionality, we may question whether the improvement in credibility is worth the effort in difficulty of assessment. Of course, if we are interested in the whole joint probability distribution represented by the knowledge map, we have no choice but to do the complete assessment. But if we are interested only in a subset of the map, a set of target quantities, then we may be able to obtain an assessment based on limited conditioning that is superior to direct assessment and much easier to perform.

Suppose, for example, that we have three quantities in our knowledge map, x, y, and z, and that the decision depends only on x, the target quantity. Suppose further that the author has drawn a knowledge map with arrows from both y and z to x and an arrow from z to y. This means that he is going to assess x conditional on y and z, then y conditional on z, and finally assign a marginal distribution on z. All these assessments (or those from another possible assessment order) would be required to obtain the joint distribution. However, in this case the only distribution desired is that on x. It could be assessed directly, possibly with misgivings that some relationship had been missed. The author could assess the entire knowledge map needed for the joint distribution, but this would require the usually difficult multiple assessment of x given both y and z.

An alternative between these extremes is to assign a distribution on z, one on y given z, and derive the distribution of y. Then this distribution can be combined with an assigned distribution on x given only y, to produce the distribution on x. In knowledge map terms what we have done is to represent the assessment by two disjoint knowledge maps, one containing y and z with an arrow from z to y and the other containing x and y with an arrow from y to x. We could connect the two nodes labeled y with a dotted arrow from the first knowledge map to the second to show that the marginal distribution derived from one is to be used as an input to the other. We could alternatively show the two nodes y as overlapping circles slightly displaced from each other.

What do we gain and lose from this procedure? We gain over assessing the complete knowledge map the advantage of replacing the multiply conditional assignment x given y and z by the simpler assignment x given y. We gain over direct assignment increased credibility because both y and z have been considered. We lose the ability to create the whole joint distribution and to answer some of the conditional probability questions that would have been determined by that distribution.

In general we can replace any expansion of a target quantity that is performed successively in terms of a, b, c, etc. by a series of pairwise expansions that link the quantity to a, then a to b, b to c, etc., with each pair representing a disjoint knowledge map. To the extent that people are much more comfortable in making conditional assessments with only one conditional, we have replaced a difficult problem with a simpler one that is logically correct and more credible than a direct assessment. While the procedure could also be used to reduce an assignment based on three conditions to one based on two conditions, it is likely that most uses of the technique will be to eliminate the necessity of making an assessment conditional on more than one variable.

1.9 VALUE DIAGRAMS

We can use the structure of decision diagrams, including deterministic nodes, to represent value functions. When several attributes affect the desirability of outcomes to the decision-maker, we can create a value node with an arrow from each of the attributes. These arrows would present deterministic functions. If there were three attributes r, s, and t, then when the level of t had been expressed, the system could derive the tradeoff curve between r and s. We can think of the value node as identifying the iso-preference surfaces among the attributes. If one of the attributes is money, the tradeoff curve between money and levels of another attribute reveals the willingness-to-pay for different levels of that attribute for given levels of the remaining attributes. To complete the specification in the case of uncertainty, we must select a numeraire to identify the different iso-preference surfaces and then construct a risk preference function on this numeraire. In many cases the numeraire will be measured in the money attribute with the values of different levels of the other attributes evaluated in money.

The usefulness of this representation is not only computational; the discussion of values can be considerably enhanced using the decision diagram framework. A common point of confusion is whether the decision-maker places a direct or indirect value on an attribute. Placing a direct value on an attribute means that it is a member of the set of attributes that collectively determine the desirability of the outcome. In terms of the discussion above, the attributes on which the decision-maker places a direct value are those that are connected by an arrow to the value node, its parents. All other nodes in the diagram will typically have indirect values: changes in them will ultimately affect nodes that have direct values.

For example, in buying a house I may place a direct value on the view because it affects desirability even when all other attributes are specified. However, I would place an indirect value on the cost of maintenance because it is only a part of total annual cost, an attribute on which I would have a direct value. I do not care in principle about the breakdown of annual cost in terms of maintenance, insurance, utilities, etc. as long as the sum remains the same. In the diagram there would be an arrow from the maintenance cost node to the total annual cost node and from there to the value node, but no arrow from the maintenance cost node to the value node. This shows graphically that I place an indirect value on maintenance cost. Note, however, that if I were very proud of having a house with low maintenance cost and it would give me great pleasure to tell my friends about it, I would place a direct value on maintenance cost and add the missing arrow.

1.10 CONCLUSION

The influence diagram paradigm can be used in special forms to frame a decision, to elicit the often confused knowledge about uncertain quantities that resides in the human mind, and to both explore and characterize the values that govern the decision. No other current tool possesses its span of application, subtlety of interpretation, and power of computation. Almost every discussion of human action or knowledge can be elucidated by its use. However, its most important feature is that its simple appearance belies both its power and its subtlety.

1.11 DISCUSSION

1.11.1 Discussion by Lawrence D. Phillips

Professor Howard portrays influence diagrams as aids for communicating about uncertainty and computing with probabilities. He shows that, because they can be used for both probabilistic and deterministic modeling, they can contribute to most stages in the cycle associated with his school of decision analysis, including multi-attribute utility or value analysis. In addition, they can represent people's cognitive maps about some area of expertise, so would be of potential interest to expert system builders.

As we know, any influence diagram can, with appropriate transformations of the arcs connecting the nodes, be expressed as a decision tree. So, is there anything new here? At the start of this conference, I was unconvinced. Yes, very complex trees can be represented more parsimoniously, usually on one page, as influence diagrams. Yes, it is far easier to see independence between events on an influence diagram then in a decision tree. For these two reasons alone, I have found influence diagrams useful. But is there a deeper use?

Certainly there are disadvantages, the most serious being the difficulty influence diagrams have with asymmetrical decision trees. When subsequent events and acts depend on the initial decision, influence diagrams run into difficulties. Often these situations occur when the decision-maker wishes to portray cause-and-effect sequences as possible scenarios. These are not times when I would choose to use an influence diagram, and they occur frequently. Look at any textbook on decision analysis and try to find a single symmetrical decision tree.

However, scenarios were rarely invoked at the conference when an illustrative use of influence diagrams was presented. Instead, the presenter described what might be called a mutual-shaping system in which causes and effects are not easily distinguished, like the readjustment a spider's web makes when one strand is broken. The description was of acts and events that were *relevant* to each other, to use Howard's helpful terminology.

This kind of modeling supports the naturalistic paradigm suggested by Lincoln and Guba (1985) as an alternative to the traditional empiricist–positivist position. The chapter by Oliver in this volume is a good example; risks posed by complex technologies are better thought of in terms of networks of relevancies than sequences of causes and effects.

By the end of the conference, this deeper representation had become obvious. If the emergent paradigm noted by Lincoln and Guba truly represents a shift in scientific thinking, then the development of influence diagrams may be a further indicant that more satisfactory ways of viewing the world are required. Professor Howard has done us a service by showing the potential richness of expression offered by influence diagrams.

REFERENCES

Lincoln, Y. S., and Guba, E. G. (1985) *Naturalistic Inquiry* Sage, Beverly Hills, Calif.

Howard, R. A. (1983) The evolution of decision analysis. In R. A. Howard and J. E. Matheson (eds) *Readings on the Principles and Applications of Decision Analysis*, vol. I, Strategic Decisions Groups, Menlo Park, Calif.

Howard, R. A. (1989) Knowledge maps *Management Science* to be published.

Howard, R. A. and Matheson, J. E. (1983) Influence diagrams. In R. A. Howard and J. E. Matheson *Readings on the Principles and Applications of Decision Analysis*, vol. II, Strategic Decisions Group, Menlo Park, Calif., pp. 719–762.

CHAPTER 2

Using Influence Diagrams to Value Information and Control

James E. Matheson, *Strategic Decisions Group, CA, USA*

ABSTRACT

The well-known concept of the value of perfect information and the recent concept of the value of control are very useful in gaining insight about decision situations. However, using the familiar tool of decision trees for this purpose can be confusing and misleading. This chapter shows how influence diagrams can clarify and correctly pose value of information questions to evaluate new opportunities for gathering information, uses the concept of the *clairvoyant* to meaningfully assess the required joint probability distributions, and introduces the value of control as a way to gain insight into the usefulness of generating new alternatives that enable more control over uncertain variables.

2.1 INTRODUCTION

Many decision analysts underestimate the difficulty of dealing with the concept of the value of perfect information. The correct analysis of the example I use in this chapter has stumped many professionals when it was first presented to them. The problem arises because a decision tree does not contain all the probabilistic information needed to answer value of information questions. In general, to answer these questions, one needs a complete description of the joint probability distribution of all the uncertain variables. This joint distribution is more easily and naturally captured in an influence diagram, which simultaneously retains the decision tree's chronological structure. In addition, ill-formed value of information questions appear as loops in the influence diagram, so they are easy to detect. Moreover, we will show how a special form of influence diagram—Howard canonical form—is created to pose and answer any value of information question.

When facing the joint probability assessments necessary to answer value of

Influence Diagrams, Belief Nets and Decision Analysis
Edited by R. M. Oliver and J. Q. Smith

information questions, we sometimes seem to be asking for the assignment of positive probability to the joint occurrence of mutually exclusive events! We will resolve this dilemma by assigning probabilities to the answers of a *clairvoyant* who foresees potential events, rather than by assigning probabilities to the events themselves.

Finally, we will introduce the concept of the value of perfect control. This might be achieved by employing a *wizard* who can control or set the value of a variable rather than simply foresee its outcome. Proper use of this concept requires that the expert, or author of the influence diagram, assert that he can contemplate controlling a set of variables without invalidating the remaining portion of his influence diagram. In this chapter, we consider only the case of controlling variables that have no predecessors.

2.2 THE PRIMARY DECISION

To create a hypothetical example, let us play the role of analysts for a space mission designed to land a remotely controlled experimental apparatus on the surface of Mars. We have thoroughly analyzed the mission and have summarized our total state of information by assigning a 0.6 probability that the Mars mission will be successful and a corresponding 0.4 probability that it will be a failure. Also, we have analyzed the values to be derived from the mission and put them in monetary units. Let us assume that the value of a successful Mars mission is 50 units and the value of an unsuccessful one is 10 units. A positive value might be attributed to a failure because attempting the mission has important social value and even a failure will provide knowledge for a better design on the next attempt.

Unexpectedly, several months before launch, another nation announces that it will attempt a similar mission to Venus in about one year. Because of the competitive nature of the space race and the important foreign policy implications of technological leadership, we realize that the value of changing our destination and successfully landing on Venus would be quite high. However, if we attempt to land on Venus and fail, we would look foolish for diverting the program, and we would set back the timetable for our extensive Martian exploration program by at least two years. When all of these factors are evaluated, we find that a successful landing on Venus is worth 100 units and a failure costs 10 units. To our surprise, when we check the feasibility of diverting the mission, we find that because of modular design only a few important, but thoroughly tested, components of the landing system need to be changed, and the mission engineers assign a 0.6 probability of success, regardless of destination.

To evaluate this problem, we must also specify a risk attitude. In this chapter, we shall assume the decision-maker is risk neutral. Therefore, we will carry out the evaluations on an expected-value basis. All the results in this chapter easily generalize to arbitrary risk attitudes.

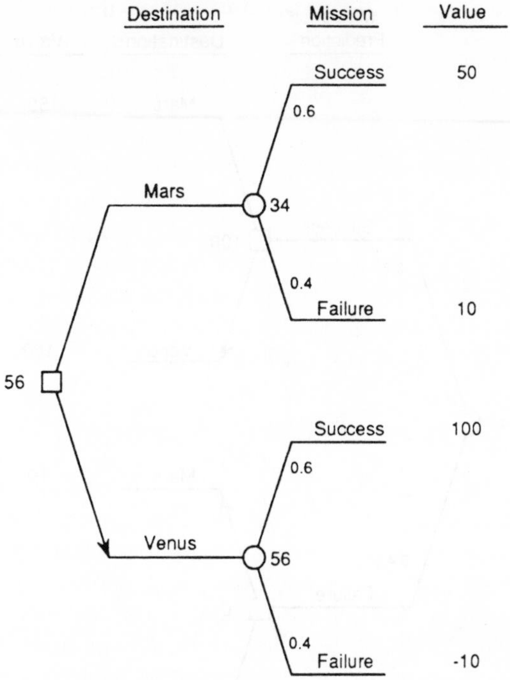

Figure 2.1 The primary decision problem.

From these assessments, we can lay out the primary decision tree of Figure 2.1. Along each outcome branch emanating from a chance node, we have written the conditional probability of following the path. Throughout this chapter, probabilities in the figures will appear in a smaller size than the values. Near each node, we have written the value, either assigned or derived, of reaching the point in the program represented by that node. We see that the expected value of going to Mars is 34, while the expected value of going to Venus is 56. Thus, to maximize the expected value of the mission, we decide to go to Venus.

2.3 VALUE OF PERFECT INFORMATION

We might wish to investigate the possibility of gathering new information before we make the final decision. To do this, we will use the value of perfect information as an upper bound for the value of less complete information-gathering programs. At this point we will show how the decision tree can be a confusing tool for this purpose, and how the introduction of the clairvoyant helps clarify the situation. Later, we will show how the use of influence diagrams avoids most of this confusion.

Most analysts, when presented with Figure 2.1 and asked to derive the value of perfect information, reverse the order of decision and chance nodes in

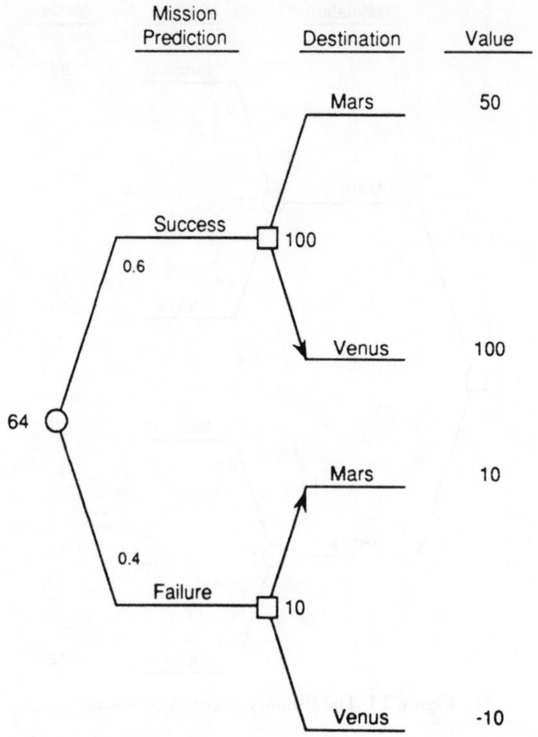

Figure 2.2 A typical, but naive, tree for determining the value of perfect information.

Figure 2.1 to produce the tree shown in Figure 2.2. With the latter tree, we learn first whether the mission will succeed or fail, and then we decide on the destination. If we know the mission will succeed, we send it to Venus for 100 units of value, and if it will fail, we send it to Mars for 10 units of value. Using the original probability of success (0.6) as the probability that the information will predict a success, we obtain an expected value of 64 units with perfect information. Subtracting 56 units for the value of the primary decision problem, we obtain 8 units for the value of perfect information.

What might be wrong with this approach? Suppose that perfect information revealed that the mission would succeed on Mars but fail on Venus, or vice versa. These possibilities do not appear in Figure 2.2. To correct this omission, we might draw a new tree for the value with perfect information, as illustrated in Figure 2.3. First, we learn one of four possible predictions consisting of the four combinations of success or failure on Mars and Venus. Then we make the best decisions given this information, as indicated in the decision tree.

To assign probabilities of these joint events, one might reason that since

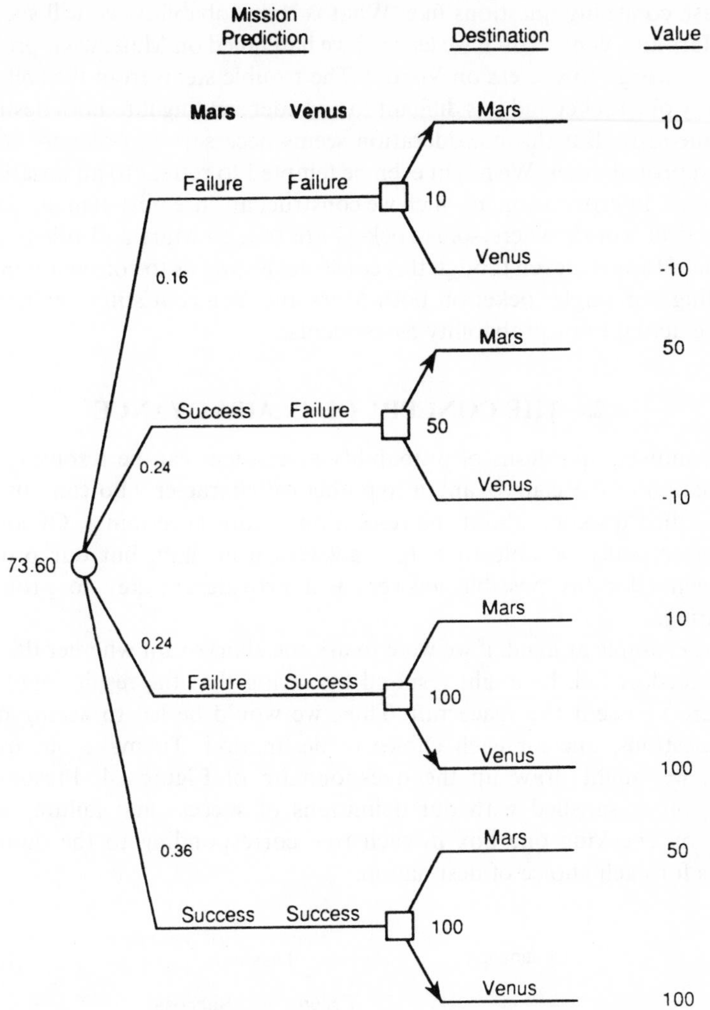

Figure 2.3 More complex tree for determining the value of perfect information.

landings on Mars or Venus appear in separate portions of the primary tree of Figure 2.1, the events must be independent and the probabilities should be multiplied with results, as shown in Figure 2.3. This will yield a value of 73.6 with perfect information. Subtracting the 56-unit value of the primary decision yields a 17.6 value of perfect information.

However, the independence assumption must be questioned. Since we can only send the mission to a single destination, might the events be mutually exclusive? If we simply try to assign the probabilities directly, we are tempted to phrase confusing questions like 'What is the probability we will succeed on both Mars and Venus?' or 'if we learned we had failed on Mars, what probability would we assign to success on Venus?' The trouble stems from the fact that we have only one rocket and it is difficult to consider sending it to both destinations simultaneously. But this consideration seems necessary in order to assign the required probabilities. We might even be tempted to retreat to an unsatisfactory 'frequentist' interpretation in which we construct an ensemble of many 'identical' hypothetical worlds where some rockets are sent to Mars and others are sent to Venus. However, even though this construct allows us to consider joint events of landing our single rocket on both Mars and Venus, it still remains difficult to make actual joint probability assessments.

2.4 THE CONCEPT OF CLAIRVOYANCE

These confusing questions of probability assessment can be resolved with the introduction of the clairvoyant, a hypothetical character who can answer any well-specified question about the resolution of any uncertainty. Of course, we shall never really be able to obtain answers from him, but our probability assignments for his possible answers will provide the key to probabilistic structuring.

In the example at hand, if we were to ask the clairvoyant whether the mission will succeed or fail, he might respond by saying that the result could depend on where you sent the spacecraft. Thus, we would be led to asking him two such questions, one for each choice of destination. To make our questions precise, we might draw up the questionnaire of Figure 2.4. Presuming the clairvoyant is satisfied with our definitions of success and failure, he could answer by checking one box in each row corresponding to the outcome he foretells for each choice of destination.

Destination	Mission	
	Failure	Success
Mars	☐	☐
Venus	☐	☐

Note: Check one box in each row

Figure 2.4 The clairvoyant's report form.

Before we engage the clairvoyant, we wish to calculate the value of his service in monetary units, i.e. the value of clairvoyance. Since the clairvoyant has two possible answers for each of the two questions, there are four possible reports for Mars and Venus together: failure, failure; failure, success; success, failure; success, success. We now must assign probabilities to receiving these reports from the clairvoyant. A possible probability assignment, compatible with our original assignments of Figure 2.1, is illustrated in Figure 2.5. This distribution implies probabilistic dependence between our assessments on the clairvoyant's two answers. For example, if he were to answer success if the spacecraft were sent to Mars, we would then assign a $0.554/0.6 = 0.923$ probability that he would also answer success if it were sent to Venus.

Philosophically, the important aspect of this formulation is that we are assigning (joint) probabilities *to events that could occur immediately*, as soon as the clairvoyant reveals his answers. We are not assigning probabilities to events that occur in different and mutually exclusive futures! We have avoided the awkward constructs of sending our single spacecraft simultaneously to both planets or of generating an ensemble of hypothetical universes.

We now apply this probability assignment by constructing the decision tree of Figure 2.6. The initial chance node represents the clairvoyant's revelation of one of the four possible reports, each indicated by the abbreviated report form on one of the following branches. The probabilities of Figure 2.6 are assigned to these reports. Following each report, we must make the best decision using the values derived in Figure 2.1. Having made the decision indicated by the arrows, we find that the expected value with clairvoyance—but before the clairvoyant reveals his answer—is 65.84. Subtracting the 56-unit value of the primary decision (without clairvoyance) yields a value of clairvoyance of 9.84.

Mars	Venus		
	Failure	Success	
Failure	0.354	0.046	0.4
Success	0.046	0.554	0.6
	0.4	0.6	

Figure 2.5 The joint probability distribution for clairvoyant's answers.

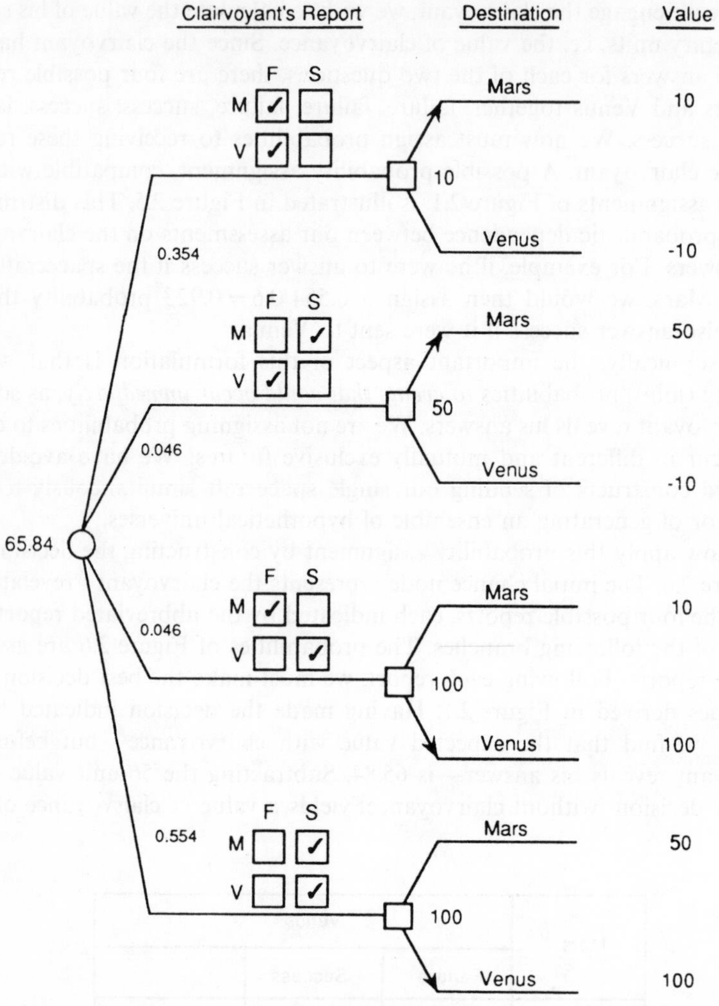

Figure 2.6 Decision tree for determining the value of complete clairvoyance.

2.5 CLARIFYING VALUE OF CLAIRVOYANCE USING INFLUENCE DIAGRAMS

The use of influence diagrams automatically avoids the pitfalls of the preceding decision tree analysis. We begin by structuring the primary decision problem with the influence diagram of Figure 2.7. We could immediately assess each node by attaching the two alternative destinations to the decision node, the probabilities of success or failure for each destination to the mission result node, and a table of values for each destination and result of the mission to the value

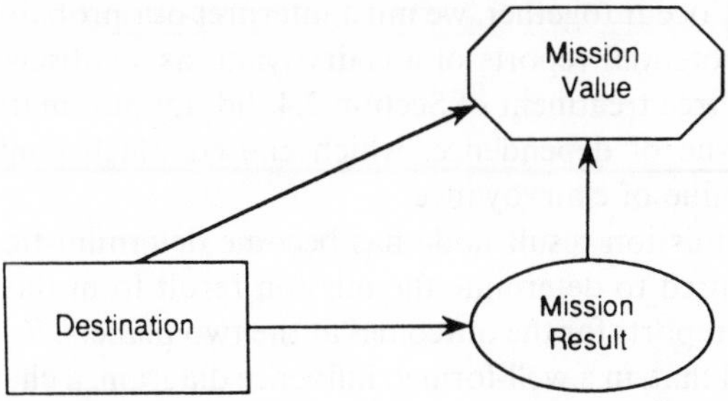

Figure 2.7 Influence diagram for the primary decision problem.

node. We could then solve for the best decision using influence diagram methods (Howard and Matheson, 1981; Olmsted, 1983).

Instead, we shall use the influence diagram to develop a deeper probabilistic model of the situation before proceeding to the evaluation stage. We are motivated to do this when we attempt to describe the situation with perfect information by adding an influence arrow from the mission result to the destination decision. We can immediately see that this modification would create a forbidden loop in the influence diagram, which indicates that the question is not well posed.

To avoid this loop, we modify the influence diagram structure to that of Figure 2.8. The Mars landing node will contain the probability of success or failure if we send the spacecraft to Mars. However, the similar probability for the Venus landing must, in general, be conditioned on the Mars landing. That is, these two events are not necessarily independent. Since these two events

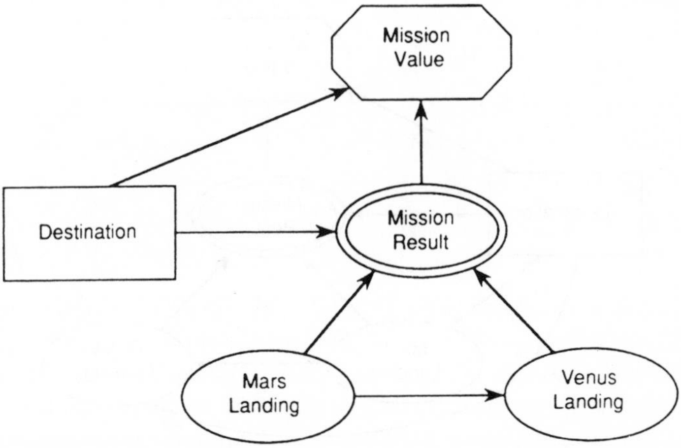

Figure 2.8 The primary decision problem in Howard canonical form.

cannot physically occur together, we must interpret our probability assignments in terms of the potential reports of a clairvoyant, as we discussed earlier. The original decision tree treatment of Section 2.4 did not automatically force us to deal with the issue of dependence, which caused much confusion when we considered the value of clairvoyance.

Note that the mission result node has become deterministic. It contains the simple logic required to determine the mission result from the destination and the clairvoyant's reports for the outcomes at the two planets. Ronald A. Howard (1988) has argued that, in a well-formed influence diagram, a chance node should never be preceded by a decision node, so we say the influence diagram of Figure 2.8 is in 'Howard canonical form', while the original diagram of Figure 2.7 was not. This form of the diagram completely separates the variables under the decision-maker's control (his free will) from the variables he cannot influence (his uncertainty). A little thought will convince you that any influence diagram can be put into Howard canonical form by introducing new chance nodes for each alternative embodied in the conditioning decision and recombining the alternatives and outcomes with a deterministic node, as we illustrate in Figure 2.9.

Because the influence diagram is in Howard canonical form, we can now add informational arrows, from either or both of the new chance nodes to the destination node, without creating loops. Therefore, each of these modifications represents a legitimate and well-formed value of information question. For example, Figure 2.9 shows an influence diagram that could be used to answer correctly the value of perfect information questions we were trying to address earlier in Figure 2.6. However, we are not yet in a position to calculate the answers, because it is unlikely we would be able to find an expert who would be comfortable directly assessing the probabilities required by Figure 2.8. So

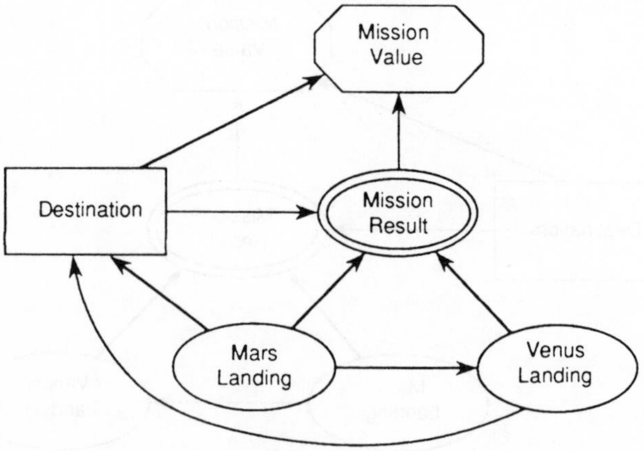

Figure 2.9 An influence diagram for complete clairvoyance.

we need to structure the influence diagram in terms of the underlying variables that experts can more easily contemplate and assess. That is, we need to expand our analysis to 'model' the expert's natural thought process.

When we discuss this problem with an expert on space flight, we might find that he is more comfortable with the influence diagram of Figure 2.10, because he has a mental model of how each of the chance nodes—launch system, Mars descender, and Venus descender—combine to produce a successful mission to either planet. And because he is more experienced in dealing with these component systems, he may be more comfortable with and better at assessing these 'more fundamental' probability distributions. After these assessments are made, we can manipulate the influence diagram into other forms for analysis, such as the one shown in Figure 2.9. We call the diagram used for assessment purposes the 'basic influence diagram'.

In this example, with his physical understanding of how the spacecraft systems produce these outcomes, the expert assesses their probabilities to be independent. This is why there are no influence arrows between the component system chance nodes. We shall assume that he assigns a 0.65 probability that the launch system will work, a 60/65 probability that the Mars descender systems will work, and an identical 60/65 probability that the Venus descender systems will work. (In general, these latter two probability assessments need not be equal, for example, because of completely different landing systems and environmental conditions

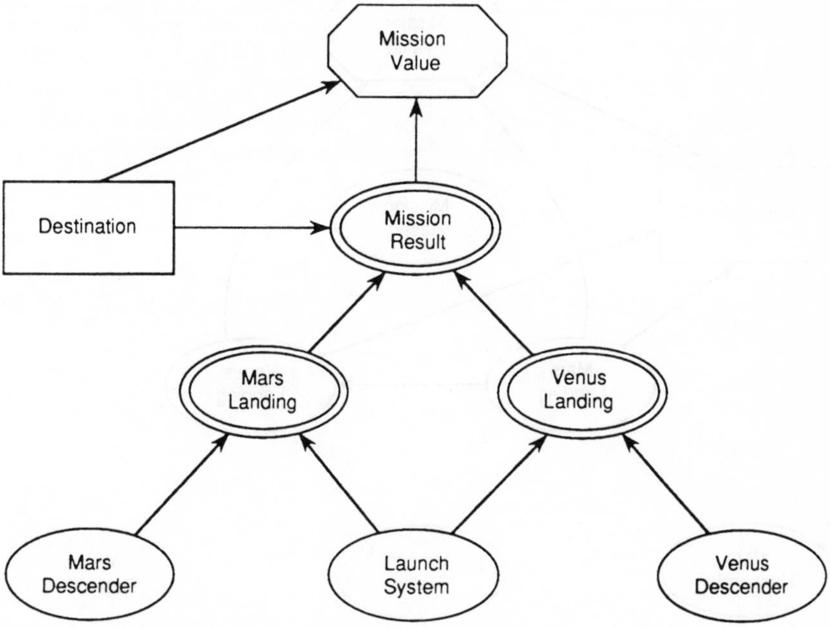

Figure 2.10 The basic influence diagram for the space mission problem.

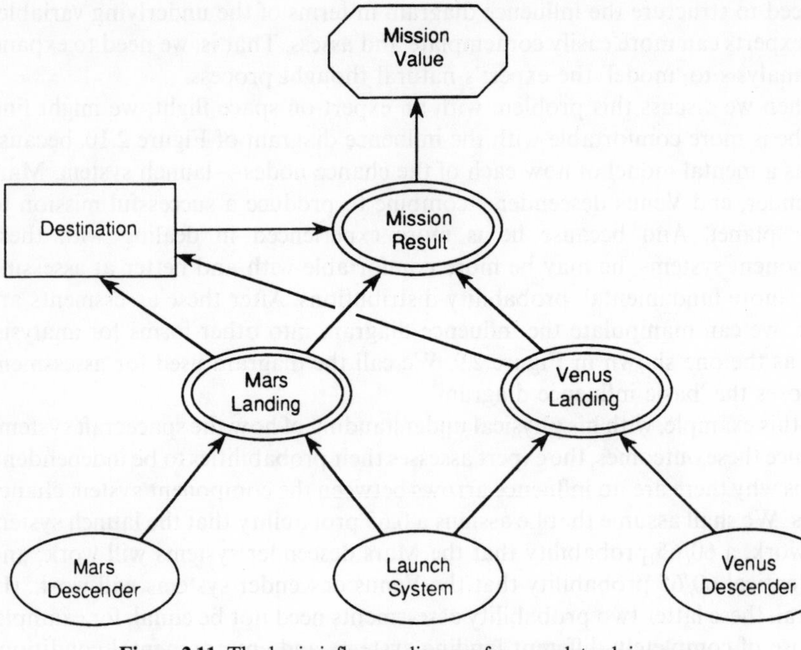

Figure 2.11 The basic influence diagram for complete clairvoyance.

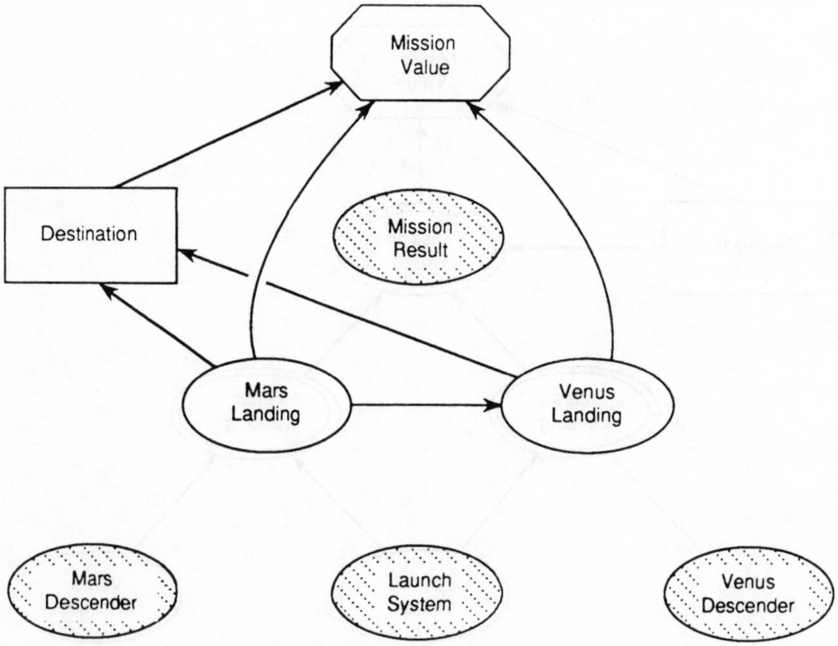

Figure 2.12 Reduced diagram for complete clairvoyance.

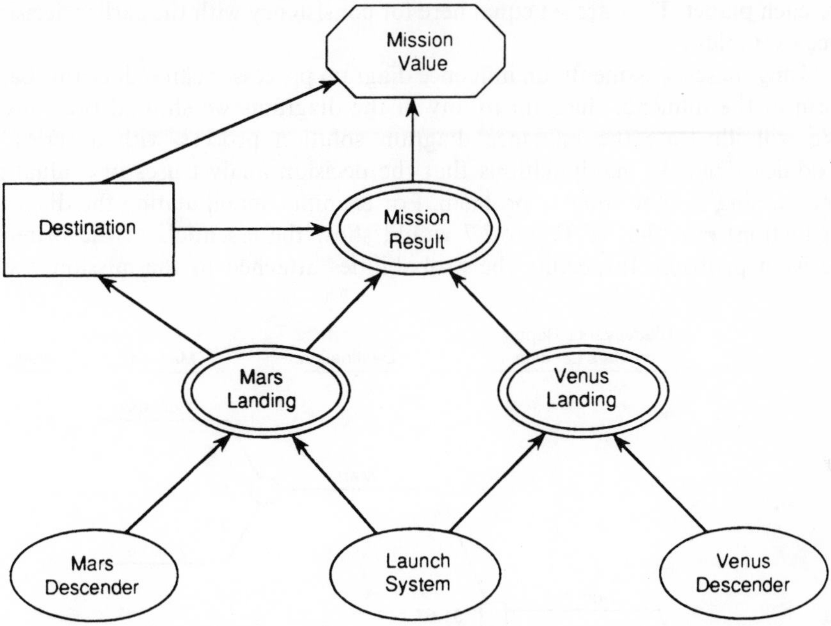

Figure 2.13 Influence diagram for determining clairvoyance on the Mars Landing.

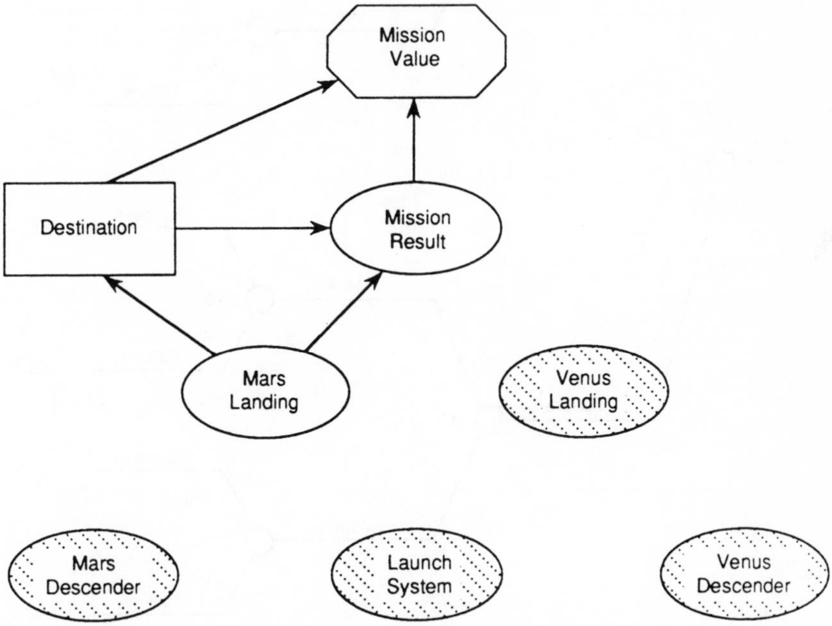

Figure 2.14 Reduced diagram for clairvoyance on the Mars Landing.

for each planet. They are set equal here for consistency with the earlier decision tree examples.)

Using these assessments, an influence diagram processor can reduce this basic form of the influence diagram to any of the diagrams we showed previously. We will illustrate the influence diagram solution process with a series of modifications and manipulations that the decision analyst uses in evaluating and gaining insight about a problem. For example, manipulating the diagram (reduction) into that of Figure 2.7 would show the essentials of the primary decision problem. Inspecting the probabilities attached to the mission result

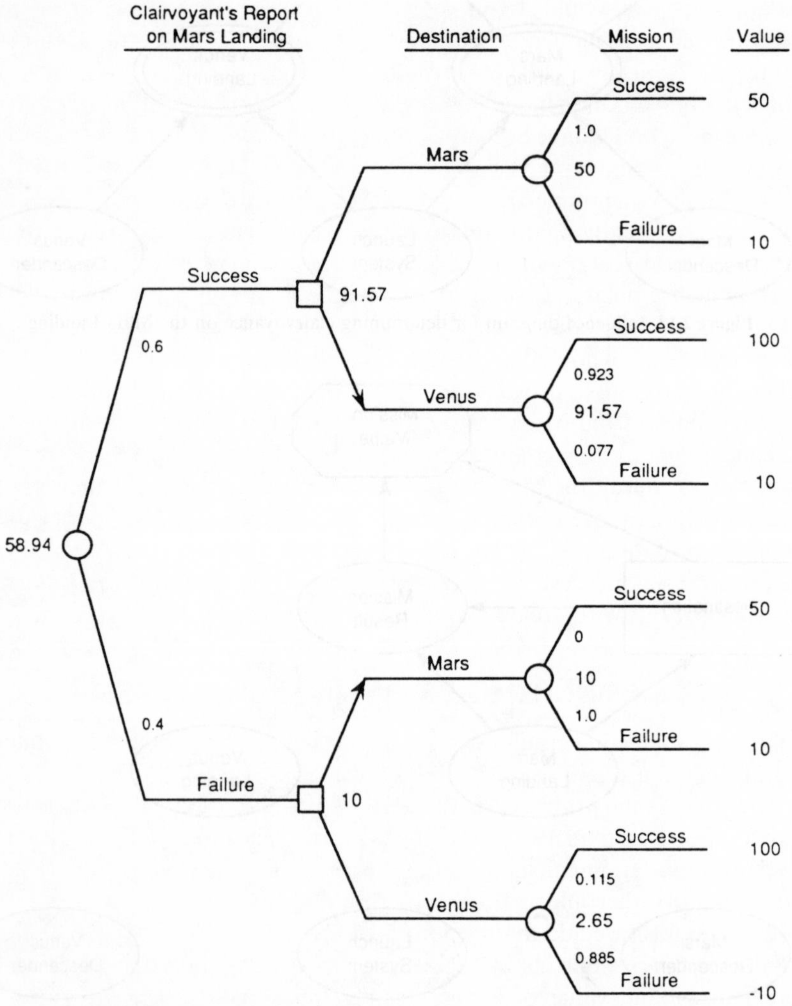

Figure 2.15 Derived decision tree for clairvoyance on the Mars Landing.

node would show a derived probability of 0.6 for success if sent to either destination. Further solution would show that the best decision is to send the spacecraft to Venus, giving an overall expected value of 56, as we saw earlier in Figure 2.1.

We are now in a position to use the influence diagram to determine the value of perfect information. Adding informational influences, as in Figure 2.11, and then reducing the diagram to the one in Figure 2.12 allows us to calculate correctly the value with free clairvoyance on mission results for both destinations. As Figure 2.6 shows, this value is 65.84, and by subtracting the primary problem value of 56, we arrive at a value of clairvoyance of 9.84. Using this approach, we would derive the decision tree representation of Figure 2.6 from the reduced influence diagram of Figure 2.12. The shaded nodes in Figure 2.12, which we call 'ghost nodes', indicate the nodes that have been removed from the influence diagram by the solution algorithm. In this case, the three ghost nodes have been integrated into the joint distribution represented by the Mars landing and Venus landing nodes.

We can also determine the value of clairvoyance on the mission result for each destination individually. For example, the influence diagram of Figure 2.13 illustrates the situation with (free) clairvoyance on the Mars landing. Reduction to the influence diagram of Figure 2.14 yields the results shown in tree format in Figure 2.15. This value of clairvoyance is 58.94 minus 56, which equals 2.94. It is interesting to note that if the clairvoyant foresees that we will have a successful Mars mission, we should send the spacecraft to Venus, and if he foresees failure of the Mars mission, we should send the spacecraft to Mars! We can rationalize this counterintuitive result by inspecting the mission value table, but we should be aware that similar unexpected results often occur in practice. Intuition, even of an experienced decision analyst, is a poor guide to valuing information.

Repeating this process, we can reproduce all the decision tree results, not only quickly, but with greater clarity of interpretation.

2.6 USING THE BASIC INFLUENCE DIAGRAM

Even more interesting results can be obtained by 'mining' the basic influence diagram of Figure 2.10 for further insights. Since this diagram captures how the expert views the problem, results derived directly from it will give the expert feedback in terms that he can readily interpret and understand. For example, values of information calculated from this diagram are often readily translated into testing or other information-gathering programs.

For example, if we add an informational arrow from the launch system node to the destination node, as in Figure 2.16, we are asking a question that may shed light on the value of further testing the launch system. Reducing the influence diagram to that of Figure 2.17 and displaying the results in tree format,

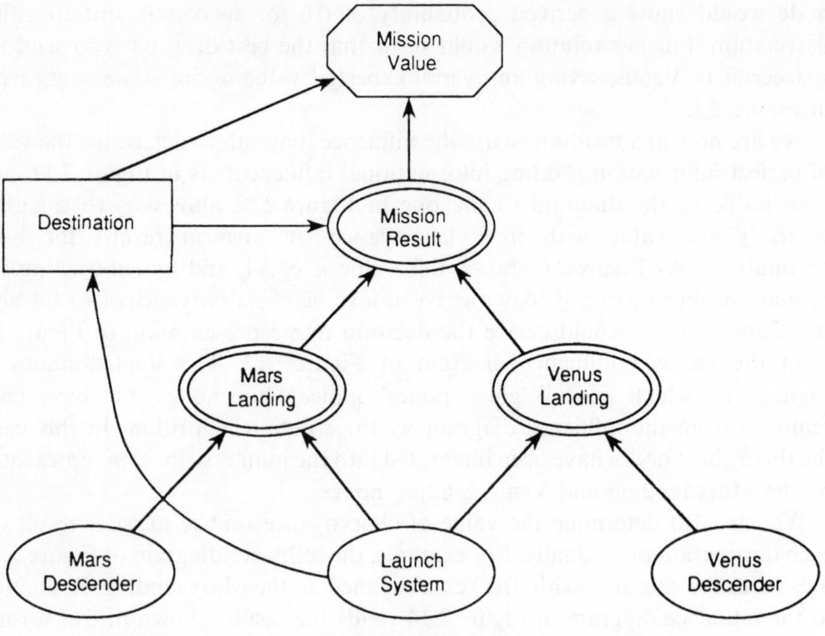

Figure 2.16 Influence diagram for clairvoyance on launch system.

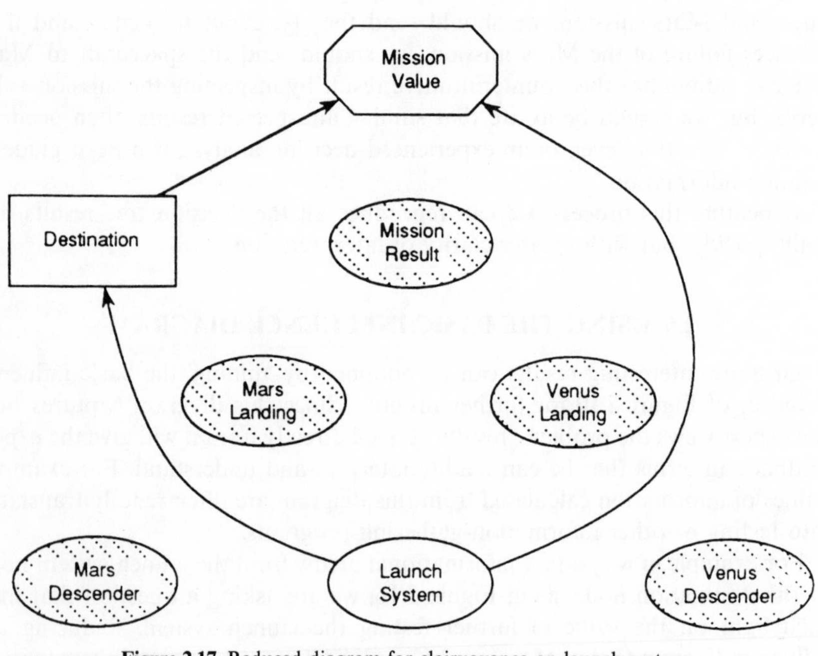

Figure 2.17 Reduced diagram for clairvoyance on launch system.

as shown in Figure 2.18, gives a value of clairvoyance of 63 less 56, or 7. This would indicate a high value of further testing. In a real case, the expert may wish to develop the launch system node into a richer set of nodes showing the physical events that combine to create a successful launch. The value of clairvoyance on these events would give the expert even more insight into information-gathering actions.

Similarly, if we add an informational arrow from the Mars descender to the

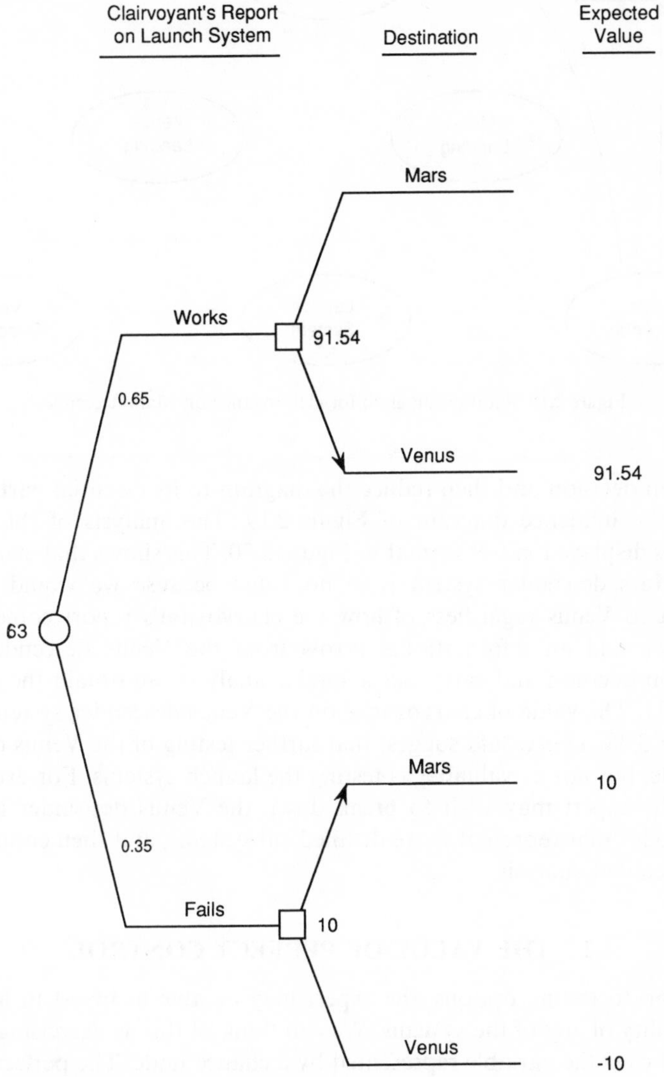

Figure 2.18 Derived decision tree for clairvoyance on launch system.

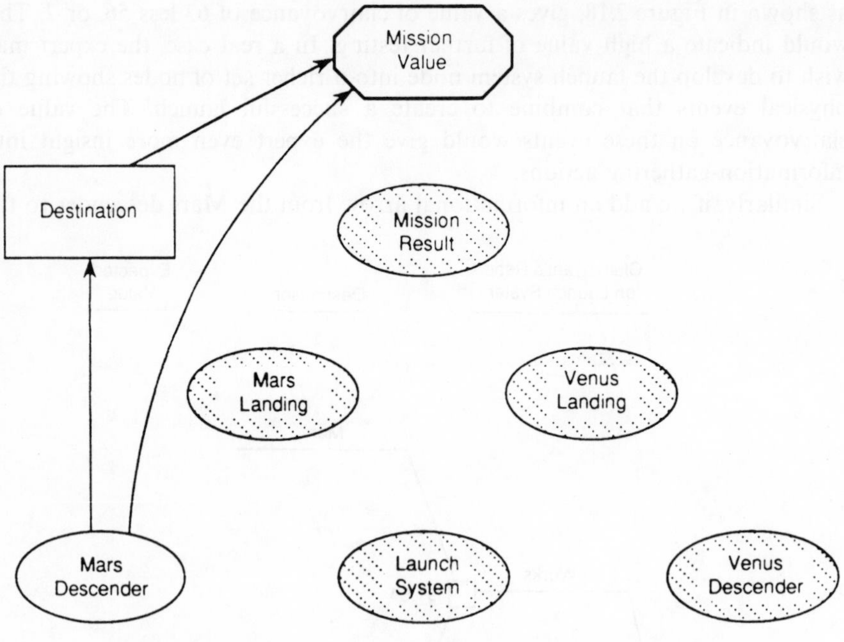

Figure 2.19 Reduced diagram for clairvoyance on Mars descender.

destination decision and then reduce the diagram to its essential parts, we will produce the influence diagram of Figure 2.19. The analysis of this reduced diagram is displayed in tree format in Figure 2.20. This shows that clairvoyance on the Mars descender system is of no value because we would send the spacecraft to Venus regardless of how the clairvoyant's report comes out. If, instead, we add an informational arrow from the Venus descender to the destination decision and carry out a similar analysis, we obtain the results of Figure 2.21. The value of clairvoyance on the Venus descender system is 59.38 less 56, or 3.38. This would suggest that further testing of the Venus descender is valuable, but not as valuable as testing the launch systems. For even deeper insight, the expert may wish to break down the Venus descender node into several nodes that represent more detailed subsystems, and then continue with a more detailed analysis.

2.7 THE VALUE OF PERFECT CONTROL

In addition to testing options, the expert may be able to invest in improving the reliability of any of the systems. We can think of this as exercising a degree of control over the variable represented by a chance node. The perfect achievement of control would be to increase the probability of the best result to 1.0.

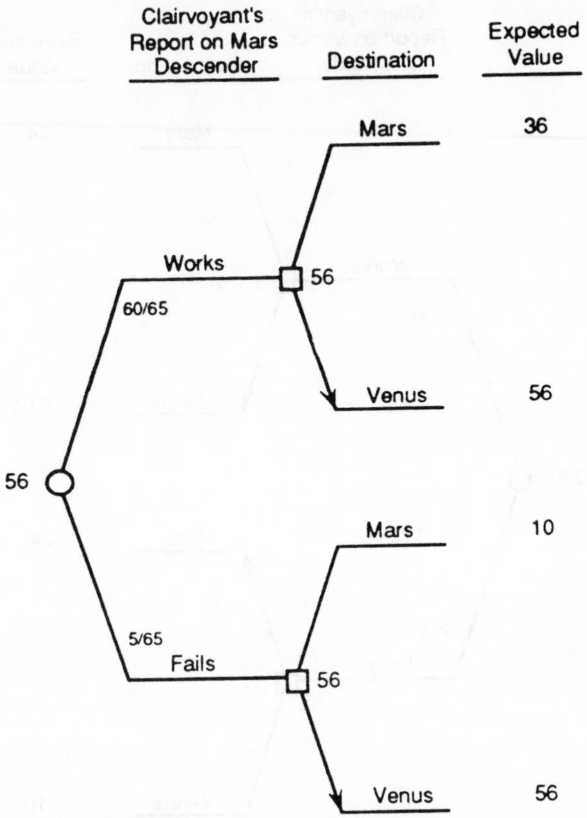

Figure 2.20 Derived decision tree for clairvoyance on Mars descender.

We can think of achieving perfect control by hiring a hypothetical wizard who will ensure that we get the best outcome.

For example, we could create the influence diagram for perfect control of the launch system by changing this chance node on Figure 2.16 to a decision node, as shown in Figure 2.22. This diagram indicates that first we choose which launch system outcome we want (by engaging the wizard), then we select the destination, and so forth. Reducing this diagram will yield the value with free wizardry.

However, we can shortcut this effort by inspecting the tree display for perfect information that was shown in Figure 2.18. Realizing that perfect control on launch system means we can choose whichever initial branch we like best, we clearly would have the launch system work for a value of 91.54. Subtracting the primary problem value of 56 yields a value of perfect control of 35.54. Of course, this value is much higher than the value of 7 we arrived at for perfect

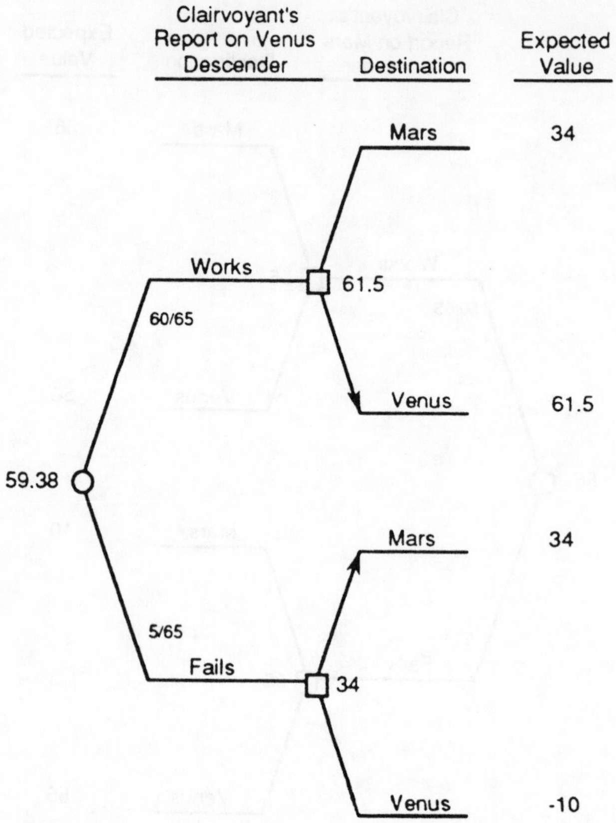

Figure 2.21 Derived decision tree for clairvoyance on Venus descender.

information—wizards are more valuable than clairvoyants. By inspecting Figures 2.20 and 2.21, we see that the value of perfect control on the Mars descender system is 0, because even if we could ensure that the Mars descender would work, we would still send the spacecraft to Venus. The value of perfect control on the Venus descender system is 61.5 less 56, or 5.5. These values would guide the expert in investing in further improvements to the spacecraft systems.

The above results are interesting to the expert because they correspond to physical systems that he can independently improve. His physical model of how these systems interrelate is represented in the influence diagram, so it is meaningful to use the influence diagram to evaluate and interpret the impact of these changes. Chance nodes on the basic influence diagram, which we defined earlier, often have a physical (or causal) interpretation. However, we should check with the expert to ensure that he can envision the possibility of controlling

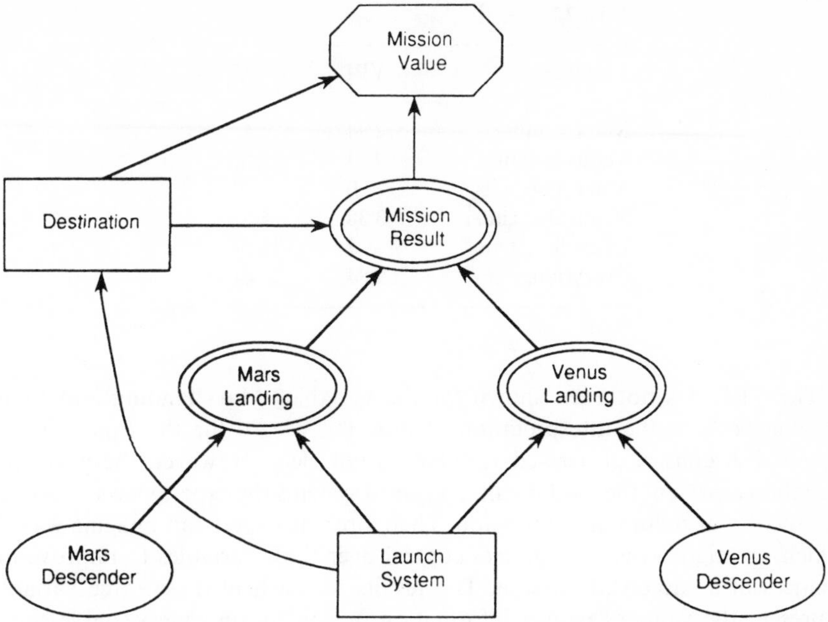

Figure 2.22 Influence diagram for wizardry on Earth Launch.

the chance node in question *without changing his structure or his assessments for the remainder of the diagram.* Otherwise, our computational results might not have any real-world interpretation.

We must be careful not to carry out these calculations on a diagram that is not basic. For example, we used the reduced diagram of Figure 2.14 to calculate the value of clairvoyance on the potential Mars landing. We might be tempted to change the Mars landing node to a decision node and calculate the value of wizardry. From the tree of Figure 2.15, we would arrive at a value of 91.57 minus 56, which equals 35.57. But because we are adjusting a derived probability, it is not clear what physical changes could create this adjustment. Usually, we cannot satisfactorily interpret the value of perfect control unless it is calculated from the basic influence diagram.

In this example, we have illustrated only the calculation of the value of perfect control for the case of chance nodes having no predecessors. While extension of this concept to the general case is possible, too much new material would need to be introduced to treat it in the present chapter.

2.8 CONCLUSIONS

Table 2.1 summarizes VPI, the value of perfect information (clairvoyance), and VPC, the value of perfect control (wizardry), for the spacecraft example.

Table 2.1

Variable	VPI	VPC
Mars landing	2.94	N.A.
Venus landing	9.84	N.A.
Mars descender	0	0
Venus descender	3.38	5.5
Launch system	7	35.54
Everything	9.84	44

The VPC has not been shown for the variables Mars landing and Venus landing because the interpretation of how the wizard (or the expert) would accomplish control of derived variables is not clear. However, the next three variables represent the fundamental physical systems the expert views as causing the success or failure of the mission. Therefore, the expert can imagine ways in which we might assert a degree of control over these variables to improve our chances of a successful mission. The results for each of these three variables represent the value of perfect information or control on each variable singly. Of course, they can also be explored in combination. The last line shows the value of gaining perfect information or control on all three variables simultaneously. Notice that these joint values are not simply the sums of the individual ones. The sum of values of clairvoyance on the three variables is 10.38, which is greater than the 9.84 value of clairvoyance on everything, while the sum of the values of wizardry on the three variables is 41.04, which is less than the value of wizardry on everything. No general conclusion can be made about these relationships; the analyst should explore all meaningful combinations.

All the calculations in this chapter have been done on an expected-value or risk-neutral basis. However, the concepts of perfect information and perfect control are meaningful for any risk attitude. If the risk attitude is constant, that is, if the utility function is exponential, then these values can be obtained by subtracting the original value (without either clairvoyance or wizardry) from the value with free clairvoyance or free wizardry. For a more general risk attitude, a cost must be attached to the use of the services of the clairvoyant or wizard. This cost must then be iteratively adjusted until the values with and without the services are equal. The final amount of this cost can then be interpreted as the most one should pay for the services.

We have shown how the influence diagram clarifies the probabilistic model of a problem and makes it easy to determine which value of perfect information questions are meaningful. If a particular question is not meaningful, because the additional required arrow would create a loop, modification of the diagram into Howard canonical form will facilitate posing well-formed informational questions.

We have also shown how influence diagrams can be used to determine the value of perfect control. However, a new subtlety arose. We defined the basic influence diagram as the one used to make the original probability assessments. Usually, but not always, the chance nodes on this diagram have a physical interpretation. Because of this interpretation, we can contemplate changing their probabilities without changing the problem captured in the rest of the diagram. In the case of chance nodes without predecessors, changing basic chance nodes to decision nodes that precede the primary decision (or decisions) allows calculating meaningful values of perfect control. The case of chance nodes that have predecessors introduces new considerations and will be treated in another paper.

Since its invention, the concept of influence diagrams has proven to be extremely useful in clearly stating and solving decision and inferential problems. For example, 20 years ago the spacecraft example would have been treated using decision trees and auxiliary 'nature's trees'. The influence diagram combines the information from both of these constructs. The influence diagram has also proven to be one of the easiest ways of communicating a decision problem to a computer. For example, all the influence diagrams in this chapter were drawn, assessed, and solved in a few hours using a properietary influence diagram processor on a personal computer.

In a decision-consulting environment, influence diagram processors allow us to focus most of our effort on capturing the structure and assessments of our client's problem and, using the concepts described in this chapter, to 'mine' the problem for maximum insight. Very little time and attention need be spent on simply getting correct answers.

ACKNOWLEDGEMENT

The author gratefully acknowledges the contributions of his colleagues Samuel Holtzman, Ronald A. Howard, and David Matheson, whose penetrating questions and ideas have greatly enhanced this chapter.

2.9 DISCUSSION

2.9.1 Discussion by Richard E. Barlow

The author has given a very clear account of the 'Howard canonical form' for influence diagrams and why it is useful and even necessary in analyzing decision problems with respect to the value of information. He provides excellent reasons why the influence diagram is preferable to the decision tree representation. Since the decision tree is basically state space enumeration, it is easier for the beginner to understand. However, it simply cannot be used effectively to understand and answer very deep decision analysis questions related to the probability calculus.

The Howard canonical form for influence diagrams is related to the foundations of Bayesian decision analysis itself. One of the principle axioms of Bayesian decision analysis is the 'sure thing principle' (cf. Savage, 1954). By the sure thing principle we mean that if decision 1 is preferred to decision 2 when a quantity θ is known to be θ_1 and also decision 1 is preferred to decision 2 when θ is known to be θ_2, then decision 1 is preferred to decision 2 when we are unsure whether θ is θ_1 or θ_2. This principle is absolutely necessary for the argument that self-consistency implies Bayesian behavior; that is, decisions should be made by maximizing expected utility (cf. Axiom 0, Rubin, 1987).

If there is an arc from a decision node to a probability node as in Matheson's Figure 2.7, then it is not clear how to interpret the sure thing principle. Since θ, i.e. mission result, depends on the decision contemplated, namely the destination, the sure thing principle does not directly apply. In fact, we might even doubt that the sure thing principle holds in this case. However, when Figure 2.7 is redrawn in Howard canonical form as in Figure 2.8, the random quantities Mars landing and Venus landing no longer depend on the destination decision and we immediately see that the sure thing principle does indeed hold. The foundations of Bayesian decision analysis are secure!

Matheson has done such a nice job of showing the power and uses of the influence diagram we might ask: why bother with the decision tree representation except as a teaching device for beginners?

REFERENCES

Howard, R. A. (1988) From influence to relevance to knowledge. Presented at the Conference on Influence Diagrams for Decision Analysis, Inference, and Prediction, 9–11 May, Berkeley, Calif. In R. M. Oliver and J. Q. Smith (eds) (1990) *Influence Diagrams, Belief Nets and Decision Analysis*, Wiley, Chichester, pp. 3–23.

Howard, R. A. and Matheson, J. E. (1984) Influence diagrams. In R. A. Howard and J. E. Matheson (eds) *Readings on The Principles and Applications of Decision Analysis*, vol. II, Strategic Decisions Group, Menlo Park, Calif., pp. 719–62.

Olmsted, S. M. (1983) On representing and solving decision problems. Ph.D. dissertation, Engineering Economic Systems Department, Stanford University, Stanford, Calif.

Rubin, H. (1987) A weak system of axioms for 'rational behavior' and the non-separability of utility from prior, *Statistics & Decisions*, **5**, 47–58.

Savage, J. L. (1954) *The Foundations of Statistics*, Wiley, New York (2nd rev. edn, Dover, 1972).

CHAPTER 3

Complexity, Calibration and Causality in Influence Diagrams

Terry Speed, *University of California, Berkeley, USA*

ABSTRACT

An influence diagram (ID) is usually part of a probabilistic model for some process taking place in the real world. All such models involve a mixture of knowledge and experience from the subject area under study, some of which statisticians would recognize as 'prior information' and some data, and an implicit or explicit objective. IDs are the most visible and most readily apprehended part of such models and in contemplating them, interpreting them and manipulating them, we can easily forget that they are only part of a model whose purpose is to assist us in addressing the questions of interest in context.

Experience from areas in which those close relatives of IDs—path diagrams—have been used for many years, suggests that care must be taken in using ID to make inferences about the real world. Do we know how the complexity of an ID trades off against its value in addressing various questions of interest? Does it (Occam's razor) matter in this context? Do we know how to calibrate an ID (i.e. provide a reality check on it)? How do we decide whether an arrow should point in one direction, or the other, or be bidirectional? Again, does it matter?

These questions will be examined using simple examples involving multinomial and Gaussian models over IDs.

3.1 INTRODUCTION

Some years ago I was very struck with the elegance and simplicity of the idea of defining independence models for random variables using undirected graphs, that is, influence diagrams. The idea grew from a study of classes of models for structured multinomial and Poisson random variables developed by L. A. Goodman, S. J. Haberman, and others, and models for normal random variables

Influence Diagrams, Belief Nets and Decision Analysis
Edited by R. M. Oliver and J. Q. Smith

introduced by A. Dempster and popularized by N. Wermuth (see Speed, 1978 for references to the work of these people). Around that time I tried to use these models in consulting work, hoping that they would lead to easily interpretable analyses which also embodied answers to substantive questions in the relevant subject-matter areas. A little latter, I entertained similar hopes in relation to causal models in the social sciences (see Kiiveri and Speed, 1982 for a review of analogues of the earlier ideas for directed graphs).

These days I am considerably less optimistic. I have come to view much of this modelling as rather sterile, and I have rarely found it to throw light on important subject-matter issues. Rather, it seems to me, we statisticians were 'modelling data', either by the use of chi-squared tests or (less often) via prior-prior posterior analyses, and we lost sight of the importance of context and of having serious and sharply focused questions to address. When we did, the models rarely provided much assistance. Of course these criticisms are not particularly directed at the use of influence diagrams; they apply to much of the linear, log-linear and other 'modelling' that went on then and still goes on today.

I might also add that around this time I first became aware of the appalling statistical practices which were to be found in the field of probabilistic risk assessment, exemplified by the various attempts to 'estimate' probabilities associated with the risks from nuclear reactor accidents. Much of this work made use of fault trees and event trees, now viewed as types of influence diagrams, and had a Bayesian flavour, but was really so bad that it could be totally discounted (see Speed, 1983 for some recent remarks on this topic).

With this background, I hope you will find my caution if not scepticism concerning the use of influence diagrams in decision analysis, expert systems, 'modelling' and so on, understandable. The influence diagram (graph, network) serves to define certain aspects of a method or model precisely, and helps us to understand what the method or model means in non-mathematical terms. At least that is how it should be. In practice influence diagrams are just like the common linear statistical models used in regression, the analysis of variance and time series, and which in a sense they generalize: they are all too easily written down and their assumptions taken as reasonable, while the hard questions are too easily ignored or quickly dismissed. The reason why a given diagram is appropriate in a particular context, and not some other diagram, or no diagram at all, and the extent to which the diagram really helps us to answer the questions of interest better than we might otherwise, are not normally seen as part of the 'theory' of diagrams. These are aspects of the relation between diagrams (models), contexts and the real world, and that is something we rarely discuss. Just as with regression, time-series analysis or anova, it seems that we all feel more comfortable focusing on issues such as the calculus of updating, proving theorems, advocating general principles of inference for others to use, suggesting applications which are little more than thought-experiments, and illustrating our work with examples which rarely involve more than replacing symbols in formulae by real numbers.

I apologize for the rather preachy nature of this chapter. It contains no theorems, indeed rather few symbols, and it reports on no case-studies. Perhaps I should have kept completely silent, but I have not. I have tried to bring to this conference the perspective of an applied statistician, of somone who is now trying to assist others in making the best use of data and, where appropriate, of the models and methods of statistics, to achieve more or less well-defined objectives in science or technology. As my current concerns involve model complexity, calibration and that old chestnut, causality, I hope you will not find my remarks about these topics in the context of influence diagrams completely without interest.

3.2 COMPLEXITY OF INFLUENCE DIAGRAMS

By the complexity of influence diagrams I mean, roughly speaking, the number of independent parameters required to specify the underlying probability model. For example, the complexity of a linear or log-linear regression model is proportional to the number of linearly independent regressors plus one (for the variance), the complexity of an ARMA (p, q) model is proportional to $p + q + 1$, while that of a spatial Markov model which might be used in image segmentation depends upon the size of the neighbourhood used. A more precise definition, cf. Rissanen (1983), would refer to the precision with which the parameters are specified; indeed what we really want is the total number of bits required to carry out the necessary probability calculations, including specifying the functional form of any probability distributions used. Such an approach would be necessary if we wished to discuss the complexity of a classification tree, cf. Breiman *et al.* (1984), Goldman *et al.* (1985), and Quinlan and Rivest (1987), for simply counting parameters in such models will not work: the tree must be described too.

When it comes to setting up an influence diagram in a particular context, the complexity of the diagram will depend to some extent on subject-matter considerations and to some extent on the amount of data available for use in evaluating the parameters of the model. There seems little point in using a diagram which involves many fine distinctions whose quantitative impact on the model can only be roughly assessed, simply because it embodies the most up-to-date knowledge about the process. Here I am assuming that we have a well-defined role for the diagram, and that we also have some measure of how well the diagram (model) performs in that role.

For simplicity and concreteness, I will now restrict my remarks to the use of influence diagrams for prediction, using this term in a general sense which includes the allocation of units to categories on the basis of ancillary information, and in particular, to the use of influence diagrams in assisting medical diagnosis

Figure 3.1

and prognosis. I further suppose that the correct category of each unit eventually becomes known to the statistician.

The simplest problem which we might have in such a context is the following: we have a binary disease class D and a binary indicant S, suggesting the influence diagrams of Figure 3.1. A standard specification of such models would be via the marginal probabilities $P(D=d)=p(d)$ and the conditional probabilities $P(S=s|D=d)=p(s|d)$, and already we have a problem since these will need to be estimated from data or evaluated in some other way. How do we know that we can do better diagnoses with $p(s|d)$ depending on d, i.e. with Figure 3.1(b), than we can with Figure 3.1(a)?

A more complex but still extremely simple example of the same issue arises when we have two indicants S_1 and S_2: we need to decide between Figure 3.2(a) and (b).

In Figure 3.2(a) we would only need to specify $p(d), p(s_1|d)$ and $p(s_2|d)$, as the figure embodies the assumption of conditional independence of S_1 and S_2 given D, while in Figure 3.2(b) we need to specify $p(d)$ and $p(s_1, s_2|d)$.

What considerations might affect our choice between Figure 3.2(a) and (b)? We might prefer Figure 3.1(b) over 3.1(a) because we know that the indicant can help our diagnosis, or Figure 3.2(a) over 3.2(b) because we know that S_1 is independent of S_2 given D. On the other hand, it is possible that we do not have much data on (D, S_1, S_2) but only on (D, S_1) and (D, S_2), and are unwilling or unable to speculate on the nature of the joint probabilistic dependence of (S_1, S_2) on D.

Do these issues arise in more realistic situations? In Lauritzen and Spiegelhalter (1988, §2) we see that the 'FORCE' node on the causal network corresponding to a single muscle in MUNIN requires 270 values for the specification of its conditional distribution. And, even in that diagram, there are many conditional independence constraints.

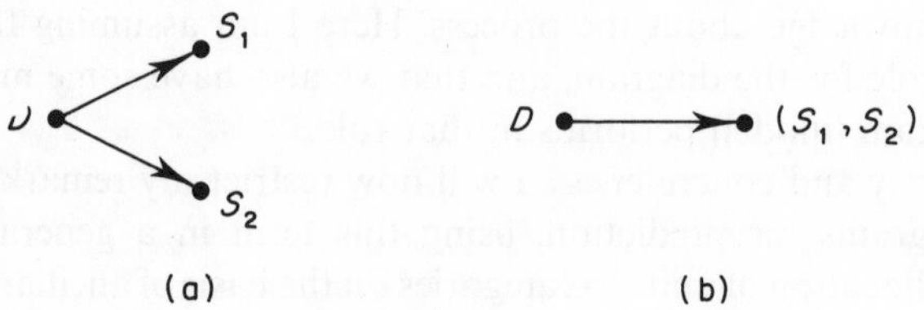

Figure 3.2

It seems to me inevitable that an influence diagram devised for use in prediction must embody, implicitly or explicitly, some trade-off between complexity and predictive power. The very reason an analyst would prefer Figure 3.1(a) to 3.1(b) would be that he or she thinks better diagnoses result; conversely often extremely complex expert systems are not permitted to become even more complex because it is not felt that better diagnoses will necessarily result.

How might we go about formalizing these ideas? We certainly need some honest measure of the performance of prediction models, but there is no shortage of these (see Dawid, 1986a for a discussion of this and related issues). Such measures would normally need to incorporate the relative importance of the different kinds of errors, cf. Titterington *et al.* (1981), Goldman *et al.* (1985), but that need not concern us here: we simply need some honest measure of the actual performance of the model. I use the term 'honest' here to draw attention to the fact that I regard performance measures based on resubstitution or cross-validation with training samples as inappropriate: they are not measures of how the model actually performed in practice, whetever asymptotic results they may have supporting their use as estimates of such a quantity, see Rissanen (1987) and the ensuing discussion of 'honesty'.

Once we have our honest performance measure and a body of data, our task is simple: we should compare alternative diagrams (models) on the basis of their performance on the data available, cf. Dawid (1984): 'Whilst there can never be any assurance that past forecasting performance will provide a reliable guide to future performance, what other reasonable guide can we follow?'

It may be that the general approach I have just described is already being widely practised. If so, I would welcome references and details as I am unaware of any major studies along these lines. The closest thing to it in the literature seems to be the classification tree methodology as outlined in Breiman *et al.* (1984). There I surmise that the models and methods are sufficiently unfamiliar to users that often inappropriate statistical methods (significance tests, Bayesian methods) and needlessly complex 'knowledge' of the processes involved can be bypassed, and the more pragmatic approach suggested here accepted. However, I would like to suggest that they reconsider their use of cross-validation: Quinlan and Rivest (1987) have shown how it can be avoided, and even their approach can be improved.

3.3 CALIBRATION OF INFLUENCE DIAGRAMS

If my future medical diagnoses are to be made by an expert system embodying some mix of epidemiological data, accepted medical knowledge and experts' personal opinions in an influence diagram, then I am going to be very interested in the performance characteristics of this system. I can talk to a human diagnostician, develop a continuing relationship if I wish, and come to my own opinions as to the degree of trust I wish to put in the diagnoses or prognoses I receive. What can I do with an expert system? Put my trust in Bayes's theorem?

A lot has been written about the calibration of probabilities, see e.g. Dawid (1986b) and references therein. Despite all this literature, I do not believe that there is any compelling reason why we should require the probability evaluation of expert diagnostic systems, for example, to be calibrated in the strict sense defined by Dawid. Nevertheless, I have to say that I do not want to have decisions relating to my medical future made by some anonymous system without first asking for some credentials. I wholeheartedly agree with the comment of Dawid (1982) made with reference to the personal aspect to Bayesian statistics: 'In its solipsistic satisfaction with the psychological self-consistency of the schizophrenic statistician it runs the risk of failing to say anything useful about the world outside.' Put another way, I do not want to hear probability assessments or evaluations by just any Bayesian, personal or impersonal, although I readily concede that they are entitled to their own opinions regarding uncertain events or quantities. I want to hear from Bayesians who assign high probability to true statements. My problem as the potential subject of a medical expert system embodying so-called Bayesian methodology is this: how do I tell the good ones from the others? As I have already indicated, I do not see the requirement of being well calibrated as especially relevant, particularly, as is well known, because analysts can be well calibrated and yet give poor predictions, and give good predictions, but be poorly calibrated. See Dawid (1986b) and De Groot and Fienberg (1983) for a discussion of the notions of refinement and resolution and their connection with scoring rules.

Where does this leave me? I am afraid that I can say nothing more on this topic than I did in Section 3.2 in my discussion of the complexity of an influence diagram: one procedure, whether or not it is based upon a diagram, or an implicit or explicit probability model, is to be preferred to another, if it is better according to a mutually agreed honest performance measure. If there are no explicit weights relating to the different outcomes which need to be considered, then my preference would be for assessments based upon the so-called logarithmic scoring rule, because of its theoretical properties and relation to likelihood, see e.g. Dawid (1984), Rissanen (1987). I would be very interested in seeing new arguments which support the use of more refined measures for the comparison of forecasters; unfortunately those which support calibration and resolution seem unconvincing, apart from the obvious fact that there is a calibration and a resolution component in total score.

3.4 CAUSALITY AND INFLUENCE DIAGRAMS

The term 'causal' is frequently mentioned in the context of influence diagrams, most commonly in their form as path diagrams in the social sciences, see the references to the work of H. M. Blalock, O. D. Duncan, L. A. Goodman, K.

Jöreskog, H. A. Simon and others in Kiiveri and Speed (1982). Usually such social scientists are investigating causal relationships, and the arrows in the diagrams indicate the direction of a hypothesized cause–effect relationship. In the probabilistic approach to expert systems exemplified by Lauritzen and Spiegelhalter (1988), 'causal networks' are viewed as encapsulating (medical) knowledge. The presence and the direction of the arrows are a reflection of assumed logical, physical, temporal or conceptual relations; they also serve to define the underlying probability model. This usage contrasts starkly with that adopted by Shachter and co-workers, see e.g. Shachter (1986), Shachter and Heckerman (1987), Shachter and Kenley (1988), where an assignment of directed edges is simply a reflection of the manner in which the probability model is currently being factorized. This approach is also (more or less) that of Barlow (1987), Smith (1988a, b) and others. For example, a joint distribution over the diagram of Figure 3.1(b) could be specified via $p(d)$ and $p(s|d)$, or via $p(s)$ and $p(d|s)$. In the former case Shachter would draw the diagram as in Figure 3.1(b), while in the latter he would reverse the arrow; indeed the notion of arc-reversal plays a central role in Shachter's theory of transformations of influence diagrams. Thus the presence of an arrow or directed arc in an influence diagram could indicate a tentatively proposed causal relation being investigated, a fully accepted causal relation or no causal relation whatsoever, depending upon the writer.

It would certainly help if some consistency of usage could be agreed upon in this field; the paper by Kiiveri and Speed (1982) was written to encourage just this within the social sciences, although it now seems unlikely that any such consistency will be achievable. Whatever one's views on the appropriate way to draw influence diagrams and the meaning of causality, one thing is clear: all that we have mathematically is a joint probability model for the random variables corresponding to the vertices (=nodes) of the diagram. The topic I want to address in this context is the following: to what extent are considerations of causality in influence diagrams—by which I mean the presence and direction of arrows—reflected in the associated probability models? My general answer to this question is: not much, most of the time, but when the presence and direction of arrows is important, some subtle issues can arise.

Firstly, when are the arrows unimportant? Consider the diagrams in Figures 3.3(a) and (b): clearly their difference cannot be reflected in probability model which everyone would assign to (U, V, W), namely one for which U and W are conditionally independent given V*. In such cases we may as well omit the arrows. A generalization of this observation was derived from the work of Wermuth (1980), and refers to what we termed configuration [>] in Kiiveri and Speed (1982), namely three vertices labelled i, j and k with $i \to k$ and $j \to k$,

*The remarks which follow can be modified to take into consideration deterministic vertices and, I think, the replacement of probability models by belief function models.

Figure 3.3

but having no edge connecting *i* and *j*. If an influence diagram has no instances of configuration [>], then its associated probability model has a particularly simple form, and all other influence diagrams with the same underlying undirected graph and no instances of configuration [>] will have the same probability model; see Kiiveri *et al.* (1984) for a detailed statement and proof of this assertion.

Having seen that in the absence of configuration [>], the probability model fails to embody the directions of the edges of an influence diagram in any meaningful way, let us turn to the simplest examples in which configuration [>] is present. We first consider the two diagrams of Figure 3.4. Here the probability models are very different: in Figure 3.4(a) we suppose that U and V are independent, and that the distribution of W given U and V is arbitrary, while in Figure 3.4(b), U and V are conditionally independent given W. As the causal interpretations of these two diagrams differ greatly, it is reasonable to ask if the probability models associated with them are readily distinguished on the basis of data. Before considering this, we should ask whether any non-trivial probabilities distributions can belong to the models associated with both Figure 3.4(a) and (b). If (U, V, W) is a triple of jointly normal random variables, or if all three are binary, then by an observation which goes back to Yule, U and V are both independent, and conditionally independent given W, if and only if, either U is independent of (V, W) or V is independent of (U, W). No such clear-cut conclusion exists for non-normal or non-binary triples, and in general we can have U and V both independent and conditionally independent given W, just as we can have (Simpson's paradox) U and V positively dependent marginally, and negatively dependent conditional on W, see e.g. Wagner (1982). In my view, the existence of such joint distributions eliminates the possibility of a clear-cut causal interpretation of joint probability distributions, and I conclude that we should regard the directed edges in an influence diagram as having no significance beyond that of helping specify a probability model for the associated random variables. The method adopted in Kiiveri *et al.* (1984)

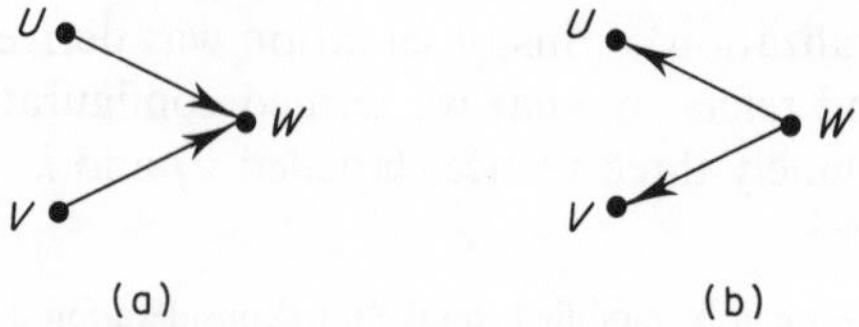

Figure 3.4

starts from the universally true factorization symbolically denoted as follows:

$$(123\ldots n) = (1)(2|1)(3|12)\ldots(n|12\ldots\overline{n-1})$$

and simplifies by writing

$$(123\ldots n) = \prod_{i=1}^{n} (i|D_i)$$

where $D_i \subseteq \{1, 2, \ldots, i-1\}, i = 1, \ldots, n$.

Summarizing, I believe that nothing would be lost if the term 'causal' was never used in the context of influence diagrams, and that the sole defensible role for directed edges is in defining the probability model. Further, I repeat that if no instances of configuration [>] are present in the diagram, the model can be defined without any directed edges at all, and is probably best done in this way, see Speed (1978) for a wide range of such model classes.

In closing let me offer for consideration two examples which do not yet seem to have arisen outside the area of social science, cf. Kiiveri and Speed (1984). Firstly, note that in Figure 3.2(b) we could require S_1 and S_2 to be conditionally independent given D equal to one value, d_1 say, but not given its other value, d_2 say, getting a model intermediate between those of Figures 3.2(a) and (b), one not adequately depicted diagrammatically. Similarly, for three arbitrary random variables U, V and W, it is clear that no distinction between the diagrams of Figures 3.5(a) and (b) can be made on the basis of the joint distribution of (U, V, W). Suppose now that U takes four values and is written as (U_1, U_2) where U_1 and U_2 are binary. Then the diagrams in Figures 3.6(a) and (b) reflect distinct probability models for the four random variables (U_1, U_2, V, W), namely Figure 3.6(a): U_1 and W conditionally independent given U_2, and U_2 and V conditionally independent given U_1 and W, and Figure 3.6(b): U_2 and V

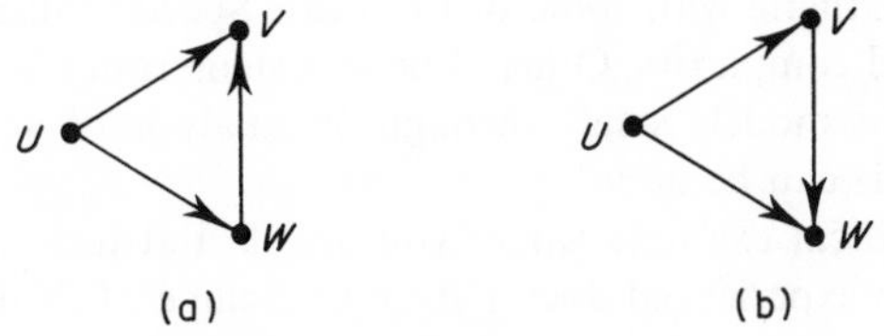

Figure 3.5

Figure 3.6

conditionally independent given U_1, and U_1 and W conditionally independent given U_2 and V. It makes no sense to put a unidirectional edge between U_1 and U_2; we can either make it bidirectional or undirected. In my mind, examples like this reiterate my point that we should treat influence diagrams as doing no more than helping define probability models; considerations of causality are highly context-dependent, and should be treated the way in which they should always be treated in statistics: preferably not at all, but if necessary, then with very great care! I invite the interested reader to construct joint distributions for (U_1, U_2, V, W) which satisfy the requirements of both of Figures 3.6(a) and (b).

ACKNOWLEDGEMENTS

I would like to thank Steffen Lauritzen, John Darroch, Harri Kiiveri and John Carlin for discussions in years gone by, and David Freedman and Phil Dawid for research which has been of great assistance to me in more recent times.

3.5 DISCUSSION

3.5.1 Discussion by J. Q. Smith

I find myself in the confusing position of simultaneously agreeing and disagreeing with Professor Speed. Most statisticians have little interest in real modelling and will tend to use modelling tools in a cavalier and inappropriate way. I do not believe that this is a good reason for not using IDs for their main purpose: to express in a simple way the structure of a proposed model so that it can be thoroughly explored, criticized and if necessary reformulated. Theorems about graph manipulation seen as an aid in this procedure can be valuable. Otherwise they are not that interesting.

I am very sympathetic with some of Professor Speed's difficulties with issues of performance and complexity. Others I believe stem from his wanting to decide objectively between models solely through an analysis of a given data set. I cannot see how this can be done.

Turn to his medical example where you are ill and have a choice between the diagnosis of an experienced doctor A or a machine M. You have previously analysed the performance over 'similar' indicants and 'similar' patients and found that although both scored well the machine M scored better than A. Should you use M?

I do not believe you can answer this question on the information given alone. Any rational person would surely want to examine other evidence pertinent to *his* symptoms *now*. How similar to you were the other patients used in the sample and how similar were their indicants? How were the algorithms in M constructed and tuned? Typically there would be little or no hard data available to answer these questions.

To a subjectivist a model is a description of the way you are currently structuring a problem. It has no objective validity. A good model is one which produces good predictions for you when you use it. Past scores might help you to assess future predictive power, but I suggest that using quadratic or log scoring rules on their own as a way to discriminate between contenders is liable to be very misleading—because it does not explicitly address important issues associated with *your* problem *now* (see Smith and Gathercole, 1986).

Again, the issue of complexity can be addressed subjectively. Simple models can usually be based on more fundamental statements about a problem. Within them it is easier to faithfully represent how you believe components of a process will be related. On the other hand, complex models tend to be more context specific and belief within them difficult to represent faithfully. For a given setting you may therefore believe, a priori, that a simple model will predict better than a complex one. But whether it does will depend again crucially on context and not just on the result of a statistical algorithm used on past data.

Finally I concur with Professor Speed on this issue of causality. Incidentally the existence of joint distributions with the property $U \perp\!\!\!\perp V$ and $U \perp\!\!\!\perp V|W$ will contain those for which the condition density/mass function of $W|U, V$ separates into the product of two functions, one a function of W and U only, the other a function of W and V only. Examples of these abound even when W is (multivariate) normal and U and V are univariate normal.

3.5.2 Discussion by Carlos Pereira

Introduction

Professor Speed stated that he expected too much from graphs. However, today he does not believe that graphs can solve all problems. In my case it happened differently. I did not believe that graphs could help me at all. Today I think

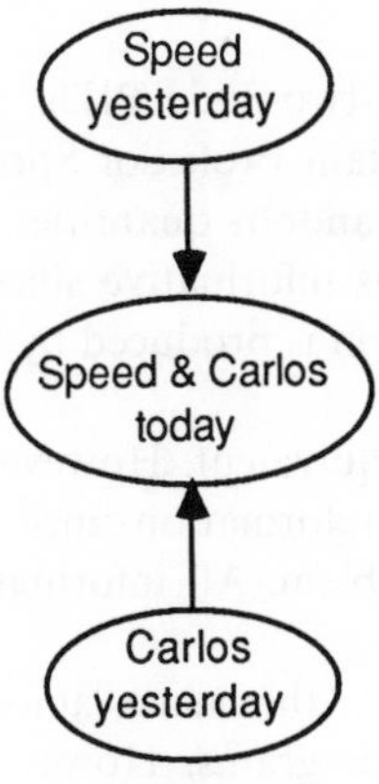

Figure 3.7

they are very useful. I also do not believe they can solve my problems. However, they can help me very much when I want to understand a problem to be solved. I imagine that Professor Speed does not completely avoid the use of graphs. In my opinion Figure 3.7 describes our positions. So, we both agree that graphs have some important uses.

Eighteen months ago when I came to Berkeley I learned for the first time about influence diagrams (IDs). That was the language that Dick Barlow and Ross Shachter were using to communicate to each other. My question was 'Who needs IDs?' and the answer was I needed IDs if I wanted to talk to them.' So, my 'interest' on IDs started. I now discover that I am taking in this new language. In my understanding, an ID for a statistical problem is the picture of the problem and a picture may replace a thousand words. I hope there will be a place for these comments in the Proceedings of this conference (this is Shachter's joke).

Definition

I think the graphs used by Professor Speed are not IDs. For this reason I will present a formal definition of ID and then reply to his critics of IDs. In fact I will need only probabilistic influence diagrams (PIDs).

A PID is an acyclic directed graph in which:

1. *Nodes* represent random (or deterministic) quantities while *directed arcs* indicate possible (not actual) dependence; and
2. Attached to each node is a conditional probability function (for the node) which depends possibly on the states of the adjacent predecessors' nodes. (The product of all these conditional probabilities function is the joint probability function of the nodes.)

PID Language

There are only three possible two-node PIDs, as given in Figure 3.8. I believe that these three diagrams explain Professor Speed's Figure 3.1.

Note now that any three random quantities can have Figure 3.2(b) as their PID. However, Figure 3.2(a) is informative since it says that s_1 ind $s_2|D$. Note that graph or visual information is produced by the lack of arcs not the presence arcs.

Figures 3.3(a) and (b) are equivalent. However, by reversing only arc $[V, W]$ in Figure 3.3(a) we lose visual information since we obtain Figure 3.9 which may represent any three-node problem. All information is represented by the joint probability function.

Figures 3.4(a) and (b) carry the same amount of information since only one arc is missing in both diagrams. However, the information is different. Figure 3.4(a) states that U and V are independent. On the other hand, U and V

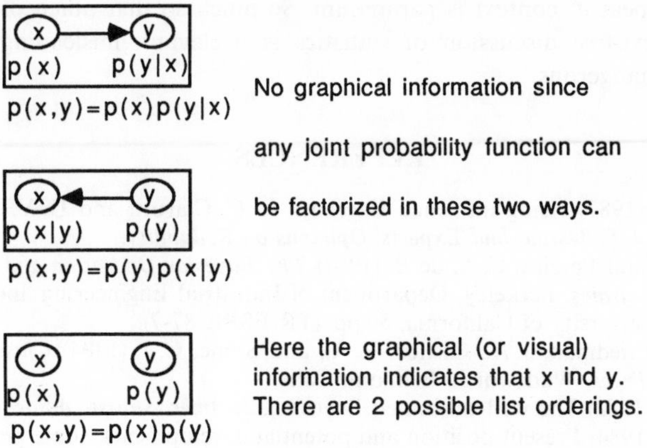
x
y
p(x) p(y|x)
p(x,y)=p(x)p(y|x)
No graphical information since
any joint probability function can
be factorized in these two ways.
x
y
p(x|y) p(y)
p(x,y)=p(y)p(x|y)
x
y
p(x) p(y)
p(x,y)=p(x)p(y)
Here the graphical (or visual)
information indicates that x ind y.
There are 2 possible list orderings.

Figure 3.8

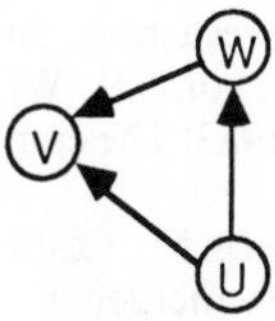
W
V
U

Figure 3.9

are conditionally independent given W in Figure 3.4(b). Note that both graphs have two possible list orderings.

Figures 3.4(a) and (b) carry no visual information since there is no lack of arcs. That is, there is only one list ordering for each diagram. Splitting node U into nodes U_1 and U_2 we gain some graph information. (Note that there must be a direction for the arc connecting U_1 and U_2 in order to be PIDs.)

Figure 3.4 introduces two informative diagrams. They are completely distinct. In order to obtain a solution satisfying both diagrams, without specifying the probability functions, we may need to eliminate one extra arc from the diagrams. Either $[U_1, V]$ or $[V, W]$ or $[U_2, W]$ or $[W, V]$ may need to be eliminated.

I hope my discussion here can help in the understanding why PIDs are pictures (maybe partial) of problems.

3.5.3 Reply

The main message I get from Carlos Pereira's comments is this: as a group we are still far from clear just what is, and what is not an influence diagram. This will not be a problem for me as I do not intend to use the term.

I thank Jim Smith for reiterating the importance of context in statistical studies. I agree, and tried to indicate this in my chapter. I have clearly failed, so let me repeat it: context is paramount. So much so that much of what goes on in context-free discussion of statistics is irrelevant, misleading and even potentially dangerous.

REFERENCES

Barlow, R. E. (1987) Using influence diagrams. In C. Clarotti and D. V. Lindley (eds) *Accelerated Life Testing and Experts' Opinions on Reliability.*

Barlow, R. E. and Pereira, C. A. de B. (1987) *The Bayesian Operation and Probabilistic Influence Diagrams*, Berkeley, Department of Industrial Engineering and Operations Research, University of California, 50 pp. (TR-ESRC 87-7).

Breiman, L., Friedman, J. A., Olshen, R. A., and Stone, C. J. (1984) *Classification and Regression Trees*, Wadsworth, Belmont, Calif.

Dawid, A. P. (1982) The well-calibrated Bayesian, *J. Amer. Statist. Assoc.*, **77**, 605–13.

Dawid, A. P. (1984) Present position and potential developments: some personal views, statistical theory, the prequential approach, *J.R. Statist. Soc.*, **B147**, 278–92.

Dawid, A. P. (1986a) Prequential data analysis, *Second Catalan International Symposium on Statistics.*

Dawid, A. P. (1986b) Probability forecasting. In N. L. Johnson and S. Kotz (eds) *Encyclopedia of Statistical Sciences*, vol. VII, Wiley, New York.

De Groot, M. H. and Fienberg, S. E. (1983) The comparison and evaluation of forecasters, *The Statistician*, **32**, 12–22.

Goldman, L. C., Weinberg, M., Weisberg, M., Olshen, R. *et al.* (1985) A computer-derived protocol to aid in the diagnosis of emergency room patients with acute chest pain, *N. Eng. J. Med.*, **307**, 588–96.

Kiiveri, H. and Speed, T. P. (1982) The structural analysis of multivariate data: a review. In Samuel Leinhardt (ed.) *Sociological Methodology*, Jossey-Bass, San Francisco, pp. 209–90.

Kiiveri, H., Speed, T. P. and Carlin, J. B. (1984) Recursive causal models, *J. Austral. Math. Soc.*, **A36**, 30–52.

Lauritzen, L. L. and Spiegelhalter, D. J. (1988) Local computation with probabilities on graphical structures and their application to expert systems (with Discussion) *J.R. Statist. Soc.*, **B50**, 157–224.

Quinlan, J. R. and Rivest, R. L. (1987) Inferring decision trees using the minimum description length principle. *Information and Computation*, **80**, 227–248.

Rissanen, J. (1983) A universal prior for integers and estimation by minimum description length, *Annals of Statistics*, **11**, 416–431.

Rissanen, J. (1986) Stochastic complexity and modeling, *Ann. Statist.*, **14**, 1080–1100.

Rissanen, J. (1987) Stochastic complexity (with discussion), *J. Roy. Statist. Soc.*, **B49**, 223–239, and 252–265.

Shachter, R. D. (1986) Evaluating influence diagrams, *Operations Research*, **34**, 874–882.

Shachter, R. D. and Heckerman, D. E. (1987) Thinking backward for knowledge acquisition, *AI Magazine*, **8**, Fall issue, 55–61.

Shachter, R. D. and Kenley, C. R. (1988) Gaussian influence diagrams, *Management Science* to appear.

Smith, J. Q., and Gathercole, R. B. (1986) Principles of interactive forecasting. In P. Goel and A. Zellner (eds) *Bayesian Inference and Decision Techniques*, pp. 405–23.

Smith, J. Q. (1988a) *Decision analysis. A Bayesian approach.* Chapman and Hall, London.

Smith, J. Q. (1988b) Statistical principles on graphs. In R. M. Oliver and J. Q. Smith (eds) (1990) *Influence Diagrams, Belief Nets and Decision Analysis*, Wiley, Chichester, pp. 89–120.

Speed, T. P. (1978) Relations between models for spatial data, contingency tables and Markov fields over graphs. In *Proceedings of the Conference on Spatial Patterns and Processes. Supplement to Adv. Appl. Prob.*, **10**, 111–122.

Speed, T. P. (1983) Probabilistic risk assessment in the nuclear industry: WASH-1400 and beyond. In L. M. Le Cam and R. A. Olshen (eds) *Proc. Berkaley Conf. in honor of J. Neyman and J. Kiefer*, vol. 1, Wadsworth, Belmont, Calif.

Titterington, D. M., Murray, G. D., Murray, L. S., Spiegelhalter, D. J., Skene, A. M., Habbema, J. D. and Gelpke, G. J. (1981) Comparison of discrimination techniques applied to a complex data set of head injured patients (with discussion), *J. Roy. Statist. Soc.*, **A144**, 145–175.

Wagner, C. H. (1982) Simpson's paradox in real life. *Amer. Statist.*, **36**, 46–48.

Wermuth, N. (1980) Linear recursive equations, covariance selection and path models, *J. Amer. Statist. Asoc.*, **75**, 963–972.

PART II

Theoretical Foundations

CHAPTER 4

The Logic of Influence Diagrams

Judea Pearl, Dan Geiger and Thomas Verma, *University of California at Los Angeles, USA*

ABSTRACT

This chapter explores the role of directed acyclic graphs (DAGs) as a representation of conditional independence relationships. We show that DAGs offer polynomially sound and complete inference mechanisms for inferring conditional independence relationships from a given causal set of such relationships. As a consequence, *d-separation*, a graphical criterion for identifying independencies in a DAG, is shown to be both correct and optimal. This criterion can be used, in lieu of arithmetic manipulations, to test the legitimacy of graphical transformations on influence diagrams, to identify relevant sources of information in the diagram and, thus, to guide the control of inferencing and information-gathering strategies.

4.1 INTRODUCTION AND SUMMARY OF RESULTS

Networks employing directed acyclic graphs (DAGs) have a long and rich tradition, starting with the geneticist Sewal Wright (1921). He developed a method called *path analysis* (Wright, 1934) which later became an established representation of causal models in economics (Wold, 1964), sociology (Blalock, 1971; Kenny, 1979) and psychology (Duncan, 1975). Good (1961) used DAGs to represent causal hierarchies of binary variables with disjunctive causes. *Influence diagrams* represent another application of DAG representation (Howard and Matheson, 1981; Olmsted, 1983; Shachter, 1986) developed for decision analysis, they contain both event nodes and decision nodes. *Recursive models* is the name given to such networks by statisticians seeking meaningful and effective decompositions of contingency tables (Lauritzen, 1982; Wermuth and Lauritzen, 1983; Kiiveri *et al.*, 1984). *Bayesian belief networks* (or *causal networks*) is the name adopted for describing networks that perform evidential

Influence Diagrams, Belief Nets and Decision Analysis
Edited by R. M. Oliver and J. Q. Smith

reasoning (Pearl, 1985, 1986, 1988). This chapter establishes a clear semantics for such networks that might explain their wide usage as models for forecasting, decision analysis and evidential reasoning.

An influence diagram* can be viewed as an economical scheme for representing conditional independence relationships and for deducing new independencies from those used in the construction of the diagram. The nodes in the diagram represent variables in some domain and its topology is specified by a list of qualitative judgements elicited from an expert in this domain. The specification list designates to each variable v a set of parents judged to have direct influence on v, and this amounts to asserting that, given its parents-set, each variable is independent of its other predecessors in some total order of the variables. This stratified list of independencies, henceforth called input list or causal list, implies many additional independence relationships that can be read off the diagram. For example, it is well known that, given its parents, each variable is also independent of all its non-descendants (Howard and Matheson, 1981). Additionally, if S is a set of nodes containing v's parents, v's children and the parents of those children, then v is independent of all other variables in the system, given those in S (Pearl, 1986). These assertions are examples of valid consequences of the input list, i.e. independencies that hold for every probability distribution that satisfies the conditional independencies specified by the input. If one ventures to perform topological transformations on the diagram (e.g. arc reversal or node removal, Shachter, 1986), caution must be exercised to ensure that each transformation does not introduce extraneous, invalid independencies, and/or that the number of valid independencies which become obscured by the transformation is kept at a minimum. Thus, in order to decide which transformations are admissible, one should have simple criteria for deciding which conditional independence statement is valid and which is not. The development of such criteria is the central theme of this chapter.

This chapter deals with the following questions:

1. What are the valid consequences of the input list?
2. What are the valid consequences of the input list that can be read off the diagram?
3. Are the two sets identical?

The answers obtained are as follows:

1. A statement is a valid consequence of the input set if and only if it can be derived from it using the following four axioms. Letting X, Y, and Z stand for three disjoint subsets of variables, and denoting by $I(X,Z,Y)$ the statement: 'the variables in X are conditionally independent of those in Y, given those in Z', the axioms state:

*Our analysis will focus on the so-called knowledge part of influence diagrams, thus excluding decision nodes.

Symmetry

$$I(X, Z, Y) \Rightarrow I(Y, Z, X) \tag{4.1a}$$

Decomposition

$$I(X, Z, Y \cup W) \Rightarrow I(X, Z, Y) \,\&\, I(X, Z, W) \tag{4.1b}$$

Weak union

$$I(X, Z, Y \cup W) \Rightarrow I(X, Z \cup W, Y) \tag{4.1c}$$

Contraction

$$I(X, Z \cup Y, W) \,\&\, I(X, Z, Y) \Rightarrow I(X, Z, Y \cup W). \tag{4.1d}$$

These axioms form a system called semi-graphoid (Pearl and Paz, 1985; Pearl and Verma, 1987) and were first proposed as heuristic properties of conditional independence by Dawid (1979).

2. Every statement that can be read off the DAG using the *d*-separation criterion represents a valid consequence of the input list (Verma, 1986).

   The *d*-separation criterion is defined as follows (Pearl, 1985): for any three disjoint subsets X, Y, Z of nodes in a DAG D, Z is said to *d*-separate X from Y, denoted $I(X, Z, Y)_D$, if there is no path from a node in X to a node in Y along which the following two conditions hold: (a) every node with converging arrows is in Z or has a descendant in Z, and (b) every other node is outside Z. (The definition is elaborated in Section 4.2.)

3. The two sets are identical, namely, a statement is valid if and only if it is graphically validated under *d*-separation in the diagram (Geiger and Pearl, 1988a).

The first result establishes the decidability of verifying whether an arbitrary statement is a valid consequence of the input set, i.e. applying axioms (4.1a)–(4.1d) on a causal input list is guaranteed to generate all its valid consequences and none other. The second result renders the *d*-separation criterion a polynomially sound inference rule, i.e. it runs in polynomial time and certifies only valid statements. The third renders the *d*-separation criterion a complete inference rule. In summary, the diagram constitutes a sound and complete inference mechanism that identifies, in polynomial time, each and every valid consequence in the system. Interestingly, result 2 holds for any semi-graphoid system, not necessarily probabilistic conditional independencies. Thus, influence diagrams can serve as effective inference instruments for a variety of dependence relationships, e.g. partial correlations (Crámer, 1946) and qualitative database dependencies (Fagin, 1977).

The results above are true only for causal input sets, i.e. those that recursively specify the dependence of each variable on its predecessors in some total order, e.g. chronological. The general problem of verifying whether a given conditional independence statement logically follows from an arbitrary set of such statements, may be undecidable. Its decidability would be resolved upon finding a complete set of axioms for conditional independence, i.e. axioms that are powerful enough to derive all valid consequences of an arbitrary input set. Until very recently, all indications were that axioms (4.1a)–(4.1d) are complete, as conjectured in Pearl and Paz (1985), but a new result of Studeny (1988) has refuted that hypothesis. Currently, it appears rather unlikely that there exists a finite set of axioms which is complete for conditional independence, thus, the general decidability problem remains unsettled. The completeness problem is treated in Geiger and Pearl (1988b) and completeness results for specialized subsets of probabilistic dependencies are summarized in Section 4.2.6.

4.2 SOUNDNESS AND COMPLETENESS

4.2.1 The *d*-separation criterion

The definition of *d*-separation is best motivated by regarding DAGs as a representation of causal relationships. Designating a node for every variable and assigning a link between every cause to each of its direct consequences defines a graphical representation of a causal hierarchy. For example, the propositions 'It is raining' (α), 'the pavement is wet' (β) and 'John slipped on the pavement' (γ) are well represented by a three-node chain, from α through β to γ. The chain indicates that either rain or wet pavement could cause slipping, yet wet pavement is designated as the direct cause; rain could cause someone to slip only by wetting the pavement, not if the pavement is covered. Moreover, knowing the condition of the pavement renders 'slipping' and 'raining' independent, and this is represented graphically by a *d*-separation condition, $I(\alpha, \beta, \gamma)_D$, showing node α and γ separated from each other by node β.

Now assume that 'broken pipe' (δ) is considered another direct cause for wet pavement, as in Figure 4.1. An informational dependency may be induced between the two causes of wet pavement: 'rain' and 'broken pipe'. Although they appear connected in Figure 4.1, these propositions are marginally independent and become dependent once we learn that the pavement is wet or that someone broke his leg. An increase in our belief in either cause would

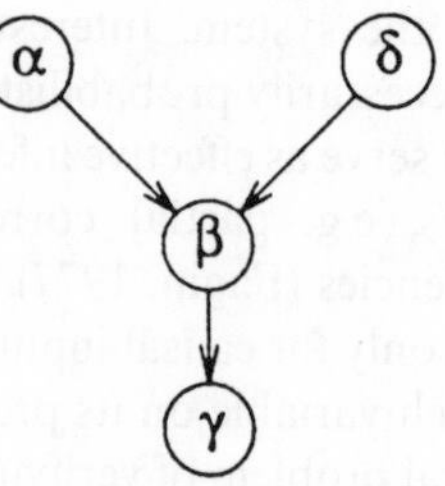

Figure 4.1

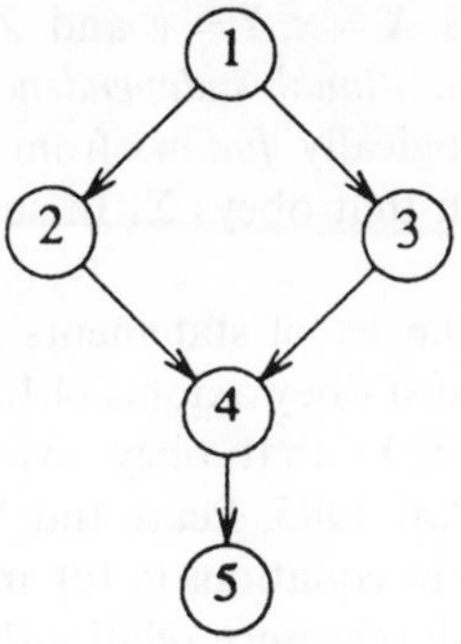

Figure 4.2

decrease our belief in the other as it would 'explain away' the observation. The following definition of d-separation permits us to graphically identify such induced dependencies from the DAG (d connoted 'directional').

Definition (*d-separation*). If X, Y, and Z are three disjoint subsets of nodes in a DAG D, then Z is said to d-separate X from Y, denoted $I(X, Z, Y)_D$, if and only if there is no path* from a node in X to a node in Y along which the following two conditions hold: (1) every node with converging arrows either is or has a descendant in Z, and (2) every other node is outside Z. A path satisfying the conditions above is said to be *active* otherwise it is said to be *blocked* (by Z). Whenever a statement $I(X, Z, Y)_D$ holds in a DAG D, the predicate $I(X, Z, Y)$ is said to be *graphically verified* (or an independency), otherwise it is *graphically unverified* by D (or a dependency). □

In Figure 4.2, for example, $X = \{2\}$ and $Y = \{3\}$ are d-separated by $Z = \{1\}$; the path $2 \leftarrow 1 \rightarrow 3$ is blocked by $1 \in Z$ while the path $2 \rightarrow 4 \leftarrow 3$ is blocked because 4 and all its descendants are outside Z. Thus $I(2, 1, 3)$ is graphically verified by D. However, X and Y are not d-separated by $Z' = \{1, 5\}$ because the path $2 \rightarrow 4 \leftarrow 3$ is rendered active. Consequently, $I(2, \{1, 5\}, 3)$ is graphically unverified by D; by virtue of 5, a descendant of 4, being in Z. Learning the value of the consequence 5, renders its causes 2 and 3 dependent, like opening a pathway along the converging arrows at 4.

4.2.2 Conditional independence and its graphical representations

Definition. If X, Y and Z are three disjoint subsets of variables of a distribution P, then X and Y are said to be *conditionally independent* given Z, denoted

$$I(X, Z, Y)_P, \quad \text{iff} \quad P(x, y \mid z) = P(x \mid z) P(y \mid z)$$

*By path we mean a sequence of edges in the underlying undirected graph, i.e. ignoring the directionality of the arcs.

for all possible assignments $X = x, Y = y$ and $Z = z$ for which $P(Z = z) > 0$. $I(X, Z, Y)_P$ is called a (*conditional independence*) *statement*. A conditional independence statement σ logically *follows* from a set Σ of such statements if σ holds in every distribution that obeys Σ. In such case we also say that σ is a *valid consequence* of Σ. □

It is easy to verify that the set of statements $I(X, Z, Y)_P$ generated by any probability distribution P must obey axioms (4.1a)–(4.1d). More generally, any system of statements $I(X, Z, Y)$ that obey axioms (4.1a)–(4.1d) is called a semi-graphoid (Pearl and Paz, 1985; Pearl and Verma, 1987). Intuitively, the essence of these axioms lies in equations (4.1c) and (4.1d), asserting that when we learn an irrelevant fact, all relevance relationships among other variables in the system should remain unaltered; any information that was relevant remains relevant and that which was irrelevant remains irrelevant. These axioms, common to almost every formalization of informational dependencies, are very similar to those assembled by Dawid (1979) for probabilistic conditional independence and those proposed by Smith (1987b) for generalized conditional independence. The difference is only that Dawid and Smith lumped equations (4.1b)–(4.1d) into one, and added an axiom to handle some cases of overlapping sets X, Y, Z. We shall henceforth call axioms (4.1a)–(4.1d) *Dawid's axioms*, or semi-graphoid axioms, interchangeably. Interestingly, both undirected graphs and DAGs conform to the semi-graphoid axioms (hence the name) if we associate the statement $I(X, Z, Y)$ with the graphical condition 'every path from X to Y is blocked by the set of nodes corresponding to Z'. (In DAGs, blocking is defined by *d*-separation.) It is this commonality, we speculate, that renders graphs such a popular scheme for representing informational dependencies.

Ideally, to employ a DAG D as a graphical representation for dependencies of some distribution P we would like to require that for every three disjoint sets of variables in P (and nodes in D) the following equivalence should hold:

$$I(X, Z, Y)_D \quad \text{iff} \quad I(X, Z, Y)_P. \tag{4.2}$$

This would provide a clear graphical representation of all variables that are conditionally independent. When equation (4.2) holds, D is said to be a perfect map of P. Unfortunately, this requirement is often too strong because there are many distributions that have no perfect map in DAGs. The spectrum of probabilistic dependencies is in fact so rich that it cannot be cast into any representation scheme that uses polynomial amount of storage (Verma, 1987). Geiger (1987) provides a graphical representation based on a collection of graphs (multi-DAGs) that is powerful enough to perfectly represent an arbitrary distribution; however, as shown by Verma, it requires, on the average, an exponential number of DAGs.

Being unable to provide perfect maps at a reasonable cost, we compromise the requirement that the graphs represent each and every independency of P, and allow some independencies to escape representation. We will require through that the set of undisplayed independencies be minimal.

Definition. A DAG D is said to be an I-map of P if for every three disjoint subsets X, Y and Z of variables the following holds:

$$I(X, Z, Y)_D \Rightarrow I(X, Z, Y)_P. \tag{4.3}$$

D is said to be a minimal I-map of P if no edge can be deleted from D without destroying its I-mapness. □

The task of finding a DAG which is a minimal I-map of a given distribution P can be accomplished by the traditional chain-rule decomposition of probability distributions. The procedure consists of the following steps: assign a total ordering d to the variables of P. For each variable i of P, identify a minimal set of predecessors S_i that renders i independent of all its other predecessors (in the ordering of the first step). Assign a direct link from every variable in S_i to i. The analysis of Verma (1986) and Pearl and Verma (1987) ensures that the resulting DAG is an I-map of P, and is minimal in the sense that no edge can be deleted without destroying its I-mapness. The input list L for this construction consists of n conditional independence statements, one for each variable, all of the form $I(i, S_i, U_{(i)} - S_i)$, where $U_{(i)}$ is the set of predecessors of i and S_i is a subset of $U_{(i)}$ that renders i conditionally independent of all its other predecessors. This set of conditional independence statements is called a causal input list and is said to define the DAG D. The term 'causal' input list stems from the following analogy: suppose we order the variables chronologically, such that a cause always precedes its effect. Then, from all potential causes of an effect i, a causal input list selects a minimal subset that is sufficient to explain i, thus rendering all other preceding events superfluous. This selected subset of variables are considered direct causes of i and therefore each is connected to it by a direct link.

4.2.3 The main results

Clearly, the constructed DAG represents more independencies than those listed in the input, namely, all those that are graphically verified by the d-separation criterion. The results reported in the previous subsection guarantee that all graphically verified statements are indeed valid in P, i.e. the DAG is an I-map of P. It turns out that the constructed DAG has another useful property: it graphically verifies every conditional independence statement that logically follows from L (i.e. holds in every distribution that obeys L). Hence, we cannot hope to improve the d-separation criterion to display more independencies, because all valid consequences of L (which defines D) are already captured by d-separation.

The three theorems below formalize these results. Proofs can be found in the references cited.

Theorem 4.1 (soundness) (Verma, 1986). Let D be a DAG defined by a causal input list L, of some dependency model obeying axioms (4.1a)–(4.1d) (e.g. probabilistic dependence). Then, every graphically verified statement is a valid consequence of L. □

Theorem 4.2 (closure) (Verma, 1986). Let D be a DAG defined by a causal input list L. Then, the set of graphically verified statements is exactly the closure of L under axioms (4.1a)–(4.1d). □

Theorem 4.3 (completeness) (Geiger and Pearl, 1988a). Let D be a DAG defined by a causal input list L. Then, every valid consequence of L is graphically verified by D (equivalently, every graphically unverified statement in D is not a valid consequence of L). □

Theorem 4.1 guarantees that the DAG displays only valid statements. Theorem 4.2 guarantees that the DAG displays all statements that are derivable from L via axioms (4.1). Theorem 4.3 assures that the DAG displays all statements that logically follow from L, i.e. the axioms in (4.1) are complete, capable of deriving all valid consequences of a causal input list. Moreover, since a statement in a DAG can be verified in polynomial time, Theorems 4.1–4.3 provide a complete polynomial inference mechanism for deriving all independence statements that are implied by a causal input list.

The first two theorems are more general than the third in the sense that they hold for every dependence relationship that obeys axioms (4.1a)–(4.1d), not necessarily those based on probabilistic conditional independence (proofs can be found in Verma, 1986 and Verma and Pearl, 1988). Among these dependence relationships are partial correlations (Crámer, 1946; Pearl and Paz, 1985) and qualitative dependencies (Fagin, 1977; Shafer *et al.*, 1987) which can readily be shown to obey axioms (4.1). The completeness of *d*-separation (Theorem 4.3) relative partial correlations has been established in Geiger and Pearl (1988b) while completeness relative to qualitative dependencies can also be shown.

4.2.4 Deterministic nodes and *D*-separation

Theorems 4.1–4.3 assume that L contains only statements of the form $I(i, S_i, U_{(i)} - S_i)$. Occasionally, however, we are in possession of stronger forms of independence relationships, in which case additional statements should be read off the DAG. A common example is the case of a deterministic variable (Shachter, 1988), i.e. a variable that is functionally dependent on its corresponding parents in the DAG. The existence of each such variable i could be encoded in L by a statement of global independence $I(i, S_i, U - S_i - i)$ asserting that conditioned on S_i, i is independent of all other variables, not merely of its predecessors. The independencies that are implied by the modified input list can be read from the DAG using an enhanced version of *d*-separation, named *D-separation.*

Definition. If X, Y, and Z are three disjoint subsets of nodes in a DAG D, then Z is said to D-separate X from Y, if and only if there is no path from a node in X to a node in Y along which the following two conditions hold: (1) every node with converging arrows either is or has a descendant in Z; (2) every other node is not in Z nor determined by Z. □

The new criterion certifies all independencies that are revealed by d-separation plus additional ones due to the enhancement of the input list. For example, if the arc $2 \rightarrow 4$ in Figure 4.2 were deterministic, then 2 D-separates 5 from 3 and, indeed, 5 and 3 must be conditionally independent given 2 because 2 determines the value of 4. In fact, the link $3 \rightarrow 4$ is redundant because $I(4, 2, 135)$ implies $I(4, 2, 13)$. It is for this reason that deterministic nodes can be presumed to receive only deterministic incoming arrows. The soundness and completeness of D-separation is stated in Theorem 4.4.

Theorem 4.4 (Geiger *et al.*, 1989b). Let D be a DAG defined by a causal input list L, possibly containing functional dependencies. Then a statement is a valid consequence of L if and only if it is graphically verified in D by the D-separation criterion. □

These graphical criteria provide easy means of recognizing conditional independence in influence diagrams as well as identifying the set of parameters needed for any given computation (Section 4.4.3). We now show how these theorems can be employed as an inference mechanism. Assume an expert has identified the following conditional independencies between variables denoted 1–5:

$$L = \{I(2, 1, \emptyset), I(3, 1, 2), I(4, 23, 1), I(5, 4, 123)\}$$

(the first statement in L is trivial). We address two questions. First, what is the set of all valid consequences of L? Second, in particular, is $I(3, 124, 5)$ a valid consequence of L? For general input lists the answer for such questions may be undecidable but, since L is a causal list, it defines a DAG that graphically verifies each and every valid consequences of L. The DAG D is the one shown in Figure 4.2, which constitutes a compact representation of all valid consequences of L. To answer the second question, we simply observe that $I(3, 124, 5)$ is graphically verified in D. A graph-based algorithm for another subclass of statements, called fixed context statements, is given in Geiger and Pearl (1988b). In that paper, results analogous to Theorem 4.1–4.3 are proven for Markov fields; a representation scheme based on undirected graphs (Isham, 1981, Lauritzen, 1982).

4.2.5 Strong completeness

Theorem 4.3 can be restated to assert that for every DAG D and any statement σ graphically unverified by D there exists a probability distribution P_σ that embodies D's causal input set L and the dependency σ. By Theorem 4.2, P_σ must embody all graphically verified statements as well because they are all derivable from L by Dawid's axioms. Thus, Theorems 4.2 and 4.3 guarantee the existence of a distribution P_σ that satisfies all graphically verified statements and a single, arbitrarily chosen, graphically unverified statement (i.e. a dependency). The question answered by Theorem 4.5 is the existence of a distribution P that embodies all independencies of D and *all* its dependencies, not merely a single dependency. Such a set of axioms is said to be strongly complete (Beeri *et al.*, 1977).

Theorem 4.5 (strong completeness) (Geiger and Pearl, 1988a). For every DAG D there exists a distribution P such that for every three disjoint sets of variables X, Y and Z the following holds:

$$I(X, Z, Y)_D \quad \text{iff} \quad I(X, Z, Y)_P. \qquad \square$$

Theorem 4.5 legitimizes the use of DAGs as a representation scheme for probabilistic dependencies; a model builder who uses the language of DAGs to express dependencies is guarded from inconsistencies.

4.2.6 Other completeness results

We conclude this section with a summary of completeness results for specialized subsets of independence statements. Proofs can be found in Geiger and Pearl (1988b) and Geiger *et al.* (1989a). The first result establishes an axiomatic characterization of marginal statements, i.e. statements of the form $I(X, Z_0, Y)$ where the middle argument Z_0 is fixed. The second result provides a complete axiomatic characterization of fixed-context statements, i.e. statements of the form $I(X, Z, Y)$ where $X \cup Z \cup Y$ sum to a fixed set of variables U.

Theorem 4.6 (completeness for marginal independence). Let Σ be a set of marginal statements closed under the following axioms:

Symmetry	$I(X, Z_0, Y) \rightarrow I(Y, Z_0, X)$
Decomposition	$I(X, Z_0, YW) \rightarrow I(X, Z_0, Y)$
Mixing	$I(X, Z_0, Y)$ and $I(XY, Z_0, W) \rightarrow I(X, Z_0, YW)$.

There exists a probability model P that obeys all statements in Σ and none other. $\square$

Theorem 4.7 (completeness for fixed-context). Let Σ be a set of fixed-context statements closed under the axioms:

Symmetry	$I(Y, Z, Y) \rightarrow i(Y, Z, X)$
Weak union	$I(X, Z, YW) \rightarrow I(X, ZY, W)$
Weak contraction	$I(XY, Z, W)$ and $I(X, ZW, Y) \rightarrow I(X, Z, YW)$.

There exists a probability model P that obeys all statements in Σ and none other. Moreover, if Σ is closed also under

$$\text{Intersection} \quad I(X, ZW, Y) \ \text{and} \ I(X, ZY, W) \rightarrow I(X, Z, YW)$$

then P can be selected to be strictly positive. □

The membership question, whether a given marginal statement follows from an arbitrary set of such statements, can be answered in quadratic time (Geiger *et al.*, 1989a). For fixed-context statements, the membership question can also be answered in quadratic time (Beeri *et al.*, 1987).

4.3 COROLLARIES

Theorem 4.1 leads to four corollaries which are the key to the construction of influence diagrams from a given distribution P. First we define influence diagrams in terms of the independencies they portray, then we justify their method of construction.

Definition. Given a probability distribution P on a set U of variables, $U = \{X_1, X_2, \ldots, X_n\}$, a DAG $D = (U, E)$ is called an influence diagram of P if and only if D is a minimal I-map of P. □

Corollary 4.1. Given a probability distribution $P(x_1, x_2, \ldots, x_n)$ and any ordering d of the variables. The DAG created by designating as parents of X_i any minimal set S_i of predecessors satisfying

$$I(X_i, S_i, U_{(i)} - S_i), \quad U_{(i)} = \{X_1, X_2, \ldots, X_{i-1}\} \tag{4.4}$$

is an influence diagram of P. Conversely, every influence diagram of P can be constructed by identifying the parent sets S_i defined in (4.4) along some ordering d. If P is strictly positive, then all the parent sets are unique (Pearl and Paz, 1985) and the influence diagram is unique as well (given d). □

Although the structure of the diagram depends strongly on the node ordering used in the construction, each diagram is nevertheless an I-map of the underlying distribution P. This means that all conditional independencies portrayed in the diagram (via d-separation) are valid in P and hence, are order independent. An immediate corollary of this observation yields an order-independent test for minimal I-mapness.

Corollary 4.2. Given a DAG D and a probability distribution P, a necessary and sufficient condition for D to be a minimal I-map (hence an influence diagram) of P is that each variable X_i be conditionally independent of all its non-descendants, given its parents S_i, and no proper subset of S_i satisfies this condition.

The necessary part follows from the fact that every parent-set S_i d-separates X_i from all its non-descendants. The sufficient part holds because X_i's independence of all its non-descendants entails X_i's independence of its predecessors in a particular ordering d (as required by Corollary 4.1). □

Corollary 4.3. If an influence diagram D is constructed from P (by the method of Corollary 4.1) in some ordering d, then any ordering d' consistent with the direction of arrows in D would give rise to an identical diagram.

The validity of Corollary 4.3 follows from that of Corollary 4.2, which ensures that the set S_i will satisfy equation (4.4) in any new ordering, as long as the new set of X_i's predecessors does not contain any of X_i's old descendants. Thus, once the network is constructed, the original order can be forgotten; only the partial order displayed in the diagram matters.

Another interesting corollary of Theorem 4.1 is a generalization of the celebrated *Markov-chain* property which is used extensively in the probabilistic analysis of random walks, time-series data and other stochastic processes (Feller, 1968; Meditch, 1969). The property states that if in a sequence of n trials $X_1, X_2, \ldots, X_n$ the outcome of any trial $X_k, k = 2, 3 \ldots n$ depends only on the outcome of its directly preceding trial X_{k-1} then, given the entire past and future trials $X_1, X_2, \ldots, X_{k-1}, X_{k+1}, \ldots, X_n$, the outcome of X_k depends only on its two nearest neighbors, X_{k-1} and X_{k+1}. Formally,

$$I(X_k, X_{k-1}, X_{k-2} \ldots X_1), \quad 2 \leqslant k \leqslant n$$
$$\Rightarrow I(X_k, X_{k-1} X_{k+1}, X_n \ldots X_{k+2} X_{k-2} \ldots X_1), \quad 2 \leqslant k \leqslant n-1$$

(The converse holds only for full graphoids, e.g. strictly positive distributions.) Theorem 4.1 generalizes the Markov-chain property to dependencies other than probabilistic and to structures other than chains. The d-separation criterion uniquely determines a *Markov blanket* for any given node X_i in an influence diagram, namely, a set $\mathbf{BL}(X_i)$ of variables that renders X_i independent of all variables not in $\mathbf{BL}(X_i)$, i.e. $I(X_i, \mathbf{BL}(X_i), U - \mathbf{BL}(X_i) - X_i)$.

Corollary 4.4. In any influence diagram, the union of the following three types of neighbors is sufficient for forming a Markov blanket of a node X_i: the direct parents of X_i, the direct successors of X_i and all direct parents of the latter. □

Thus, if the diagram consists of a single path (i.e. a Markov chain), the Markov blanket of any nonterminal node consists of its two immediate neighbors, as expected. In trees, the Markov blanket consists of the (unique) parent and the immediate successors. In Figure 4.2, however, the Markov blanket of node 3 is $\{1, 4, 3\}$. Note that in general, these Markov blankets are not minimal; alternative ordering might display X_i with a smaller set of neighbors.

The necessary part of Corollary 4.2 was stated without proof in Howard and Matheson (1981) and was later used in the derivations of Olmsted (1983) and Shachter (1986). Corollaries 4.2 and 4.3 are proven in Smith (1987) using the axioms of equation (4.1). Since Theorem 4.1 establishes d-separation a sound procedure relative to Dawid's axioms, the validity of such corollaries can now be verified by purely graphical means.

4.4 APPLICATIONS

4.4.1 Entailment and equivalence

One problem concerning influence diagrams that has received much attention is that of *entailment*, i.e. establishing a criterion for testing whether one influence diagram entails all the conditional independencies portrayed by another (Smith, 1987). We shall now establish such a criterion using the notions of d-separation.

Definition. Given two diagrams D_1 and D_2, we say that D_1 entails D_2 (equivalently, D_2 is implied by D_1) if every independency verified in D_2 is implied by D_1, i.e. if for all disjoint subsets of nodes X, Y and Z, we have

$$I(X, Z, Y)_{D_2} \Rightarrow I(X, Z, Y)_{D_1}. \tag{4.5}$$

Remark. Although the two diagrams may contain different sets of nodes, we require only that equation (4.5) holds for all triplets (X, Y, Z) that appear in both diagrams.

Based on Theorem 4.1 and its corollaries, condition (4.5) is equivalent to stating that every independency displayed in D_2 must hold in any probability distribution P_1 that underlies D_1, i.e. D_2 is an I-map of any distribution P of which D_1 is an I-map. Moreover, Theorem 4.1 also yields a simple procedure for verifying (4.5) without testing all triplets (X, Y, Z) in D_2. According to Corollary 4.2, it is sufficient to verify equation (4.5) for just one causal list of D_2.

Corollary 4.5. A necessary and sufficient condition for a diagram D_1 to entail D_2 is that, for every node X_i having parents S_i^2 in D_2 we have: $I(X_i, S_i^2, U_{(i)}^2 - X_i - S_i^2)_{D_1}$, where $U_{(i)}^2 = \{X_1, X_2, \ldots, X_{i-1}\}$ is a set of predecessors of X_i in some ordering that is consistent with the arrows of D_2. □

As an example, consider the four DAGs shown in Figure 4.3. Here D_1 does not entail D_2 because $I(5, 1, 24)$ is not verified in D_1. Similarly D_1 does not entail D_3; $I(5, 14, 2)$ is not verified in D_1. However, D_1 entails D_4 because the following three statements:

$$I(2, 1, \varnothing), \qquad I(5, 1, 2), \qquad I(4, 125, \varnothing)$$

are verified in D_1. These statements correspond to the causal list of D_4 in the ordering $(1, 2, 5, 4)$. In general, the question of entailment can be decided in $O(n^2)$ time, because each d-separation statement can be verified in linear time (Geiger *et al.*, 1989b).

Definition. Two DAGs, D_1 and D_2, are said to be equivalent, written $D_1 \equiv D_2$, if each entails the other, namely, if they portray the same set of conditional independence relationships over the set of nodes common to both.

The equivalence of two DAGs can be established either by entailment both ways or, more transparently, by invoking the notions of adjacency and conditional adjacency. □

Definition. Two nodes a and b in an influence diagram are adjacent, written $a - b$, if there is a direct arc connecting them. They are conditionally adjacent

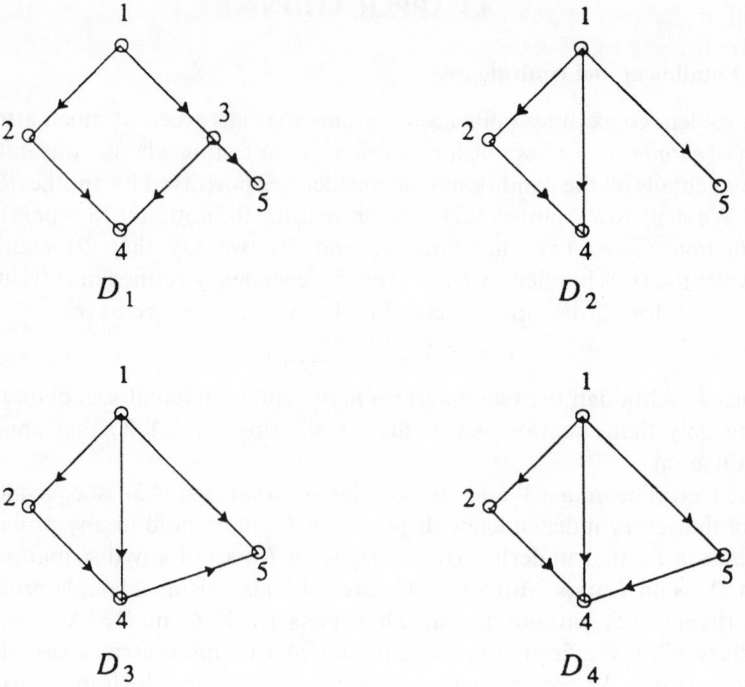

Figure 4.3

given c, written $a - b|c$, if a and b are adjacent or, alternatively, if a and b are the common parents of c or some ancestor of c. □

Adjacency is equivalent to non-d-separability, namely, $a - b$ if and only if there is no set S (not containing a or b) which d-separates a from b. Using these notions, the criterion for entailment can be stated concisely:

Corollary 4.6 (equivalence). For any two influence diagrams, D and E, over the set of nodes $N, D \equiv E$ if and only if both of the following hold:

1. $\forall_{a,b \in N} \; a \underset{D}{-} b \Leftrightarrow a \underset{E}{-} b$

2. $\forall_{a,b,c \in N} \; a \underset{D}{-} b|c \Leftrightarrow a \underset{E}{-} b|c.$ □

Thus, $D \equiv E$ if and only if D and E share the same set of undirected edges and the same set of converging arrows emanating form non-adjacent sources; non-converging arrows may have arbitrary orientations (as long as they do not form directed cycles). Typical examples of equivalent networks are those created by different orientations of a tree or, more generally, different orientations of a chordal graph. Chordal graphs can be oriented in such a way that all converging arrows emanate from adjacent sources (Pearl, 1988), thus satisfying the conditions of Corollary 4.6.

4.4.2 Arc reversal and node removal

A transformation on an influence diagram is said to be sound if it does not introduce extraneous independencies, namely, if the resulting diagram is entailed by the original one. The importance of guarding against the introduction of new independencies lies in the requirement that the numerical parameters which label the arcs of the transformed diagram be sufficient to reconstruct the distribution that underlies the original diagram. It can be shown that as long as the transformed diagram does not display new independencies, the original distribution can be reconstructed by taking the product of the conditional probabilities of all child–parents families in the transformed diagram (e.g. see Pearl, 1988). Thus, no information is lost in the transformation.

Corollary 4.5 defines the conditions for sound transformations of influence diagrams. For example, the conditions forbid the removal of any arc from the diagram; only reorientation and addition of arcs are permitted. Additionally, if a reorientation destroys a pair of non-adjacent common parents, then the pair must be joined by a link; otherwise a conditional independence between such parents (given the child and the grandparents) would be created by the transformation. Finally, if a reorientation creats a new pair of common parents, then the pair must be joined by a link.

From this corollary, most results concerning manipulations of influence diagrams are immediate. Consider, for example, the rule for *arc reversal* (Olmsted, 1983): if an arc from node X to node Y is reversed, X and Y should inherit each other's parents. The reason for this rule is shown in Figure 4.4, where S_X, S_Y and S_{XY} represent sets of parent nodes of X, Y and X and Y respectively. The transformed diagram without parent inheritance (Figure 4.4b) displays two independencies which are not verified by the original diagram (Figure 4.4a), namely, $I(X, YS_{XY}S_X, S_Y)$ and $I(Y, S_{XY}S_Y, S_X)$. To destroy these two independencies it is necessary (and sufficient) to add the two arcs $S_Y \rightarrow X$

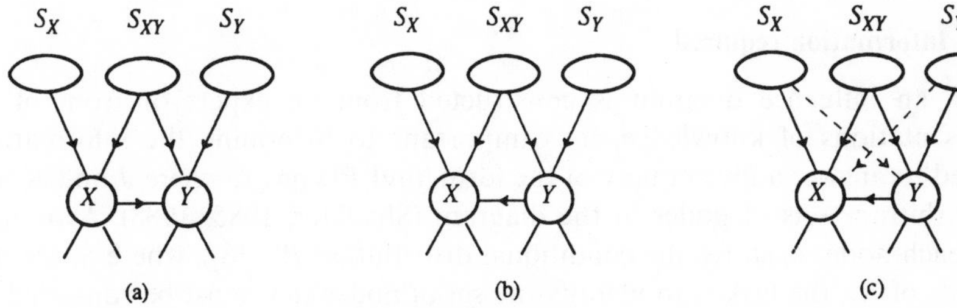

Figure 4.4

and $S_X \to Y$, as shown in Figure 4.4(c). Figure 4.4(c) satisfies the condition of Corollary 4.5, because the final parent set $YS_XS_{XY}S_Y$ of X is shown in Figure 4.4(a) to *d*-separate X from all its non-descendants (as defined by Figure 4.4(c)). A similar *d*-separation condition is verified for the final parent set of Y. Note, that the added arc $S_X \to Y$ can be reversed and still satisfy the condition for X but, then, additional arcs might be required between Y and the ancestors of S_X.

The parent-inheritance rule for arc reversal was alluded to in Howard and Matheson (1981) and later was proven for probability models by Olmsted (1983), assuming that Corollary 4.2 holds. This arc-reversal rule was also used by Shachter (1986) as the basic mechanism for evaluating influence diagrams, and was proven by Smith (1987) using axioms (4.1a)–(4.1d). The soundness of the *d*-separation criterion (Theorem 4.2) offers a graphical proof for the soundness of the arc-reversal rule, once we verify that every parents–child statement verified in the transformed diagram is graphically verified in the original diagram. Moreover, the completeness of the *d*-separation criterion implies that the parent-inheritance rule for arc reversal is also minimal, i.e. every arc added in this process in necessary, when only one arc is to be reversed; it may not be minimal when several arcs are reversed.

The rule for *node removal* also follows from Corollary 4.5. Traditionally, node removal was executed using a two-step process: to remove an arbitrary node, simply reorient any links directed out of it (using arc reversal), then remove the node and all its links. If several arcs are to be reversed, then sequentially applying the arc-reversal method to the individual arcs may result in the addition of unnecessary links, as is shown in Figure 4.3; eliminating node 3 by first reversing arc $3 \to 4$ then reversing $3 \to 5$ requires the addition of an arc between 2 and 5. Yet, diagram D_4 shows that the arc $2 \to 5$ is superfluous.

The rule for direct node removal is stated in the following corollary.

Corollary 4.7 (node removal). A node X can be eliminated from the diagram if every child of X inherits X's parents, every pair of X's children is connected by an arc and every child that receives an arrow from another child inherits all parents of the latter. These added arcs are both necessary and sufficient. Accordingly, a leaf node can be removed without adding any arcs. □

4.4.3 Information required

When an influence diagram is constructed from an expert or from other representations of knowledge, it is important to determine the information needed to answer a given query of the form 'find $P(x_J \mid x_K)$', where J and K are two arbitrary sets of nodes in the diagram (Shachter, 1985, 1988). Assuming that each node X_i stores the conditional distribution $P(x_i \mid S_i)$, where S_i are the parents of X_i, the task is to identify the set of nodes that must be consulted in the process of computing $P(x_J \mid x_K)$ or, alternatively, the set of nodes that can be assigned arbitrary conditional distributions without affecting the quantity $P(x_J|x_K)$.

The required set can be identified by the *d*-separation criterion if we represent the parameters of the distribution $P(x_i|S_i)$ as a dummy parent of node X_i. From Theorem 4.1, all dummy nodes that are *d*-separated from J by K represent variables that are conditional independent of J given K and so, the information stored in these nodes can be ignored. Thus, the information required to compute $P(x_J|x_K)$ resides in the set of dummy nodes which are not *d*-separated from J given K.

Shachter (1985, 1988) has devised an algorithm for finding a set of nodes M guaranteed to contain sufficient information for computing $P(x_J|x_K)$. The outcome of Shachter's algorithm can now be stated declaratively: M contains every ancestor of $J \cup K$ that is not *d*-separated from J given K and none other. The completeness of *d*-separation further implies that M is minimal; no node in M can be excluded on purely topological grounds (i.e. without considering the numerical values of the probabilities involved). If the diagram contains deterministic nodes, then the *D*-separation criterion should be used instead of *d*-separation.

Identifying the set of nodes that are *d*-separated from J by K is also important for limiting the scope of the subnetwork that should undergo restructuring (e.g. clustering or conditioning, Pearl, 1988) as well as for controlling evidence-gathering strategies. Knowing the precise values of variables represented by such nodes will have no effect on the outcome $P(x_J|x_K)$ and, therefore, these variables need not be tested or examined. A linear time algorithm for identifying the set of relevant nodes is reported in Geiger *et al.*, (1989b).

ACKNOWLEDGEMENTS

We thank N. Dalkey, R. Fagin, S. Lauritzen and A. Paz for many valuable discussions. This work was partially supported by the National Science Foundation Grant No. IRI-8610155. 'Graphoids: a computer representation for dependencies and relevance in automated reasoning (computer information science)'.

4.5 DISCUSSION

4.5.1 Discussion by Herman Rubin

The chapter seems to be a well-written mathematical paper, providing methods for deducing whether independence relations and conditional independence relations hold from the graph structure alone of influence diagrams. I have not had the time to look over the details, but this looks like a contribution to the mathematical treatment of the objects involved.

On the other hand, is it anything more? The author seems obsessed in reducing everything to those independence relations. He even wishes to define causality in those terms. I find this difficult to understand. In thinking over his remarks to my discussion at the meeting, I was pushed to derive a definition of causality. It is not definable from the joint distributions.

> A is a cause for B if an extranatural agent interfering only with A can affect B.

This is not symmetric in A and B; we have what the philosophers and physicists call time's arrow to contend with. I maintain that the sun shining is a cause of the grass growing but not vice versa. I will continue to maintain this asymmetry until the barely conceivable possibility of infants returning to the womb to render their mothers pregnant occurs with reasonable frequency.

Furthermore, independence relationships are not all that important. Sewall Wright essentially expressed his ideas in a manner very similar to influence diagrams. However, he was interested not so much in which variables affected which, but in how. He had no hesitation in introducing unobservable variables, and while his statistical knowledge was necessarily limited, he made good use of it. Of course he used British statistics; what else was there?

The mathematical economists, who were attempting to devise a method for extraeconomic agents, in particular governments, to take actions which would have 'beneficial' effects on the economy, were essentially using influence diagrams and decision trees, although these formal concepts had not even been invented. I do not think the terminology would have been particularly helpful to them. In any case, the problem was not that of which variables could be ignored, but of the quantitative nature of the relations involved.

4.5.2 Reply

The thrust of Dr. Rubin's comment is that one cannot infer causal relationships from dependence information or even from a complete joint distribution function.

There is no dispute over this observation. Our claim is that every causal relationship implies a certain pattern of independencies, not the converse. This gives rise to two interesting questions:

1. What, precisely, is the pattern of independencies that accompanies causal assertions?
2. If the bulk of human knowledge is derived from empirical observations, and if empirical observations are summarized by statistical averages, how is it that people manage to cast this knowledge in standard causal schemata whose structures are uniformly accepted and rarely disputed?

In our chapter we have attempted to address question 1. We have treated causal assertions as a shorthand notation for certain patterns of conditional independencies, and we have provided graphical and logical machinery for extracting the sum total of other independencies that follow from such assertions.

One can argue (as Jim Matheson did in one cocktail party) that our results bear no relation whatsoever to causation, since the theorems and corollaries retain their validity for any stratified list of independencies, not necessarily those ordered chronologically*. Our answer is as follows. Of course, if in some peculiar domain we find experts conversing with weirdly stratified dependencies that are ordered contrary to the flow of causation—so be it. Our contention is only that whenever people do choose to communicate in causal sentences, the independencies portrayed in our so-called 'causal lists' or 'input lists' offer a reasonable probabilistic interpretation of such sentences, and that the graphs that ensue offer an efficient mechanism for representing the ramifications of these sentences.

The mathematical properties of the graphical representation are verifiable analytically, while the reasonableness of our interpretation is an empirical claim about the psychology of causation, and should be verified by behavioral experiments. The claim is that, if we apply our proposed probabilistic interpretation to sentences about causation, and draw their probabilistic consequences, then these consequences will never clash with the intuition or intent of the sentence provider.

Let us now address question 2, regarding the empirical basis of causal knowledge. Temporal information undoubtedly provides important clues for causal directionality and causal organization, but, are these clues essential or merely an expedient convention? In the social sciences, for example, we often seek causal models for events that cannot be temporally ordered. We say, for example, that the attitude of a person is a cause for a certain behavior, though it is impossible to determine which comes first. It is important, therefore, to give a nontemporal probabilistic interpretation of causation, based on the notion of dependence, viewing the temporal component of causation merely as a convenient indexing standard chosen to facilitate communication and prediction. The first author has attempted to develop such an interpretation in Pearl (1988) and it seems appropriate to quote some related passages:

> The asymmetry conveyed by causal directionality is viewed as a notational device for encoding still more intricate patterns of relevance relationships, such as *nontransitive* and *induced* dependencies... . Two events do not become relevant to each other merely by virtue of predicting a common consequence, but they do become relevant when the consequence is actually observed. The opposite is true for two consequences of a common cause (pages 18–19).
>
> Note that the topology of a Bayesian network can be extremely sensitive to the node ordering *d*. What is a tree in one ordering might become a complete graph if the ordering is reversed.... This sensitivity to order may seem paradoxical at first; *d* can be chosen arbitrarily, whereas people have fairly

*Similar objections can be raised against the use of the term 'influence', and no doubt were raised centuries ago when the word 'dependence' was first introduced into probability theory.

uniform conceptual structures, e.g. they agree on whether two propositions are directly or indirectly related. This consensus about the structure of dependencies shows the dominant role causality plays in the formation of these structures. In other words, the standard ordering imposed by the direction of causation indirectly induces identical topologies on the networks that people adopt to encode experiential knowledge. Were it not for the social convention of adopting a standard ordering of events that conforms to the flow of time and causation, human communication as we know it might be impossible. Why, then, do we use temporal ordering to organize our memory? It may be because information about temporal precedence is more readily available than other indexing information, or it may be that networks constructed with temporal ordering are inherently more parsimonious (i.e. they display more independencies). Experience with expert systems applications does not entirely rule out the second possibility (Shachter and Heckerman, 1987) (pages 125–126).

REFERENCES

Beeri, C., Fagin, R., and Howard, R. A. (1977) A complete axiomatization of functional dependencies and multi-valued dependencies in database relations, *Proceedings, 1977 ACM SIGMOD Int. Conf. on Mgmt of Data* Toronto, Canada, pp. 47–61.

Blalock, H. M. (1971) *Causal Models in the Social Sciences*, Macmillan, London.

Cramér, H. (1946) *Mathematical Methods of Statistics*, Princeton University Press, Princeton, N.J.

Dawid, A. P. (1979) Conditional independence in statistical theory, *J.R. Statist. Soc.*, **B41**(1), 1–31.

Duncan, O. D. (1975) *Introduction to Structural Equation Models*, Academic Press, New York.

Fagin, R. (1977) Multivalued dependencies and a new form for relational databases, *ACM Transactions on Database Systems*, **2**(3), 262–78, September.

Feller, W. (1968) *An Introduction to Probability Theory and its Applications*, 3rd edn, Wiley, New York, Chapter XV.

Geiger, D. (1987) *Towards the Formalization of Informational Dependencies*, UCLA Cognitive Systems Laboratory, Technical Report 880053 R-102, December.

Geiger, D., Paz, A., and Pearl, J. (1989a) *Axioms and Algorithms for Inferences Involving Probabilistic Independence*, Technical Report (R-119), to appear in *Information and Computation.*

Geiger, D., and Pearl, J. (1988a) On the logic of causal models, *Proc. of the 4th Workshop on Uncertainty in AI*, St Paul, Minn., August, pp. 136–47.

Geiger, D., and Pearl, J. (1988b) *Logical and Algorithmic Properties of Conditional Independence*, UCLA Cognitive Systems Laboratory, Technical Report 870056 (R-97), February, 1989.

Geiger, D., Verma, T. and Pearl, J. (1989b) *Identifying Independence in Bayesian Networks*, UCLA Cognitive Systems Laboratory, Technical Report R-116, to appear in *Networks.*

Good, I. J. (1961) A causal calculus, *Philosophy of Science*, **11**, 305–18. Also in I. J. Good, *Good Thinking: The Foundations of Probability and its Applications*, University of Minnesota Press, Minneapolis (1983).

Howard, R. A., and Matheson, J. E. (1981) Influence diagrams. In *Principles and Applications of Decision Analysis*, Strategic Decisions Group, Menlo Park, Calif.

Isham, V. (1981) An introduction to spatial point processes and Markov random fields, *International Statistical Review*, **49**, 21–43.

Kenny, D. A. (1979) *Correlation and Causality*, Wiley, New York.

Kiiveri, H., Speed, T. P., and Carlin, J. B. (1984) Recursive causal models, *Journal of Australian Math Society*, **36**, 30–52.

Lauritzen, S. L. (1982) *Lectures on Contingency Tables*, 2nd edn, University of Aalborg Press, Aalborg, Denmark.

Meditch, J. S. (1969) *Stochastic Optimal Linear Estimation and Control*, McGraw-Hill, New York.

Olmsted, S. M. (1983) On representing and solving decision problems, Ph.D. thesis, EES Dept, Stanford University.

Pearl, J. (1985) A constraint propagation approach to probabilistic reasoning, *Proc. of the First Workshop on Uncertainty in AI*, Los Angeles, August, pp. 31–42.

Pearl, J. (1986) Fusion, propagation and structuring in belief networks, *Artificial Intelligence*, **29**(3) September, 241–88.

Pearl, J. (1988) *Probabilistic Reasoning in Intelligent Systems: Networks of Plausible Inference*, Morgan Kaufmann, San Mateo, Calif.

Pearl, J., and Paz, A. (1985) *GRAPHOIDS: a graph-based logic for reasoning about relevance relations*, UCLA Computer Science Department Technical Report 8500038 (R-53), October; also, *Proceedings, ECAI-86*, Brighton, UK, June 1986.

Pearl, J., and Verma, T. (1987) The logic of representing dependencies by directed acyclic graphs, *Proc. AAA-I*, Seattle, Wash., July, pp. 374–79.

Shachter, R. D. (1985) Intelligent probabilistic inference, *Proc. of the First Workshop on Uncertainty in AI*, Los Angeles, August, pp. 237–44.

Shachter, R. D. (1986) Evaluating influence diagrams, *Operations Research*, **34**(6), 871–82.

Shachter, R. D., and Heckerman, D. (1987) Thinking backward for knowledge acquisition, *AI Magazine*, **8**, 55–62.

Shachter, R. D. (1988) Probabilistic inference and influence diagrams, to appear in *Operations Research.*

Shafer, G., Shenoy, P. P., and Mellouli, K. (1987) Propagating belief functions in qualitative Markov trees, *International Journal of Approximate Reasoning*, **1**(4), 349–400.

Smith, J. Q. (1987) *Influence Diagrams for Statistical Modelling*, Technical Report No. 117, Department of Statistics, University of Warwick, Coventry, England, June.

Studeny, M. (1988) Attempts at axiomatic description of conditional independence, *Workshop on Uncertainty processing in Expert Systems*, Alsovice, Czechoslovakia, 20–23, June.

Verma, T. S. (1986) *Causal Networks: Semantics and Expressiveness*, UCLA Cognitive Systems Laboratory Technical Report R-65.

Verma, T. S. (1987) *Some Mathematical Properties of Dependency Models*, UCLA Cognitive Systems Laboratory Technical Report R-103.

Verma, T., and Pearl, J. (1987) The logic of representing dependencies by directed acyclic graphs, *Proc. AAAI*, Seattle, Wash., July, pp. 374–79.

Verma, T. and Pearl, J. (1988) Causal networks: semantics and expressiveness, *Proceedings of the 4th Workshop on Uncertainty in AI*, St Paul, Minn., pp. 352–9.

Wermuth, N. and Lauritzen, S. L. (1983) Graphical and recursive models for contingency tables, *Biometrika*, **70**, 537–52.

Wold, H. (1964) *Econometric Model Building*, North-Holland, Amsterdam.

Wright, S. (1921) Correlation and causation, *Journal of Agricultural Research*, **20**, 557–85.

Wright, S. (1934) The method of path coefficients, *Ann. Math. Statist.*, **5**, 161–215.

CHAPTER 5

Statistical Principles on Graphs

J. Q. Smith, *University of Warwick, UK*

ABSTRACT

Many statistical and decision theoretic principles and concepts can be represented diagramatically using a directed graph. Such a graph, called an influence diagram, is designed to capture dependences between variables—both random variables (through conditional independence) and through decisions. It is possible to use the laws of conditional independence to augment a directed graph to represent other sets of implied dependences.

The algebra for manipulating these graphs is simple. On the other hand many statistical principles and concepts can be translated to be manipulation rules on influence diagrams. How statistical principles should guide inferences arising from a given model is often transparent once an appropriate graph is drawn.

In this paper a formal definition of an influence diagram if given. It is shown, through illustrative examples, how basic statistical ideas like observability and sufficiency act on these graphs. We shall discuss some of the advantages and disadvantages of graphical methods as compared with the more conventional algebraic methods of manipulation for analysing the implications of a given statistical model.

5.1 INTRODUCTION

In my opinion the uses of graphical representatives of conditional independence (c.i.) fall into three broad categories. The first is to enable the efficient propagation of probabilities, providing a framework for the quick and efficient calculations of various marginal and conditional distributions. An excellent survey of the work done in this area is given in Lauritzen and Spiegelhalter (1988). A second use is to help in the elicitation of a model structure from a client. A graph can be used to represent statements by a client about how various variables and decisions in a problem are related. The properties of c.i.

Influence Diagrams, Belief Nets and Decision Analysis
Edited by R. M. Oliver and J. Q. Smith

can then be used to derive other relationships implicit in the client's model which can be fed back to him to be checked and if necessary modified. In this way models can be adjusted and elaborated without needing to confront the client with numerical evaluations of uncertainty (e.g. probabilities) early in an analysis—a process about which many clients harbour great suspicion. This second use of graphs is implicit in Howard and Matheson (1981) and has been formalized for probabilistic systems by Pearl (1986b) and Smith (1989b) and for decision systems first by Olmsted (1984) by Shachter (1986a) and then more generally by Smith (1989a). The software of Shachter (1987) is designed to both help elicit and then efficiently propagate probabilities and expected utilities.

The third use of graphical methods, which is linked but distinct from the second, is to help the decision analyst or statistician to understand and use a model's c.i. structure. He uses graphs directly to derive rigorously both the relationships embedded between variables and the forms of optimal policies implicit within a given model structure. Since this analysis works only on the conditional independence statements in a model, the results obtained will be completely distribution-free. Indeed they apply whether this conditional independence is defined probabilistically or is based on some other form of inference as will be shown below.

This chapter concerns, in the main, this third use of graphical methods, giving special emphasis to fundamental ideas of statistical inference. I shall use the axioms of c.i. to show how both probabilistic and non-probabilistic graphs can be constructed to give insight into a particular choice of model.

In Section 5.2. I introduce a general form of conditional independence and proceed with some examples of this structure. In Section 5.3 I address some of problems that arise when an inferential system fails to satisfy the c.i. axioms. In Section 5.4. I motivate and define the influence diagram (ID) as a graphical representation of c.i. and list some rules to manipulate this structure informatively. In Sections 5.5–5.8 I give example of how ID can help to elucidate and to generalize useful statistical constructions. In Section 5.9 I outline some of the possible extensions of current graphical method which might make them into an even more useful tool.

5.2 GENERAL CONDITIONAL INDEPENDENCE ON GRAPHS

Essential to any inferential system, be it based on full probability specifications as in Bayesian statistics or some other structure, are rules by which information is propagated through the system. Informally, graphical methods depict whether or not there exists an information link between variables in a problem.

In Smith (1989) I define a tertiary operator $\cdot \perp\!\!\!\perp \cdot | \cdot$ called *generalized conditional independence* defined on all triples $\mathbf{X}, \mathbf{Y}, \mathbf{Z}$ where $\mathbf{X}, \mathbf{Y}, \mathbf{Z}$ are subsets of a set of uncertain quantities. In agreement with Dawid's (1979) notation $\mathbf{X} \perp\!\!\!\perp \mathbf{Y} | \mathbf{Z}$ reads 'given the information in a measurement $\mathbf{Z}$, the measurement

X is uninformative about the measurement **Y**, whatever the actual value of **Z** is'. Here we do not restrict $\mathbf{X} \in A$ to be sets of random vectors, but just elements satisfying the three properties given below.

P1: for all subsets $\mathbf{X}, \mathbf{Y}, \mathbf{Z}, \mathbf{X} \perp\!\!\!\perp \mathbf{Y} \mid \mathbf{Y} \cup \mathbf{Z}$ which reads: Once **Y** (and anything else, **Z**) is known, **X** conveys no further information about **Y**.

P2: for all subsets $\mathbf{X}, \mathbf{Y}, \mathbf{Z}, \mathbf{X} \perp\!\!\!\perp \mathbf{Y} \mid \mathbf{Z} \Leftrightarrow \mathbf{Y} \perp\!\!\!\perp \mathbf{X} \mid \mathbf{Z}$ which reads: Once **Z** is known, if **X** is uninformative about **Y**, then **Y** is uninformative about **X**.

P3: for all subsets $\mathbf{W}, \mathbf{X}, \mathbf{Y}, \mathbf{Z}$

$$\left.\begin{array}{l}\mathbf{X} \perp\!\!\!\perp \mathbf{Y} \mid \mathbf{Z} \cup \mathbf{W} \\ \mathbf{X} \perp\!\!\!\perp \mathbf{Z} \mid \mathbf{W}\end{array}\right\} \text{ together imply and are implied by } \mathbf{X} \perp\!\!\!\perp \mathbf{Y} \cup \mathbf{Z} \mid \mathbf{W}$$

which reads: Once **W** is known '**X** uninformative about **Y** and **Z**' is equivalent to the two statements '**X** is uninformative about **Z**' and '**X** is uninformative about **Y** once **Z** is also known'.

It is easily checked that these three properties hold when $\{X_1, \ldots, X_m\} = A$ are random vectors and $\cdot \perp\!\!\!\perp \cdot \mid \cdot$ is, in Dawid's (1979) notation, conventional c.i. Henceforth I shall write $\mathbf{X} \perp\!\!\!\perp \mathbf{Y}$ as shorthand for $\mathbf{X} \perp\!\!\!\perp \mathbf{Y} \mid \varnothing$ where $\varnothing$ is the empty set.

All the results given in this chapter hold for any inferential system in which a generalized c.i. operator $\cdot \perp\!\!\!\perp \cdot \mid \cdot$ satisfying P1, P2 and P3 can be defined, though of course in different systems theorems may have different interpretations. As well as the usual probabilistic defnition of $\cdot \perp\!\!\!\perp \cdot \mid \cdot$, there are at least three other important operators on which graphical methods and manipulations can be based.

5.2.1 Linear systems, weak $\cdot \perp\!\!\!\perp \cdot \mid \cdot$

If A is a set of random vectors and $\mathbf{X} \perp\!\!\!\perp \mathbf{Y} \mid \mathbf{W}$ reads a best linear estimate (under quadratic loss) of elements in **X** based on elements in **W** and **Y** need only include elements in **W**'. This relation also satisfied P1, P2 and P3. To see this, first assume that **X**, **Y** and **W** are jointly normally distributed. Then the statement above is equivalent to the statement that **X** is independent of **Y** given **W**. So when A contains only jointly normal variates then P1, P2 and P3 hold. But the formulae for best linear estimates for arbitrarily distributed random variables always agree with those for normal variates with the same covariance structure. Henceforth call this $\cdot \perp\!\!\!\perp^{W} \cdot \mid \cdot$.

The operator $\cdot \perp\!\!\!\perp^{W} \cdot \mid \cdot$ is widely used in time-series analysis and control theory which traditionally has been closely associated with linear systems and second-order processes. Notice that its definition allows for the analogous interpretation of P1, P2, P3 given above provided that we are working in linear structures. Although inferences in such systems are necessarily weaker than is possible under probabilistic c.i. there are some major advantages to working with this operator.

First, it may be very difficult to elicit or specify with any degree of certainty that two variables are independent given **W**. It may, on the other hand, be much easier to assert they are uncorrelated in the sense above. Indeed in very complex systems this may be the best we can hope for.

Second, there are additional properties of this c.i. which can enable stronger inferences to be made than under the more usual c.i. For example, if $\mathbf{X} \not\perp\!\!\!\perp^{W} \mathbf{Y}$ then **Y** can be partitioned into $\mathbf{Y} = (f(\mathbf{X}), \varepsilon)$ where $f(\mathbf{X})$ is a function of **X** and

$$\mathbf{Y} \overset{W}{\perp\!\!\!\perp} \varepsilon.$$

This is not a property shared by probabilistic $\cdot \perp\!\!\!\perp \cdot | \cdot$.

Third, graphs of these systems can have arcs which are labelled by a single coefficient (corresponding to a function of a regression coefficient) to encapsulate the magnitude of a relationship between two variables, when such a relationship exists.

Properties of possibly non-linear systems under this type of linear operator have been studied extensively by Goldstein (1981).

5.2.2 Parametrized family conditional c.i.

Suppose $A = \{X_1(\boldsymbol{\theta}), X_2(\boldsymbol{\theta}), \ldots, X_n(\boldsymbol{\theta}) : \boldsymbol{\theta} \in \Theta\}$ and for each fixed value of $\boldsymbol{\theta}^* \in \Theta$ a c.i. operator $\cdot \perp\!\!\!\perp^{x} \cdot | \cdot$ over $\{X_1(\boldsymbol{\theta}^*), \ldots, X_n(\boldsymbol{\theta}^*)\}$ can be defined satisfying P1, P2, P3. Then a new c.i. operator

$$\cdot \overset{\hat{x}}{\perp\!\!\!\perp} \cdot | \cdot$$

can be defined over A by stipulating

$$\mathbf{X}_i(\boldsymbol{\theta}) \overset{\hat{x}}{\perp\!\!\!\perp} \mathbf{X}_j(\boldsymbol{\theta}) | \mathbf{X}_k(\boldsymbol{\theta})$$

if and only if

$$\mathbf{X}_i(\boldsymbol{\theta}) \overset{x}{\perp\!\!\!\perp} \mathbf{X}_j(\boldsymbol{\theta}) | \mathbf{X}_k(\boldsymbol{\theta}) \quad \text{for all values } \boldsymbol{\theta} \in \Theta.$$

For m parameters $(\theta_1, \theta_2, \ldots, \theta_m)$ the score random variable for a differentiable (in $\boldsymbol{\theta}$) family of densities $p(\mathbf{y}|\boldsymbol{\theta})$ of **Y** is defined by $\mathbf{S} = (S_1, S_2, \ldots, S_m)$ where

$$S_i = \frac{\partial}{\partial \theta_i} \log p(\mathbf{Y}|\boldsymbol{\theta}) \qquad \boldsymbol{\theta} = (\theta_1, \theta_2, \ldots, \theta_m).$$

If, for all values of $\boldsymbol{\theta}$, $B_1 \perp\!\!\!\perp B_2$ probabilistically where B_1 and B_2 are disjoint subsets of $\{S_1, \ldots, S_m\}$, then the likelihood will separate. This suggests that 'information' from **Y** about θ_i whose score component is in B_1 is separate from 'information' from **Y** about θ_j whose score is in B_2. So, in particular, inferences about θ_j need only be based on S_j. Unfortunately not many experiments will separate 'information' in this way. Cox and Reid (1987) loosen the structures of separation of information used in the above, employing $\cdot \perp\!\!\!\perp^W \cdot | \cdot$ rather than probabilistic $\cdot \perp\!\!\!\perp \cdot | \cdot$. Because of well-known problems (see Jeffreys, 1961) they can only separate one parameter from the rest in general, but our c.i. structure here allows a conditional separation which helps to break down a likelihood. This enables us to construct parametrizations with just a small number of linkages of dependence.

5.2.3 Decision system c.i.

It is not difficult to define a c.i. system over a client's decision problem. The simple trick, as outlined in Smith (1989a), is to treat a client's decision variable as an uncertain quantity from the decision analyst's point of view. Similarly his client's utility function can also be thought of as an uncertain quantity by the decision analyst. The client's random variables $C_1, \ldots, C_{n_1}$, the client's decisions $D_1, \ldots, D_{n_2}$ and the client's utility function U comprise the elements of A, the set of uncertain quantities. We can now choose a c.i. operator $\cdot \perp\!\!\!\perp \cdot | \cdot$ across subsets of this set. To date probabilistic $\cdot \perp\!\!\!\perp \cdot | \cdot$ has been used as the underlying information operator, but there is no reason why one should not use any other. For example, $\cdot \perp\!\!\!\perp^W \cdot | \cdot$ might be more appropriate if the decision analyst believed that his client would always use decision rules that were linear functions of random variables and decisions that had been observed or acted upon so far and that the utility function was quadratic in a linear combination of the other variables.

Because of their special nature, we can prove useful theorems valid only on c.i. structures over decision systems. Some of these theorems are quite subtle (see Smith, 1989a). Note that in many applications both decision variables and the utility variables will be degenerate, i.e. just functions of other variables in the system. Technically this was a nuisance since many of the relevant results in Markov field theory and IDs implicitly or explicitly assumed positivity across just distributions of variables. Hence the results from these areas needed to be reproved in this slightly more general context before theorems on c.i. relevant to these types of systems could be regarded as formally correct, even for probabilistic c.i. (see Smith, 1989b).

Treating decisions and utilities as uncertain quantities has several advantages. It extends usual forms of decision analysis to allow both for the formal processing of information when utilities have elicitation errors and for the possibility that our client may act irrationally in the future.

An example of how to use graphical methods based on decision c.i. to prove a result in game theory was presented in Smith (1988a).

5.3 WHY WE SHOULD EXPECT THE INFORMATION HYPOTHESES TO HOLD

I will now discuss some of the implications of P1, P2 and P3 and how once accepted they provide the basis of the definition of a flow of information through a system.

5.3.1 Violating P1 by estimating variables you already know

That P1 should hold I think is self-apparent. To violate this rule and believe that anything more than the measurement y itself could help to improve an estimate of some feature of y appears ridiculous.

However, people still misapply classical statistics and violate this principle, especially when selecting models. Suppose we have observed variables $\mathbf{x} = (x_1, x_2, \ldots, x_n)$ and believe that they come from a joint distribution parametrized by a set of parameters $\boldsymbol{\theta}(i)$ for one of a set of m models $\mathcal{M}_i$. Under each model $\mathcal{M}_i$ we can obtain estimates of $\boldsymbol{\theta}(i)$, for example by methods of maximum likelihood, and hence estimates $\hat{\mathbf{x}}(i)$ of a different random vector $\mathbf{X}^*$ whose distribution conditional on $\boldsymbol{\theta}(i)$ is the same as $\mathbf{X}$ whose values were $\mathbf{x}$. By confusing $\mathbf{X}^*$ with $\mathbf{X}$ and basing his choice of model on the value of $\mathbf{x} - \hat{\mathbf{x}}(i)$ alone the unsuspecting practitioner will be led to believe that models with more parameters are better. This is because, when the dimension of $\boldsymbol{\theta}(i)$ is large compared with n there is likely to be at least one vector of values of $\boldsymbol{\theta}$ which allows $\hat{\mathbf{x}}(i)$ to be close (if not equal to) $\mathbf{x}(i)$. This is not a paradox but nonsense arising from the implicit assumption that we can obtain more information about something already known—thus violatingP1. By using Bayesian methods which treat $\hat{\mathbf{x}}(i)$ as an estimate of future observation $\mathbf{x}^*$ or other methods which exclude x_j from $\hat{x}_j(i)$ such as cross-validation (see Stone, 1974) or prequential statistics (see Dawid, 1984) this apparent paradox is avoided.

5.3.2 Violating P2 and symmetry of information

In many ways P2 is the most contentious axiom of c.i. because it imposes a symmetry on measures of information which may not seem necessary in some applications. In this section it is sufficient to consider only independence $\cdot \perp\!\!\!\perp \cdot$, since analogous comments for c.i. also hold true.

Consider the following example. Suppose we need to make a decision d_1 about X_1 and a decision d_2 about X_2 and are penalized in these decisions by a common loss function $L(d_i, x_i)$, $i = 1, 2$. It is natural in this situation to say that X_1 is uninformative about X_2 if, when d_2 is allowed to depend upon $X_1 = x_1$, our expected loss associated with our optimal decision d_2^* with no information about X_1 is equal to our expected loss when given perfect information about X_1, i.e. the 'value of information' of x_1 for estimating X_2 is zero. If under this condition we were to write $X_1 \perp\!\!\!\perp X_2$ we could not conclude that $X_2 \perp\!\!\!\perp X_1$ as the following example shows.

Example 5.1. Let

$$X_1 = \begin{cases} -1 & \text{w.p.} \frac{1}{4} \\ 0 & \text{w.p.} \frac{1}{2} \\ 1 & \text{w.p.} \frac{1}{4} \end{cases} \qquad X_2 = |X_1| \quad \text{and} \quad L(x_i, d_i) = (x_i - d_i)^2 \qquad i = 1, 2$$

Then since given $X_2 = x_2$ the optimal decision $d^*(x_2) = \mathsf{E}(X_1 | X_2 = x_2) = 0$ whether or not $X_2 = 0$ or 1, the value of X_2 for estimating X_1 under L is 0. On the other hand, since X_2 is a function of X_1, once X_1 is known we can estimate X_2 perfectly. The expected value of information $= \frac{1}{2} \neq 0$. So $X_1 \not\perp\!\!\!\perp X_2$ but $X_2 \perp\!\!\!\perp X_1$, Thus 'value of information', in its usual sense, cannot be used in this way to construct $\perp\!\!\!\perp$ satisfying P2. Before the graphical manipulations which are given later are relevant, $\perp\!\!\!\perp$ must be symmetrized so that if X is informative about Y then Y is informative about X and vice versa.

Another apparent violation of P2 appears to occur even within conventional c.i.

5.3.3 Assumption P2 and paradoxes of non-conglomerability

A probability P is said to be *conglomerable* in a partition Π (see de Finetti, 1972 p. 99) when for all events E such that $P(E|h_i)$ is defined and for all constants k_1 and k_2

$$k_1 \leqslant P(E|h_i) \leqslant k_2 \qquad \text{for all events } h_i \in \Pi,$$
$$k_1 \leqslant P(E) \leqslant k_2.$$

When $\mathscr{H}$ is a countable partition and P is σ-additive then conglomerability must hold over $\mathscr{H}$, for

$$P(E) = \sum_{i=1}^{\infty} P(E|h_i)P(h_i) \qquad \text{where} \sum_{i=1}^{\infty} P(h_i) = 1,$$

is a weighted average of $P(E|h_i)$, $1 \leqslant i \leqslant \infty$.

However, in Bayesian inference and decision theory it is now fairly widely accepted that although the axiomatic case for subjective beliefs being expressed in terms of a finitely additive distribution is a compelling one, the stipulation that subjective probabilities should be countably additive is not really supported by a convincing axiomatic system (see Kadane *et al.*, 1986; Hill and Lane, 1986). And in finitely additive distributions paradoxes of non-conglomerability over countably infinite partitions can appear as the following example shows.

Example 5.2. Suppose Y takes the values 0 or 1 each with probability $\frac{1}{2}$ and that

$$P(X = x | Y = y) = 2^{-(x+y)}. \tag{5.1}$$

It is easily checked that this is consistent with a finitely additive joint distribution over (X, Y). Also notice that definition (5.1) implies that X is not independent of Y, in the sense that knowing the value of y of Y affects our probability assessment of X. $P(Y=0|X=x)$ is non-conglomerable in the partition on the margin of X, since using the conventional definition of conditional probability, for any integer x,

$$P(Y=0|X=x)=\tfrac{2}{3} \quad (5.2)$$

and yet $P(Y=0)$ has been defined as $\frac{1}{2}$. The reason that non-conglomerability is important to an understanding of independence is that since $P(Y=0|X=x)$ (and also $P(Y=1|X=x)$ do not depend on x we would like to say $Y \perp\!\!\!\perp X$ but, we have argued $X \not\perp\!\!\!\perp Y$, so P2 is violated. Whenever we have non-conglomerability, if we read $X \perp\!\!\!\perp Y$ as 'the distribution of X given Y does not depend on Y for all possible values of Y' then there is always the possibility that P2 will be violated in finitely additive probability distributions.

Even if you insist on σ-additivity over your subjective probabilities, if a partition Π is not countable, as happens when considering conditional distributions over continuous variables, non-conglomerability can still appear, even in fairly common models like one involving ratios of independent normal variates (see Kadane *et al.*, 1986). It is thus possible to construct joint densities $p(x, y)$ over variables X and Y such that $p(x|y)$ is an explicit function of y but $p(y|x)$ does not depend on x, i.e. under the interpretation of independence above $X \not\perp\!\!\!\perp Y$ but $y \perp\!\!\!\perp X$.

Now, although at first sight this is disturbing, in fact this paradox is a mirage. Independence and c.i. actually exist on the sigma-field of events on the joint distributions of all the variables and is not a function of densities and mass. By redefining independence and c.i. properly the sorts of paradox above just evaporate. Thus we say that two uncertain quantities **X** and **Y** are independent of each other if

$$\mathsf{E}(\tau(\mathbf{X})\sigma(\mathbf{Y})) = \mathsf{E}(\tau(\mathbf{X}))\mathsf{E}(\sigma(\mathbf{Y}))$$

for all bounded functions τ, σ of **X** only and **Y** only respectively on which the expectation operator E is defined on terms either side of this equation. In heuristic terms we say that $\mathbf{X} \perp\!\!\!\perp \mathbf{Y}$ if and only if all bounded functions of **X** are uncorrelated with all bounded functions of **Y**. Conditional independence can be defined similarly.

The moral of this discussion is that even when dealing with probabilistic c.i., once we move away from finite event spaces we must be very careful how we choose to define c.i. In particular we cannot conclude that $\mathbf{X} \perp\!\!\!\perp \mathbf{Y}$ just because at all possible values of **Y** a distribution of **X** conditional on $\mathbf{Y}=\mathbf{y}$ does not depend on **y**.

On the other hand, in my opinion if we say that **X** is uninformative about **Y** it is extremely counter-intuitive not to conclude that **Y** is uninformative about

X. For example I contend conglomerability paradoxes are paradoxes precisely because we seem to be led to an interpretation of probabilistic information which violates P2.

5.3.4 Simpson's paradox and P3

A common misinterpretation of independence and c.i. over joint probability distributions is, in my opinion at the root of several statistical controversies. One such controversy arises from the so-called Simpson's paradox which can be summarized in two statements.

1. If **X** is uninformative about **Y** this does not mean that **X** is uninformative about **Y** once we have been given further information from a measurement **Z**. In our notation

$$\mathbf{X} \perp\!\!\!\perp \mathbf{Y} \not\Rightarrow \mathbf{X} \perp\!\!\!\perp \mathbf{Y} \mid \mathbf{Z}. \tag{5.3}$$

   The fact that this implication is not clear is probably due to the heuristic interpretation of the word 'independence'. If we read $\mathbf{X} \perp\!\!\!\perp \mathbf{Y}$ as '**X** is independent of **Y**' there seems to be an implicit statement that whatever the circumstances **X** is uninformative about **Y**. In fact $\mathbf{X} \perp\!\!\!\perp \mathbf{Y}$ means **X** used on its own is useless for predicting anything about **Y**.
2. If for any given measurement **z** of **Z**, **X** is uninformative about **Y** we cannot conclude that **X** is uninformative about **Y**, i.e.

$$\mathbf{X} \perp\!\!\!\perp \mathbf{Y} \mid \mathbf{Z} \not\Rightarrow \mathbf{X} \perp\!\!\!\perp \mathbf{Y}.$$

It is fairly clear that we cannot expect this c.i. to carry over if we think of these statements in the right way. Just because we are given **Z**, to state that **X** provides us with no further information about **Y** obviously does not imply **X** (on its own) provides no information about **Y**. The measurement **X** may provide information about **Z** which itself is informative about **Y**, even though, were **Z** known, no further information about **Y** could be gleaned from **X**. On the other hand, interpreting this equation in terms of 'independence' encourages the sort of confusion mentioned in the last paragraph. Notice here that if $\mathbf{X} \perp\!\!\!\perp \mathbf{Z}$ or $\mathbf{Y} \perp\!\!\!\perp \mathbf{Z}$ then by using P2 and P3 we can deduce that $\mathbf{X} \perp\!\!\!\perp \mathbf{Y}$ does hold, for in this case the information chain from **X** to **Y** through **Z** is broken. The property P3 ensures that under our definition of c.i. if **X** is 'uninformative' about **Y** given **Z** then you can only learn about **Y** from **X** through the information **X** provides about **Z**. In particular if **X** and **Y** can only be 'linked' through **Z** and **X** and **Z** are not linked, then **X** and **Y** cannot be linked.

This type of explanation leads us quite neatly into a graphical representation of c.i. because it essentially uses the idea of information flow through a network. By representing $\mathbf{X}_1 \perp\!\!\!\perp \mathbf{X}_2 \mid \mathbf{Z}_1 \cup \mathbf{Z}_2 \cdots \cup \mathbf{Z}_m$ by a graph with nodes labelled $\mathbf{X}_1$, $\mathbf{X}_2$, $\mathbf{Z}_1, \ldots, \mathbf{Z}_m$ and directed arcs joining all nodes $\mathbf{Z}_i$ to $\mathbf{X}_j$, $1 \leqslant i \leqslant m$, $1 \leqslant j \leqslant 2$, but with no arc between $\mathbf{X}_1$ and $\mathbf{X}_2$, we can represent a set of c.i. statements by a graph of links between variables. Algebraic rules P1, P2 and P3 can then be shown to imply less opaque rules about flow of information on this graph. Here then is the influence diagram—the structure which provides the vital link between c.i. algebra and geometry.

5.4 INFLUENCE DIAGRAMS AND GRAPHS OF CONDITIONAL INDEPENDENCE

Let $A = \{\mathbf{X}_1, \mathbf{X}_2, \ldots, \mathbf{X}_m\}$ be a set of uncertain quantities over which a c.i. operator is defined. First list the element in A in some arbitrary order $\mathbf{X}_{(1)}$, $\mathbf{X}_{(2)}, \ldots, \mathbf{X}_{(m)}$ where $\{\mathbf{X}_{(i)} = X_j : 1 \leqslant i \leqslant m\}$ determines a permutation of indices (called Λ). An influence diagram (ID) over A and Λ has a directed graph I with nodes labelled $\mathbf{X}_1, \ldots, \mathbf{X}_m$ for which the $m-1$ c.i. statements

$$\mathbf{X}_{(r)} \perp\!\!\!\perp \mathbf{X}^{(r-1)} | P(\mathbf{X}_{(r)}) \qquad r = 2, 3, 4, \ldots, m \tag{5.4}$$

hold, where

$$\mathbf{X}^{(r-1)} = (\mathbf{X}_{(1)}, \mathbf{X}_{(2)}, \ldots, \mathbf{X}_{(r-1)})$$

and $P(\mathbf{X}_{(r)}) \subseteq X^{(r-1)}$ is the set of nodes/uncertain quantities connected by one directed edge to $X_{(r)}$.

By properties P1 and P3 an equivalent set of c.i. statements to the set given above is

$$\mathbf{X}_{(r)} \perp\!\!\!\perp \bar{P}(\mathbf{X}_{(r)}) | P(\mathbf{X}_{(r)}) \qquad r = 2, 3, 4, \ldots, m, \tag{5.5}$$

where $\bar{P}(\mathbf{X}_{(r)})$ is the complement of $P(\mathbf{X}_{(r)})$ in $\mathbf{X}^{(r-1)}$. Thus we can read from the graph of I that $\mathbf{X}_{(r)}$ is independent of all those nodes in our list (defined by permutation Λ) introduced before $\mathbf{X}_{(r)}$ which are not attached to it by edges, conditional on those nodes listed before $\mathbf{X}_{(r)}$ which are attached to it by an edge. Before we discuss why such a graph is useful note the following. Since the c.i. relationships between variables in an ID for which $P(\mathbf{X}_{(r)}) = \mathbf{X}^{(r-1)}$ (whose graph is completely connected) are vacuous, there always exists at least on ID over uncertain quantities A. Normally we would try to choose a permutation Λ so that a corresponding ID could be found whose graph had least arcs, since graphs with the smallest numbers of arcs contain the strongest statements about c.i. among the elements of A. Note that if $X_{(i)}$ is attached to $X_{(k)}$ then $i < k$ so that the graph of I must be acyclic.

If in the graph of I, $\mathbf{X}$ is connected by an arc to $\mathbf{X}_r$ so that so that $\mathbf{X} \in P(\mathbf{X}_r)$, it is called a *direct predecessor* (d.p.) of $\mathbf{X}_r$ and $\mathbf{X}_r$ is called a direct successor (d.s.) of $\mathbf{X}$. The set of all d.s.s. of $\mathbf{X}$ is denoted by $S(\mathbf{X})$. Here $\mathbf{X}$ is called a *predecessor* of $\mathbf{Y}$ and $\mathbf{Y}$ called a *successor* of $\mathbf{X}$ if there exists a directed path from $\mathbf{X}$ to $\mathbf{Y}$ in an ID labelled I. The set of all successors of $\mathbf{Y}$ is denoted by $\sigma(Y)$ and $\bar{\sigma}(\mathbf{Y})$ denotes the complement of $\sigma(\mathbf{Y})$ in $A \backslash \mathbf{X}$. An ID labelled I_1 is said to be implied by an ID labelled I_2 (written $I_2 \Rightarrow I_1$) if all c.i. statements implied in the graph of I_1 can be deduced from those implied by the graph of

I_2. If $I_1 \Rightarrow I_2$ and $I_2 \Rightarrow I_1$, I_1 and I_2 are said to be equivalent (written $I_1 \equiv I_2$). Unless otherwise stated, IDs will be assumed to be on the same set of m nodes and the d.p.s and d.s.s of Z in I_i will be denoted by $P_i(Z)$ and $S_i(Z)$ respectively, $i = 1, 2, 3 \ldots$.

An ID is said to be decomposable if the d.p.s $P(X)$ of any node X in I have the property that if X_1, $X_2 \in P(X)$ then $X_1 \in P(X_2)$ or $X_2 \in P(X_1)$. Two IDs I_1 and I_2 defined in the same set of nodes are called similar (written $I_1 \sim I_2$) if the undirected graph obtained by deleting arrows from the directed graph of I_1 is identical to that obtained by deleting arrows from I_2.

Perhaps the most important reason why IDs can be useful is that it can be proved (see Smith, 1989b) that any two listings of variables corresponding to permutations of variables Λ_1 and Λ_2 consistent with the partial order induced by a directed graph I of an ID over elements in A expresses equivalent set of c.i. statements (equation 5.4). Therefore the graph on its own (without the permutation) is sufficient to define an equivalent set of $m-1$ c.i. statements in the quantities labelled by its nodes.

Because of this one-to-one relationship between ID graphs and sets of c.i. statements (5.4) we can translate the c.i. rules P1, P2 and P3 on to rules about ID representations. Because the ID gives us a global picture of the relationships between the variables in our problem it is often easy to calculate the global c.i. implications of a list of (local) c.i. statements from its graph. So although nothing can be proved or explained using IDs which cannot be proved directly from c.i. they do help to explain and motivate otherwise opaque results, just as a geometrical argument explains and motivates the proof of an algebraic identity.

Below I give a short, and very incomplete list of some rules on how to read implied c.i. statements from the graph of an ID. These results and many others are variously proved by Pearl (1986b) and Smith (1988, 1989a, 1989b). Earlier forms of proofs are also given in Olmsted (1984) and Shachter (1986b).

Rule 1 (partial order characterization). In any ID labelled I for all nodes $\mathbf{X} \in I$

$$\mathbf{X} \perp\!\!\!\perp \bar{\sigma}(\mathbf{X}) \mid \mathbf{P}(\mathbf{X}),$$

i.e. given its d.p.s $\mathbf{X}$ is independent of the set of all variables which are not its successors.

□

Rule 2 (addition of arcs). If, for two IDs I_1 and I_2,

$$P_1(X_i) \subseteq P_2(X_i) \qquad 1 \leqslant i \leqslant m$$

then

$$I_1 \Rightarrow I_2. \qquad \square$$

Rule 3 (node elaboration). Let ID labelled I_1 have nodes $\{\mathbf{X}_1, \ldots, \mathbf{X}_m\}$ introduced in that order. Let ID labelled I_2 have nodes $\{\mathbf{X}_1, \ldots, \mathbf{X}_{k-1}, \mathbf{X}_k(1), \ldots,$

$\mathbf{X}_k(n)\}$ introduced in that order and let ID labelled I_3 have nodes $\{\mathbf{X}_1, \ldots, \mathbf{X}_{k-1}, \mathbf{X}_k(1), \ldots, \mathbf{X}_k(n), \mathbf{X}_{k+1}, \ldots, \mathbf{X}_m\}$ introduced in that order where

$$P_1(\mathbf{X}_i) = P_2(\mathbf{X}_i), \qquad 1 \leqslant i \leqslant k-1$$

and

$$\mathbf{X}_k = \bigcup_{i=1}^{n} \mathbf{X}_k(i).$$

If

(a) $P_3(\mathbf{X}_k(r)) = P_2(\mathbf{X}_k(r)) \qquad 1 \leqslant r \leqslant n$

(b) $$P_3(\mathbf{X}_i) = \begin{cases} P_1(X_i) \cup \{\mathbf{X}_k(1), \ldots, \mathbf{X}_k(n)\} \backslash \mathbf{X}_k & \text{if } \mathbf{X}_k \in P(\mathbf{X}_i) \\ P_1(\mathbf{X}_i) & \text{otherwise} \end{cases}$$

$1 \leqslant i \leqslant m,\ i \neq k$

then I_1 and I_2 together are equivalent to I_3. □

Rule 4 (the Howard–Matheson theorem). If ID labelled I_1 has $X_j \in P_1(X_k)$ and ID labelled I_2 has its d.p.s defined by

$$P_2(X_i) = P_1(X_i) \qquad i \neq j, k$$
$$P_2(X_j) = \{P_1(X_j) \cup P_1(X_k) \cup X_k\} \backslash X_j$$
$$P_2(X_k) = P_1(X_k) \cup P_1(X_j)$$

then $I_1 \Rightarrow I_2$. □

Rule 5 (the decomposition theorem). If I_1 and I_2 are both decomposable IDs and $I_1 \sim I_2$ then $I_1 \equiv I_2$. □

Sometimes inherent in a problem are additional pieces of information about c.i. not represented in the graph of an ID. A node $\mathbf{W}$ is said to be identified by a set of nodes $(\mathbf{X}_1, \ldots, \mathbf{X}_r)$ if

$$\mathbf{W} \perp\!\!\!\perp B \mid (\mathbf{X}_1, \ldots, \mathbf{X}_r),$$

where B is any set of nodes not containing $\mathbf{W}$. For example if $\mathbf{W}$ is a function of $\mathbf{X}_1, \ldots, \mathbf{X}_r$ it is identified by these variables.

Rule 6 (reduction by identification). If $\mathbf{W}$ is identified by $(\mathbf{X}_1, \ldots, \mathbf{X}_r)$ and $(\mathbf{W}, \mathbf{X}_1, \ldots, \mathbf{X}_r) \subseteq P_1(\mathbf{Y})$ for some $Y \in I_1$ then $I_1 \Rightarrow I_2$, where I_2 has all nodes with the same d.p.s as I_1 except the d.p.s of Y which are given by $P_2(\mathbf{Y}) = P_1(\mathbf{Y}) \backslash \mathbf{W}$. □

5.5 CONDITIONAL INDEPENDENCE AND THE ROLE OF PARAMETERS IN BAYESIAN INFERENCE

We know that we will observe that values of random variables $\mathbf{X} = (X_1, X_2, \ldots, X_n)$ and on the basis of this information be required to make

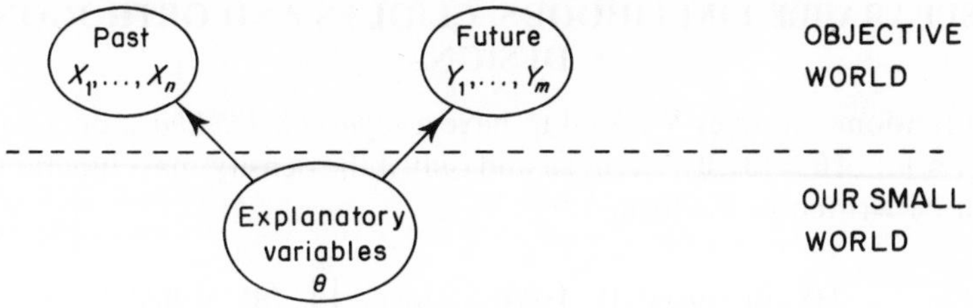

Figure 5.1

decisions about the future random variables $\mathbf{Y} = \{Y_1, Y_2, \ldots, Y_m\}$. Let $\mathbf{Z} = (\mathbf{X}, \mathbf{Y})$. Because all these variables are related to one another, to construct a model based on the conditional breakdown of **Z** is arduous. However, in many cases there is considerable symmetry about our information about **Z**. We can usually create another small dimensional vector of explanatory variables (or parameters) $\boldsymbol{\theta}$ such that conditional on $\boldsymbol{\theta}$ it is reasonable to assume that

$$\coprod_{i=1}^{m+n} Z_i \mid \boldsymbol{\theta}.$$

The variable $\boldsymbol{\theta}$ is the construct, based on ideas of partial causality and analogy, we use in our personal and subjective small world to help us understand and categorize how supposedly objective measurement events are linked together. This idea is encapsulated in the graph of Figure 5.1.

Since $\mathbf{X} \perp\!\!\!\perp \mathbf{Y} \mid \boldsymbol{\theta}$ we can only learn about the future, **Y**, from the way the past, **X**, influences our beliefs about $\boldsymbol{\theta}$. For us, personally, knowing how **x** has influenced our beliefs about $\boldsymbol{\theta}$ is sufficient for predicting **Y**. In our current state of mind (model) all information other than **x** is irrelevant for this forecast. Not only does this clarify what we are trying to do in a Bayesian model but also provides a useful tool for propagating information. Once we accept $\mathbf{X} \perp\!\!\!\perp \mathbf{Y} \mid \boldsymbol{\theta}$ it is clearly much more computationally efficient to construct our forecast distribution using $\boldsymbol{\theta}$ than not, since $\boldsymbol{\theta}$ gives a small dimensional summary of information relevant to our task of predicting **Y** from **x**.

Thus in my opinion, a good statistical model needs to introduce explanatory variables $\boldsymbol{\theta}$ which have both sparse c.i. relationships between their components and the observables in the system. If this feature does not exist it will be difficult to both understand (and so initialize) the model and also to communicate to others through it. Note that the creation of these explanatory variables is a complementary one to decision analytic techniques of removing irrelevant variables in the model description (e.g. barren node reduction, Shachter, 1986a). More details of these ideas are given in Smith (1987).

I shall now give some simple examples of how statistical constructs can be manipulated and extended by using graphs to synthesize sets of c.i. statements they implicitly impose.

5.6 SEPARABLE LIKELIHOODS, CLIQUES AND ORTHOGONAL DESIGNS

A set of random variables $\mathbf{Y}$ is said to have a *separable likelihood* on a partition $\{\boldsymbol{\theta}_i : 1 \leqslant i \leqslant m\}$ of $\boldsymbol{\theta} = (\boldsymbol{\theta}_1, \boldsymbol{\theta}_2, \ldots, \boldsymbol{\theta}_m)$ if and only if the density/mass function $f(\mathbf{y}\,|\,\boldsymbol{\theta})$ of $\mathbf{y}$ can be written in the form

$$f(\mathbf{y})|\boldsymbol{\theta}) = g(\mathbf{y}\,|\,\mathbf{t}_1(\mathbf{y}), \mathbf{t}_2(\mathbf{y}), \ldots, \mathbf{t}_m(\mathbf{y})) \prod_{i=1}^{n} f_i(\mathbf{t}_i(\mathbf{y})\,|\,\boldsymbol{\theta}_i), \tag{5.6}$$

where $\mathbf{t}_1, \ldots, \mathbf{t}_m$ are functions of $\mathbf{y}$, g is a function of $\mathbf{y}$, and $\mathbf{t}_1, \ldots, \mathbf{t}_m$ only and $f_i(\mathbf{t}_i(\mathbf{y})\,|\,\boldsymbol{\theta}_i)$ is a function of $\mathbf{t}_i(\mathbf{y})$ and $\boldsymbol{\theta}_i$ only.

It is easy to see that, assuming conglomerability, an absolutely continuous/discrete random vector $\mathbf{Y}$ has a separable likelihood on $\{\boldsymbol{\theta}_i : 1 \leqslant i \leqslant m\}$ if and only if when $\perp\!\!\!\perp_{i=1}^{m} \boldsymbol{\theta}_i$ a priori, we expect after observing $\mathbf{Y}$ that this independence will be preserved. Thus to a Bayesian statistician separability corresponds to a type of preservation of independence in a prior to posterior analysis, the two graphs I and K given in Figure 5.2 being equivalent. (Separability implies graph I can be expanded to J_1. Here J_1 is equivalent to J_2 using Rule 4 on $(t_i, \boldsymbol{\theta}_i)$ over $1 \leqslant i \leqslant m$; J_2 implies K by Rule 3.) Notice that our experiment has to be structured in a special way for this equivalence to hold. The properties of c.i. do not on their own imply separability, see Simpson's paradox in Section 5.3.4 and the Howard–Matheson theorem of Section 5.4.

It is often very helpful, both analytically and interpretatively, if experiments can be designed to have observables with a likelihood separable in some components of $\boldsymbol{\theta}$. The well-used orthogonal design in linear models uses this idea as the following example illustrates.

Example 5.3. The two-way analysis of variance. Our model states

$$Y_i\,|\,x_i(1), x_i(2) = \mu + \mathbf{x}_i(1)\boldsymbol{\theta}_1 + \mathbf{x}_i(2)\boldsymbol{\theta}_2 + \varepsilon_i \ \perp\!\!\!\perp_{i=1}^{m} \varepsilon_i \qquad \varepsilon_i \sim n(0, \sigma^2), \tag{5.7}$$

where $\mathbf{x}_i(1) \sim$ the indicator vector on the lathe used in experiment i;
$\mathbf{x}_i(2) \sim$ the indicator vector on the lathe operator conducting experiment i;
$y_i\,|\,\mathbf{x}_i(1), \mathbf{x}_i(2) \sim$ the job time given the lathe and operator indicated by $(\mathbf{x}_i(1), \mathbf{x}_i(2))$.
The parameters $(\mu, \boldsymbol{\theta}_1, \boldsymbol{\theta}_2)$ are usually interpreted thus:
$\mu \sim$ mean time over all jobs
$\boldsymbol{\theta}_1 \sim$ effect of each lathe on the speed of doing the given task
$\boldsymbol{\theta}_2 \sim$ effect of each operator on the speed of completing the task

Usually we want to learn about $\boldsymbol{\theta}_1$, $\boldsymbol{\theta}_2$ and are free to choose the 'design matrix' $(\mathbf{x}_i(1), \mathbf{x}_i(2))$, $1 \leqslant i \leqslant n$ as we think best.

I contend that for the additive model (5.7) to be compelling we are likely to believe that the parameters $\mu, \boldsymbol{\theta}_1, \boldsymbol{\theta}_2$ are mutually independent prior to the experiment. If, for example, we had reason to believe that some structural

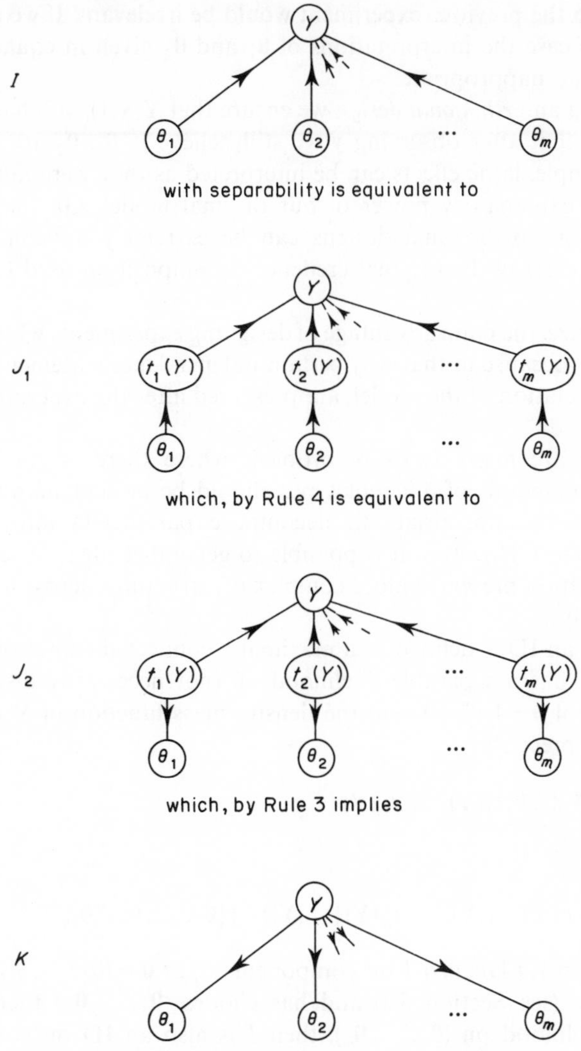

Figure 5.2 Separability and presentation of independence.

relationship existed between $\boldsymbol{\theta}_1, \boldsymbol{\theta}_2$ (for example better operators in the past worked on better machines) then surely we should reparametrize the model so as to lable this effect directly.

Furthermore, by separating our beliefs into the partition $(\mu, \boldsymbol{\theta}_1, \boldsymbol{\theta}_2)$ we implicitly relate the marginal lathe 'effect' $\boldsymbol{\theta}_1$ to predictions about other experiments. Hence if we repeated the experiment with a disjoint set of operators of a similar profile, the lathe effect $\boldsymbol{\theta}_1$ should be preserved, whereas information about $\boldsymbol{\theta}_2$ from the previous experiment would be irrelevant. If we did not want this to be the case the interpretations of $\boldsymbol{\theta}_1$ and $\boldsymbol{\theta}_2$ given in equation (5.8) are misleading and inappropriate.

By choosing an *orthogonal design* we ensure that $\mathbf{Y}|\mathbf{x}(1), \mathbf{x}(2)$ has a separable likelihood so that after observing $\mathbf{y}$ we still believe μ, $\boldsymbol{\theta}_1$, $\boldsymbol{\theta}_2$ are independent. Thus for example, lathe effects can be interpreted as they were initially and we maintain the explanatory power of our original model. On the other hand, results from non-orthogonal designs can be extremely difficult to interpret because they destroy the original credence decomposition used to structure a model.

To summarize, the main advantage of designing experiments whose likelihood is separable is because in that way certain independence statements, important on the interpretation of the model, are preserved after the experiment has been performed.

Now there are many types of problem where there is good reason for supposing interactions of different types should be evident between effects. It will not then be appropriate to decompose parameters into independent partitions *a priori*. However, it is possible to generalize ideas of separability so that our definition preserves more complex c.i. structures across $\boldsymbol{\theta}$ in a prior to posterior analysis.

A clique of an ID is defined as a maximally connected subset of nodes of its graph. Say $\mathbf{Y}$ has a separable likelihood on (not necessarily disjoint) cliques $(\boldsymbol{\theta}_1, \boldsymbol{\theta}_2, \ldots, \boldsymbol{\theta}_m)$ if $\boldsymbol{\theta} = \bigcup_{i=1}^{m} \boldsymbol{\theta}_i$ and the density mass function of $\mathbf{Y}$ has the form (5.6) or more precisely:

$$\mathbf{Y} \perp\!\!\!\perp \boldsymbol{\theta} | t_1(\mathbf{Y}), \ldots, t_m(\mathbf{Y}) \tag{5.9}$$

and

$$t_i(\mathbf{Y}) \perp\!\!\!\perp (\{t_1(\mathbf{Y}), \ldots, t_m(\mathbf{Y})\} \backslash t_i(\mathbf{Y})) \cup (\{\boldsymbol{\theta}_1, \ldots, \boldsymbol{\theta}_m\} \backslash \boldsymbol{\theta}_i)/\theta_i. \tag{5.10}$$

Claim. If an ID labelled I on components θ_j of $\boldsymbol{\theta} = (\theta_1, \ldots, \theta_k)$, $1 \leqslant i \leqslant k$, is decomposable (see Section 5.4) and has cliques $(\boldsymbol{\theta}_1, \ldots, \boldsymbol{\theta}_m)$ then if $\mathbf{Y}$ has a separable likelihood on $(\boldsymbol{\theta}_1, \ldots, \boldsymbol{\theta}_m)$, then I is also an ID on components θ_j, posterior to observing $\mathbf{Y}$.

Proof. Without loss assume the cliques $(\boldsymbol{\theta}_1, \ldots, \boldsymbol{\theta}_m)$ are ordered consistently with the partial order induced by an ID J_0 similar to I so that $(\boldsymbol{\theta}_1,\ldots,\boldsymbol{\theta}_m)$ have the *running intersection property*, i.e. $\forall_j \geqslant 2$, $\exists i < j$ such that

$$\boldsymbol{\theta}_i \supseteq \boldsymbol{\theta}_j \cap (\boldsymbol{\theta}_1 \cup \cdots \boldsymbol{\theta}_{j-1}) = S_j.$$

It has been proved (see Beeri *et al.*, 1981, 1983; Lauritzen *et al.*, 1984; Tarjan and Yannakakis, 1984) that such J_0 always exist if I is decomposable. By Rule 5, $J_0 \equiv I$. □

We now condition on our observations in the order $\{\mathbf{t}_1(\mathbf{y}), \mathbf{t}_2(\mathbf{y}), \ldots \mathbf{t}_m(\mathbf{y}), \mathbf{y}\}$. To introduce the node $\mathbf{t}_1(\mathbf{y})$ as to J_0, by equation (5.10), we need only connect $\mathbf{t}_1(\mathbf{y})$ by arcs from $\boldsymbol{\theta}_1 = (\theta_1(1), \ldots, \theta_1(k_1))$. Call this new ID J_0^+. Now use Rule 4 on J_0^+ to reverse arcs between $\mathbf{t}_1(\mathbf{y})$ and the components of $\boldsymbol{\theta}_1$ in the order $(\theta_1(k_1), \mathbf{t}_1(\mathbf{y})), (\theta_1(k_1 - 1), \mathbf{t}_1(\mathbf{y})), \ldots, (\theta_1(1), \mathbf{t}_1(\mathbf{y}))$. Since $\boldsymbol{\theta}_1$ completely connected, no new arcs between components of $\boldsymbol{\theta}_1$ can be added and $\mathbf{t}_1(\mathbf{y})$ now precedes $\boldsymbol{\theta}_1$. Relabel the components $\theta_1(i)$ of $\boldsymbol{\theta}_1$ as $\theta_1(i)|\mathbf{t}_1(\mathbf{y})$ to condition out this node to obtain the ID J_1, representing our beliefs after observing $\mathbf{t}_1(\mathbf{y})$. Note that $J_1 = J_0$. We can now proceed by induction.

Let $\mathbf{t}_r(\mathbf{y})$ be the lowest ordered statistic introduced and then conditioned out, as described above such that we cannot deduce that $J_r = J_{r-1}$, where J_k, $k = 1, 2, \ldots$ is defined inductively by the ID obtained from adding $\mathbf{t}_k(\mathbf{y})$ to J_{k-1} to obtain J_{k-1}^+ in the analogous way to the above and then conditioning out $\mathbf{t}_k(\mathbf{y})$, using Rule 4 so that $\mathbf{t}_k(\mathbf{y})$ has no predecessors. By equation (5.10), $\mathbf{t}_r(\mathbf{y})$ is only connected by arcs from $\boldsymbol{\theta}_r = (\theta_r(1), \ldots, \theta_r(k_r))$ in J_{r-1}^+. Now use Rule 4 to reverse arcs from components of $\boldsymbol{\theta}_r$ to $\mathbf{t}_r(\mathbf{y})$ in the order

$$(\theta_r(k_r), \mathbf{t}_r(\mathbf{y})), (\theta_r(k_r - 1), \mathbf{t}_r(\mathbf{y})), \ldots, (\theta_r(1), \mathbf{t}_r(\mathbf{y})).$$

By the running intersection property $\mathbf{t}_r(\mathbf{y})$ is now a direct successor only of the elements of the clique $\boldsymbol{\theta}_{r-1}$. Since $\boldsymbol{\theta}_r$ is a clique no additional arcs have been added between its components under these operations. Also by running intersection property $\theta_r(j) \in \boldsymbol{\theta}_{r-1}$ iff $j < l$ for some l. It follows that the sequence of operations above joints nodes $\theta_r(j) \in \boldsymbol{\theta}_r$ to those in $\boldsymbol{\theta}_{r-1}$ only if $\theta_r(j) \in \boldsymbol{\theta}_{r-1}$. Since $\boldsymbol{\theta}_{r-1}$ is a clique, no arcs have been added between its components. By the running intersection property no node in J_{r-1} not in $\boldsymbol{\theta}_{r-1}$ or $\boldsymbol{\theta}_r$ has arcs attached from it by this operation. The node $\mathbf{t}_r(\mathbf{y})$ now has as direct predecessors only nodes in $\boldsymbol{\theta}_{r-1}$. Call this ID $\tilde{J}_{r-1}$. Since Rule 4 adds on additional arcs into successors of both nodes involved in an arc reversal by treating $\mathbf{t}_r(\mathbf{y})$ as $\mathbf{t}_{r-1}(\mathbf{y})$ we can conclude from the inductive hypothesis that $\tilde{J}_{r-1}$ implies $\hat{J}_{r-1}$ where $\hat{J}_{r-1}$ is equal to J but with an additional node $\mathbf{t}_r(\mathbf{y})$ conditioning the components of cliques $(\boldsymbol{\theta}_1, \boldsymbol{\theta}_2, \ldots, \boldsymbol{\theta}_r)$. Conditioning out $\mathbf{t}_r(\mathbf{y})$ and relabelling the components of $\theta_j(i)|t_{r-1}(\mathbf{y}), \ldots, t_j(\mathbf{y})$ of $\boldsymbol{\theta}_j$ in J_{r-1} by $\theta_j(i)|t_r(y) \cdots t_j(y)$, $1 \leqslant i \leqslant m$, proves that $\hat{J}_{r-1} \Rightarrow J_r$. This contradicts our inductive hypothesis. It follows that we must conclude that J_m can be deduced from J_0.

Finally since conditional on $\mathbf{t}_1, \ldots, \mathbf{t}_m$, $\mathbf{Y}$ is independent of $\boldsymbol{\theta}$ by equation (5.9), a node representing $\mathbf{Y}$ is attached to no node in J_m. It follows that $\mathbf{Y}$ can be marginalized out with no new arcs needing to be added to J_m. Our result now follows. □

This type of idea is examined extensively in Speigelhalter and Lauritzen (1988) although there the emphasis is on efficiency of calculation rather than the maintenance of explanatory power of a model prior and posterior to likely experiments.

5.7 NON-LINEAR EXTENSIONS OF STATE-SPACE FORECASTING

In 1976, Harrison and Stevens adapted the Kalman filter so that it was a viable tool for business forecasting. Explicitly a sequence of observables $\{Y_t : t = 1, 2, \ldots\}$, for example Y_t representing sales in week t, were related to each other through m-dimensional state vectors $\{\boldsymbol{\theta}_t : t = 1, 2, 3, \ldots\}$ whose components represented variables like the mean level of demand at time t or the seasonal effect on demand at week t. Thus $\{Y_t, \boldsymbol{\theta}_t : t = 1, 2, \ldots\}$ were related as follows:

$$Y_t = \mathbf{F}\boldsymbol{\theta}_t + v_t, \tag{5.11}$$

$$\boldsymbol{\theta}_t = \mathbf{G}\boldsymbol{\theta}_{t-1} + \omega_t; \tag{5.12}$$

where $v_t \sim n(0, V_t)$, $\omega_t \sim n(0, W_t)$, $\perp\!\!\!\perp_{t=1,2} v_t$, ω_t, $\perp\!\!\!\perp \boldsymbol{\theta}_t, \boldsymbol{\theta}_{t+1}, \ldots, | \boldsymbol{\theta}_{t-1} \omega_{t-1}$, $\{v_t\}_{t \geq 1} \perp\!\!\!\perp \{\omega_t\}_{t \geq 1}$ are error terms, $\mathbf{F}$ is a $(1 \times m)$ vector of constants and $\mathbf{G}$ is an $m \times m$ matrix of constants. Note that the observations $\{Y_t\}_{t \geq 1}$ are related to each other in time only though their relationship to $\{\boldsymbol{\theta}_t\}_{t \geq 1}$, the 'true levels' of the process at various times, where $\{\boldsymbol{\theta}_t\}_{t \geq 1}$ is a Markov chain. This process has an ID whose graph I_1 is represented in Figure (5.3).

Although the Kalman filter (or dynamic linear model) has been shown to be a very useful model, it has drawbacks in the sense that it is essentially a linear normal process and most business series are neither linear or normal. There is therefore a need to extend the dynamic linear model defined by equations (5.11) and (5.12) so that it can model non-linear as well as linear time series.

Now the graph given in Figure 5.3 represents none of the distributional and linearity assumptions embedded in equations (5.11) and (5.12). It represents only a set of c.i. statements implied by the equations. It would be natural therefore to demand that an extension of the dynamic linear model away from normal processes would produce the same graph of c.i. statements in Figure 5.3. In this way the role of the state space variables $\boldsymbol{\theta}_t$ is most likely to be preserved.

The first point to notice here is that the role of the error variables $\{v_t\}_{t \geq 1}$ and $\{\omega_t\}_{t \geq 1}$ are only useful in the way they relate $\{\boldsymbol{\theta}_t\}_{t \geq 1}$ and $\{Y_t\}_{t \geq 1}$. They are certainly not variables which we have a direct interest in forecasting. Our losses will involve only future values of $\{Y_t\}$ and possibly the states $\{\boldsymbol{\theta}_t\}_{t \geq 1}$. In a non-normal extension we can therefore safely marginalize out these variables.

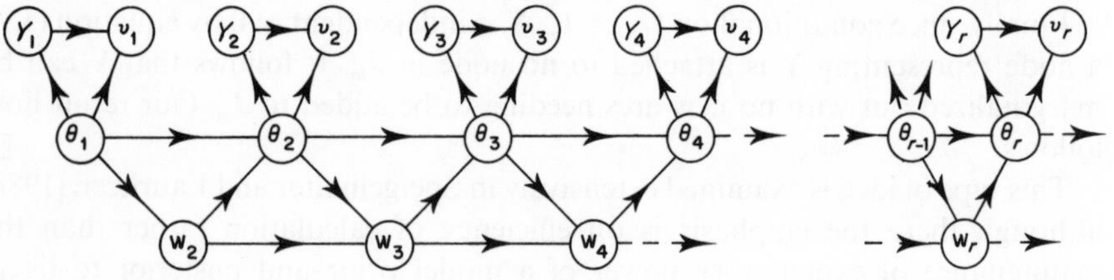

Figure 5.3 I_1

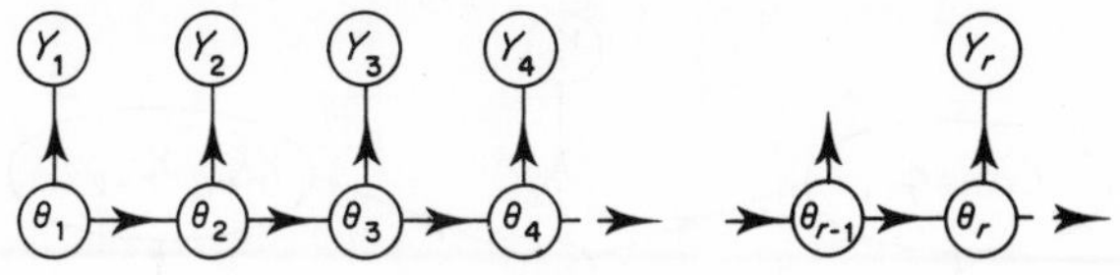

Figure 5.4 I_2

Well Rule 4 allows us to reverse all $(\omega_i, \boldsymbol{\theta}_i)$ arcs without addition in Figure 5.3 and obtain an equivalent influence diagram. After these arc reversals by Rule 1 all error terms could have been introduced last into the graph of the ID. So in particular all error terms could be dropped from the original ID J_1 giving us the implied ID I_2. Figure 5.4 can be interpreted as saying:

1. The distribution of Y_T depends only on $\{\boldsymbol{\theta}_t, \{Y_t\} \backslash Y_T\}$ only through $\boldsymbol{\theta}_T$.
2. The vector $\boldsymbol{\theta}_T$ depends only on $\{\boldsymbol{\theta}_1, \ldots, \boldsymbol{\theta}_{T-1}\}$ through its relationship to $\boldsymbol{\theta}_{T-1}$.

Any specification of distributions on $\mathbf{Y}_T | \boldsymbol{\theta}_T$ and $\boldsymbol{\theta}_T | \boldsymbol{\theta}_{T-1}$ which respect the c.i. statements in (1) and (2) can be thought of as a non-normal/non-linear extension of equations (5.11) and (5.12) preserving the essential advantage of using a state-space characterization of a time series rather than using a model on the observables directly. Some of these advantages are listed below.

(a) The existence of compact forecasting distributions

Rule 3 implies that for any value $\tau = T, I_2$ implies the ID given in Figure 5.5. By Rule 4 an equivalent ID is obtained by reversing the arc labelled $(*)$ and then employing Rule 3 again gives I_4 whose graph is given in Figure 5.6.

This implies, in particular that the distribution of the future observables $Y_{T+1}, Y_{T+2}, \ldots$ is independent of the past $Y_1, \ldots, Y_T$ conditional on the explanatory state variable $\boldsymbol{\theta}_T$. So the variable $\boldsymbol{\theta}_T$ contains all relevant knowledge gained from the past with which to predict the future. Thus any generalization of equations (5.11) and (5.12) which respects (1) and (2) shares the desirable property that to predict the future we need only store a small dimensional (m) distribution of relevant information.

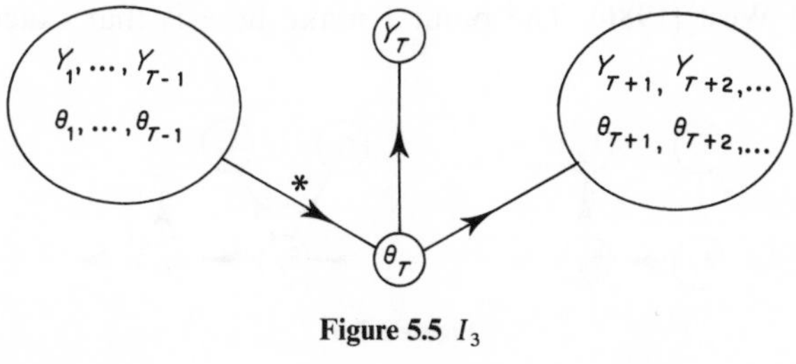

Figure 5.5 I_3

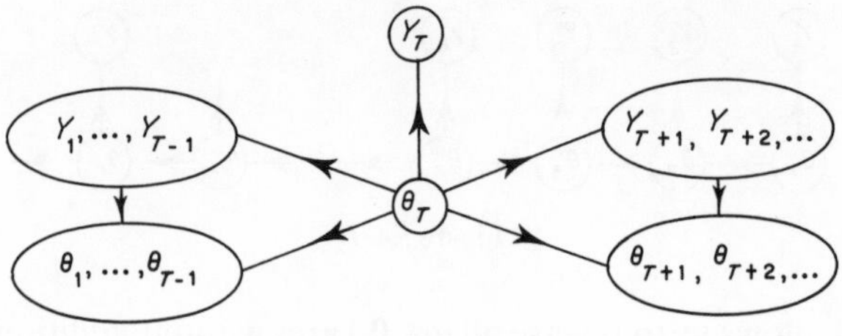

Figure 5.6 I_4

(b) Simple accounting for outliers

In many interesting series, data points at certain times are missing. Non-state-space forecasting models are greatly disrupted by this, state-space models are not. The reason why state-space models can cope with this problem can be seen immediately from the graphs of Figure 5.6. Suppose the observation Y_T is missing. Because of the c.i. structure of I_4 we still have that the past $Y_1, \ldots, Y_{T-1}$ is independent of the future, $Y_{T+1}, Y_{T+2}, \ldots$. So $\boldsymbol{\theta}_T$ contains all relevant information from the past about the future. The same forecasting equations hold except that the margin $\boldsymbol{\theta}_T | y_1, \ldots, y_{T-1}$ is substituted for $\boldsymbol{\theta}_T | y_1, \ldots, y_T$. And, we know that this will be the case for any generalization to non-normal models retaining the same c.i. structure.

(c) The accommodation of external information

Typically business series are not closed but regularly disrupted by unexpected external events. For example the sales of a product may be disrupted by a competitor ceasing to manufacture. If the state space $\boldsymbol{\theta}_T$ has been chosen wisely we are often able to assume that the effect of these external events, measured by Z_T will only disrupt our beliefs about the future sales $Y_{T+1}, Y_{T+2}, \ldots$ though its effect on the causal variables $\boldsymbol{\theta}_T$, i.e.

$$Z_T \perp\!\!\!\perp Y_{T+1}, Y_{T+2}, \ldots | \boldsymbol{\theta}_T.$$

When this is not possible it may still be possible to modify $\boldsymbol{\theta}_T$ for example, by adding new components ψ_T. Graphs of the IDs of these two cases are given in Figure 5.7 and 5.8. An example of how to modify $\boldsymbol{\theta}_T$ in practice is given in Harrison and West (1986). The point I make here is that exactly analogous

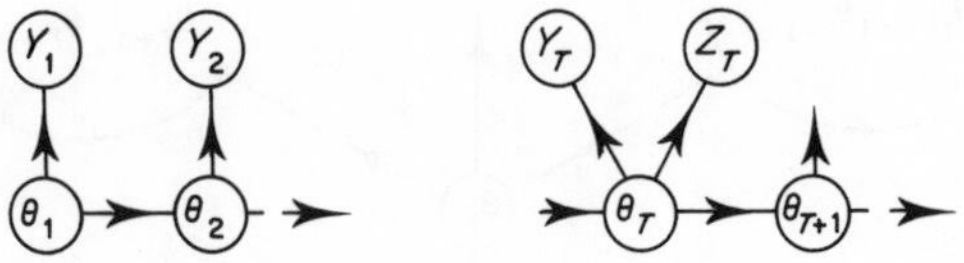

Figure 5.7 I_5

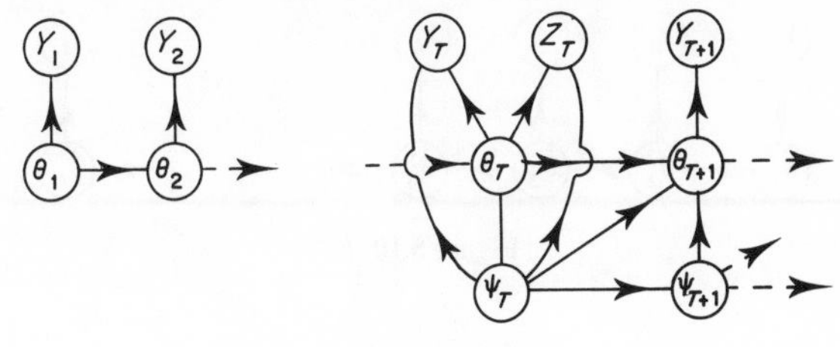

Figure 5.8 I_6

procedures are available for non-linear processes. So after minor adjustments made in the light of this new information Z_T the algorithm for computing forecasting distributions can proceed more or less as usual.

Notice that all the comments made here are equally valid under different definitions of the c.i. operator $\cdot \perp\!\!\!\perp \cdot | \cdot$. In particular $\cdot \perp\!\!\!\perp^{W} \cdot | \cdot$ gives the analogous results for second-order processes, more in keeping with the Kalman original model.

5.8 OBSERVABILITY OF STATE-SPACE MODELS—A GENERAL DEFINITION

A linear system as defined by equations (5.11) and (5.12) is said to be observable if the matrix $(\mathbf{F}, \mathbf{FG}, \mathbf{FG}^2, \ldots, \mathbf{FG}^{m-1})$ is of full rank. The idea stems from the study of deterministic systems where all error terms appearing in equations (5.11) and (5.12) are identically zero. Then, in an observable system, after the first m observations $Y_1, \ldots, Y_m$ we can solve for the values $\boldsymbol{\theta}_t$ in the system $t \geqslant m$.

In a normal linear stochastic system it is easy to check that the state space $\boldsymbol{\theta}_t$ is unobservable if and only if there exists a linear reparametrization of

$$\boldsymbol{\theta}_t \to \boldsymbol{\theta}_t^* = (\bar{\boldsymbol{\theta}}_t^*, \boldsymbol{\theta}_t^*(m)) \qquad \bar{\boldsymbol{\theta}}_t^* = (\theta_t^*(1), \ldots, \theta_t^*(m-1))$$

such that the ID I_1 holds with $\bar{\boldsymbol{\theta}}_t^*$ replacing $\boldsymbol{\theta}_t$ and $\theta_t^*(m)$ replacing $\boldsymbol{\omega}_t$. The state $\theta_r^*(m)$ is called an unobservable state in this system. In general then it would seem natural in non-linear systems to say that $\theta_t^*(m)$ were an unobservable state if I_7 of Figure 5.9 held. Notice that we can remove $\{v_t\}$ from the

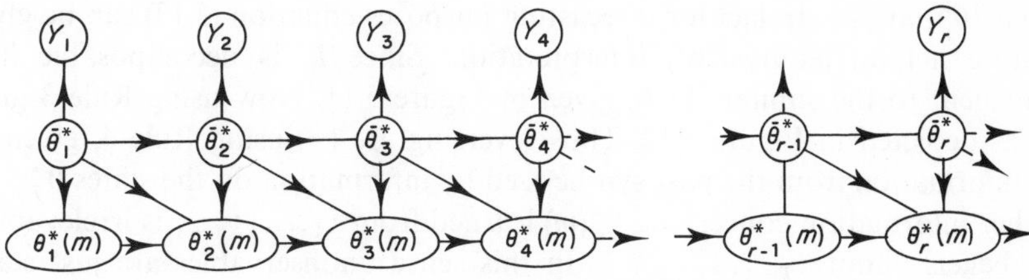

Figure 5.9 I_7

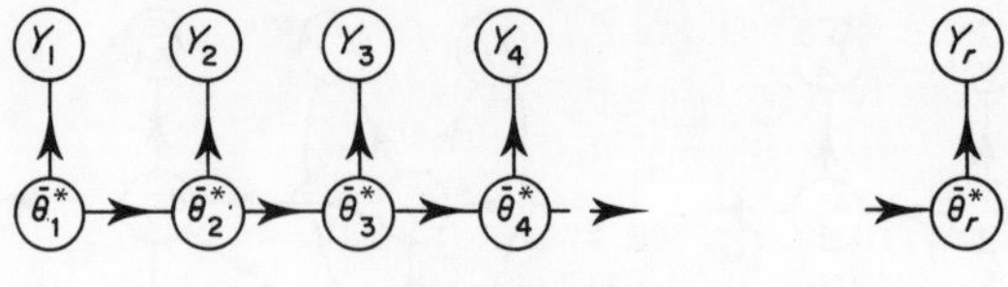

Figure 5.10 I_8

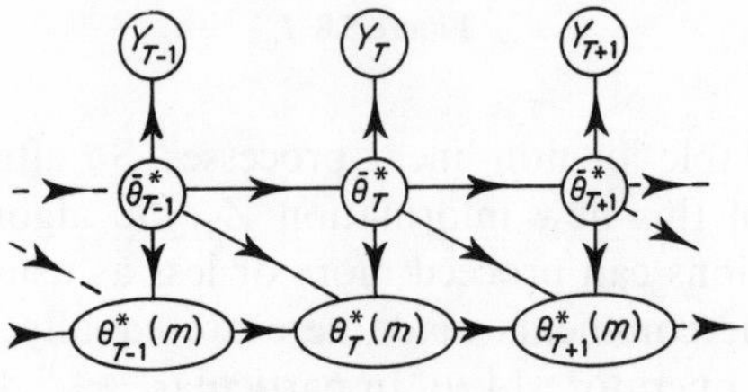

Figure 5.11 I_9

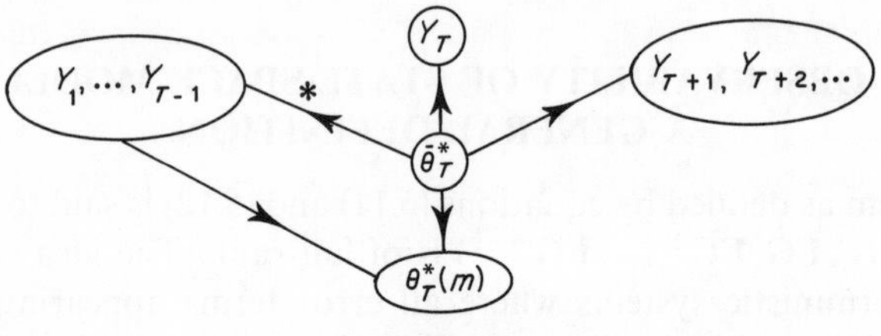

Figure 5.12 I_{10}

corresponding diagram I_1 because these are not quantities that we will be required to forecast.

The same argument we used on I_1 can now be used on I_7 to obtain the implied ID I_8 of Figure 5.10. So $\{\bar{\boldsymbol{\theta}}_t^*\}_{t \geq 1}$ can fulfil the role of the states of the process on their own, without regard to $\{\theta_t^*(m)\}_{t \geq 1}$. In particular we have that

$$\{Y_t\}_{t \geq 1} \perp\!\!\!\perp \{\theta_t^*(m)\}_{t \geq 1} | \{\bar{\boldsymbol{\theta}}_t^*\}_{t \geq 1}. \tag{5.13}$$

This redefines the unobservability of $\{\theta_t^*(m)\}_{t \geq 1}$ by saying—given the other states $\{\bar{\boldsymbol{\theta}}_t^*\}_{t \geq 1}$—that the observations $\{Y_t\}_{t \geq 1}$ provide no further information about $\{\theta_t^*(m)\}_{t \geq 1}$. In fact for forecasting purposes equation (5.13) can be given a more helpful (equivalent) interpretation. Since I_7 is decomposable it is equivalent to the similar ID I_9 given in Figure 5.11. Now using Rule 3 gives I_{10} as depicted in Figure 5.12. Thus reversing arc (*) using Rule 4, given all the information from the past synthesized by information on the states $\bar{\boldsymbol{\theta}}_T^*$, any further information concerning $\theta_t^*(m)$ learned from $y_1, \ldots, y_{T-1}$ is irrelevant to our beliefs about $y_{T+1}, y_{T+2}, \ldots$. In this sense unobservables are just states which are uninformative about the future given information on the other states.

The fact that these two different interpretations of observable states are equivalent is due entirely to the properties of c.i.

5.9 PROBLEMS AND LIMITATIONS IN USING INFLUENCE DIAGRAMS AND BEYOND

Although I hope I have succeeded in illustrating how ID can be useful in expressing and manipulating quite difficult statistical constructions we must not lose sight of the fact that IDs are just one of many tools that help us understand the implications of a given c.i. structure on a set of uncertain quantities. Some of their limitations are listed below.

(i) The necessity for more than one graph

We may have a model which contains a package of c.i. statements that cannot have all its implications expressed in a single ID. The simplest of these are the two statements $X_1 \perp\!\!\!\perp X_2 | X_3, X_4$ and $X_3 \perp\!\!\!\perp X_4 | X_1, X_2$ where a minimum of two IDs, those given in Figure 5.13, are required to express these statements. To my knowledge the question of how many IDs are required to express all the c.i. statements in a given set and what rules can be used to manipulate these sets of graphs are as yet unanswered. In any case if we insist on using only one ID to manipulate a given set of c.i. statements then, because of this inadequacy, we may not be able to make proper use of all given information in our system.

Because sets of statements like the two given above do not define any one breakdown of a joint probability into its conditionals, combination of statements like these often imply strong distributional forms over the variables (X_1, X_2, X_3, X_4) provided we know that in our given problems neither of the IDs of Figure 5.13 can be simplified. This is really hardly surprising. Full distributions can often be characterized in terms of a c.i. statement and some other property. A well-known example is when $X_1 \perp\!\!\!\perp X_2$ and $f(X_1, X_2) \perp\!\!\!\perp g(X_1, X_2)$ where $f(X_1, X_2) = X_1 + X_2$ and $g(X_1, X_2) = X_1 - X_2$. Unless X_1 and X_2 are both degenerate, this property characterizes the normal distribution. If we are given functional relationships, for example, as well as c.i. statements, because these functional relationships are not fully integrated within an ID we may not realize the full implication of a given model.

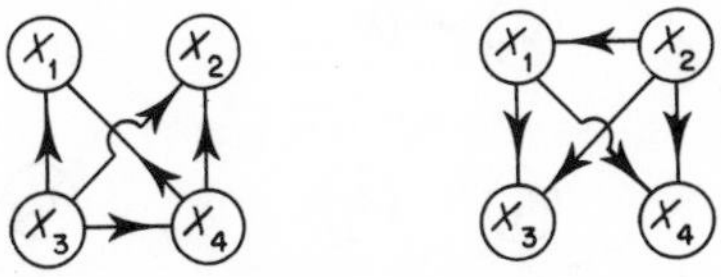

Figure 5.13

(ii) Inefficient descriptions of c.i. statements

As Howard and Matheson (1981) point out, IDs are most useful when there exists symmetry in the c.i. statements in a system. Explicitly, for any set of nodes, we require that the c.i. statements hold for all possible values of their predecessors. Although most conventional models contain this type of symmetry, in many practical problems such symmetry does not exist. Here is a formal example of the sort of structure I am considering.

Example 5.4. Identically distributed random variables $X_1, \ldots, X_n$ conditional on $X_0 = nk + r$, $1 \leqslant k \leqslant n$, $1 \leqslant r \leqslant n$, $r \neq k$ have the property $X_k = X_r$ and

$$\coprod_{i \in \{1,2,\ldots,n\} \backslash r} X_i.$$

Here, the ID of $X_0, X_1, \ldots, X_n$ is completely connected. On the other hand the ID of $X_1, \ldots, X_n$ conditional on each value of X_0 is extremely simple, each including just one of the $n(n-1)/2$ arcs between nodes on the original diagram. Posterior to observing X_0 most of the dependencies in the system disappear.

Although the example above is idealized, this type of structure is not unusual. For example, once the type of fault, X_0, in a system is identified, which components of the system that will interfere with each other's proper working can often be identified. In these circumstances it seems necessary to develop

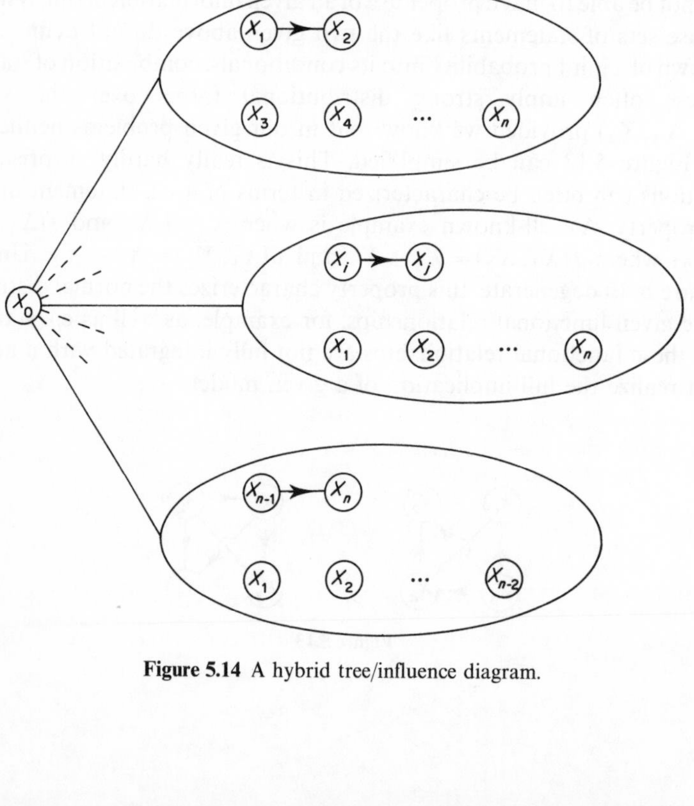

Figure 5.14 A hybrid tree/influence diagram.

hybrid probability tree ID representations. Such a hybrid representation is given in Figure 5.14. Rules for manipulating these hybrid systems can obviously be developed in a similar way to those for IDs.

(iii) Depicting 'weak' relationships

Sometimes relationships between the variables in a problem are weak but they do exist. An ID cannot represent the strength of information flow through a system. This is unfortunate since it will often be crucial to know whether a variable is significantly informative about another.

Under $\cdot \perp\!\!\!\perp^{W} \cdot | \cdot$ the arcs of an ID can be indexed with their corresponding (scaled) regression coefficients, larger indices corresponding to more information flow in some sense. One way of labelling strength of relationship on arcs of an ID when we employ conventional c.i. is to base this index on possible observed c.i. that was discussed in (ii). Thus label the (i, j) arc of an ID by a number $a(i, j)$, $0 < a(i, j) \leqslant 1$, where

$$a(i,j) = \int \#_{i,j}\{R(X_j) = \mathbf{x}\}\, \mathrm{d}F(\mathbf{x}),$$

where $R(X_j)$ are the predecessors of X_j, $\#_{i,j}(\mathbf{x})$ is the indicator function on the event that $X_j \perp\!\!\!\perp X_i | R(X_j) = \mathbf{x}$ and $a(i,j)$ is the expectation of this quantity, the expectation being taken over the distribution $F(\mathbf{x})$ on the predecessors $R(X_j)$ of X_j. In Example 5.4 each arc on the complete graph would be labelled by n^{-1}.

Such indexing is practically significant especially if $(X_1, X_2, \ldots, X_n)$ are ordered in the ID consistent with when it will be possible to observe them. For then it will be possible to at least obtain an indication of whether or not it is worth commiting yourself *a priori* to buy the value of a given measurement at X_i.

5.10 CONCLUSIONS

Despite some of the difficulties mentioned above I believe that the use of IDs will become more and more common as an aid to help a client and a decision analyst/statistician to arrive at a mutual understanding of the client's problem. This will be in large part due to the fact that IDs are linked not to complicated probabilistic structures but directly to the syntax of explanations. I look forward to the time when probabilities will be considered only as artefacts of logical measuring and not primitive quantities to be elicited directly.

5.11 DISCUSSION

5.11.1 Discussion by David J. Spiegelhalter

This chapter presents a full and attractive description of the way that statistical models can be related to directed graphs, and then graph-theoretic ideas exploited

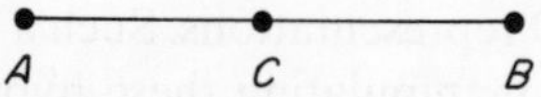

Figure 5.15 Markov field in which $A \perp\!\!\!\perp B|C$ since C separates A from B.

to clarify and extend properties of those models. Manipulations on directed graphs form the basis for much work in IDs, but in recent work with Steffen Lauritzen (Lauritzen and Spiegelhalter, 1988) we have found it advantageous to also use undirected graphs to represent certain independence properties, and I would like to show how this may provide further simplification. (Only strictly positive probability densities are considered here, and for detailed discussions see, for example, Speed, 1979, Lauritzen, 1982 and Lauritzen *et al.*, 1988.)

We say that our beliefs form a Markov field on an undirected graph G if, for any node subsets A, B, C in which C separates A from B in G, we have that $A \perp\!\!\!\perp B|C$. Figure 5.15 shows an example.

For strictly positive densities such a property is equivalent to the density factorizing into functions on the cliques of G.

This immediately leads to some simplified analysis on directed graphs. In an ID it can be shown by induction that the density is equal to the product of terms

$$p(X_{(r)}|P(X_{(r)}))$$

and hence factorizes into terms containing each node and its direct predecessors (parents). Hence it forms a Markov field on what Lauritzen and Spiegelhalter (1988) call the 'moral' graph (formed by joining parents and dropping directions). This is illustrated in Figure 5.16.

This has immediate consequences in terms of reading off conditional independencies from an ID. Suppose in Figure 5.16 we wish to establish whether $A \perp\!\!\!\perp B|D, E, S$. First, we note that (by Smith's Rule 2) we also have an influence

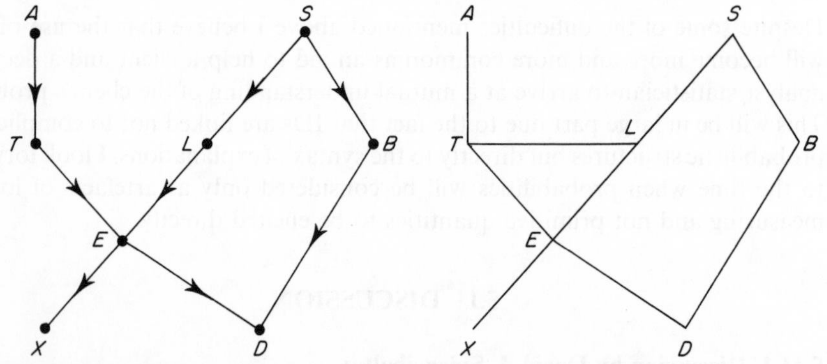

Figure 5.16 Influence diagram and corresponding 'moral' graph formed by joining 'parents' and dropping directions: our beliefs form a Markov field on this graph.

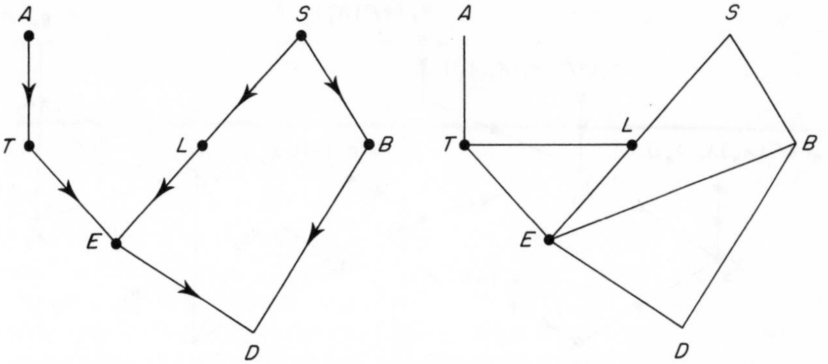

Figure 5.17 Is $A \perp\!\!\!\perp B \mid D, E, S$ in Figure 5.16. By forming the moral graph of the ancestral set of the nodes in question, we need only check whether E, S separate A from B in this graph.

diagram on the ancestral graph of A, B, D, E, S, which contains the nodes under scrutiny and all their predecessors. Second, by the preceding argument, we have a Markov field on the moralized ancestral graph, shown in Figure 5.17.

Now, since D, E, S separate A from B in this graph, we can immediately see that the independence relation holds. Lauritzen *et al.* (1988) discuss this attractively simple procedure in detail and show it exhausts all the implied conditional independencies, and is equivalent to Pearl's (1986a) *d*-separation criterion.

Another application concerns conditional independence of parameters under separable likelihoods. Figure 5.18 shows a decomposable ID on $\theta_1, \ldots, \theta_4$ and statistics $T_1, \ldots, T_4$ with separable likelihood. The moral graph in such a situation consists of just dropping directions, and since the joint distribution of $\mathbf{T}, \boldsymbol{\theta}$ is

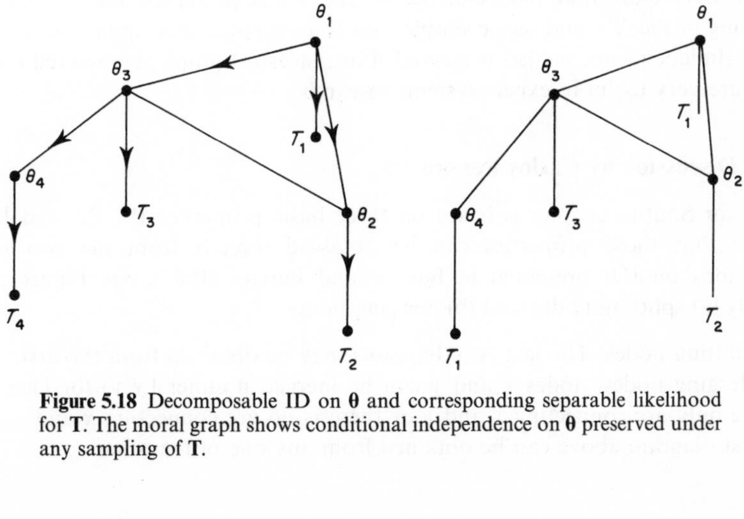

Figure 5.18 Decomposable ID on $\boldsymbol{\theta}$ and corresponding separable likelihood for $\mathbf{T}$. The moral graph shows conditional independence on $\boldsymbol{\theta}$ preserved under any sampling of $\mathbf{T}$.

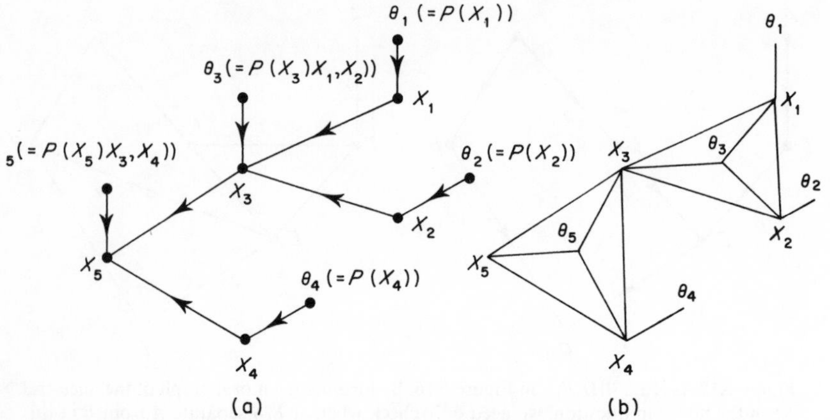

Figure 5.19 Influence diagram with conditional probability tables **θ** considered as independent random quantities. Moral graph shows that independence of θ's is preserved under complete sampling of X's (also under sampling an ancestral set).

Markov on this, it is clear that conditional independence properties on **θ** are unaffected by sampling on **T**.

Finally, we consider the problems of learning about the conditional probabilities in IDs by collecting data on many cases. In Figure 5.19(a) we show a four-variable diagram, and indicate how we can formally consider the conditional probabilities relative to this diagram as random quantities which are additional direct predecessors, in which the full ID corresponds to a joint density

$$p(X, \boldsymbol{\theta}) = \prod_{i=1}^{5} p(X_{(i)} \mid P(X_{(i)}), \theta_i) p(\theta_i).$$

Figure 5.19(b) shows the moral graph of this diagram, and from this it is immediately clear that independence of the θ's is preserved under complete sampling of the X's, and hence simple case-to-case parameter updating can take place. (Independence is also preserved if an ancestral graph is observed.) Such ideas are very useful in expert systems research.

5.11.2 Discussion by Carlos Pereira

Professor Smith's chapter is based on three basic properties P1, P2, and P3. I believe that those properties can be obtained directly from the two basic operations on IDs presented in Barlow and Pereira (1987), (see Figure 5.20) namely (a) splitting nodes and (b) merging nodes.

(a) Splitting nodes: The last two diagrams may be obtained from the first.
(b) Merging nodes: nodes x and y can be merged if either $[x, y]$ (or $[y, x]$) is the only arc connecting x and y or there is no arc connecting x and y. The first diagram above can be obtained from any one of the others.

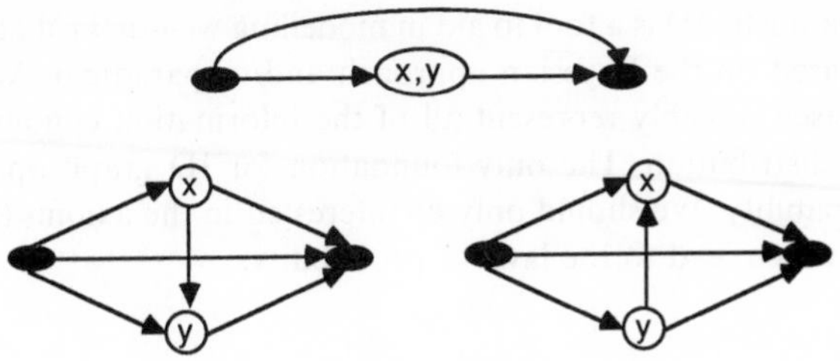

Figure 5.20

To show P1 and P2 we only need to take out some arcs of the picture, to eliminate the black node of the right-hand side, and to replace the left-hand black node by node z. Note also we can add to the graph a deterministic node $f(y)$ connected from a directed arc $[y, f(y)]$.

To prove P3, we need only to draw the diagrams for the statements of P3. The equivalence of the diagrams is a direct consequence of the above operations (Figure 5.21).

A more complete discussion on these conditional independence properties can be found in Barlow and Pereira (1987, Section 6).

5.11.3 Discussion by R. E. Barlow

As Rubin (1987) points out, we have to start somewhere and we have to assume something. I think we start with a decision problem—then we define (through axioms) 'rational' behaviour. This leads to Bayesian decision analysis and inductive inference based on probability and the laws of probability. To meet a deadline and in desperation, a man by the name of Miller invented the ID (at least constructed the acyclic directed graph with decision and probabilistic nodes). He did this because he just could not convey his ideas adequately using decision trees. Now you, Pearl, etc. come along and try to abstract the ID from its original purpose. You concentrate on conditional independence as if this somehow existed apart from any well-defined probability structure. To paraphrase Howard, you have substituted mathematics for clarity.

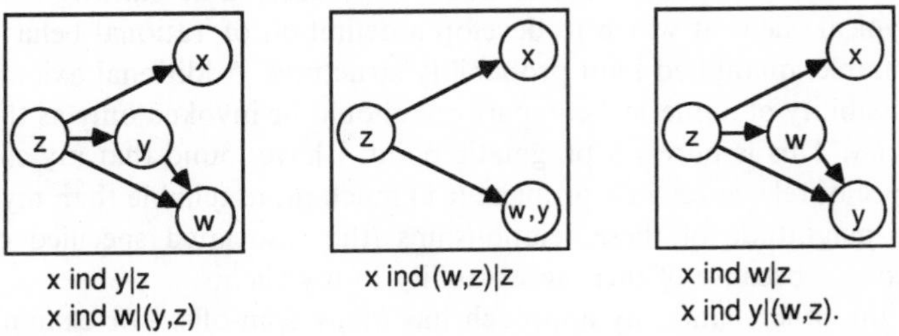

Figure 5.21

The probabilistic ID is a tool to aid in modelling well-defined decision theoretic problems based on the Bayesian approach and/or paradigm. An ID (or graph) cannot by itself possibly represent all of the information contained in the joint probability distribution. The only foundation for ID graph operations are the laws of probability. We should only be interested in the axioms (namely rational behaviour) which lead to the laws of probability.

5.11.4 Reply

I would first like to thank David for politely illustrating how mixed and undirected graphs are useful and in some cases more powerful representations of a model than the directed graph of an ID. I have always been wary of these representations in the past because of their apparent reliance on positivity of joint densities mass functions although the recent work of Lauritzen *et al.* (1988) appears to show that this condition is often not necessary (see also Verma, 1987).

Professor Pereira's comments are very interesting especially in the way they helpfully elucidate P3. Of course, the justification of his operations (a) and (b) must be derived from the c.i. properties of his underlying probability distribution. It is not clear to me that the formal establishment of these rules (stated in a more complete way) from the laws of probability are any easier than P1, P2 and P3.

I find Professor Barlow's position perplexing. He begins by stating 'we define through the axioms, "rational" behaviour'. I would like to know to which axioms he refers: there are many such sets and they do not all arrive at the same conclusions. For example, de Finetti's axioms emphatically support the use of finitely additive distributions, whereas DeGroot (1971) arrives at countably additive distributions. Both these positions are labelled Bayesian. Again 'rational behaviour' can apply to groups of individuals and various axiomatic systems, many of them Bayesian (see e.g. Harsanyi, 1977; Raiffa, 1968), justify different definitions of rationality. So which axiomatic systems should we support?

When constructing axiomatic systems I suggest we go back to basics. Whereas many of us have found it difficult to elicit specific probabilities from a client early in an analysis we have found that he is happy with an ID representation of his problem. To me this suggests that sets of c.i. statements, on which these diagrams solely depend, form a much more acceptable starting-point (axiom) to a typical client at which to develop a definition of 'rational behaviour' than, for example, quantified joint probability structures. Additional axioms implying the possibility of numerical comparisons should be invoked only as they become necessary. This is purely a pragmatic point. I have found that my elicitation of existence of relevances in a problem (c.i.) much more reliable than my elicitation of the magnitude of these relationships (the associated specified conditional probability) because of their acceptability to my clients.

On the other hand, my approach has many spin-offs. For example, a client who feels very uncomfortable in specifying a specific set of probabilities can still have his problem analysed coherently. Of course we cannot be explicit as to the decision rule he should use, but we can be explicit about on what variables that decision should depend. This may be all he needs. If we have several clients all with different beliefs about magnitudes of relationships they may all agree as to the existence or otherwise of those relationships and again the analysis of the structure of the problem implied by these relevances may well be enough for them. Furthermore, as I have tried to illustrate in my chapter, many implications of the most profound aspects of a client's model in fact will lie in the c.i. statements he chooses to make. For c.i. structures a model; probabilities fill this model out in our specific instance and enable us to identify optimal decisions in the Bayesian sense.

Of course we all realize one ID 'cannot by itself possibly represent all probability assessments made'. Indeed it cannot express all c.i. statements made. On the other hand, by concentrating on a client's c.i. statements and by using tools like IDs we are often able to produce a faithful representation of a problem as that client perceives it, uncontaminated by often arbitrarily joint distributional assumptions imposed by the analyst/statistician.

REFERENCES

Barlow, R. E. and Pereira, C. (1987) *The Bayesian Operational and Probabilistic Influence Diagrams*, TR ESRC87-7, University of California, Berkeley.

Beeri, C., Fagin, R., Maier, D., Mendelzon, A., Ullman, J. and Yannakakis, M. (1981) Properties of acyclic database schemes. *Proc. 13th Annual ACM Symposium on the Theory of Computing, Milwaukee*, J. Assoc. Comput. Mach., New York.

Beeri, C., Fagin, R., Maier, D. and Yannakakis, M. (1983) On the desirability of acyclic database schemes, *J. Assoc. Comput. Mach.*, **30**, 479–513.

Cox, D. R. and Reid, N. (1987) Parameter orthogonality and approximate conditional influence (with discussion), *J.R. Statist. Soc.*, **B49**, 1–39.

Dawid, A. P. (1979) Conditional independence in statistical theory (with discussion), *J.R. Statist. Soc.*, **B41**, 1–31.

Dawid, A. P. (1984) *Tomorrow and Tomorrow and Tomorrow ... the Prequential Approach to Statistical Theory*, Research Report No. 32, Dept. of Statistical Science, University College, London.

de Finetti, B. (1972) *Probability, Induction and Statistics*, Wiley, New York.

DeGroot, M. H. (1971) *Optimal Statistical Decisions*, McGraw-Hill, New York, Ch. 6.

Goldstein, M. (1981) Revising previsions: a geometric approach (with discussion), *J.R. Statist. Soc.*, **B43**, 105–30.

Harrison, P. J. and Stevens, C. F. (1976) Bayesian forecasting (with discussion), *J.R. Statist. Soc.*, **B38**, 205–47.

Harrison, P. J. and West, M. (1986) *Bayesian Forecasting in Practice*, Bayesian Statistics Study Year Research Report No. 13, Dept. of Statistcs, University of Warwick.

Harsanyi, J. (1977) *Rational Behaviour and Bargaining Equilibrium in Games and Social Situations*, Cambridge University Press.

Hill, B. M. and Lane, D. (1986) Conglomerability and countable additivity. In P. Goel and A. Zellner (eds) *Bayesian Inference and Decision Techniques*, Elsevier, Amsterdam. pp. 45–57.

Howard, R. A. and Matheson, J. E. (1981) Influence diagram. In R. A. Howard and J. E. Matheson (eds) *Readings on the Principles and Applications of Decision Analysis*, vol. II, Strategic Decisions Group, Menlo Park, Calif. pp. 719–762.

Jeffreys, H. (1961) *Theory of Probability*, Oxford University Press.

Kadane, J. B., Schervish, M. J. and Seiderfield, T. (1986) *Statistical implications of finite additivity*. In P. Goel and A. Zellner (eds) *Bayesian Inference and Decision Techniques*, Elsevier, Amsterdam. pp. 59–76.

Lauritzen, S. L. (1982) *Lectures on Contingency Tables*, 2nd edn Aalborg University Press.

Lauritzen, S. L., Speed, T. P. and Vijayan, K. (1984) Decomposable graphs and hypergraphs, *J. Austral. Math. Soc.*, **A36**, 12–19.

Lauritzen, S. L., Dawid, A. P., Larsen, B. N. and Leimer, H. G. (1988) *Independence Properties of Directed Markov Fields*, Working Paper, Aalborg University.

Lauritzen, S. L. and Spiegelhalter, D. J. (1988) Local computations with probabilities on graphical structures and their application to expert systems (with discussion), *J.R. Statist. Soc.*, **B50**, 157–224.

Olmsted, S. M. (1984) *On representing and solving decision problems*, Ph.D. Thesis, Dept. of Engineering-Economic Systems, Stanford University, Stanford, Calif.

Pearl, J. (1986a) Fusion, propagation and structuring in belief networks, *Artificial Intelligence*, **29**, 241–88.

Pearl, J. (1986b), *Markov and Bayes Networks: A Comparison of Two Graphical Representations of Probabilistic Knowledge*, Technical Rep. CSD 860024, Cognitive Systems Lab, Comp. Sci. Dept., University of California, Los Angeles.

Raiffa, H. (1968) *Decision Analysis*, Addison-Wesley, New York, Ch. 7.

Rubin, H. (1987) 'A weak system of axioms for 'rational' behavior and the non-separability of utility from prior,' *Statistics and Decisions*, **5**, 47–58.

Shachter, R. D. (1986a) Evaluating influence diagrams. In A. P. Basu (ed.) *Reliability and Quality Control*, Elsevier, North-Holland, pp. 321–44.

Shachter, R. D. (1986b) Intelligent probabilistic inference. In L. N. Kanal and J. Lemmer (eds) *Uncertainty and Artificial Intelligence*, North-Holland, Amsterdam, pp. 371–382.

Shachter, R. D. (1987) *David: Influence Diagram Processing System for the Macintosh*, Documentation Centre for Health Policy Research and Education, Duke University.

Smith, J. Q. (1987) *Decision Analysis: A Bayesian Approach*, Chapman & Hall, London.

Smith, J. Q. (1988) Models, optimal decisions and influence diagrams. In J. M. Bernardo, M. H. DeGroot, D. V. Lindley and A. F. M. Smith (eds) *Bayesian Statistics 3*, Oxford, Oxford University Press, pp. 765–776.

Smith, J. Q. (1989a) Influence diagrams for Bayesian decision analysis, *European Journal of Operations Research*, **40**, No. 3, pp. 363–376.

Smith, J. Q. (1989b) Influence diagrams for statistical modelling, *Annals of Statistics*, **17**(2), 654–72.

Speed, T. P. (1979) A note on nearest-neighbour Gibbs and Markov probabilities, *Sankhya*, **A41**, 184–97.

Stone, M. (1974) Cross-validatory choice and assessment of statistical predictions (with discussion), *J.R. Statist. Soc.*, **B36**, 111–47.

Tarjan, R. E. and Yannakakis, M. (1984) Simple linear-time algorithms to test chordality of graphs, text acyclicity of hypergraphs, and selectively reduce acyclic hypergraphs, *SIAM J. Comput.*, **13**, 566–79.

Verma, T. (1987) *Causal Networks: Semantics and Expressiveness*, Technical Report, R-65-1, Cognitive Systems Lab., University of California at Los Angeles.

CHAPTER 6

Construction and Local Computation Aspects of Network Belief Functions

A. P. Dempster, *Harvard University, USA*

ABSTRACT

Belief function methodology extends Bayesian reasoning about uncertainty to situations where the available evidence does not support a full probabilistic specification. A network belief function is constructed by combining independent component belief functions representing logical and probabilistic knowledge assessed locally across the network. Computational methods for combining (propagating and fusing) the local knowledge assessments are sketched. Examples are introduced without detailed formulas, including the Kalman filter, fault trees, and sampling inference.

6.1 FORMULATION

The theory of belief functions is a theory of reasoning from uncertain evidence to uncertain conclusions. The theory is expressed through applications to specific situations, where in each instance the evidence is given a precise formal representation, typically as a system of independent belief functions over the possible states of a network. Hence the formal reasoning consists of well-defined computations, whose inputs are the assumed belief functions, and whose outputs are uncertain conclusions in the form of implied marginal belief functions.

In addition, preceding every application of the formal theory, there is an essential and often extensive informal process of model construction, also involving reasoning and evidence. Since the informal activities effectively control the final conclusions of the formal processes, they too deserve open recognition and study. Hence the organization of the chapter into Section 6.2 on model construction and Section 6.3 on inferential computation.

The idea that the term 'reasoning' accurately describes the activity of computing with belief functions goes back to the title of the first attempt

Influence Diagrams, Belief Nets and Decision Analysis
Edited by R. M. Oliver and J. Q. Smith

(Dempster, 1966) to define the theory. More broadly, I think of applied statistics as a process of reasoning about real-world phenomena. For example, the key feature of applied Bayesian statistics, as I practice it, is not the cachet of strict fidelity to coherent Bayesian principles that advocates (e.g. Lindley, 1987) tirelessly emphasize. I take for granted a need for adherence to an explicit, precise, and well-tested formalism. Instead, the key feature is the rich mixture of informal model construction and formal inference (i.e. informal and formal reasoning) that constitutes Bayesian thinking about the real world. Due to its enhanced power and flexibility, the belief function extension of Bayesian thinking offers prospects for improved performance.

A popular alternative to reasoning about real problems is attempting to solve them by following correct rules of procedure. The result of a procedure is a reported outcome, together with a list of (assumption, property) pairs that characterize the meaning of the procedure. There is no place in the behaviorist outlook for assessing the meaning of the outcome of a specific application of a procedure. While such a restriction may suffice for repetitive practice, a basic requirement of much applied work is a credibly calibrated numerical assessment of uncertainty that is tailored to, and can be trusted for subsequent analysis of, the phenomenon under study. Credibility derives from quality of evidence and reasoning about specific circumstances, not knowledge about procedures. Moreover, since the credibility is as much our business as that of our clients, it follows that statisticians need to be full partners in situation-specific reasoning, and to be deeply cognizant of as much relevant evidence, both hard (e.g. data) and soft as can be brought to bear.

The formal theory of belief functions is described in Glenn Shafer's remarkable 1976 book *A Mathematical Theory of Evidence* (Shafer, 1976) that developed from my narrowly aimed efforts of the 1960s. Shafer introduced many terms, including belief function. He conceived the theory broadly, unlike my early descriptions that were limited to sampling inference, and he substantially enriched the mathematical content. A major development of the 1980s has been the recognition that the theory applies naturally and effectively to network-type knowledge structures. See, for example, Dempster and Kong (1988) and Shenoy and Shafer (1986). A key aspect of the new viewpoint is that units of evidence represented by logical belief functions, by Bayesian belief functions, or by more general belief functions, should be conceived and treated in a unified way.

Statisticians usually define the mathematics of a probability model in terms of one or more distributions over a sample space. In a subjectivist interpretation, the sample space defines the possible states of some small world under study, and the probabilities measure an analyst's uncertainty concerning the true state. In belief function theory, the space of possible states of the world is called the frame of discernment (or simply frame), but is no longer the sample space of a probability measure. Instead, the mathematical measure space is the class of focal elements, where a focal element is a subset of the frame of discernment.

Thus the probability distribution that defines a belief function is, formally, a distribution over subsets of the frame. Heuristically, this means that the uncertainty measure is spread over subsets instead of being necessarily attached to single points of the frame.

Several familiar formulations of incomplete knowledge are obvious special cases. For example, a Boolean (or logical) belief function asserting that the truth lies in a certain subset of the frame is defined by a (trivial) probability distribution that assigns probability one to the subset and zero to all other subsets. A more general belief function construction takes the class of focal elements to be mutually exclusive subsets that partition the frame, then specifies a probability distribution over the partition. Such a partition is often called a margin in common statistical parlance. If the partition consists of all singleton subsets of the frame, then the margin is equivalent to the full frame, and the marginal belief function reduces to an ordinary Bayesian distribution over the frame.

When it is recognized that for any application all the structures (i.e. frame, class of focal elements, and probability assignment) of a belief function must be constructed, it makes sense to allow reduction of a possibly onerous task by permitting assignment of distributions to margins, instead of requiring, as does Bayesian methodology, that a full probability distribution be forced on the user. All modelers do this routinely in the Boolean case, and do so implicitly in the probabilistic case by effectively restricting analysis to formal small worlds that are margins of more realistic small worlds. So it hardly seems a radical step to move to probability models over more general classes of focal elements. Belief function theory should be construed as a flexible generalization of the theory of subjective probability, allowing removal of restrictions that users should through practice come to see as unnecessary.

A major consequence of allowing probabilities to be assigned to subsets of the frame is that some (often all) of the possible statements about the true state of the world fail to have precise numerical probabilities. Again, in the case of Boolean belief functions we do not find this surprising, for we are accustomed to inferring only Boolean consequences. A general belief function model specifies that any statement has a lower probability or belief, and an upper probability or plausibility, each determined in an obvious way from the distribution over the focal elements. In the semantics of belief function theory, both beliefs and plausibilities are instances of ordinary probabilities, with the twist that a belief is the minimum amount of probability committed to the assertion in question by the formal model, while the plausibility is the maximum amount. For example, a Boolean assertion that is known to be true has belief and plausibility both unity, that is known to be false has belief and plausibility zero, and that may be true or false has belief zero and plausibility unity. The reader is directed to belief function references such as Dempster (1966, 1967), Shafer (1976) and Pearl (1988) for more details.

Given a belief function over a frame, it follows that beliefs and plausibilities are determined for all subsets of a margin of the frame, and it is easily seen that these beliefs and plausibilities in fact define a marginal belief function for the marginal frame. In the case of a Bayesian belief function, the marginal belief function is simply the familiar marginal distribution of elementary probability theory. Unlike the ordinary theory of probability, however, where a given marginal distribution can result from many different choices of a distribution over the full sample space, and no one of these has any special claim to attention based on its margin alone, a given marginal belief function does correspond to a unique belief function over the full frame, called the minimal extension, which conveys exactly the same state of uncertainty as conveyed by the original marginal. In fact, because belief functions on networks are typically built up from specifications on local margins of the network, whereas it is otherwise more convenient to consider these local beliefs to be defined over the full frame, the operation of minimally extending a marginal belief function to a more inclusive frame is typically used many times, almost subconsciously, in any application of the theory.

Any given pair of independent belief functions over a common frame may be combined into a single belief function over the same frame, the interpretation being that the combination represents the pool of the independent sets of evidence represented by the original belief functions. The combination mechanism or rule (Dempster, 1967; Shafer, 1976) has been called the direct sum operator or Dempster's rule by Shafer (1976). I have recently been calling it the product–intersection rule because it simultaneously generalizes two fundamental devices, namely, (i) the multiplication of Bayesian probabilities of independent events to find the probability that both events occur, and (ii) the intersection of Boolean sets to find the set representing conjunction. In mathematical settings, the term 'direct sum' (written $\oplus$) is appropriate because the rule can be expressed analytically as the sum of the logs of certain quantities called unnormalized communalities. Again, more details may be found in basic references.

The product–intersection rule is the most frequently criticized aspect of the theory of belief functions, by authors who are skeptical about the theory (e.g. Aitchison, 1968), or who wish to propose different combination rules (e.g. Walley, 1987). But since the rule is so firmly established in special cases, there is obviously no way to reject it entirely. The cited authors object to applications involving parametric inference from random samples, where the consequences of using certain belief function models are perceived by them to be either counterintuitive or to violate familiar principles. I am less disturbed by the criticisms than I might be, perhaps, if I did not find the authors themselves repeatedly using the rule to calculate likelihood functions and Bayesian posteriors, and if I could see a coherent explanation of why one application is right and another wrong. Unexpected findings from unfamiliar processes are not in themselves valid objections, especially in the general area of statistical inference, where even the most fundamental notions (e.g. probability itself) are regarded as controversial.

The key safeguard against misuse of the theory is insistence that each independence assumption introduced while constructing a model for a specific application must be explicitly recognized as a judgement call based on all available theoretical and empirical knowledge of the subject-matter field. In the case of Boolean assertions we rarely feel compelled to question independence. Specifically, if we receive convincing evidence that the truth lies in A, and we subsequently receive convincing evidence that the truth lies in B, we do not ordinarily feel compelled to question the validity of the first statement given the second piece of evidence, and hence we feel comfortable applying the rule of combination which in this case asserts that the truth lies in the intersection of the subsets of the frame representing A and B. But as statisticians, we are trained to question all probabilistic judgements when information arrives, say, about A and B. For example, a key question in many important applications concerns whether or not tiny probabilities (those too small to be directly assessed empirically), can ever be judged trustworthy, because all such probabilities depend on independence assumptions made in product–intersection rule applications. The type of independence judgements that are both commonplace and critical in the construction of ordinary Bayesian models carries over virtually intact to the theory of belief functions.

The rule of combination is the centerpiece of the theory, as shown by its central place in both model construction and inferential computation.

6.2 MODEL CONSTRUCTION

The construction of statistical models is a skilled activity that draws on three fundamental varieties of knowledge and understanding. The first is the ability to recognize and name entities, such as a system (e.g. a nuclear power plant) or a set of statistical units (e.g. a human population), that are the basic building blocks of the model, and to understand how such entities and subentities relate and interlock to form the complete system or mosaic represented by a model. The second variety reflects perception and understanding of variables or attributes that characterize similarities and differences among entities of the same type, such as the myriad properties of a nuclear plant recognized by nuclear engineers, or similarly diverse properties of populations and persons recognized by epidemiologists. Thirdly, a skillful modeler has a substantial body of theoretical and empirical knowledge, including causal understanding about factors that govern or influence temporal processes of change. The third variety of knowledge makes possible credible choices of the mathematical representations of deterministic and probabilistic relations that give inferential power to a model.

I sketch in this section a view of how these elements enter the development of major classes of belief function models. One class consists of Kalman filter models. Another concerns fault tree models used in reliability studies of complex systems. A third class goes back to the sampling inference models of Dempster (1966). Each of these types is reviewed from the new perspective of belief function network models.

The Kalman filter (Kalman and Bucy, 1961) is concerned with a sequence of states $\mathbf{T}_t$ progressing through times $t = 0, 1, \ldots, n$ where $\mathbf{T}_t$ is typically a vector of unknowns describing a system at time t. The random evolution of states is described by a set of state equations

$$\mathbf{T}_{t+1} = \mathbf{G}_{t+1}\mathbf{T}_t + V_{t+1}, \tag{6.1}$$

where the $\mathbf{G}_{t+1}$ for $t = 0, 1, \ldots, n-1$ are assumed known matrices describing systematic linear behavior of the system, and the $\mathbf{V}_t$ are independent random disturbances with assumed known distributions for $t = 0, 1, \ldots, n-1$. A corresponding sequence of observable Y_t is available, satisfying the observation equations

$$Y_t = \mathbf{F}_t\mathbf{T}_t + U_t, \tag{6.2}$$

where the $\mathbf{F}_t$ for $t = 1, 2, \ldots, n$ are also known matrices, and the U_t are independent random observation errors. The forward filter is a computational process that can be interpreted (e.g. Meinhold and Singpurwalla, 1983) as successively updating the Bayesian posterior of $\mathbf{T}_t$ given $Y_1, Y_2, \ldots, Y_t$, as t increases by steps of one.

The Kalman filter is a precise illustration of a belief function network, in form, construction, and computational aspects. The basic knowledge structure is defined by the system of variables

$$(T_0, V_1, T_1, Y_1, U_1, V_2, T_2, Y_2, U_2, \ldots, V_n, T_n, Y_n, U_n), \tag{6.3}$$

whose product space of possible values defines the frame of discernment for the belief function to be constructed. These variables correspond one-to-one with the nodes of the network.

The edges or hyperedges of the network correspond to independent belief functions specified on margins of the frame that correspond to subsets of the variables. The partial knowledge built into the component belief functions is of three types. First, there is deterministic knowledge, captured in the logical belief functions described by equations (6.1) and (6.2). Second, there is probabilistic knowledge described by distributional assumptions on the U_t and V_t sequences. These distributional assumptions are Bayesian belief functions on margins of the full frame of discernment. Finally, it is a given that the values of $Y_1, Y_2, \ldots, Y_n$ are known by observation, thus providing a third set of logical belief functions. The full Kalman filter model results from judging all these component belief functions to be independent, and combining them into a single belief function according to the product–intersection rule, as further explained in Section 6.3. The key property is that the component belief functions are all specified locally in a network whose nodes are the variables listed in equation (6.3). Network aspects are described and pictured more explicitly in Section 6.3.

The fact that Kalman filter models have proven useful in many situations is partly a consequence of their mathematical and computational tractability, but a more fundamental reason is that the models portray with sufficient accuracy the phenomena under analysis. In the case of a physical system, the state equations often derive in part from well-studied physical phenomena, and similarly the observation equations derive from understanding of measurement processes. The associated Gaussian distribution assumptions for the terms U_t and V_t are sometimes made without much thought, but can often be empirically substantiated, and of course should be treated as potential sources of mistaken conclusions if their validity and consequences are not analyzed. In other situations with softer theoretical and empirical support, the assumed equations, distributions, and independence properties may be deliberately speculative, but still may improve substantially on nonstochastic thinking.

The success of least-squares methods in the nineteenth century, and their extensions to Markov (e.g. Kalman filter) methods in the twentieth century can be seen as limited to circumstances where the available measurements fix the values of the desired estimands with sufficient accuracy for intended uses. In another class of important estimation problems, the so-called ill-posed inverse problems, successful solution may be explained Bayesianly (Turchin *et al.*, 1971) as depending on the ability of the problem-solver to supply prior knowledge that makes up for the deficiencies of the data. It is interesting that the original, and I believe still dominant, Russian school associated with Tikhonov (Tikhonov and Arsenin, 1979) uses prior knowledge in the form of bounds rather than distributions, i.e. uses prior logical belief functions in place of Bayesian prior distributions. The mindset of most Tikhonov regularizers (e.g. Tikhonov and Goncharsky, 1987), and indeed of most Kalman filter users, is point estimation, i.e. getting close enough with a single numerical answer rather than providing a Bayesian or belief function posterior distribution of possible answers. It is an interesting open issue how far belief function assessments of uncertainty will contribute to the solution of the ill-posed inverse problems that are widespread in many areas of modern science, but it is surely important to stress the potential contribution, in line with the statistician's credo that point estimates are worth little without corresponding measures of uncertainty.

The task of assessing possible failure modes and associated uncertainty measures for a complex system, such as a nuclear power plant, is both similar to and different from the error-correction tasks described above. A basic similarity is that system failure modes are commonly represented in network terms. Figure 6.1 shows a fault 'tree' representation of an artificial system where the nodes are variables indicating the presence or absence of certain causes of,

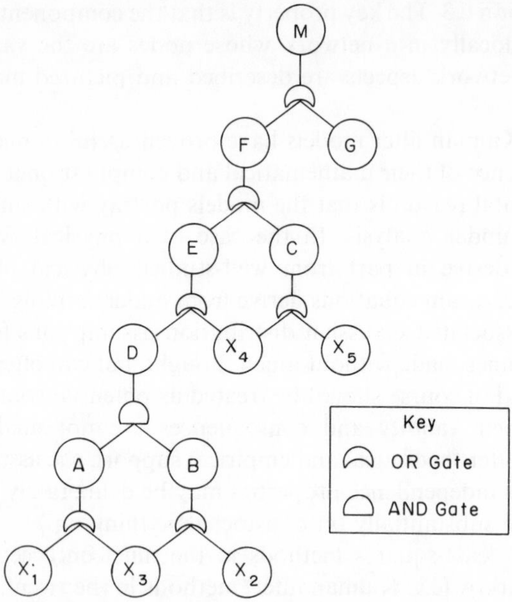

Figure 6.1 An illustrative fault tree for a hypothetical machine *M* as described in Dempster and Kong (1988). (*Reproduced by permission of Elsevier Science Publishers, Physical Sciences & Engineering Div.*)

or factors in, failure. The special AND-gate and OR-gate relations derive from physical knowledge of how the system operates, and are immediately expressible as logical belief functions (Dempster and Kong, 1988). Similarly, it is natural to introduce stochastic uncertainty in the form of belief functions that quantify failure probabilities in localized parts of the system. Note, however, that risk assessment is carried out from knowledge that is strictly *a priori* relative to the particular system under analysis, as contrasted with the repeated on-line observations assumed in Kalman filtering. For this reason, a reliability assessment is inherently more problematic even than an ill-posed inverse physical problem where the dependence on prior assumptions is restricted to a subset of dimensions.

Another complicating element is that the ultimate probability of failure may be known to be quite small. For example, in the reactor case meltdown rates are typically 10^{-4} per year or less, so that direct estimation of probabilities from counting failures of complete reactor units is impractical (as well as unacceptable). Consequently, final answers become dependent on and sensitive to inputs of specific numerical probabilities and independence assumptions.

Two factors that work against the development of a scientific and trustworthy consensus on risk assessments are, first, that the leadership and overall design of risk assessment studies are generally in the hands of subject matter specialists who are not deeply schooled in matters relating to probability judgement, and, second, that those statistical professionals who are so schooled are deeply divided on how to proceed. Consequently, what is done is often *ad hoc* in nature and lacks a defensible overall statistical posture. For example, Speed (1985) can make a devastating critique of the statistical methodology, used in the 'WASH 1400' reactor safety study (Reactor Safety Study, 1975), and evidently only marginally improved since 1975, but there is no consensus within the profession on what can realistically be achieved (cf. Singpurwalla, 1984; Abramson, 1984).

In my view, there are three strongly differentiated statistical strategies towards probabilistic risk assessment, whose differences involve attitudes both to inference and to model construction. Most prominent is the 'objectivist' or 'frequentist' school who interpret the long strings of conditional probability factors that appear in event trees as parameters that in principle should be estimated only from frequency counts based upon repeated empirical trials under appropriately relevant circumstances. The task of the statistician therefore is to assemble the best available data on component failures, relevant human factors studies, and so on, and then define a statistical estimation procedure capable of producing a final answer of acceptable quality. Statisticians of this school would often conclude, for example, in the reactor safety context, that no adequate analysis is possible given the available data, and suggest putting little credence in current methodology.

The second school are 'subjectivists' or 'Bayesians' who insist that every unknown value of variables specified in a model shall be assigned a joint prior distribution, so that gaps in knowledge due to the absence of relevant empirical data do not inhibit formal inference. Their essential tool is elicitation of expert opinion about the unknown values. In situations where hard data are scarce, as in reactor safety studies, elicitation is the primary source of formal uncertainty measures put into formal models. Again long strings of multiplied conditional probabilities are identified and their logs are assigned prior distributions, usually log normal for tractability. The technology requires extensive independence assumptions to be effective, but dependence both across and within event trees can be introduced. As with frequentists, the final goal of traditional Bayesians is estimation of an overall risk parameter, typically represented as a subjectivist probability distribution computed by Monte Carlo sampling. For example, in current reactor safety studies there is much reference to arcane computer 'codes' and 'Latin hypercube' sampling schemes (Reactor Risk Reference Document, 1987). Reviews of the current procedures from within the community of nuclear engineers and scientists appraise them as 'state of the art' (Kouts *et al.*, 1987), but are sharply critical (Kastenburg *et al.*, 1988) of the failure to design and carry out elicitation processes that capture high-quality 'engineering judgement'.

My position is that neither of the two traditional strategies is adequate for risk assessment of complex systems with low failure rates, and I suggest that belief function network modeling makes possible a third, as yet undeveloped, strategy combining elements of both 'objectivist' and 'subjectivist' approaches. Since belief function formulations do not require that all unknowns be assigned a joint prior distribution, the modeler is freed from the need to expend energy on the creation of priors derived from soft or even nonexistent information sources. Instead, model construction can concentrate on sources of evidence that are objective in the sense of incorporating hard data or experienced engineering judgement. Rather than be driven by the full spectrum of all unknowns defined by the structure of the given system, the focus can be on assembling the best evidence available into packets that can reasonably be represented by independent component belief functions. The result will be 'subjectivist' probability judgements about prospective failures, but in the benign, and in my opinion necessary, sense of credible mathematical representations of uncertainty, and not in the sense of speculative guesses that 'objectivist' analysts find offensive.

The distinction here is not between Bayesian models and belief function models, for the former are special types of the latter, and are in fact the most desirable types when justifiable through a credible modelling process. The problem is that Bayesian practice often follows the frequentist habit of attempting to assess hypothetical population frequencies, whereas a more relevant strategy is to focus on causal processes affecting the specific system under analysis, and thence attempt to develop probabilistic and logical information directly applicable to that system (cf. Dempster and Kong, 1988). Sometimes the process may lead to a Bayesian model, or something very close thereto, but in other cases the desired risk assessments may be represented by (belief, plausibility) pairs that are close to (0, 1). Under the latter condition, which is certainly a mainline intended feature of belief function theory, the belief function conclusion requires unBayesian interpretation to the effect that the evidence does not support useful statements about risk. The analyst must either seek new evidence or abandon the analysis.

Finally, consider the most basic of all applied statistical questions: how to draw inferences from an observed sample to the unobserved population from which it was randomly drawn? All the modelling issues and difficulties discussed in the preceding two example types arise also in this third type, including the issue of assessing how ill-posed the inference task might be in a specific application. The belief function model that I prefer for sampling inference remains the simplex model of Dempster (1966, 1968, 1972). The model assumes an observed random sample $X_1, X_2, \ldots, X_n$ where X_i specifies which of the k categories of a multinomial population the ith sample individual occupies. The task is to draw inferences about the unknown vector of population proportions

$$p = (p_1, p_2, \ldots, p_k). \tag{6.4}$$

In belief function network terms, the representation may be pictured as in Figure 6.2, where the edges joining pairs of nodes indicate that independent

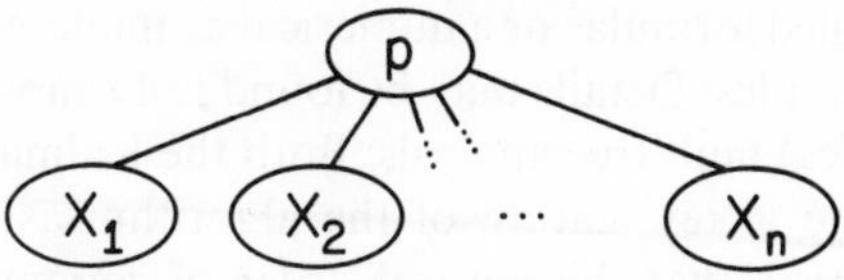

Figure 6.2 Network representation of a random sample $(X_1, X_2, \ldots, X_n)$ from a multinomial population p.

component belief functions are associated with the pairs (p, X_i) for $i = 1, 2, \ldots, n$. The essential property of the belief function assigned to the marginal frame (p, X_i) is that, when conditioned by the determination of the vector $\mathbf{p}$, the belief function on X_i must specialize to the ordinary probability measure of a single multinomial observation from population distribution p. The model is a simple and elegant way to satisfy the requirement. For detailed definitions, see the original references cited above.

The simplex model is capable of representing inferences from all types of populations, including high dimensional multivariate populations, and can handle both parametric and nonparametric specifications. It has never been implemented for complex populations, however, in part because the computations appeared forbidding 20 years ago, and in part because it is clear from what little is known about the results that the inferential message is a pessimistic one, namely, that very large samples are typically required before the data alone, without help from very strong *a priori* knowledge, can yield better than weakly informative inferences. In other words, the pure sampling problem in most realistic situations is ill-posed. For small k it does make sense to formally bind sampling inference models to network models such as Kalman filter models or fault tree models, but the message for a wide range of statistical sampling situations is that the need for external sources of credible prior knowledge should be taken seriously.

6.3 INFERENTIAL COMPUTATION

The remainder of the chapter is devoted to describing the remarkable computational process of propagation and fusion associated with belief function networks. The Bayesian special case was discovered and publicized within artificial intelligence (AI) by Pearl (1986), and more recently in statistics (Lauritzen and Spiegelhalter, 1988). The direct generalization to belief function network models was independently formulated by Kong (1986) and Shenoy *et al.* (1988). My approach here will be to sketch the key ideas, using an example where they have been widely known and used for many years. The example is provided by the Kalman filter, where the standard forward filtering and backward smoothing algorithms are exact illustrations of the general processes.

I do not provide detailed formulas or a numerical example, either for the Kalman filter or for other examples. Details may be found in Dempster and Kong (1988) for a small hypothetical fault tree analysis. Both the Kalman filter and the fault tree analysis illustrate a key feature of the algorithm as applied to network belief functions, namely, that the natural order of computations intermingles Boolean and Bayesian belief function components. Recognition of this feature enhances understanding of alternative computing strategies, and may help to speed the development of realistic examples.

Several different types of network representation are likely to appear in a single example. As illustrated above in Figures 6.1 and 6.2 and below in Figure 6.3, the nodes may represent single attributes of a system. Another type pictured below in Figure 6.4–6.6 creates nodes from clusters of attributes that are in effect generalized attributes. Edges joining nodes in the first type of network represent relations among the nodes, in the general sense of causal, logical, and probabilistic relations. Some discussions of networks distinguish between directed edges and undirected edges, correspondingly pictured with and without arrows, where the direction normally indicates either causal or inferential flow. In my exposition, causation is important mainly in the informal processes of model construction, and as noted above I tend to be critical of modelling processes that attempt to elicit subjective judgements concerning long chains of conditional probabilities. Hence arrows do not appear on my edges.

Two basic network representations of the Kalman filter model will be described. The first of these is more familiar, but the second is more illuminating and convenient for explaining the propagation and fusion algorithm. In Figure 6.3, each of the variables in the complete list (equation 6.3) is represented by a node. The network may be conceived as a hypergraph whose hyperedges consist of all the triples appearing in the equations (6.1) and (6.2). A variant of the hypergraph, also shown in Figure 6.3, is the ordinary graph whose edges simply join pairs of points. More precisely, the graph shown in Figure 6.3, whose edges are dotted lines, is called the 2-section of the hypergraph and is defined by edges contained in at least one hyperedge of the hypergraph.

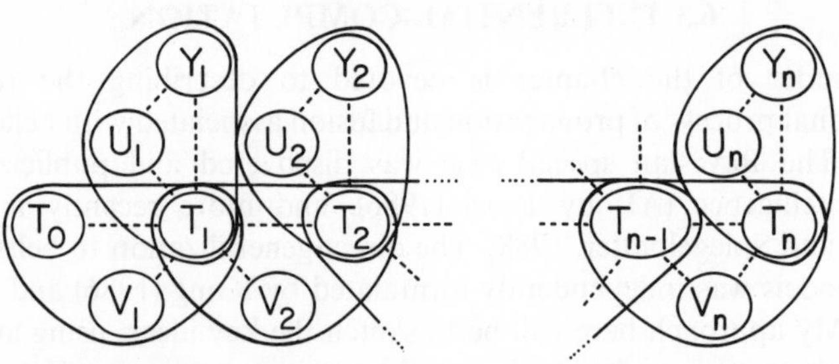

Figure 6.3 A network representation of the Kalman filter. The thick solid lines enclose hyperedges. The dashed lines are the edges of the 2-section of the hypergraph.

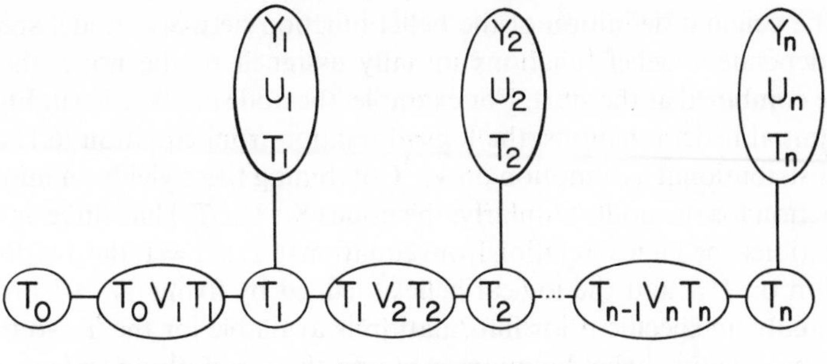

Figure 6.4 A tree of cliques representation of the Kalman filter.

The second network representation shown in Figure 6.4 is more immediately useful. The cluster-nodes in Figure 6.4 consist of cliques in the sense of containing variables mutually joined by the 2-section edges. The edges in Figure 6.4 indicate a nonempty intersection among the variables of the joined nodes. The tree of cliques representation illustrated in Figure 6.4 is not unique for a given example. The set of nodes is required to be large enough so that each component belief function in the network model refers to a margin contained in at least one node. Not all the edges indicating a common variable are included in the graph. A choice of edges is made in such a way that the result is a tree, i.e. is minimally connected with no loops, but it is clear from inspection of Figure 6.4 that other choices of edges would do the same. Another essential property is that each variable is allowed to appear in only one connected set of nodes. The cited literature discusses how to construct such a tree of cliques in general. It can always be done in a trivial way by declaring that the variables all belong to a single node, but the trick is to do it so as to disperse the input belief functions as widely as possible into small nodes. The particular choice shown in Figure 6.4 was made largely for convenience of exposition.

The propagation and fusion algorithm has many variants, in the sense of performing the computations in different orders. A variant is defined by selecting a node in the tree of cliques, propagating all beliefs inwards to that node, then reversing the flow to propagate outwards to the leaves of the tree. For example, the standard forward Kalman filter algorithm propagates in terms of Figure 6.4 from left to right and downwards, ending at the rightmost node with posterior beliefs for T_n given observations $Y_1, Y_2, \ldots, Y_n$. Then the standard backward smoothing algorithm propagates back from right to left, obtaining full posterior marginal beliefs about each of $T_n, T_{n-1}, \ldots, T_0$. As stated above, however, one could equally well propagate to any preselected node of the tree, and back out, obtaining the same posteriors.

The definitions of the terms propagation and fusion, and the rules specifying their use, are very simple. Fusion refers to product–intersection combination within a node of the tree of cliques. Fusion is carried out for two basic reasons.

First, if the original definition of the belief function network model specified a set of independent belief functions initially assigned to the node, then these should be combined at the start. For example, the node (T_0, V_1, T_1) in Figure 6.4 has two initial belief functions: the logical relation from equation (6.1) at $t = 1$, and the distributional assumption on V_0. Combining these yields an initial prior belief function for the node. Similarly, the node (Y_1, U_1, T_1) has three associated belief functions: the logical relation from equation (6.2) at $t = 1$, the distributional assumption on U_1, and the logical beliefs defined by fixing Y_1 at its observed value. Usually no specific prior information is available for the T_t, so these are assigned the vacuous belief function, meaning there is nothing to fuse. Second, if information is received at a node due to propagation from a neighboring node, then such information is fused or combined with the pool of the initial beliefs assigned to the node. For example, the node labeled T_1 may receive belief functions propagated from each of the triple nodes (T_0, V_1, T_1) and (Y_1, U_1, T_1), which would then be combined with each other and with the initial vacuous belief function at the node.

Propagation refers to sending a belief function from one node to a neighboring node. Propagation involves either marginalization, or minimal extension, or sometimes both. For example, if a belief function is sent from the node (T_0, V_1, T_1) to the node T_1, the task is to marginalize the original belief function from the triple variable to the single variable, since then it fits the recipient node. Alternatively, when transmitting from T_1 to (T_1, V_2, T_2), the task is to minimally extend from the single variable to the triple variable, again because the propagated belief function then fits the recipient node. Although not illustrated in Figure 6.4, it is in general necessary to both marginalize from the sending node to the intersection subset and then minimally extend to the receiving node.

The rules governing propagation and fusion are as follows. First select a node in the tree of cliques to be the recipient node for the inward propagation process. Then from each extremal node propagate its fused initial beliefs to its neighbor. When an interior node, excepting the recipient node, has received inward directed beliefs from all inward directed edges, these beliefs are fused with each other and with the initial beliefs at that node, and are then propagated inward. The process continues until beliefs are received at the recipient node from all directions, and finally fused with each other and with the initial beliefs at that node. The result, as shown by the algorithmic formulas of Kong (1986), is the combined marginal belief function at the recipient node.

By roughly doubling the amount of propagation and fusion, marginal beliefs can be obtained for all the nodes. This is achieved by outward propagation from the original recipient node, using the rules governing inward propagation except that beliefs originally propagated inwards along any edge must be excluded when fused beliefs are propagated back along that edge. The validity of this claim is immediate, assuming the validity of the inward process, because

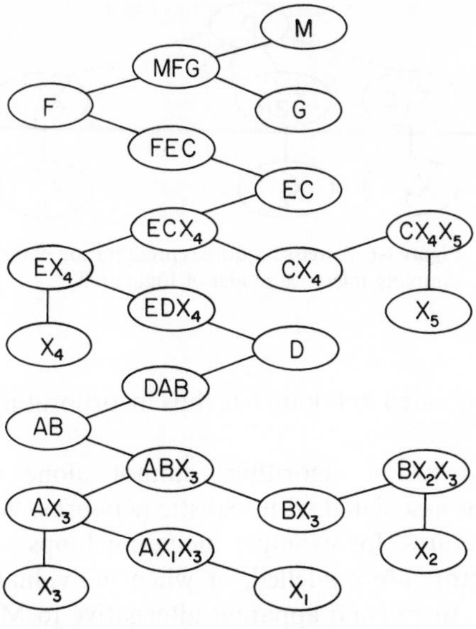

Figure 6.5 An extended tree of cliques representation of the fault tree of Figure 6.1.

the final answer at each node is obviously the same as it would have been had the node been chosen as initial recipient.

The inward and outward processes just described require a measure of global control, since the system must know which edge from each node leads to the recipient node. These could of course be computed in advance, in which case the algorithm can operate totally under local control. In another variant, there is no need even to establish an initial recipient node. Instead, each node simply receives and records beliefs from any incoming edge, and when beliefs are received from any selection of all but one edge, the beliefs are fused with each other and with the initial belief at that node, and then are propagated along the remaining edge. This operation continues until each node has transmitted beliefs (even if vacuous) along each associated edge. The result is an algorithm completely under local control that computes marginal belief functions at every node.

Although the foregoing description of propagation and fusion was nominally tied to the Kalman filter example, it is in fact quite general. The worked hypothetical reliability example in Dempster and Kong (1988) uses a tree of cliques representation of the fault tree of Figure 6.1 that is shown in Figure 6.5. Similarly, the hypergraph representation of Figure 6.2 can be cast as a tree of cliques representation as in Figure 6.6. Although the network pictured in Figure 6.6 may appear trivially simple, it represents an important type of component for extended model construction. For example, each of the X nodes in Figure 6.5 denotes an event that could be connected to an (X, p) node in an inference

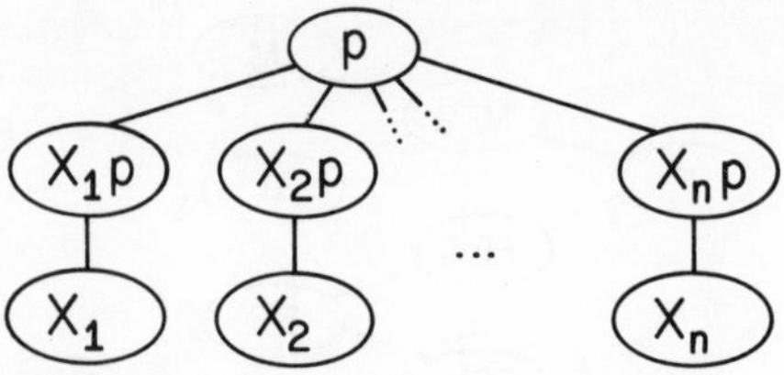

Figure 6.6 A tree of cliques representation of the sampling inference model of Figure 6.2.

model like that represented in Figure 6.6, thus incorporating sample information into the model.

Propagation and fusion algorithms cannot alone resolve the massive computing problems associated with realistic networks. When the hypergraphs have complex structures, for example including loops such as appear when common causal factors are modelled, or when individual hyperedges become highly multivariate, there is no apparent alternative to Monte Carlo sampling of belief functions. Even here, however, judicious introduction of propagation and fusion into Monte Carlo schemes will greatly extend the range of feasible computation.

ACKNOWLEDGEMENT

This work was facilitated by support from ARO Contract DAAL03-86-K-0042.

6.4 DISCUSSION

6.4.1 Discussion by Lotfi A. Zadeh

The seminal work of Professor Dempster (Dempster, 1967) became widely known after the publication of Glenn Shafer's book *A Mathematical Theory of Evidence* (Shafer, 1976). Today, it would be hard to find a person within the community of those who are concerned with the management of uncertainty in expert systems who is unfamiliar with what has come to be known as the Dempster–Shafer theory.

To some, the Dempster–Shafer theory is a very effective collection of concepts and techniques which have the capability of addressing the important issue of information granularity, that is, the association of probabilities with sets rather than with singletons. To others, it is basically a theory of interval-valued probabilities. And there are some who claim that anything which can be done with the Dempster–Shafer theory can be done equally well through the use of techniques which lie within the framework of classical probability theory.

I would classify myself as a strong believer in the Dempster–Shafer theory, albeit with some reservations.

My first reservation, which I have discussed elsewhere (Zadeh, 1979a, 1986), relates to the validity of the Dempster rule of combination. In more specific terms, I believe that the use of normalization in the Dempster rule may lead to counterintuitive results. Furthermore, there may be a limitation on the applicability of the Dempster rule which is not considered within the current version of the theory (Zadeh, 1986).

The second reservation relates to the fact that in the Dempster–Shafer theory there is no provision for degrees of containment and intersection. Thus, if A is a focal element and B is a given set, then either $A \subset B$ or $A \not\subset B$, implying that if A is almost contained in B it will not contribute to the measure of belief of B. One way of dealing with this problem is to assume that the focal elements are fuzzy sets (Zadeh, 1979b, 1981), which makes it possible to associate with A and B the conditional possibility and necessity of B given A, with both taking values in $[0, 1]$.

My third reservation, which is actually more of a question than a reservation, relates more directly to the subject of the chapter under discussion. Specifically, before the publication of the thesis of Kong (1986) and the paper by Shenoy and Shafer (1986), the Dempster–Shafer theory did not provide a complete set of rules of combination for dealing with uncertainty in expert systems or with the closely related problem of propagation of belief in networks. The question is: do the techniques described in Professor Dempster's chapter and a recently updated version of the Shafer–Shenoy paper (Shafer and Shenoy, 1988) provide a definitive solution to the problem?

As a first test, let us consider for simplicity the propagation of Bayesian belief functions in the context of MYCIN (Shortliffe and Buchanan, 1975). More specifically, consider the following rules (H = hypothesis, E = evidence, CF = certainty factor (which for simplicity is assumed to be nonnegative)):

(a) conjunction of hypotheses

$$CF[H_1 \text{ and } H_2 | E] = \min(CF[H_1 | E], CF[H_2 | E]),$$

(b) conjunction of evidence

$$CF[H | E_1 \text{ and } E_2] = CF[H | E_1] + CF[H | E_2] - CF[H | E_1]CF[H | E_2],$$

(c) chaining

$$CF[H | E_1] = CF[H | E_2]CF[E_2 | E_1].$$

In MYCIN, these relations and their duals are the basic rules that are needed for the propagation of certainty factors in inference networks. The question is: What would be the corresponding rules in Professor Dempster's approach to the computation of network belief functions? (A discussion of this question in the context of the standard Dempster–Shafer theory may be found in Grosof, 1986.)

To make the question more explicit it is necessary to relate the definition of *CF* to the underlying probabilities. In the original version of MYCIN, *CF* is defined by

$$CF[H|E] = \frac{P(H|E) - P(H)}{1 - P(H)}, \qquad P(H|E) \geqslant P(H). \tag{6.5}$$

As shown by Heckerman (1986), this definition is inconsistent with (b). An alternative expression suggested by Heckerman is

$$CF[H|E] = \frac{P(H|E) - P(H)}{P(H|E)(1 - P(H))}, \qquad P(H|E) \geqslant P(H). \tag{6.6}$$

In this perspective, the question posed above may be restated as follows: What would be the rules of combination for (a), (b) and (c) (and their duals) in Professor Dempster's approach and how would they relate to the MYCIN rules under the assumptions (6.5)–(6.7), $CF[H|E] = P(H|E)$?

The reasons for posing this question are the following. First, the MYCIN rules are *ad hoc* and it is known that (b) is inconsistent with equation (6.5); (b) is consistent with equation (6.6); and (c) can be deduced from equation (6.5). Thus, it is of interest to find out if, under appropriate assumptions regarding conditional independence, the MYCIN rules are consistent or inconsistent with the rules implied by Professor Dempster's approach in the special case of Bayesian belief functions. If they are inconsistent, then what are the correct forms of these rules?

My second question centers on the adequacy of Professor Dempster's approach when one is confronted with the problem of propagation of belief functions when the relations between node variables are imprecise in nature—as is frequently the case in expert systems.

For example, assume that a variable Z is related to X and Y by the conditional propositions:

$$Z \text{ is small if } X \text{ is large and } Y \text{ is medium}, \tag{6.7}$$

$$Z \text{ is medium if } X \text{ is small and } Y \text{ is large}, \tag{6.8}$$

$$Z \text{ is large if } X \text{ is very small and } Y \text{ is very large}, \tag{6.9}$$

in which small, large, medium, etc. play the role of given possibility or, alternatively, probability distributions. Now suppose that we would like to compute the value of Z when, say, X is not large and Y is not very small. The question is: How can one compute the desired value of Z given X and Y through the use of the rules of combination in Professor Dempster's approach?

As a somewhat more complicated example, assume that in addition to propositions (6.7)–(6.9) we have the following information:

$$(X \text{ is small and } Y \text{ is large}) \text{ is likely,} \tag{6.10}$$

$$(X \text{ is medium and } Y \text{ is small}) \text{ is unlikely,} \tag{6.11}$$

$$(X \text{ is large and } Y \text{ is medium}) \text{ is not very likely.} \tag{6.12}$$

Given propositions (6.7)–(6.12), the problem is to compute the probability that, say, Z is not very small. The problem in question is in the Dempster–Shafer spirit, except that the underlying probabilities and predicates are assumed to be fuzzy rather than crisp. Can the theory described in Professor Dempster's chapter deal with problems of this type?

In conclusion, Professor Dempster's chapter presents a number of important ideas regarding the propagation of belief functions in networks and influence diagrams. So far as my discussion is concerned, it is intended in the main to raise two issues: (a) How would the rules for propagation of Bayesian Belief functions compare with the corresponding rules in, say, MYCIN; and (b) how would Professor Dempster's approach apply in situations in which the relations between the node variables are expressed imprecisely—e.g. in linguistic terms—rather than as Boolean belief functions. The same applies to the underlying probabilities.

ACKNOWLEDGEMENT

Research supported in part by NASA Grant NCC-2-275 and NSF Grant DCR-8513139

REFERENCES AND RELATED PUBLICATIONS

Abramson, L. G. (1984) Introduction-Part II. In R. A. Waller and V. T. Covello (eds) *Low-Probability, High-Consequence Risk Analysis*, Plenum Press, New York, pp. 183–5.

Adams, E. W. (1981) Transmissible improbabilities and marginal essentialness of premises in inferences involving indicative conditionals, *Journal of Philosophical Logic*, **10**, 149–77.

Aitchison, J. (1968) Discussion, *Journal of the Royal Statistical Society*, **B30**, 234–7.

Dempster, A. P. (1966) New methods for reasoning toward posterior distributions based on sample data, *Annals of Mathematical Statistics*, **37**, 355–74.

Dempster, A. P. (1967) Upper and lower probabilities induced by a multivalued mapping, *Annals of Mathematical Statistics*, **98**, 325–39

Dempster, A. P. (1968) A generalization of Bayesian inference (with discussion), *Journal of the Royal Statistical Society*, **B30**, 205–47.

Dempster, A. P. (1972) Random convex polytopes, *Annals of Mathematical Statistics*, **43**, 260–72.

Dempster, A. P. and Kong, Augustine (1988) Uncertain evidence and artificial analysis. *Journal of Statistical Planning and Inference*, **20**, 355–68.

Dong, W.-M. and Wong, F. S. (1987) Propagation of evidence in rule based systems, *Int. J. Man–Machine Studies*, **26**, 551–66.

Dubois, D. and Prade, H. (1988) *Possibility Theory: An Approach to Computerized Processing of Uncertainty*, Plenum Press, New York.

Grosof, B. N. (1986) Evidential combination as transformed probability. In L. N. Kanal and J. F. Lemmer (eds) *Uncertainty in Artificial Intelligence*, North-Holland, Amsterdam, pp. 153–66.

Heckerman, D. (1986) Probabilistic interpretations for MYCIN's certainty factors. In L. N. Kanal and J. F. Lemmer (eds) *Uncertainty in Artificial Intelligence*, North-Holland, Amsterdam, pp. 167–97.

Hummel, R. A. and Landy, M. S. (1986) *A Statistical Viewpoint on the Theory of Evidence*, Robotics Research Technical Report No. 194/57, Computer Science Division, Courant Institute of Mathematical Sciences, New York University.

Kalman, R. E. and Bucy, R. S. (1961) New results in linear prediction and filtering theory, *Transactions ASME*, Series D, *J. Basic Eng.*, **83**, 95–107.

Kastenberg, W. F. *et al.* (1988) *Findings of the Peer Review Panel on the Draft Reactor Risk Reference Document, NUREG-1150*, NUREG/CR-5113, US Nuclear Regulatory Commission.

Kong, A. (1986) Multivariate belief functions and graphical models, Ph.D. thesis, Department of Statistics, Harvard University.

Kouts, Herbert *et al.* (1987) *Methodology for Uncertainty Estimation in NUREG-1150: Conclusions of a Review Panel*, NUREG/CR-5000, US Nuclear Regulatory Commission.

Kyburg, H. (1985) *Bayesian and non-Bayesian Evidential Reasoning*, Computer Science Technical Report No. 139, University of Rochester, New York. (To appear in *Artificial Intelligence*.)

Lauritzen, S. L. and Spiegelhalter, D. J. (1988) Local computations with probabilities on graphical structures and their application to expert systems (with discussion), *J. Royal Statistical Society*, **B50**, 157–224.

Lawrance, J., Garvey, T. and Strat, T. (1986) Framework for evidential reasoning systems, *Proc. AAAI Conference*, Philadelphia, Pa, pp. 896–903.

Levi, I. (1984) *Decisions and Revisions*, Cambridge University Press, Cambridge.

Lindley, Dennis V. (1987) The probability approach to the treatment of uncertainty in artificial intelligence and expert systems, *Statistical Science*, **2**, 17–24.

Liu, S.-H. G. (1987) Causal and plausible reasoning in expert systems, Ph.D. dissertation, Computer Science Division, University of California, Berkeley.

McCarty, P. (1988) The management of uncertainty in expert systems, Ph.D. dissertation, Computer Science Division, University of California, Berkeley.

Meinhold, Richard J. and Singpurwalla, Nozer D. (1983) Understanding the Kalman filter, *American Statistician*, **37**, 123–7.

Oblow, E. M. (1985) *O-theory—A Hybrid Uncertainty Theory*, ORNL Report TM-9759, Oak Ridge National Laboratory, Oak Ridge, Tenn.

Pearl, J. (1986) Fusion, propagation and structuring in belief networks, *Artificial Intelligence*, **29**, 241–88.

Pearl, Judea (1988) *Probabilistic Reasoning in Intelligent Systems*, Morgan Kaufmann.

Reactor Risk Reference Document (draft NUREG-1150) (1987) US Nuclear Regulatory Commission.

Reactor Safety Study (WASH-1400) (1975) US Nuclear Regulatory Commission.

Shafer, G. (1976) *A Mathematical Theory of Evidence*, Princeton University Press, Princeton, NJ.

Shafer, G. and Shenoy, P. (1988) *Bayesian and Belief-Function Propagation*, School of Business Working Paper-Series No. 192, University of Kansas, Lawrence.

Shenoy, P. P. and Shafer, G. (1986) Propagating belief functions with local computations, *IEEE Expert*, **1**, 43–52.

Shenoy, Prakash P., Shafer, Glenn and Mellouli, Khaled (1988) Propagation of belief functions: a distributed approach. In J. F. Lemmer and L. N. Kanal (eds) *Uncertainty in Artificial Intelligence*, Vol. 2, North-Holland, Amsterdam.

Shortliffe, E. H. and Buchanan, B. (1975) A model of inexact reasoning in medicine, *Mathematical Biosciences*, **23**, 351–79.

Singpurwalla, N. D. (1984) Introduction—Part I. In R. A. Waller and V. T. Covello (eds) *Low-Probability, High-Consequence Risk Analysis*, Plenum Press, New York, pp. 181–2.

Speed, T. P. (1985) Probabilistic risk assessment in the nuclear industry: WASH-1400 and beyond. In C. A. LeCam and R. A. Olshen (eds) *Proceedings of the Berkeley Conference in Honor of Jerzy Neyman and Jack Kiefer*, **1**, 173–200.

Tikhonov, A. N. and Arsenin, Y. Va. (1979) *Methods for Solving Ill-Posed Problems*, Nauka, Moscow.

Tikhonov, A. N. and Goncharskey, A. V. (eds) (1987) *Ill-Posed Problems in the Natural Sciences*, Mir, Moscow.

Turchin, V. F., Kozlov, V. P. and Malkevich, M. S. (1971) The use of mathematical statistics methods in the solution of incorrectly posed problems, *Soviet Physics Uspekhi*, **13**, 681–703.

Walley, Peter (1987) Belief function representations of statistical evidence, *Annals of Statistics*, **15**, 1439–65.

Yen, J. (1986) Evidential reasoning in expert systems, Ph.D. dissertation, Computer Science Division, University of California, Berkeley.

Zadeh, L. A. (1979a) *On the Validity of Dempster's Rule of Combination of Evidence*, ERL Memorandum M79/24, University of California, Berkeley.

Zadeh, L. A. (1979b) Fuzzy sets and information granularity. In M. Gupta, R. Ragade, and R. Yager (eds) *Advances in Fuzzy Set Theory and Applications*, North-Holland, Amsterdam, pp. 3–18.

Zadeh, L. A. (1981) Possibility theory and soft data analysis. In L. Cobb and R. M. Thrall (eds) *Mathematical Frontiers of the Social and Policy Sciences*, Westview Press, Boulder, Colo., pp. 69–129.

Zadeh, L. A. (1983) The role of fuzzy logic in the management of uncertainty in expert systems, *Fuzzy Sets and Systems*, **11**, 199–227.

Zadeh, L. A. (1986) A simple view of the Dempster–Shafer theory of evidence and its implications for the rule of combination of evidence, *AI Magazine*, **7**, 85–90.

CHAPTER 7

Influence and Belief Adjustment

Michael Goldstein, *University of Hull, UK*

ABSTRACT

We discuss the construction of influence-style diagrams for partially specified stochastic systems, using the general methodology of 'belief adjustment'. We pay particular attention to the rearrangement of a given diagram to uncover simple underlying structure. We discuss the quantification of the influence exerted across each arc of the diagram, and suggest an associated collection of diagnostics for the specification. We give two illustrations. The first is qualitative and concerns the simplification of a diagram with coexchangeable specifications. The second is quantitative and relates our general influence measures to influence measures used for regression diagnostics.

7.1 INTRODUCTION

Knowledge maps can depict 'information flow' between random objects. There are various ways to measure such information flow. We will discuss one particular approach to influence representation, based upon 'belief adjustments', for analysing partially specified stochastic systems.

We need such methods because, due to the complexity of large stochastic systems, partial specification often is the most that we can achieve. Further, even when we can, in principle, achieve a more complete specification, filling in unneccessary aspects of the specification may be very time consuming and lead to inefficient methods of analysis. Partial specifications are often both easier and safer to analyse, due to various interpretive and diagnostic measures for such analysis which offer safeguards against potential misuse.

A series of papers, Goldstein (1981–1988b), develop a systematic approach to the analysis of partial specifications. A prototype computer implementation is available. The program is called [B/D]—an acronym for Beliefs adjusted by Data.

Influence Diagrams, Belief Nets and Decision Analysis
Edited by R. M. Oliver and J. Q. Smith

There is a close analogy between the qualitative information displayed visually in an influence diagram and the quantitative information which [B/D] provides. In this chapter we investigate the relationship between the two approaches. Potential benefits from generating influence-style diagrams within our formulation are as follows:

1. We can construct informative diagrams, based on modest specification requirements, for which the nodes represent meaningful collections of objects.
2. There is an automatic methodology for identifying potential simplifications to a given diagram.
3. There are natural labellings for the arcs of the diagram quantifying the amount of 'influence' exerted across each arc.
4. There are simple diagnostics for assessing the performance of the diagram.

In this chapter, we summarize those aspects of belief adjustment methodology which are most directly relevant to the construction and interpretation of influence-style diagrams. We emphasize the process of systematic simplification of influence diagrams, and discuss the quantification of influence, suggesting an associated system of diagnostics. We give two illustrations. The first is a qualitative analysis of 'coexchangeability' judgements and the second is a quantitative analysis which relates our general influence measures to those used for regression diagnostics.

7.2 BELIEF STRUCTURES AND INFLUENCE DIAGRAMS

7.2.1 Belief specification: prevision

When working within a framework of explicit partial belief specification, it is very useful to be able to make expectation statements directly, if we wish, rather than requiring prior specification of probability distributions over some limiting partition of possibilities. (We can still work exclusively within a probability framework where appropriate, as the probability of an event is equivalent to the expectation of the indicator function corresponding to the event.) The gains from taking expectation as primitive are both logical and practical. Our theory will fit more closely to our actual specification abilities and it is often easier to extract the useful features from an expectation-based analysis than from a probability-based analysis.

Thus, we take as primitive the expectation of a random quantity, Our development is in terms of an individual describing beliefs about aspects of some system, so that, for example, a random quantity is simply any quantity whose value is not at present known to the individual. (The formal theory is essentially the same if we recast the development in less 'personalistic' terms, though the interpretation will differ.) The most detailed treatment of expectation as 'primitive' is given in de Finetti (1974). In de Finetti's development, which we shall follow, a direct expectation statement for a random quantity X is termed a 'prevision', with notation $P(X)$. This notation abolishes any distinction between expectation and probability; when E is an event, we write the probability of E as $P(E)$, where E is now used as the indicator function for the corresponding event. In de Finetti's approach, prevision is given various direct operational definitions, and simple coherence conditions are used to justify the basic expectation properties;

$$\sup(X) \geqslant P(X) \geqslant \inf(X);$$
$$P(X+Y) = P(X) + P(Y).$$

We now describe how to organize a collection of expectation statements.

7.2.2 Belief structures and nodes of influence diagrams

Collections of probability statements are organized into probability spaces. This is the natural mathematical structure to exploit Boolean operations over event spaces, but is not well suited to expectation-based structures, whose essential features derive from the linearity of the expectation operator.

If a prevision is specified for each member X_i of some collection, $C = [X_1, \ldots, X_k]$, of random quantities, then this implicitly specifies the prevision for each linear combination $a_1X_1 + \cdots + a_kX_k$ (and in general for no other quantities). Thus, the collection of all such linear combinations is a natural object of interest. We view the structure as a vector space, L, in which each X_i is a vector and linear combinations of vectors are the corresponding linear combinations of the random quantities.

We now introduce multiplication of the elements of L as an inner product over L, namely, for each X, Y in L

$$(X, Y) = P(XY).$$

We call an inner product space, B, built up in this way a *belief structure* (which is shorthand for a minimally structured collection of beliefs). The generating set, C, is referred to as the *base* of the belief structure, which we write as $C = b(B)$. Unless otherwise stated, the unit constant $X_0 = 1$ is automatically a member of each base.

The reason for choosing the belief structure as the fundamental object in our system is that many of the operations that we require in a stochastic analysis have their fundamental form in such structues. (Particular aspects of this argument are covered in the papers given in the references.)

In this formulation, a full probability specification becomes the corresponding Hilbert space of square integrable functions over the probability space with respect to the underlying probability measure. The two representations are equivalent for fully specified structures. However, because the Hilbert space representation has shifted the emphasis to the vectors, i.e. the random variables, it is easier in this formulation to identify how we may reduce our input specifications. We can restrict inputs to just those sufficient to determine some (typically small) subspace of the full space which is adequate to answer the questions at issue. In this sense, rather than choosing between a probability or expectation framework, we simply decide on a level of detail appropriate to the problem at hand.

The analysis of beliefs is carried out, in principle, within a single belief structure, collecting all of the belief statements. To help us to understand the flow of information, we can split the full structure into a collection of smaller belief structures. We now seek to construct influence-style diagrams visualizing the information flow between the substructures.

Our pictorial representation begins with a collection of nodes. Each node represents a belief structure, and thus indexes a base of random quantities, and a collection of belief statements, namely the prevision (or expectation) for each element, and for the product of each pair of elements in the base. The influence diagram will supply a qualitative visual representation of the links between the various structures.

7.2.3 Belief adjustment

The simplest way to measure the information flow between belief structures is by assessing the effects of linear fitting. Two belief structures, B and D, may be viewed as substructures of the combined structure $A = B + D$. (The sum of two inner product spaces is the space spanned by all of the elements of both spaces.) A simple quantification of the uncertainty about an element b of B is the distance from b to b_0 (the unit constant), i.e. the length $\| b - P(b)b_0 \|^2$, i.e. the variance of b. We can split the variance of b into two parts as

$$\operatorname{var}(b) = \operatorname{var}(P_D(b)) + \operatorname{var}([b/D]),$$

where $P_D(b)$ is the orthogonal projection of b into D (equivalently the linear fit of b on the elements of the base of D), and $[b/D] = b - P_D(b)$. The first variance term on the right of the above equation may be characterized as the portion of the variance of b which can be 'explained' by the variation in D, while the second variance term may be characterized as the 'residual' variation in b, given D. Intuitively, D is informative for b in B if the ratio

$$D(b) = \operatorname{var}([b/D])/\operatorname{var}(b)$$

is small.

It is useful to have a notation for the effect of stripping out from the belief structure B that portion of the variation which can be explained by the structure D. For any set $C = \{U_1, U_2, \ldots\}$ of random quantities, $[C/D]$ denotes the set $\{[U_1/D], [U_2/D], \ldots\}$.

Definition. If B is a belief structure with base $b(B)$, then the belief structure $[B/D]$ (the belief structure B adjusted by D), is the belief structure with base $b([B/D]) = [b(B)/D]$. □

(An equivalent definition which better identifies the underlying construction is as follows: the adjusted belief structure $[B/D]$ is the orthogonal complement of the subspace D in the composite space $B + D$.)

Thus, the adjusted belief structure expresses all of the residual variation in B after linear fitting on D.

In Goldstein (1988a), the following properties of adjusted belief structures are described, for any spaces B, C, D:

(P1) $[B/D] = 0$, the zero space, if and only if B is contained in D, $[B/D] = B$ if and only if B and D are orthogonal spaces;

(P2) $[(B + C)/D] = [B/D] + [C/D]$;

(P3) Any sum of belief structures $B_1 + \cdots + B_k$ can be decomposed as

$$B_1 + \cdots + B_k = D_1 + \cdots + D_k,$$

where $D_1 = B_1$, $D_i = [B_i/(B_1 + \cdots + B_{i-1})]$, $i = 2, \ldots, k$ and $D_1, \ldots, D_k$ are mutually orthogonal subspaces.

(P4) $[B/(D + E)] = [[B/D]/[E/D]]$.
Note, in particular, that if D and E are orthogonal, then

$$[B/(D + E)] = [[B/D]/E] = [[B/E]/D].$$

In this case, the adjustment can be partitioned into a series of individual adjustments which can be performed in either order. (The quantification of the effect of the adjustment will similarly separate the effects of such adjustments in an additive fashion.)

7.2.4 Separation by belief adjustment

Smith (1987) demonstrates that many of the properties that we usually associate with the property of conditional independence will apply equally to any generalized conditional independence property, namely a tertiary property $(. \perp\!\!\perp . | .)$ defined over a collection of objects with the following characteristics for each B, C, D, E:

(C1) $B \perp\!\!\perp C | (C + D)$;

(C2) $B \perp\!\!\perp C | D$ if and only if $C \perp\!\!\perp B | D$;

(C3) $B \perp\!\!\perp (C + D) | E$ implies and is implied by the pair of conditions $B \perp\!\!\perp C | (D + E)$ and $B \perp\!\!\perp D | E$.

Any tertiary property obeying (C1)–(C3) will behave computationally as a conditional independence property. For adjusted belief structures the natural form is as follows.

Definition. For any three belief structures, B, C, D we say that B is separated from C by D, written $B \perp\!\!\!\perp C/D$, if $[B/D]$ is orthogonal to $[C/D]$. □

Note from properties P1 and P4 above that

$$B \perp\!\!\!\perp C/D \text{ if and only if } [B/(D+C)] = [B/D].$$

Therefore B is separated from C by D if there is no further information about B which can be gained by adjusting by C, if we have already adjusted both structures by D and, symmetrically, there is no information about C from adjusting by B after adjusting for D.

(Compare the usual notion of conditional independence, where there is no information about the random quantity X to be gained by (probabilistic) conditioning on the random quantity Y, if we have already conditioned all quantities on the random quantity Z. In our formulation, conditional independence corresponds to the orthogonality of the associated adjusted L_2 spaces.)

Claim. $(.\perp\!\!\!\perp./.)$ satisfies properties (C1)–(C3) of the general conditional independence relation above.

Proof. (C1) is immediate as $[C/(C+D)]$ is the zero space by property (P1); (C2) is immediate from the symmetry of the orthogonality relation; (C3) is shown as follows:

The condition $B \perp\!\!\!\perp (C+D)/E$ is equivalent to the condition that

$$[B/(E+C+D)] = [B/E]$$

which is equivalent to the condition that

$$[[B/(E+D)]/[C/(E+D)]] = [B/E],$$

which condition both implies and is implied by the pair of conditions

$$[[B/(E+D)]/[C/(E+D)]] = [B/(E+D)] \quad \text{and} \quad [B/(E+D)] = [B/E]$$

i.e. by the pair of conditions $B \perp\!\!\!\perp D/E$ and $B \perp\!\!\!\perp C/(D+E)$. □

Thus, separation of belief structures acts as a generalized conditional independence property, and diagrams based upon such a definition may be manipulated by the rules governing the usual diagrams.

We have motivated the separation of belief structures informally, as a pragmatic method for analysing partially specified beliefs. However, as we have noted above, these methods are not special to partial specifications and may be applied equally to probability spaces, represented as spaces of indicator functions for various partitions, for which orthogonal projection is equivalent to conditional expectation. General discussion of the foundational status of our operations is given in Goldstein (1985, 1986a, 1987b, 1988a), where such projections are formulated as a natural extension of the basic operational definition of prevision. Also considered are the fundamental temporal arguments for relating current belief analysis to future beliefs (for example, when and why are we justified in converting conditional beliefs, specified *a priori*, into actual posterior beliefs?), and the case is made for regarding the projection-operation as used here (of which conditioning is simply a special case) as providing the full (stochastic) relationship between *a priori* and *a posteriori* beliefs.

7.2.5 Influence diagrams for belief structures

The construction of influence diagrams on collections of belief structures is formally similar to the usual construction over random quantities.

Each of the k nodes is a belief structure, B_j say, with base $b(B_j) = \{X_{0j}, X_{1j}, X_{2j}, \ldots\}$, consisting of m_j elements. The belief structure $B_1 + \cdots B_v$ combining the first v nodes is denoted by $B(v)$. The belief structure combining all of the structures except B_r and B_s is denoted by $B_{\{r,s\}}$.

We may construct, for example, the usual type of directed influence graph on the ordered nodes B_1 to B_k, namely any directed graph which expresses the collection of statements $B_r \perp\!\!\!\perp B(r-1)/P(B_r)$, where $P(B_r)$ is that subcollection of the nodes $B_1, \ldots, B_{r-1}$ which is connected by one directed edge to B_r.

An equivalent requirement is that $[B_r/P(B_r)] = [B_r/B(r-1)]$, for each r. Thus, for example, for each node B_r in the system, we can uniquely partition the influence of nodes $B_1, \ldots, B_{r-1}$ on B_r into the influence from the $r-1$ mutually orthogonal structures $D_i = [B_i/P(B_i)]$.

We can form an interesting alternative diagram to express 'cumulative' effects, which we term a cumulative knowledge map. In this diagram, a directed arc is drawn between each structure B_r and all nodes B_i, $i = 1, \ldots, r-1$, for which $D_i = [B_i/P(B_i)]$ is not orthogonal to B_r.

The graph is formed recursively. First, we check if B_1 is joined to B_2. If so, we replace B_2 by D_2. Then we check if B_3 should be joined to B_1 and/or D_2. We then replace B_3 by D_3, and so forth.

This type of diagram has the property that, for $i < j$, $B(i-1) \perp\!\!\!\perp B_j/P(B_i)$.

Further, $B_r \perp\!\!\!\perp D(r-1)/P_{[D]}(B_r)$, where $D(s)$, $P_{[D]}(B_s)$ replaces each structure B_i in $B(s)$, $P(B_r)$ respectively by the corresponding D_i.

A simple undirected graph on the k (unordered) nodes is defined as follows. Each pair of nodes B_i, B_j is joined by an arc unless

$$B_i \perp\!\!\!\perp B_j/B_{\{i,j\}}.$$

All properties of the usual influence diagrams which can be deduced from the general form of the property $B \perp\!\!\!\perp C|D$ will apply to these constructions. Many of the qualitative properties of the diagrams will therefore be similar to those of more familiar diagrams. Thus, we will instead focus attention upon certain aspects of our diagrams which are specifically suggested by our formulation, namely the simplification of belief adjustment diagrams and the quantification of influence.

7.3 BELIEF TRANSFORMS AND THE SIMPLIFICATION OF KNOWLEDGE MAPS

7.3.1 Example

To motivate our development, consider the following example. Suppose that we intend to select individuals for training for certain abilities. Suppose that at the time of selection we have access to various test scores for candidates, plus various relevant background variables. The eventual success of the candidate will be measured by further variables, such as subsequent test outcomes.

Specifically, suppose that we wish to predict two quantities S, T. We have two batteries of tests $s = [s_1, \ldots, s_m]$ and $t = [t_1, \ldots, t_n]$. We also have a profile for each individual, $Z = [Z_1, \ldots, Z_k]$, where each Z_i is a possibly relevant quantity such as 'age', 'experience', etc. The overall belief structure has base C which contains S, T, all of the elements in s, t, Z, and the unit constant.

We must specify all of the means, variances and covariances between the variables. (In a detailed analysis, we would include various functional forms and products of the variables in our choice of bases, which would involve certain higher-order specifications, but for this purely qualitative example it will be sufficient to work with the variables themselves.) A natural simplifying assumption would be that, in the terminology of Goldstein (1986b), all of the test items in s, t have been standardized to be exchangeable within s, t and coexchangeable across C.

In terms of the belief specification, we require that

1. The prior mean and variance is the same for each s_i, and the prior covariance between each pair s_i and s_j is the same, i.e. that s_1, $s_2, \ldots$ is a (finite second-order) exchangeable sequence;
2. For any further element x in C, the prior covariance between x and s_i is the same for each i, i.e. the sequence $s_1, s_2, \ldots$ is coexchangeable across C.

We impose similar conditions on beliefs over t.

We might collect our beliefs into four structures, 's-tests', 't-tests', 'profile' and 'targets', where s-tests, t-tests and profile index the belief structures derived from the bases s, t and Z and targets is generated by the base containing S and T. Neither directed or undirected diagrams appear to be very interesting, as given our specifications each node is connected to all other nodes.

However, suppose that we introduce two new structures in our diagram. The first, which we denote by 'tests' is the combined test structure containing all test items in s and t. The second new belief structure, which we term 'means' has a base containing the two quantities s_m and t_m, where s_m is the average of the tests $s_1, s_2, \ldots$ and t_m is the average of the tests t_1, $t_2, \ldots$

It follows, from the results on coexchangable belief structures in Goldstein (1986), that the test scores are separated from profile and target by the test

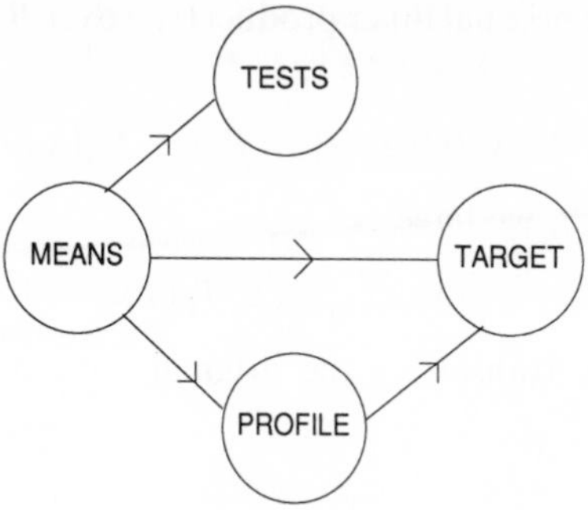

Figure 7.1

averages, i.e. that tests $\perp\!\!\!\perp$ (profile + target)/means. The directed diagram on the ordered nodes means, tests, profile, target, is as given in Figure 7.1.

Figure 7.1 represents the implication that, once the test averages have been reported, then there is no further use to be made of the individual test scores in the analysis.

The above example illustrates in simple form various basic ideas. In particular, notice that to obtain an interesting diagram we will often need to (1) regroup various of the quantities (in this example, we combine *s*-tests and *t*-tests), and (2) compute new quantities (in this example, the means).

How do we reach simple representations? In many problems, we will have ideas about 'simple structure' that can be directly incorporated into our diagrams. But often, as here, the simple structure emerges after some preliminary analysis. Thus, in Section 7.3.2, we describe a systematic approach to the uncovering of a simple structure, based on the eigenstructure of the 'belief transform'.

7.3.2 Belief transforms: notation

The effects of an adjustment of beliefs are summarized by the induced belief transform, defined as follows.

Definition. The belief transform T_D associated with the adjustment of the belief structure B by the belief structure D is the operator T_D over B defined by $T_D = I - P_B P_D$, where I is the identity operator on B and P_B, P_D are respectively the orthogonal projections from D to B, and from B to D. When we want to explicitly designate T_D as operating over B, we write the transform as $T_D\{B\}$. □

The belief transform is a bounded self-adjoint linear operator on B. The basic property of the belief transform is as follows. If we view a belief adjustment

$[B/D]$ as replacing the original inner product $(\cdot,\cdot)$ over B by a new inner product $(\cdot,\cdot)_{[D]}$, defined by

$$(X, Y)_{[D]} = P([X/D][Y/D]) = P(X - P_D(X))(Y - P_D(Y))$$

then, for any X, Y in B, we have

$$(X, Y)_{[D]} = (X, T_D(Y)),$$

so that T_D essentially transforms the original structure B into the adjusted structure $[B/D]$. □

7.3.3 Belief transforms: eigenstructure

All of the information about the belief adjustment is carried by the corresponding transform. A simple way to exhibit this information is by extracting the eigenstructure of the transform. For this discussion, it will be sufficient to consider the special case in which B is finite dimensional. (The infinite case is essentially the same, but there are various technicalities which require attention.)

If the dimension of B is $v + 1$, then T_D will have precisely $v + 1$ orthogonal eigenvectors, $Z_0, Z_1, \ldots, Z_v$ with ordered eigenvalues, $1 \geqslant m_v \geqslant \cdots \geqslant m_1 \geqslant m_0 = 0$. By convention, we set Z_0 to be the unit constant (as the unit constant will always be an eigenvector of T_D, with eigenvalue 0). All eigenvectors except Z_0 will have zero prior prevision. We standardize each of these eigenvectors to have prior variance 1.

As any element x in B can be written as

$$X = P(X)Z_0 + \mathrm{cov}(X, Z_1)Z_1 + \cdots + \mathrm{cov}(X, Z_v)Z_v,$$

the influence ratio, $D(X) = \mathrm{var}([X/D])/\mathrm{var}(X)$, is given by

$$D(X) = \frac{(m_1 \mathrm{cov}^2(X, Z_1) + \cdots + m_v \mathrm{cov}^2(X, Z_v))}{(\mathrm{cov}^2(X, Z_1) + \cdots + \mathrm{cov}^2(X, Z_v))}.$$

(Values of $D(X)$ near zero identify elements for which the adjustment is informative.) From the above equation,

1. $\mathrm{Min}\{D(X)\text{: } X \text{ in } B\text{:}P(X) = 0\} = m_1$ (this minimum occurring for Z_1);
2. $\mathrm{Min}\{D(X)\text{: } X \text{ in } B \text{ with } P(X) = 0, \mathrm{cov}(X, Z_1) = 0\} = m_2$ (for Z_2), and so forth.

Therefore, the eigenstructure of T_D lays an orthogonal grid over B, with the property that the adjustment has most effect for Z_1, next most effect for Z_2, etc., and the effect of the adjustment on a general element, X, of B is determined by resolving X along each element of the grid. Thus, the eigenvectors of T_D summarize the nature of the expected information, while the eigenvalues summarize the expected strength of the information.

7.3.4 Belief transforms: further properties

1. There is a natural symmetry between $T_D\{B\}$, the transform over B induced by D, and $T_B\{D\}$, the transform over D induced by B. We call the eigenvalues/vectors of the transform for which m_i is less than one the proper eigenvalues/vectors of the transform. (A proper eigenvalue corresponds to an element of B for which there is positive 'expected information'.) The proper eigenvalues of $T_B\{D\}$ and $T_D\{B\}$ are the same. Further, the proper standardized eigenvectors, W_i, of T_B are related to the proper standardized eigenvectors, Z_i of T_D by the relation
$$W_i = m_i^{-1/2} P_D(Z_i).$$
Denote by $H\{D/B\}$, the belief structure with base $[W_0, W_1, \ldots, W_s]$, consisting of all of the proper eigenvectors of $T_B\{D\}$. We call $H\{D/B\}$ the *Heart* of the transform.
The basic property of the heart of the transform is that
$$T_D\{B\} = T_{H\{D/B\}}\{B\}, \tag{7.1}$$
i.e. the adjustment is performed strictly over the space generated by the proper eigenvectors of the transform.
Note the following property of the heart of the transform. If $H\{B/D1\} = H\{B/D2\} = \cdots = H\{B/Dk\} = H$, then H is also equal to $H\{B/(D1 + \cdots + D_K)\}$.
2. Belief transforms can be built up in stages. For example, the transform $T_{(C+D)}$ over the structure B can be decomposed as
$$T_{(C+D)} = T_D T_{\{C/D\}},$$
where $T_{\{C/D\}}$ is the belief transform over $[B/D]$ induced by the space $[C/D]$, but viewed as an operator over B. That is, T_D transforms B to $[B/D]$, and $T_{\{C/D\}}$ transforms $[B/D]$ to $[B/(C+D)]$, as follows:
$$\begin{aligned}(X, T_{C+D} Y) &= (X, T_D T_{\{C/D\}} Y)\\ &= (X, T_{\{C/D\}} Y)_{[D]}\\ &= (X, Y)_{[C+D]}.\end{aligned}$$
(Compare the usual probabilistic conditioning of W on pieces of evidence U and V. We can proceed in two stages, by conditioning out U on V and W, and then conditioning W on V under the induced conditional distribution.)
In particular, $[B/D]$ is orthogonal to $[C/D]$ if and only if $T_{\{C/D\}}$ is the identity operator, so that
$$T_{(C+D)}\{B\} = T_D\{B\}$$
if and only if $[B/D]$ is orthogonal to $[C/D]$.

Notice, from this property and equation (7.1), that $[B/H\{D/B\}]$ and $[D/H\{D/B\}]$ are orthogonal.

7.3.5 Separating belief structures by eigenvectors

We now translate the above results into the belief separation terminology of Section 7.2.

We have said that belief structure D separates structures B and C if $[B/D]$ is orthogonal to $[C/D]$. In the notation of (2) of Section 7.3.4, $B \perp\!\!\perp C/D$, if and only if $T\{C/D\}$ is the identity operator. Thus, an alternative characterization of the separation is as follows:

Corollary 7.1. (i)$B \perp\!\!\perp C/D$ if and only if $T_{\{C+D\}}\{B\} = T_D\{B\}$; (ii) for any B, D, we have $B \perp\!\!\perp D/H\{D/B\}$ (and, equivalently, $B \perp\!\!\perp D/H\{B/D\}$). (In particular, if $H\{D/B1\} = \cdots = H\{D/Bk\} = H$, then $D \perp\!\!\perp (B1 + \cdots B_K/H.)$ □

This corollary shows how we may automatically identify possible reductions of the diagram—we may simplify our influence diagrams when various of the associated belief transforms have a heart which is simpler than the full structure.

7.3.6 Example continued

In our example on the influence of test scores on various remaining variables, we can identify the redrawn diagram, Figure 7.1, automatically as follows.

First, we observe that each of the transforms T_{target} {s-tests}, T_{profile} {s-tests}, $T_{t\text{-tests}}$ {s-tests}, have a common heart generated by the single element s_m.

(We may realize this directly from theoretical considerations. As problems become more involved, we would instead reach this realization simply by running a program such as [B/D], which finds and then compares the heart of each transform.)

We may immediately conclude, from the corollary of Section 7.3.5, that s_m separates s-tests from t-tests + target + profile. We may similarly conclude that t-tests $\perp\!\!\perp$ (s-tests + profile + target)/t_m. Finally, we may conclude that the structure tests, which contains s_m and t_m, separates (s-tests + t-tests) from (target + profile), which is the relationship expressed in Figure 7.1.

This is a special case of general results on belief structures expressing finite exchangeable systems, for which the 'average' structures quite generally separate the observed values from any related belief structures. There are similar results which enable us to automatically update our judgements upon the beliefs specified in the diagram when we make repeated observations upon trials of the diagram under 'exchangeable' conditions.

For example, if we view the individual realizations as coming from an (in principle, infinite) exchangeable sequence of such structures, then, by the results of Goldstein (1986b), we may extract a 'limiting' or 'population' structure, analogous to 'means' in the present example, which separates the 'future performance' of the diagram from the 'past observations' on the diagram. Thus, our updating algorithm is that past data updates the population structure which influences future realizations.

Thus, examining the eigenvectors of the various adjustments in the diagram provides us with systematic approaches to the redrawing of our diagrams. The eigenstructure summarizes the types of information which pass between the various nodes of the diagram.

To get a more interesting diagram, we need to feed in some more symmetries. As an example, suppose that there are k test items in s and in t and that each pair (s_i, t_i) of test responses is matched, $i = 1, \ldots, k$, forming k exchangeable belief structures—one for each pair of questions. Suppose that the combined collection of questions is coexchangeable with profile, as are the pair S, T and that s is related to T as t is related to S. Formally, suppose that

$$\operatorname{var}(s_i) = \operatorname{var}(t_j) \quad \text{for all } i,j;$$

$$\operatorname{corr}(s_i, s_j) = \operatorname{corr}(t_u, t_v) \quad \text{for all } i,j \text{ and } u,v \text{ not equal};$$

$$\operatorname{corr}(s_i, t_i) \text{ does not depend on } i;$$

$$\operatorname{corr}(s_i, t_j) \text{ does not depend on } i,j \text{ not equal};$$

$$\operatorname{corr}(s_i, Z_k) = \operatorname{corr}(t_j, Z_k) \quad \text{for all } i,j,k;$$

$$\operatorname{var}(S) = \operatorname{var}(T);$$

$$\operatorname{corr}(s_i, S) = \operatorname{corr}(t_j, T) \quad \text{for all } i,j;$$

$$\operatorname{corr}(s_i, T) = \operatorname{corr}(t_j, S) \quad \text{for all } i,j;$$

$$\operatorname{corr}(S, Z_k) = \operatorname{corr}(T, Z_k) \quad \text{for all } k.$$

By the same types of automatic procedure which generated the above reduction of the diagram, namely partitioning the eigenvectors of the transforms from the 'mean' structure, we are led to construct the four quantities, $\text{sum} = s_m + t_m$, $\text{diff} = s_m - t_m$, $\text{SUM} = S + T$, $\text{DIFF} = S - T$, from which we generate Figure 7.2.

Thus, we have two distinct influence systems, the first involving overall score, dependent on individual test scores and profile, the second involving test difference alone. This simplifies both our analysis and also the diagnostic checking of the network.

Such simplifications are so useful that it is important to be able to quantify how much our diagrams are distorted if we pretend such strict exchangeability in circumstances for which there are actually certain small differences between individual specifications. More generally, we need to consider how much information passes between the various nodes, (so that, for example, we can assess whether an arc that we have not drawn in an idealized diagram such as Figure 7.2 can pragmatically be considered to carry 'negligible influence'). We now consider how to quantify the influence across the arcs of our diagrams.

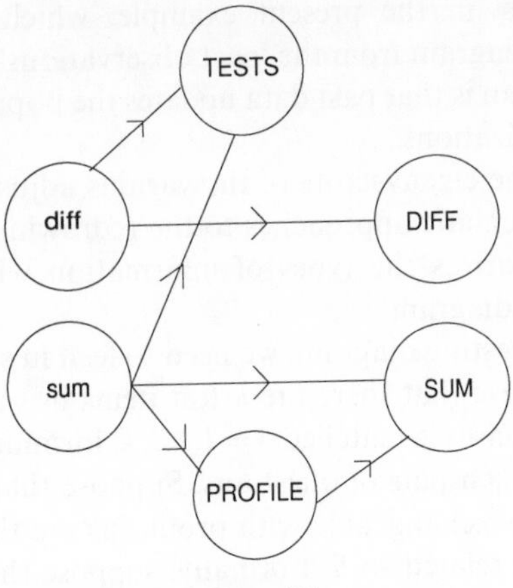

Figure 7.2

7.4 QUANTIFYING INFLUENCE

7.4.1 Trace quantification

A simple quantification of the adjustment of a single random quantity Y by a structure D is the ratio $D(Y) = \text{var}([Y/D])/\text{var}(Y)$. It is more convenient to have zero representing no influence, so we use the quantification $R(Y) = \text{var}(P_D(Y))/\text{var}(Y)$ for the effect of the adjustment. Here $R(Y)$ is analogous to the familiar coefficient of determination in multiple regression, and is equivalent to $D(Y)$ as, from Section 7.2.3, $R(Y) + D(Y) = 1$.

The belief transform T_D over the structure B with base Y in this case has a single eigenvector, namely Y, with eigenvalue equal to $D(Y)$. Similarly, $R(Y)$ is the single eigenvalue of the transform $S_D = I - T_D$, where I is the identity operator. For general structures, B, the transform S_D is more convenient if we wish to analyse changes in belief, while the transform T_D is preferable for analysing the residual uncertainty. Thus, we make the following definition.

Definition. The adjustment transform over B induced by D is

$$S_D\{B\} = I - T_D\{B\}. \qquad \square$$

Note that the eigenvectors of S and T are the same, and eigenvalues of S are one minus the eigenvalues of T.

When we adjust a general belief structure B by structure D, then the eigenstructure of S_D provides the full quantification of the adjustment. However, it is useful to reduce the collection of eigenvalues to a single number quantifying the 'overall' effect of the adjustment (for example, to label all of the arcs in our influence diagrams).

Each large (i.e. near one) eigenvalue of $S_D\{B\}$ identifies a direction in B for which the adjustment is influential. A simple measure of the overall influence of the adjustment on B is the sum of all of the eigenvalues of S_D, i.e. the trace of S_D. The trace of S does not depend on any particular basis representation and does not 'double-count' quantities which are in S twice under different names. For example, the trace of S would be the same over the belief structure with base $\{X, Y\}$, and the base $\{X, Y, X + Y\}$, which allows us to introduce average quantities (such as s_m, t_m in the example of Section 7.3) into the belief structures, without distorting our influence measures.

Useful trace properties of S are:

1. Trace $(S_D\{B\}) = \text{trace } (S_B\{D\})$;
2. If $D1$ is orthogonal to $D2$, then trace $(S_{D1+D2}\{B\}) = \text{trace } (S_{D1}\{B\}) + \text{trace } (S_{D2}\{B\})$.

Although any single numerical summary of influence will lose some information, and may need to be modified in specific applications, there is a particular characterization of the trace of S, which we shall now describe, which provides both a natural interpretation of the trace as an influence measure and also suggests a systematic system of diagnostics for the influence diagram.

7.4.2 The length of a belief adjustment

The belief transform summarizes expected effects of a belief adjustment of a structure B by a structure D. After the values of the elements of D become known, we may implement the adjustment by replacing the initial prevision $P(X)$, for each X in B, by the adjusted prevision $P_D(X)$, i.e. the linear fit of X on the base of D. We use the following notation: if d is the collection of observed values for the elements of D, then $P_d(X)$ is the value of $P_D(X)$ when we substitute d for D.

A simple and revealing summary of the effect of the observations is the length of the adjustment, defined as follows.

Definition. Consider the collection, B_U, of all elements of B which have prevision zero and variance one, *a priori*. For each element X in B_U, evaluate the squared change in prevision $(P_d(X))^2$. The length $L\{B/d\}$, of the revision is defined to be the supremum of the values $(P_d(X))^2$, for X in B_U. □

Thus, if the length of the revision is small then all of the revisions of belief are small, whereas if the length is large then at least some aspects of the beliefs have changed by a substantial amount. How large would we expect the largest change to be? The answer is given by the following result.

Theorem 1.1. For any belief structures B, D we have

$$P(L\{B/D\}) = \text{trace } (S_D\{B\}).$$

(Note as a consequence that $P(L\{B/D\}\} = P(L\{D/B\}))$.

This result is proved for finite dimensional belief structures in Goldstein (1988b). (Proof in full generality will be reported elsewhere.) In that paper, we gave the following simple algorithm to evaluate $L\{B/d\}$.

1. Evaluate the eigenvalues $m_1, \ldots, m_r$ and the standardized eigenvectors $Z_1, \ldots, Z_r$ of $S_B\{D\}$.
2. From the observed values $D = d$, evaluate the corresponding observed values z_i for each Z_i.
3. Now we can compute $L\{B/d\}$ as the sum

$$m_1 z_1^2 + \cdots + m_r z_r^2.$$

As the trace of the adjustment transform may be interpreted as the prior expectation for the maximal change in belief in the system, a simple diagnostic for the adjustment is to compare the actual value of the largest change in belief, $L\{B/d\}$, with the trace of the transform $S_B\{D\}$. If L is much larger than the trace, then this suggests either that something 'surprising' has happened, which we will presumably be interested to investigate, or alternately that our prior assessments were wrong. (A typical wrong specification would occur if the user, instead of carefully quantifying his uncertainties, simply entered an arbitrary collection of values for belief inputs. Having observed how people use most standard statistical packages, it is safe to assume that there will be plenty of abuses of this type.) If L is much smaller than the trace, then this might suggest that we have been too cautious in our prior specifications (i.e. by down-weighting our beliefs too much in favour of the observational data). Of course there are many other types of diagnostic that we may employ. However, because this diagnostic directly concerns the features of primary interest, namely the change in beliefs, it will usually be of intrinsic interest.

In Section 7.4.1, we suggested the trace as a simple summary of influence. We now have a substantitive interpretation of the trace, plus an informative diagnostic system based on the use of this influence measure. We now interpret the trace of the transform between adjusted belief structures, and introduce further diagnostic quantities.

7.4.3 Data trajectories: stepwise trace

Knowledge maps display changes in belief by separating and analysing the information flow between the various evidential components of the system. The simplest extended relation is to suppose that there are two evidential nodes, D and E, say, which influence node B. Each influence statement can be quantified as a 'length' or expected change in belief. A natural measure of composite influence is to quantify the progressive changes in belief over B as each node is introduced.

Suppose that we will first evaluate the previsions for the elements of B adjusted by D alone, and then evaluate the effects of adjustment by $(D + E)$. There are three 'lengths' of interest:

1. $L\{B/d\} = \max(P_d(X))^2$;
2. $L\{B/(d+e)\} = \max(P_{d+e}(X))^2$;
3. $L\{B/[e/d]\} = \max(P_{[e/d]}(X))^2 = \max(P_{e+d}(X) - P_d(X))^2$.

As before, all maximizations are over those elements of B with zero prevision and unit variance, *a priori*.

Here $L\{B/(d+e)\}$ expresses the effect of the overall adjustment, while the two quantities $L\{B/d\}, L\{B/[e/d]\}$ express the effects of the first-stage adjustment by D and the change when E is introduced. The expected value of each of the above lengths is equal to the trace of the corresponding adjustment transform.

We may now compare $L\{B/(d+e)\}$ with the sum $L\{B/d\} + L\{B/[e/d]\}$. If the combined length $L\{B/(d+e)\}$ is substantially greater than the sum of the two composite lengths then the combined effect of the two pieces of evidence is substantially larger than the separate effects of the individual items, so that the two collections of evidence are complementary. If $L\{B/\{d+e)\}$ is substantially less than $L\{B/d\} + L\{B/[e/d]\}$, then the two pieces of evidence are contradictory. (If adjustment by D causes large changes then $L\{B/d\}$ will be large. If adding E contradicts all these changes then $L\{B/[e/d]\}$ will also be large, as the revised previsions will return towards their prior values. However, the effect of the combined adjustment, $L\{B/(d+e)\}$, will be small.) Finally, if the sum of the stepwise changes equals the overall change, then the pieces of evidence are providing information about 'unrelated' aspects of B, in which case we can meaningfully resolve the overall change in belief due to adjustment by D and then E into portions corresponding to the adjustment by D and the subsequent adjustment by E.

A simple measure of the degree of support/conflict between D and E in adjusting beliefs over B is the *path correlation* $C\{d, e{:}B\}$, defined by

$$C\{d, e{:}B\} = \frac{(L\{B/(d+e)\} - (L\{B/d\} + L\{B/[e/d]\})}{2(L\{B/d\}L\{B/[e/d]\})^{1/2}}.$$

Here $C\{d, e{:}B\}$ lies between -1 and $+1$, $+1$ corresponding to full agreement between D and E, -1 being full conflict, while 0 represents unrelated pieces of evidence. Our description of $C\{d, e{:}B\}$ as a correlation is based upon the development in Goldstein (1988b), where each of the lengths in the above formula is derived as the variance of a particular quantity, termed the bearing of the adjustment.

(The bearing Y_d of an adjustment of B by observed values $D = d$ is the (essentially unique) element of B with the property that for any X in B, $P_d(X) - P(X) = \text{cov}(X, Y_d)$. The bearing summarizes the direction and magnitude of all the adjustments over B as a single vector, and the above path correlation is essentially the correlation between Y_d and $Y_{[e/d]}$. Existence, construction, properties and interpretation of bearings, with examples, is provided in Goldstein, 1988b.)

In summary, the trace of a transform over adjusted spaces expresses the influence (i.e. expected change) when the new nodes are added to pre-existing nodes. These expected changes can be compared diagnostically to the observed changes. The degree of support/conflict between aspects of the data can be expressed through the path correlations.

7.4.4 Splitting the heart of a transform

In general, the effect on the trace of a transform when we add or subtract belief structures depends on the full structure that we are analysing. However, there is an important class of transforms for which it is straightforward to combine or separate substructures.

We have noted that the heart of a belief transform separates the individual structures, i.e. that $B \perp\!\!\!\perp D/H\{B/D\}$. We now observe that any partition of the heart resolves the length of an adjustment into the corresponding components, as follows. (Full details are given in Goldstein, 1988b.)

Theorem 7.2. Let $bH[1], \ldots, bH[r]$ be any partition of the proper eigenvectors of $T_B\{D\}$ into disjoint subsets. Let $H_1, \ldots, H_r$ be the belief structures with bases $bH[1], \ldots, bH[r]$ respectively.

1. For any observed outcomes, and each $j, L\{B/H_j\} = L\{B/[H_j/(H_1 + \cdots + H_{j-1})]\}$;
2. $P(L\{B/H_j\})$ is equal to the sum of the eigenvalues of $S_B\{D\}$ corresponding to the eigenvectors in $bH[j]$;
3. Each path correlation $C\{(H_1 + \cdots + H_{j-1}, H_j{:}B\}$ is zero.

Therefore, if we split a knowledge map through the heart of a transform, then we can study observed and expected effects of adjustments separately along each arm of the diagram which greatly simplifies our problems of analysis and interpretation. In Section 7.3, we implicitly treated a simple case of this, namely separating the proper eigenvectors from the zero eigenvectors of S, as the zero eigenvectors have no effect on the adjustment. Theorem 7.2 provides the full generalization of this property.

7.4.5 Labelling knowledge maps

In our construction of knowledge maps, each arc joining two nodes implicitly represents an adjusted belief transform between the two nodes. We now quantify the influence across each arc by labelling the arc by the trace of the corresponding adjustment transform. This labelling will differ between the various directed and undirected graphs that we have defined, due to the differences between the various transforms.

Labelling the undirected graph

For the undirected graph, the arc between two nodes B_i and B_j represents the adjustment of the adjusted node $[B_i/B_{\{i,j\}}]$ by the adjusted node $[B_j/B_{\{i,j\}}]$, with corresponding transform S_{ij}. Let $s_{ij} = \text{trace}(S_{ij})$. Each arc (i,j) can be uniquely labelled by s_{ij}, as $s_{ij} = s_{ji}$. This label will be zero if and only if there is no arc between B_i and B_j. Each arc label may be interpreted as the expected maximal squared change in adjusted prevision over the standardized elements of B_i, when the adjustment by B_j is made, following the adjustment by $B_{\{i,j\}}$.

An equivalent definition of the undirected graph would follow from joining all pairs of nodes and then deleting all arcs with $s_{ij} = 0$. A method for simplifying a knowledge map with a small number of important arcs and a large number of arcs which are non-zero but whose effect is negligible, is to define an undirected knowledge map at level c to be a diagram constructed as before but for which a pair of nodes is only joined if $c < s_{ij}$.

However, for a visual display, it may be better to present the information as follows. First, we treat the arc labels as a pseudo-metric (i.e. the larger the label, the closer the pair of nodes in our diagram), and construct the diagram through some non-metric multidimensional scaling routine. Second, we can colour code the arcs, so that instead of deleting arcs with small labels, we can draw them in a lighter colour. We can either choose absolute cut-offs to determine labels or choose cut-offs on the quantiles of the observed label values.

Labelling the directed graph

For the directed graph, each directed arc from B_i to B_j, $(i < j)$, can be interpreted in various ways. One interpretation, which relates to our approach above, is to view each arc from B_r to B_s as expressing the transform from $[B_r/P(B_s/r)]$ to $[B_s/P(B_s/r)]$, where $P(B_s/r)$ is the collection $P(B_s)$, but with B_r removed.

The adjustment transform between these two spaces is R_{rs}, say, with trace r_{rs}, which value can be used to label each arc. As before, the arc length is the expected change in the length of the adjustment over B_s, when B_r is added after $P(B_s/r)$.

A useful property of this labelling system is that if an individual arc has a zero label, the arc can be deleted without affecting the validity of the influence diagram. (If several arcs have zero labels, then they must be erased sequentially, recalculating labels at each stage to see which labels remain zero. This process will ultimately result in a graph with no surplus arcs.)

Labelling the cumulative graph

For the cumulative knowledge map, the arcs represent the adjustment of $[B_j/B(i-1)]$ by $[B_i/B(i-1)]$ (equivalently by $[B_i/P(B_i)]$). The adjustment transform between these two spaces is S_{ij}, say, with trace s_{ij}, which value can be used to label each arc. As before, the arc length is the expected length of the adjustment.

This labelling has the property that if we replace all of the nodes B_1 to B_r with the single node $B(r)$, and join $B(r)$ to any other node B_s, $r < s$, then the label on the arc from $B(r)$ to B_s is the sum of the labels on all of the arcs from B_s to $P(B_s)$ in the original graph.

An alternative definition of the cumulative knowledge map would be to join each node to each higher numbered node, and then delete all arcs with $s_{ij} = 0$. As for the undirected graph, we can eliminate, or shade lightly arcs with small s_{ij} values, and present our diagrams, via multi-dimensional scaling routines, so that nodes with large s_{ij} values are close together.

We may also use this labelling system to induce an ordering over an initially unordered set of nodes. The upward ordering is determined by the influence of each node upon some target or value node B_0. Choose $B_{(1)} = B_j$, where j is chosen to maximize s_{0j} over $j = 1, 1, k$. Choose $B_{(2)}$ to maximize the trace s_{20} of the operator from $[B_{(2)}/B_{(1)}]$ to $[B_0/B_{(1)}]$. At each step choose $B_{(i)}$ to maximize s_{i0} under this labelling.

We may instead create a downward ordering by choosing$B_{(k)}$, $B_{(k-1)}, \ldots$ sequentially to minimize the value of s_{0j} at each stage.

The upward ordering is often more natural. At each stage we add the node which has the most (adjusted) influence on the value node. However, we know from experience that 'backward elimination' style approaches to identifying important combinations of 'predictors' may avoid some of the pathologies of forward selection.

7.4.6 Example

To illustrate the quantification of influence, we give an example which relates the types of influence measures used in regression diagnostics to the more general types of influence measures that we have discussed.

We proceed as follows. First, we introduce a data set which has been used several times in the statistical literature to illustrate the need for influence measures in regression diagnostics. Next, we establish the relevant belief structure analysis. We then draw the associated graph, and show that the labelling that we have suggested identifies the 'influential' point in fitting the regression. We evaluate the diagnostic labelling of the graph, given the data, and discuss the similarities between this labelling and adjusted residual plots. Finally, by analysing the associated belief transform, we refine the diagram to identify the features of the regression which are most affected by the influential case.

This example illustrates the following ideas. First, the information conveyed by traditional influence diagrams lies in the missing arcs. However, strong relationships are also important, and a good diagram should be able to display

Table 7.1 Data on body weight (x_1), liver weight (x_2), drug dose (x_3) and percentage of dose in liver (y) on a sample of 19 rats (*source*: Weisberg, 1985)

y	x_1	x_2	x_3
0.42	176	6.5	0.88
0.25	176	9.5	0.88
0.56	190	9.0	1.0
0.23	176	8.9	0.88
0.23	200	7.2	1.0
0.32	167	8.9	0.83
0.37	188	8.0	0.94
0.41	195	10.0	0.98
0.33	176	8.0	0.88
0.38	165	7.9	0.84
0.27	158	6.9	0.80
0.36	148	7.3	0.74
0.21	149	5.2	0.75
0.28	163	8.4	0.81
0.34	170	7.2	0.85
0.28	186	6.8	0.94
0.30	146	7.3	0.73
0.37	181	9.0	0.90
0.46	149	6.4	0.75

both 'strong' and 'absent' relations. Second, just as we must evaluate diagnostics for statistical models, so also should we evaluate the corresponding diagnostics for our knowledge maps. Finally, the interest in the diagram will often concern relationships between groups of quantities. Often, we will not identify the important groupings *a priori*, so that we need routine methods to improve our groupings.

For our illustration, we shall use a data set described in Weisberg (1985, p. 121–4), as an illustration of case statistics in regression diagnostics. The data, as given in Table 7.1, concerns the relationship between x_1 (the body weight), x_2 (the liver weight) and x_3 (the relative dose of a particular drug) and y (the percentage of the dose present in the liver) on a sample of nineteen rats.

Weisberg observes that none of the simple regressions of y on any of the independent variables appears significant. Fitting both x_1 and x_3 together, both regression coefficients appear significant, which is surprising as a feature of the experiment was that dose was approximately selected to be a multiple of weight, suggesting that x_1 and x_3 should be measuring the same thing. Weisberg evaluates the leverage for each observation. Case 3 has a very high leverage indicating that the response vector for case 3 is different from the others. In the fitted regression with case 3 deleted, all the regression coefficients are small, so that the relationship can be ascribed to the third case alone. Inspecting case 3 reveals that the third rat received a much larger dose for its weight than did any of the other rats, with various possible implications for the interpretation of the experimental results.

The model underlying the analysis supposes that each observation Y_i is of form

$$Y_i = b_0 + b_1 x_{1i} + b_2 x_{2i} + b_3 x_{3i} + e_i,$$

where b_0, b_1, b_2, b_3 are the regression coefficients, x_{rs} is the value of x_r for individual s, and $e_1, e_2, \ldots$ is a sequence of uncorrelated random quantities, with zero mean and the same unknown variance.

For the purpose of this illustration, suppose that we specify $P(b_i) = 0$, for each i, and specify $\text{var}(b_0) = 0.5$, $\text{var}(b_1) = 0.01$, $\text{var}(b_2) = 0.01$, $\text{var}(b_3) = 400$, and all covariances zero. These values are chosen so that before the experiment we consider that the effect on Y of each standardized variable is likely to be of a similar, moderate order of magnitude. Finally, we specify an expected 'error' variance of 0.01. (Our output will change with the prior specification. The influence of case 3 upon the belief adjustment emerges in a similar fashion for most belief specifications.)

We draw the undirected influence diagram for which each observation Y_i is the base for a belief structure node O_i, and the coefficients $\{b_0, b_1, b_2, b_3\}$ form the base for a belief structure which we call 'coef'. The elements of coef are the linear combinations of the regression coefficients, so that coef represents the collection of predictions for possible sets of predictor values.

Under our belief specifications above, there are no arcs between any of the observation nodes O_i andO_j. Each O_i is joined to coef.

In our knowledge map, we label each arc between O_i and coef by two numbers. The first number expresses the influence of Y_i on coef and is the trace, s_i, of the belief transform over $[\text{coef}/O_{(i)}]$ induced by $[O_i/O_{(i)}]$, where $O_{(i)}$ is the belief structure generated by all of the observations except Y_i. Each s_i can be interpreted as the expected maximal change in prevision for an element of coef, i.e. the maximal change over the standardized collection of predictions, caused by observing Y_i after observing the remaining observations. The maximal value of each s_i is one. The second number, l_i, is the actual maximal change in prevision over coef when the observation Y_i is added last.

The undirected graph is as given in Figure 7.3. In Figure 7.3, O_3 is by far the most influential point. Fitting Y_3 leads to a moderate change in maximal prevision, compared to our expectations. In a more complicated diagram we would probably suppress explicit labels, but convey influence by differential colour coding of the arcs or, as here, by treating influence as a pseudo-metric. We might place 'warning' marks on arcs for which the diagnostic ratio $d_i = l_i/s_i$ was larger than some fixed amount. (In Figure 7.3, the maximum value of d_i is 2.3, so there are no particular diagnostic warnings.)

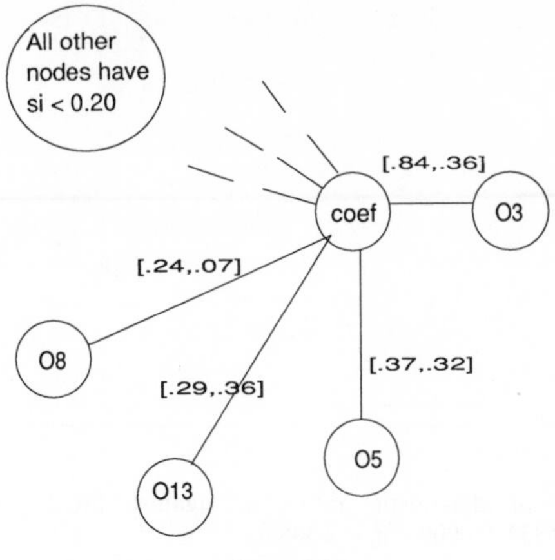

Figure 7.3

In the regression model that we are analysing in this example, the diagnostic ratios d_i may be thought of as somewhat analogous to the standardized 'fit' of the data points to the model. To illustrate this similarity, we evaluate, for each i, the 'residual' value r_i, which is the adjusted prevision for the ith error term e_i in the above regression model. (This adjusted prevision is evaluated based upon all of the data.) We now plot the pairs of values $(d_i, |r_i|)$, with plot as in Figure 7.4.

In Figure 7.4, d_i is monotonically related to $|r_i|$. Another obvious plot is the diagnostic ratio against the 'fitted' value $f_i = y_i - r_i$, which plot is as given in Figure 7.5.

Note in particular, that Y_3 has a very large fitted value, and a comparatively modest diagnostic ratio, as is often the case for very influential values.

To clarify Figure 7.3, we examine the eigenstructure of the transform

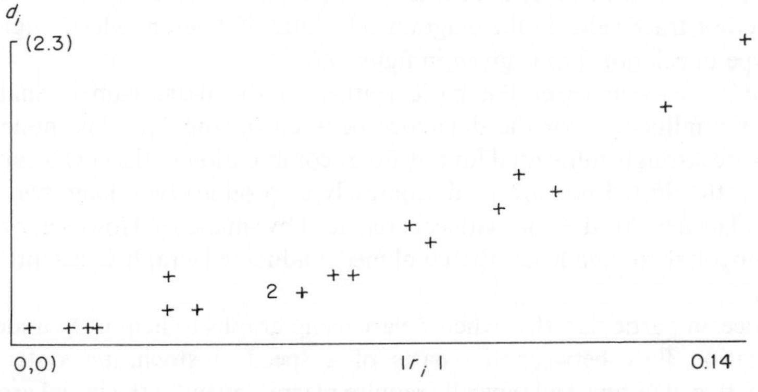

Figure 7.4 Plot of diagnostic ratios d_i against absolute adjusted residuals $|r_i|$.

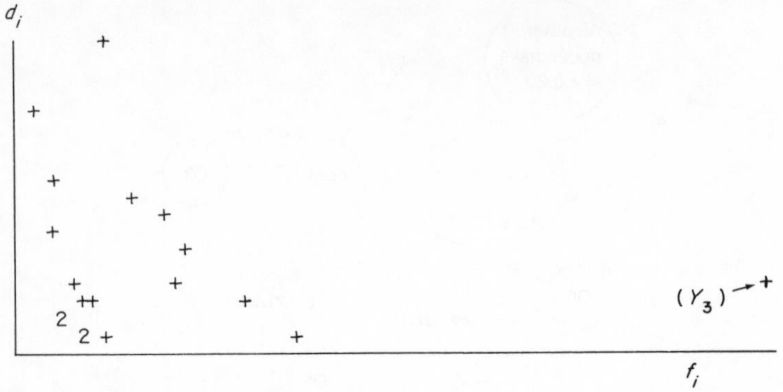

Figure 7.5 Plot of diagnostic ratios d_i against fitted values f_i (scale: $0.2980 < f_i < 0.5334$; $0.0000 < d_i < 2.3053$).

associated with the influential value, i.e. the transform over $[\text{coef}/O_{(3)}]$ induced by $[O_3/O_{(3)}]$. The eigenvector of this transform is almost orthogonal to the eigenvectors corresponding to the transforms for each remaining point. The eigenvector for point 3 is essentially the standardized difference between coefficients b_1 and b_3, namely $b13$ diff $= 10b_1 - 0.05b_3$. (The coefficients are the inverses of the standard deviations of b_1 and b_3.) We split $b13$ diff from coef, leaving the reduced coef with base b_0, b_2, $b13$ sum $(= 10b_1 + 0.05b_3)$. Call the structure with this reduced base 'smallcoef.'

The influence diagram in which coef is replaced by $b13$ diff has a single 'influential' arc, as the trace of the adjustment transform over $[b13\text{diff}/O_{(3)}]$ induced by $[O_3/O_{(3)}]$ is 0.82, while the largest trace for any of the other transforms, $[b13\text{diff}/O_{(i)}]$ induced by $[O_i/O_{(i)}]$, is 0.07.

The influence diagram in which coef is replaced by smallcoef has no single 'influential' arc. The trace of the adjustment transform for each point varies between a minimum value of 0.05, for point 10, and a maximum value of 0.37 for point 5. The previously influential point, Y_3, has trace 0.17, which is around the average trace value in the diagram. The kind of diagram which might show this type of relationship is given in figure 7.6.

Figure 7.6 summarizes the basic feature of the data, namely that Y_3 is extremely influential for the difference between b_1 and b_3, while none of the points are strongly influential for any other combination of the coefficients. (The items in the dotted box are used separately to produce two diagrams, one for coef replaced by $b13$diff, one with coef replaced by smallcoef. However, as $b13$diff is orthogonal to smallcoef, the combined undirected graph is essentially the same.)

Notice, in particular, that when we are using graphs to help us to understand information flow between the parts of a specified stochastic system, then informative diagrams will typically require rearrangment of the initial groupings

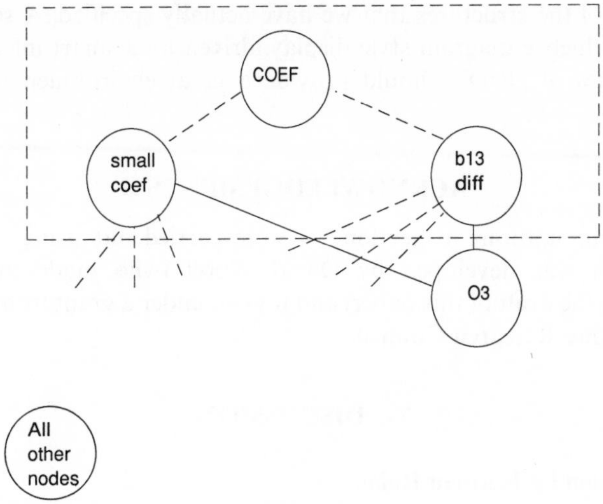

Figure 7.6

and computation of various interesting new quantities. Further, it will usually be essential to distinguish strengths of relationships in our diagrams.

These calculations may be fully automated. All of the computations for the above diagrams were written as short macros in the program [*B/D*]. The macros (apart from the special macro to declare and organize the regression beliefs/data) are fully general.

In Section 7.1 we briefly addressed the advantages of partial belief specifications. The above types of calculation illustrate one of the great benefits of such limited specifications, namely that we can subject our analysis to much more detailed scrutiny than is typically the case for fully specified structures. In principle, the calculations that we have described may be applied in identical fashion to any fully specified probabilistic structure. However, in practice, precisely because the analysis is so searching, it will usually uncover many features and discrepancies which are due to the artificial simplifications of the model, so that it becomes difficult to distinguish between real effects and simple artefacts of the analysis. One way to avoid such analytical overkill is to run an analysis of the type we have discussed purely on certain substructures, extracted from the fully specified structure, which are intended to form the substantive core of the specification.

7.5 CONCLUDING COMMENTS

Problem-solving is an iterative process of heuristics, specificaton, analysis, further heuristics, revised specification and so forth. Just as knowledge maps may help us to structure problems initially, they may also help us to explore the implications of the structures that we have actually specified. A sophisticated, interactive influence diagram style display, driven by a smart inference engine along the lines of [B/D], should provide a great environment for problem analysis.

ACKNOWLEDGEMENTS

All of the computations in Section 7.4 were carried out using the program [B/D], which was developed by David Wooff (who made many helpful comments on the draft of this paper) and myself, under a grant from the Science and Engineering Research Council.

7.6 DISCUSSION

7.6.1 Discussion by Herman Rubin

I believe that this chapter has much formalism, but limited usefulness. He starts out with De Finetti's use of expectation (prevision) as the fundamental property rather than probability; while that this can be done is obvious, what is gained? It is much easier to update probabilities rather than previsions. However, he immediately jumps to the inner product as the thing to be looked at. Having spoken Matrix for almost 45 years, I find the terminology quite understandable. However, neither inner products nor similar constructs as characteristic functions provide any insight into the probability questions involved. Furthermore, he compounds the obfuscation by the Humpty Dumptyism of defining a 'belief structure' to be a covariance structure. Beliefs are things to be operated on by Bayes' theorem, not by projection operators.

The preoccupation with linear models and second-order effects is all too common in statistics. Just as the preoccupation with conjugate priors causes rash Bayesian behaviour when the prior matters, the preoccupation with linear models causes one to restrict attention to second-order properties. I find that in few situations are linear models appropriate; and that while one may profitably use linear models in some of those situations, the non-linearity usually must be considered. Of course, if there are not enough data to come to reasonable conclusions (a failing of far too many studies), there may not be any important effect. However, usually non-linearity shows its head long before too much else does.

Another point to which I object is the cavalier use of exchangeability. In the examples given, I would not consider it reasonable. The author mentioned in the discussion the unreasonable agonies that some test designers go through to make the questions of comparable difficulty. I agree with him that this is undesirable. However, I doubt that they ever attempt to make pairs of questions of comparable joint difficulty.

It may be that some of the procedures mentioned in the chapter are of use in the field of linear statistical inference. Not enough information is given to verify this. The connection with influence diagrams seems to lie mainly in the correspondence between correlation and dependence.

7.6.2 Reply

It is unfortunate that Professor Rubin was so distressed by the divergences between my approach and the formal Bayes analysis that it prevented him from addressing any of the actual content of the chapter, as many of the issues are common to both approaches. However, as the discussion emphasizes the differences between the conventional Bayesian approach and a more flexible subjectivist approach, perhaps we should consider how you might tackle the examples in the chapter within the standard Bayes paradigm.

Start with the example on test scores. Suppose that you have 30 questions in each battery of tests, each scored as an integer from 1 to 25. You also have, say, 20 relevant profile variables, some continuous and some discrete. Suppose that you work strictly within the language of probability. You must describe your beliefs about the collection of 80 variables, by specifying an 80-dimensional probability distribution.

Professor Rubin objects to the 'cavalier use of exchangeability' in specifying beliefs over this collection, even when dealing with a purely hypothetical illustrative example in which we have split the tests into two non-exchangeable groups, and matched test responses across labelled pairs. Thus, in any real example, you had best suppose that each pair of questions will be considered separately and each triple of questions and each group of four and so forth. Somehow you must place a joint probability distribution on the 60 test scores. You must extend this distribution to include the 20 profile quantities, so, in addition to a 20-dimensional probability distribution over the profile quantities, you must specify, for each possible collection of 20 profile scores, the conditional joint distribution of the 60 test scores. Your aim is to predict two 'target' quantities. Therefore, for every possible combination of values of the 80 'predictors', you must specify a full two-dimensional distribution over the target variables.

Do not ask me how you could really make all of these specifications. Any individual probability judgement about the system can be a given a meaningful value after careful reflection. The problem lies entirely in giving careful consideration to billions of probability judgements! (There are 25^{60} numbers required to specify a probability distribution over the test questions alone.)

Let us suppose that somehow you manage to write down a meaningful probability distribution over all of the quantities above. You have barely scratched the surface. As information comes in, you must update your beliefs. Here, the simplest form of information is to update beliefs from data as you observe test, profile and target values for candidates. In the probability framework, you must consider every possible joint probability distribution that you could possibly construct over the 80-odd variables, and somehow place a probability distribution over all these possible probability distributions. (Again, do not ask me how—you might like to evaluate the order of magnitude of the number of individual probability judgements that are involved.)

Actually, it is far worse than this. You can only get away with specifying a single probability distribution over all 80 dimensional probability distributions under the assumption that all candidates are exchangeable. However, candidates are even less exchangeable than are test questions. I am not sure whether, even in principle, you can handle this specification problem within a strict probability framework, as probabilistic specification requires a full categorization of possibilities a priori. However, let us suppose that you can somehow list out all of the nuances which might in principle distinguish the individual candidates and incorporate these as, say, another 20 dimensions tacked on to this monster probability specification. Now take a further step back, and consider any system within which this type of problem might be analysed. For example, you might wish to design a better test battery within an improved evaluation system. Now all of the probability evaluations that we detailed above sit in a small corner of the large structure of actual and potential tests, political and economic uncertainties constraining the choice and so forth.

Where do you stop? At which level of complexity do you draw the line and admit that, even in principle, it is ridiculous to pretend that you can make all of the probability specifications which are required to describe the system? What do you do when you reach that level of complexity?

If we are being completely honest, most of us will probably admit that, except for very special problems, this level of complexity is pretty low—at around three or four dimensions, we are usually struggling. To describe a probability distribution over all of the probability distributions over 80-odd variables as representing anybody's 'beliefs' seems so bizarre as to defy serious consideration.

Of course, provided that you gloss over the difference between making a single probability judgement and making a billion such judgements, and put some arbitrary probability structure in place, then everything becomes strictly a matter of technique. In this purely formal sense it is easy to update probabilities. However, updates to largely arbitrary probability specifications are themselves largely arbitrary, which is why, in practice, probability systems are very hard to update. Somehow, you must ensure that the genuine insights which formed the basis for the specification are used in a sensible fashion and are not swamped by all of the essentially arbitrary features of the specification. I do not know any reliable way to do this, and I do not think any one else does (let alone how you could convince a sceptical outsider of the validity of your analysis).

Is there any compelling reason why you must build such gigantic structures? Not that I know of. People do manage to make inferences for large systems, because almost all of the probability structure is irrelevant.

Is there an alternative formulation that does not force us to embed meaningful judgements within such an artificial framework? An alternative system must involve less specifications than the probabilistic approach (which is why approaches which are based on 'uncertain', 'fuzzy', 'interval' probabilities, and so forth, for all their intrinsic interest, may not help us fundamentally in handling complexity).

This is why expectation is useful as a basic quantity. You can make exactly that collection of expectation statements about a problem which seem to you to be both directly relevant and also within your capability (remembering always that just because you can make any single expectation statement, this does not mean that you can make millions of such statements).

I have never understood why many people who are prepared to accept probability as basic find it so hard to treat expectation as basic. What they appear to be saying is that if they have to specify the expectation of every single function of a collection of random quantities (i.e. to specify the whole joint probability distribution) then they can do this. However, if they have to specify the expectation of only a small number of functions of the random quantities then this is beyond their ability.

Here are some methods that you might combine to specify a collection of expectations: symmetry considerations; prior experimental results in which sample means, correlations and standard errors were quoted; evaluation of covariances for a variety of plausible prior specifications—choosing values somewhere in the middle of your range of evaluations. (Note that if the expectation is the quantity of interest you can specify it more thoughtfully as a primitive than by forcing it to be the evaluation for a single prior measure.)

Any collection of expectation statements has certain implications. Increasing the collection of expectation statements increases the number of implications, but often a small collection of implication statements will suffice.

For example, in the regression problem discussed in the chapter, a small number of variance and convariance statements are sufficient to suggest that the data do not support belief in a regression model, and to identify the single point which has produced an apparent regression effect. (A strict 'subjectivist' analysis would be concerned with conditional expectations for the observable Y given the X values. The analysis is essentially the same, and reaches similar conclusions without involving hidden parameters, but requires about a page of explanatory notation.)

I chose this example partly for simplicity, because influence and residuals for regression are sufficiently well known that readers can easily understand why the case that is identified should be so influential, and how the 'residual-type' diagnostics should be interpreted. However, any Bayes-type analysis will profit from a similar influence/residual style analysis.

This analysis may be adequate for the problem. If the person with the data believed in the regression simply because they had found a least-squares regression fit, then it only needs a small number of uncertainty specifications to convince them otherwise. You do not need to go through the charade of pretending to place a probability measure over all probability measures on the four random quantities, and then struggling to somehow duplicate the results of the simple analysis within the extended structure.

But suppose that you really do have a sensible full Bayes analysis to carry out.

You have a meaningful likelihood about whose (meaningful!) parameters the scientist has detailed beliefs which can be sensibly cast in probabilistic form. Then go ahead and make all of the specifications—the theory and analysis will be exactly the same as I have described. The probability specification defines expectations over the class of all indicator functions. Bayes analysis is simply the projection into this linear space. Any Bayes analysis is carried out by projection operators—what is at issue is simply whether the projection should be restricted to depend on your actual specifications, or whether it should mostly depend upon arbitrary additional inputs.

Surely this should be a matter of judgement not of dogma. If you can find a satisfactory probability formulation, then use it. But otherwise, you have a choice, and the expectation-based system will typically be more flexible. For example, if you dislike having test questions exchangeable in my example, then it is much easier to modify the appropriate variance structures than to modify all of the full probability measures over probability measures. Include whatever specifications you feel are relevant—the raw quantities, polynomials on products of the quantities, indicator functions for important ranges of combinations of the variables—build small or large spaces as the situation requires. Just do not feel uniquely constrained to work in the maximal linear space which it is possible to imagine for the problem.

Professor Rubin summarizes the orthodox Bayes position very succinctly, when he writes that, for him, 'beliefs are things to be operated on by Bayes' theorem,...'. I would suggest that the term 'beliefs' should be reserved for evaluations in which you actually believe, whether expressed as probabilities, expectations or in some more fundamental form. To define beliefs as things which are handled by a particular, rigid methodology has the paradoxical effect of ensuring that the methodology hardly ever applies to beliefs at all.

Some of the practical advantages of working with expectations in influence diagram style analysis are discussed in the chapter. Because probabilities are expectations, all of the structure and methods that I describe in the chapter are available in just the same way to analyse the usual probabilistic influence diagrams. However, because the theory applies equally to what, in my opinion, is the more natural and general formulation, I described it in full generality instead. (The theory used in the chapter has, of necessity, been covered only briefly. For both technical details, and a proper description of the subjectivist rationale underlying the analysis, refer to the papers quoted in the references.) If you want to impose the restriction that all variables are indicators defined over enormous product partitions, the theory will be just the same (with a few technicalities needed for bookkeeping over infinite collections of apparent belief statements).

The issues will be unchanged. First, pictures only communicate if they are understandable. A picture which runs over several pages of similar seeming boxes, whose information is carried by the arcs which are not there somehow does not seem very communicative. We need to summarize the diagrams down into their essential ingredients, by grouping similar quantities into large boxes, and, where possible, introducing new quantities which express all of the influence. It is easy to draw complicated diagrams. The trick is to draw simple diagrams, and this trick can, to a large extent, be automated, as explained in the chapter.

Having summarized the relationships which a diagram expresses, it seems interesting to distinguish the important from the inconsequential. The information in an influence diagram may be contained in the missing arcs, but the action is happening on the arcs which are there. (For example, the influence diagram for the regression has no interest, until we identify the one important arc.) In the chapter I suggest a natural way to quantify this influence. It is natural because it directly summarizes changes in belief and leads to a simple diagnostic system.

Diagnostics are crucial to the development. During the conference, I saw large diagrams, intended for control of complex processes, with little attention being paid to what seems the most important question of all, namely validating the diagram. Again, one of the reasons for choosing the regression example was to emphasize that diagnostics for general influence diagrams can be systematically tackled by methods that have much in common with those which are commonplace in evaluating statistical models.

I do not know whether the discussant objects, in principle, to the operations of organizing, summarizing, labelling and critically validating influence-style diagrams, or only to such a treatment within a subjectivist as opposed to a Bayesian framework. I would be very interested to see an intrinsically different approach to these operations within the strict Bayes paradigm, but my feeling is that while it is natural and straightforward to address such questions in an expectation-based system, in the probabilistic framework the issues seem much harder to formulate.

REFERENCES

De Finetti, B. (1974) *Theory of Probability*, vol. 1, Wiley, Chichester.

Goldstein, M. (1981) Revising previsions: a geometric interpretation, *J. Roy. Statist. Soc.*, **B43**, 105–30.

Goldstein, M. (1983) The prevision of a prevision, *J. Amer. Statist. Assoc.*, **78**, 817–19.

Goldstein, M. (1985) Temporal coherence. In J. H. Bernardo *et al.* (eds) *Bayesian Statistics*, **2**, Elsevier, Amsterdam.

Goldstein, M. (1986a) Separating beliefs. In P. K. Goel and A. Zellner (eds) *Bayesian Inference and Decision Techniques*, North-Holland, Amsterdam.

Goldstein, M. (1986b) Exchangeable belief structures, *J. Amer. Statist. Assoc.*, **81**, 971–6.

Goldstein, M. (1987a) Systematic analysis of limited belief specifications, *The Statistician*, **36**, 191–9.

Goldstein, M. (1987b) Can we build a subjectivist statistical package? In R. Viertl (ed.) *Probability and Bayesian Statistics*, Plenum Prss, New York.

Goldstein, M. (1987c) [*B*/*D*]: *Introduction and Overview*. (Technical report.)

Goldstein, M. (1988a) Adjusting belief structures. *J. Roy. Statist. Soc.*, **B50**, 133–154.

Goldstein, M. (1988b) The data trajectory. In J. H. Berndo *et al.* (eds) *Bayesian Statistics 3*, Oxford University Press, Oxford, pp 189–209.

Smith, J. Q. (1987) *Influence Diagrams for Statistical Modelling*, Research Report 117, Dept of Statistics, University of Warwick.

Weisberg, S. (1985) *Applied Linear Regression*, Edn, Wiley, New York.

Wooff, D.A. (1987) [*B*/*D*]: *Reference Manual.* (Technical report.)

PART III

Problems and Applications: Industrial

CHAPTER 8

Informative Sampling Methods: The Influence of Experimental Design on Decision

Richard E. Barlow, Telba Z. Irony and S. W. W. Shor, *University of California, USA*

ABSTRACT

Influence diagrams are used to clarify the role of the experimental design in data analysis. In particular, we analyze in some detail the records on frequency and duration of forced outages of fossil fuel electrical power plants. Data were collected in order to predict future downtimes of similar electrical power units. The method of data collection suggested that downtimes on some units were missed. When used in the data analysis, this information turned out to be quite useful. Influence diagrams were used to model our problem and were also of considerable aid at the data analysis stage.

8.1 INTRODUCTION

In recording or extracting information for statistical analysis, some rule or set of instructions must be employed either implicitly or explicitly in order to determine the recording procedure. By *sampling method* or *sampling rule*, we mean the rule that defines the information extraction procedure (the way by which the sample is collected and recorded).

In most decision problems, the method of sampling is irrelevant to the form of the likelihood function relative to the model parameters. Therefore it is usually accepted that the sampling method is itself irrelevant in the Bayesian approach.

This chapter makes use of influence diagrams to clarify the role of informative sampling methods in data analysis.

A sampling rule is said to be *non-informative* if the Bayesian or likelihood

Influence Diagrams, Belief Nets and Decision Analysis
Edited by R. M. Oliver and J. Q. Smith

inferences about the parameters of interest are the same whether or not the sampling probability is introduced explicitly into the statistical analysis. For an *informative* sampling rule, the inferences differ depending upon whether or not the sampling probability is introduced. In calculating the likelihood function, the sampling probability is usually assumed not to be a function of the unknown parameters of interest. In those cases, it can be absorbed into a proportionality constant and therefore, it is non-informative and irrelevant to the inference. On the other hand, if the sampling rule is a function of the unknown parameters, we must explore further to see whether the rule is informative or non-informative. In Bayesian terms, we will consider a prior distribution to determine if the sampling probability is reflected in the posterior distribution. This means that we will compare the posterior distribution in which explicit consideration of the sampling method is made with the posterior distribution that would be reached if the sampling method were ignored.

For further discussion of informative and non-informative sampling methods and stopping rules, see Raiffa and Schlaifer (1961) and Roberts (1967).

8.2 RECORD SEARCH PROCEDURE

This chapter deals with records on fossil fuel electrical power plants. Records relative to the frequency and duration of forced outages exceeding 60 days were searched and extracted from a huge data base. A forced outage is an outage which is not a planned maintenance one but is due to an unforeseen failure. The records were tabulated by quarter of a year and all outages exceeding 30 days in a quarter were extracted from the data base. If an outage exceeding 30 days was still in effect at the end of a quarter, the following quarter was searched to complete the record for that particular outage. If an outage exceeded 30 days at the start of a quarter, the previous quarter was searched to also complete the record for that particular outage. By following this procedure, we could be sure that no 60 days or greater outage was missed (see Figure 8.1).

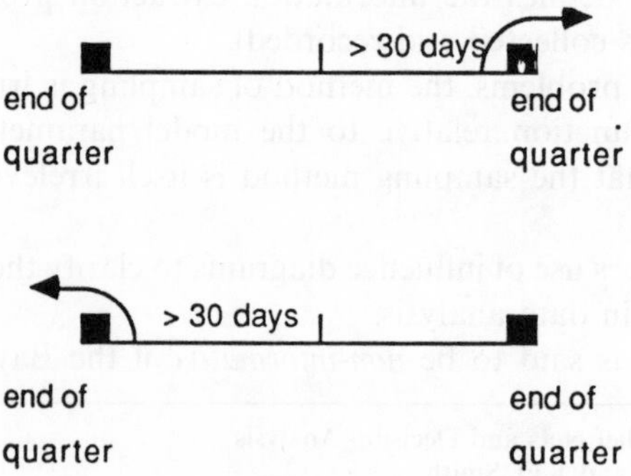

Figure 8.1

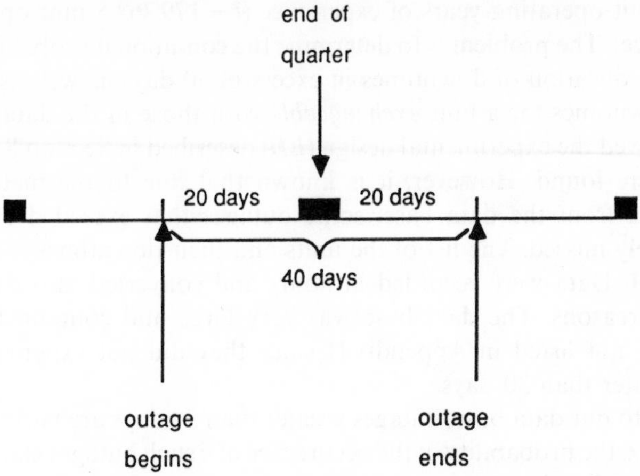

Figure 8.2 Example of a missed outage exceeding 30 days.

Relative to 60 days or greater outages, this particular sampling method, given the data, was non-informative with respect to the expected frequency and expected duration of such outages (the model parameters). All information was contained in the recorded data. None of the 60 days or greater outages were missed.

However, it subsequently became necessary to use the same extracted data to estimate the expected frequency (λ) and the expected duration (θ) of 30-day or greater forced outages. With respect to these outages, our search procedure, and consequently our sampling method, almost surely missed some such outages (see Figure 8.2).

The missed outages constitute an unobserved random quantity whose distribution depends on both the experimental design and the unknown parameters of interest. Given the observed 30-day or greater outages and the knowledge that some could have been missed, based upon the way data were collected, the sampling rule may be informative with respect to the unknown model parameters.

In the following we compare the model that takes into account the sampling method with the one that does not. In other words we try to find out if in this case the experimental design is informative or not. Here we are interested in the parameter (θ) that is related to the duration of the forced outages exceeding 30 days.

8.3 CONDITIONAL PROBABILITY DISTRIBUTIONS

The database consisted of quarterly records for forced outages of fossil fuel electrical power units recorded during $p = 1613$ days. This period covered

$t = 492.5$ unit operating years of experience ($t = 179\,762.5$ unit operating days of experience). The problem is to determine the conditional probability distribution for the duration of downtimes in excess of 30 days as well as the number of such downtimes for a unit *exchangeable* with those in the data base.

We followed the experimental design (*ED*) described in Section 8.2 and $k = 72$ outages were found. However, it is known that due to the method used to extract data from the data base, some outages that exceeded 30 days were almost surely missed. The list of the units and their downtimes is contained in Appendix II. Data were recorded in hours and converted into days for computational reasons. The data base was very large and contained records for many units not listed in Appendix II since they did not experience a forced outage greater than 30 days.

Relative to our data base, outages greater than 30 days are fairly rare. Hence, given t and λ, the probability of the occurrence of n such outages can be described by a Poisson distribution, i.e.

$$P[N(t) = n \mid t, \lambda] = \frac{(\lambda t)^n \exp(-\lambda t)}{n!}$$

where λ is the expected number of such outages per unit day. We assume that any exchangeable collection of additional units will have this same unknown rate.

Suppose that n downtimes exceeding 30 days had actually occurred. Then, the probability of recording k such downtimes given n, θ, and the experimental design (*ED*), will be

$$p(k \mid n, \theta, ED) = \binom{n}{k} C(\theta)^k (1 - C(\theta))^{n-k}$$

where $C(\theta)$ is the probability that an occurred excess over 30 days will not be missed.

Let y denote the excess over 30 days of a forced outage greater than 30 days. Then, $f(y \mid \theta)$ will be the conditional density of y given that the corresponding forced outage has exceeded 30 days. In this case, we have assumed (see Appendix III) that the distribution of y given θ is exponential with mean θ, i.e.

$$f(y \mid \theta) = \frac{1}{\theta} \exp\left(\frac{-y}{\theta}\right).$$

Here θ is the mean of the excess (over 30 days) of downtimes exceeding 30 days.

Under this assumption, it can be shown (see Appendix I) that

$$C(\theta) = 1 - \frac{m\Delta}{p} + (1 - \exp(-\Delta/\theta))\frac{m\theta}{p}$$

where $\Delta = 30$ days, $m = 17$ and $p = 1613$ days.

If $\mathbf{y}$ is the vector of observed excesses over 30 days then, given that we have observed k downtimes that exceeded 30 days, θ and ED, the density of $\mathbf{y}$ will be

$$p(\mathbf{y}|k, \theta, ED) = [\theta \cdot C(\theta)]^{-k} \exp\left(-\frac{T}{\theta}\right) \prod_{i=1}^{k} \left[1 - \frac{m(\Delta - y_i)^+}{p}\right]$$

where $T = \sum_1^k y_i$ (sum of recorded excesses).

This probability is derived in Appendix I.

If the experimental design (ED) had not missed any 30-day or longer forced outages, then the density for recorded excesses over 30 days would be

$$p(\mathbf{y}|k, \theta) = (\theta)^{-k} \exp\left(-\frac{T}{\theta}\right)$$

and the experimental design would not have been informative for θ.

8.4 THE INFLUENCE OF THE EXPERIMENTAL DESIGN ON DECISION: INFLUENCE DIAGRAMS

A suitable way to study the influence of the experimental design on inference (decision) about the parameters of interest is via an influence diagram (see Howard and Matheson, 1981).

Figure 8.3 is an influence diagram corresponding to our problem when the fact that the sampling method influences the recorded data is not ignored.

Since the number of actual downtimes exceeding 30 days, n, depends on λ and the experimental design, ED (through t), there are arrows originating at those two nodes and ending at node n. The influence of the sampling method

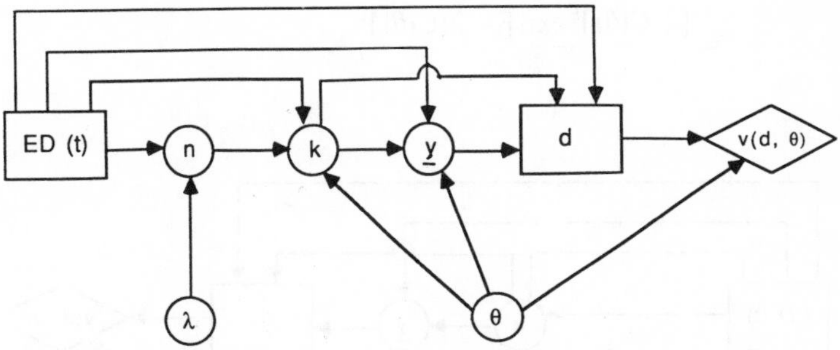

Figure 8.3 ED: experimental design; n: number of downtimes exceeding 30 days; t: 179 762.5 unit operating days; λ: expected number of outages exceeding 30 days per unit day; θ: mean of the excess of downtimes exceeding 30 days; k: number of recorded downtimes exceeding 30 days; $\mathbf{y}$: vector of recorded excesses of downtimes exceeding 30 days; d: decision (in this case, the estimate of θ); $v(d, \theta)$: value function; $C(\theta)$: probability that a downtime exceeding 30 days will not be missed.

on k and $\mathbf{y}$ is explicitly pointed out by the arrows originating at the ED node and ending at nodes k and $\mathbf{y}$. Note that, given n and ED, k and λ are conditionally independent (there is no arrow from λ to k). Here $\mathbf{y}$ also depends on k and θ and so there are arrows from these nodes to node $\mathbf{y}$; ED, k and $\mathbf{y}$ are known at the time of decision. Hence, arrows go from these nodes to the decision node. In this case, the decision will be an estimate of θ.

Associated with each circle node there is a conditional probability distribution of the random quantity represented by that node given the quantities in the adjacent predecessor nodes. In this instance we have assessed:

1. $p(\lambda)$ and $p(\theta)$—flat priors;

2. $p(n|\lambda, ED) = \dfrac{(\lambda t)^n \exp(-\lambda t)}{n!}$;

3. $p(k|n, \theta, ED) = \dbinom{n}{k} C(\theta)^k [1 - C(\theta)]^{n-k}$;

4. $p(\mathbf{y}|k, \theta, ED) = [\theta \cdot C(\theta)]^{-k} \exp\left(-\dfrac{T}{\theta}\right) \displaystyle\prod_{i=1}^{k} \left[1 - \dfrac{m(\Delta - y_i)^+}{p}\right]$.

The second step of the procedure will be the 'node elimination' operation. In this case we are interested in the elimination of the node n, since its value was not observed.

After node elimination, we have (Figure 8.4):

1. $p(\lambda)$ and $p(\theta)$—flat priors;

2. $$p(k|\lambda, ED, \theta) = \sum_{n=k}^{\infty} p(k|\lambda, ED, \theta, n) \cdot p(n|\lambda, \theta, ED)$$
$$= \frac{[\lambda t C(\theta)]^k \exp[-\lambda t C(\theta)]}{k!};$$

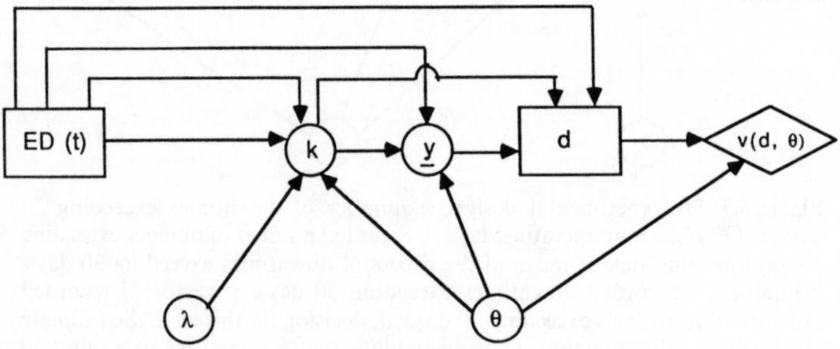

Figure 8.4

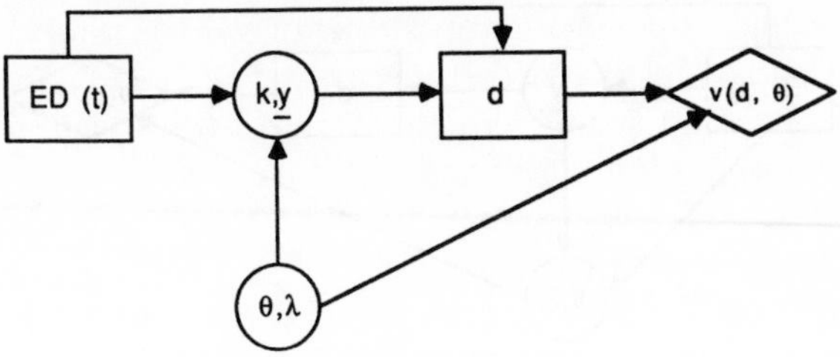

Figure 8.5

3. $p(\mathbf{y}|k,\theta,ED)=[\theta\cdot C(\theta)]^{-k}\exp\left(-\frac{T}{\theta}\right)\prod_{i=1}^{k}\left[1-\frac{m(\Delta-y_i)^+}{p}\right].$

The third and fourth step will be to merge nodes θ and λ and nodes k and $\mathbf{y}$. In our opinion, λ and θ are *a priori* independent because we are dealing with a huge database and, therefore, $p(\theta,\lambda)=p(\theta)\cdot p(\lambda)$. On the other hand,

$$p(k,\mathbf{y}|\lambda,\theta,ED)=p(k|\lambda,\theta,ED)\cdot p(\mathbf{y}|k,\lambda,\theta,ED).$$

The next influence diagram is shown in Figure 8.5.

1. $p(\lambda,\theta)$—flat prior (product of flat priors);
2. $p(k,\mathbf{y}|\lambda,\theta,ED)\propto\left(\frac{\lambda}{\theta}\right)^k\exp\left[-\lambda tC(\theta)-\frac{T}{\theta}\right]$

$$\propto L(\theta,\lambda|k,\mathbf{y},ED)$$
$$=L(\theta,\lambda|k,T,ED)$$

where $T=\sum_1^k y_i$.

Now, we use Bayes' formula to compute the posterior distribution for λ and θ given k, T, and the experimental design. This procedure is equivalent to the operation of reversing the arc from the node λ,θ to the node $k,\mathbf{y}$ (see Figure 8.6). By Bayes' formula:

$$p(\lambda,\theta|k,\mathbf{y},ED)\propto L(\lambda,\theta|k,T,ED)\cdot p(\lambda,\theta).$$

Here, $p(\lambda,\theta)$ is the joint prior distribution for λ and θ. Since, in our opinion, θ and λ are independent *a priori* and their prior distributions are flat, their posterior distribution is proportional to the likelihood, that is,

$$p(\lambda,\theta|k,T,ED)\propto L(\lambda,\theta|k,T,ED)$$
$$\propto\left(\frac{\lambda}{\theta}\right)^k\exp\left[-\lambda tC(\theta)-\frac{T}{\theta}\right].$$

After arc reversal, the influence diagram becomes as shown in Figure 8.6.

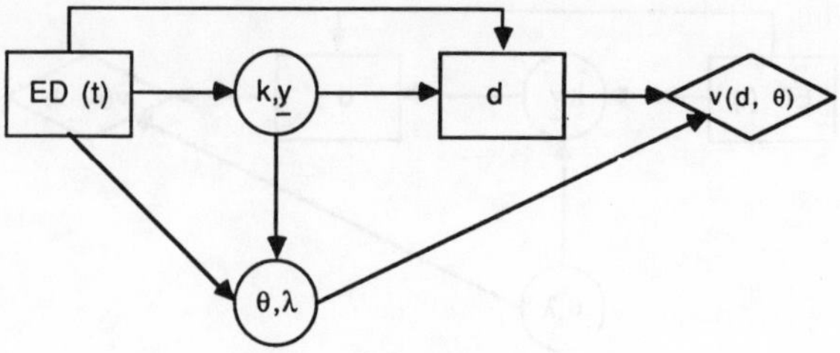

Figure 8.6

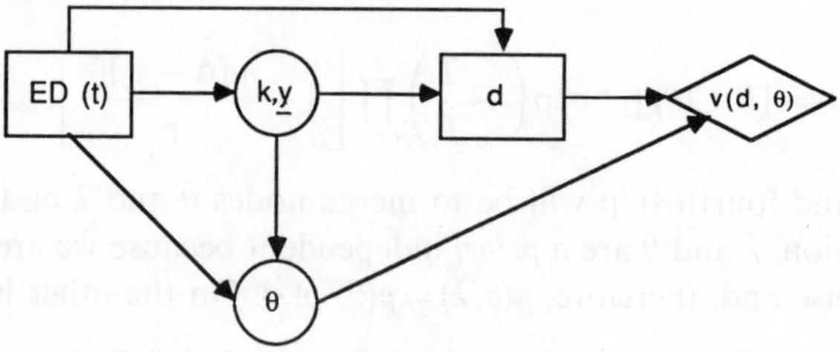

Figure 8.7

The next operation is the elimination of λ, the nuisance parameter. To do this, we integrate the joint posterior distribution of λ and θ in λ, that is,

$$p(\theta|k, T, ED) = \int_0^\infty p(\lambda, \theta|k, T, ED)\,d\lambda$$

i.e.

$$p(\theta|k, T, ED) \propto \frac{\theta^{-k}\exp(-T/\theta)}{C(\theta)^{k+1}}.$$

The last influence diagram is as shown in Figure 8.7.

Finally, we should choose the decision d that optimizes the value $v(d, \theta)$:

$$\text{If } v(d, \theta) = \begin{cases} 1 \text{ if } |d-\theta| < \varepsilon \text{ where } \varepsilon > 0 \text{ is arbitrarily small;} \\ 0 \text{ otherwise} \end{cases}$$

then the optimal decision is the posterior mode.

The analysis is completed.

If the experimental design, given the data, were non-informative, we would start with the influence diagram shown in Figure 8.8.

1. $p(\lambda)$ and $p(\theta)$—flat priors;

2. $p(k|\lambda) = \dfrac{(\lambda t)^k e^{-\lambda t}}{k!}$;

3. $p(\mathbf{y}|k, \theta) = \theta^{-k} e^{-T/\theta}$.

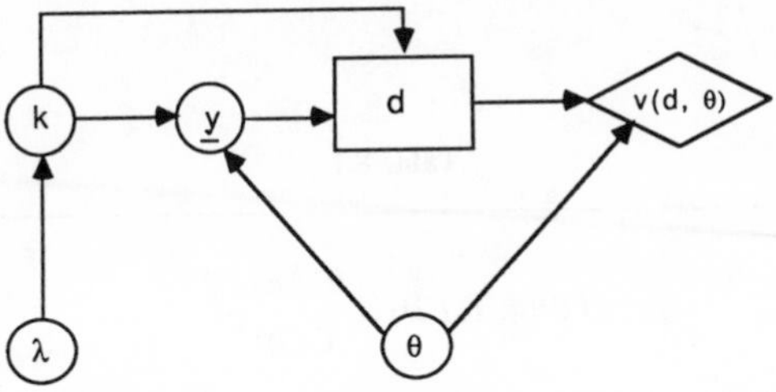

Figure 8.8

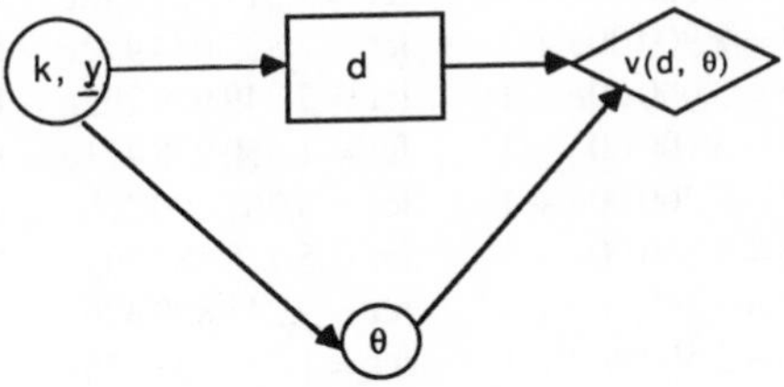

Figure 8.9

Going through the analysis of this influence diagram, it can be easily verified that, if λ and θ were judged independent *a priori*, they would remain independent *a posteriori*. In other words, the joint posterior distribution of λ and θ would be

$$p(\lambda, \theta | k, T) \propto L(\lambda, \theta | k, T) \propto \lambda^k e^{-\lambda t} \theta^{-k} e^{-T/\theta}.$$

Our problem is to make inferences about θ. Therefore, λ can be regarded as a nuisance parameter. In this case, $L(\lambda | k, T)$ can be absorbed into a proportionality constant and the last influence diagram is as shown in Figure 8.9. Here $p(\theta | k, T) \propto \theta^{-k} e^{-T/\theta}$ and the posterior mode will be $\theta^* = T/k$.

8.5 NUMERICAL RESULTS

In this section we compare the posterior distribution in which explicit consideration of the sampling probability is made with the posterior distribution that would be reached if the sampling rule were ignored.

Both priors for λ and θ are judged flat. Therefore, for our purposes, it suffices to compare the integrated likelihood that takes into account the sampling rule (*ED*) with the likelihood that does not

$$L(\theta | k, T, ED) \propto \frac{\theta^{-k} e^{-T/\theta}}{C(\theta)^{k+1}}$$

where

$$C(\theta) = 1 - \frac{m\Delta}{p} + [1 - \exp(-\Delta/\theta)] \frac{m\theta}{p}$$

Table 8.1

$$L(\theta|k, T, ED) \propto \frac{\theta^{-k} e^{-T/\theta}}{C(\theta)^{k+1}}$$

$$((\theta^{(-72)})*(e^{(-3157/\theta)}))/((0.6838 + 0.010539*\theta*(1 - e^{(-30/\theta)}))^{73})$$

at $\theta = 2.80000e+1$	fct $= 9.319925798e-150$
at $\theta = 2.90000e+1$	fct $= 2.837044442e-149$
at $\theta = 3.00000e+1$	fct $= 7.349909292e-149$
at $\theta = 3.10000e+1$	fct $= 1.650978421e-148$
at $\theta = 3.20000e+1$	fct $= 3.267600268e-148$
at $\theta = 3.30000e+1$	fct $= 5.778387599e-148$
at $\theta = 3.40000e+1$	fct $= 9.241809426e-148$
at $\theta = 3.50000e+1$	fct $= 1.351138776e-147$
at $\theta = 3.60000e+1$	fct $= 1.822595588e-147$
at $\theta = 3.70000e+1$	fct $= 2.287170088e-147$
at $\theta = 3.80000e+1$	fct $= 2.689539534e-147$
at $\theta = 3.90000e+1$	fct $= 2.982786467e-147$
at $\theta = 4.00000e+1$	fct $= 3.137707151e-147$
at $\theta = 4.10000e+1$	fct $= 3.146721600e-147$
at $\theta = 4.20000e+1$	fct $= 3.022264843e-147$
at $\theta = 4.30000e+1$	fct $= 2.791258660e-147$
at $\theta = 4.40000e+1$	fct $= 2.487962715e-147$
at $\theta = 4.50000e+1$	fct $= 2.147268072e-147$
at $\theta = 4.60000e+1$	fct $= 1.799732704e-147$
at $\theta = 4.70000e+1$	fct $= 1.468805971e-147$
at $\theta = 4.80000e+1$	fct $= 1.170037030e-147$
at $\theta = 4.90000e+1$	fct $= 9.117141043e-148$
at $\theta = 5.00000e+1$	fct $= 6.963034901e-148$
at $\theta = 5.10000e+1$	fct $= 5.221526357e-148$
at $\theta = 5.20000e+1$	fct $= 3.850907201e-148$
at $\theta = 5.30000e+1$	fct $= 2.797304602e-148$
at $\theta = 5.40000e+1$	fct $= 2.004080897e-148$
at $\theta = 5.50000e+1$	fct $= 1.417836177e-148$
at $\theta = 5.60000e+1$	fct $= 9.916596039e-149$
at $\theta = 5.70000e+1$	fct $= 6.863944805e-149$
at $\theta = 5.80000e+1$	fct $= 4.706203671e-149$
at $\theta = 5.90000e+1$	fct $= 3.199119870e-149$
at $\theta = 6.00000e+1$	fct $= 2.157741011e-149$

Table 8.2

$L(\theta|k, T) \propto \theta^{-k} e^{-T/\theta}$
$((\theta^{(-72)}))*(e^{(-3157/\theta)})$

at $\theta = 2.800\,00e+1$	fct $= 6.885\,237\,236e-154$
at $\theta = 3.000\,00e+1$	fct $= 8.810\,766\,284e-153$
at $\theta = 3.200\,00e+1$	fct $= 6.072\,067\,360e-152$
at $\theta = 3.400\,00e+1$	fct $= 2.558\,574\,054e-151$
at $\theta = 3.600\,00e+1$	fct $= 7.261\,170\,406e-151$
at $\theta = 3.800\,00e+1$	fct $= 1.495\,728\,053e-150$
at $\theta = 4.000\,00e+1$	fct $= 2.371\,311\,679e-150$
at $\theta = 4.200\,00e+1$	fct $= 3.030\,976\,261e-150$
at $\theta = 4.400\,00e+1$	fct $= 3.241\,740\,273e-150$
at $\theta = 4.600\,00e+1$	fct $= 2.989\,626\,967e-150$
at $\theta = 4.800\,00e+1$	fct $= 2.436\,244\,783e-150$
at $\theta = 5.000\,00e+1$	fct $= 1.789\,806\,580e-150$
at $\theta = 5.200\,00e+1$	fct $= 1.205\,238\,834e-150$
at $\theta = 5.400\,00e+1$	fct $= 7.542\,356\,954e-151$
at $\theta = 5.600\,00e+1$	fct $= 4.437\,216\,694e-151$
at $\theta = 5.800\,00e+1$	fct $= 2.477\,915\,791e-151$
at $\theta = 6.000\,00e+1$	fct $= 1.324\,286\,595e-151$

and

$$L(\theta|k, T) \propto \theta^{-k} e^{-T/\theta}.$$

The following values were extracted from the data in Appendix II: $k = 72$ outages, $T = 3157$ days, $m = 17$ quarters, $p = 1613$ days, $\Delta = 30$ days.

Both integrated likelihoods are tabulated (see Tables 8.1 and 8.2). The values of θ are given in days/outage (the range goes from 28 days/outage to 60 days/outage).

The integrated likelihood that does not take into account the experimental design overestimates θ, as should be expected. This happens because it does not consider the outages that could have been missed. The smaller the outage, the greater the probability that it could be missed. Therefore, the second likelihood almost surely did not consider some outages whose duration was short and hence, overestimated θ.

The posterior modes are:

1. Considering the experimental design in the likelihood: $\theta^* = 41$ days/outage;
2. Not considering the experimental design in the likelihood: $\theta^* = 43.85$ days/outage;
3. Percentage of change in the posterior mode: 7 percent.

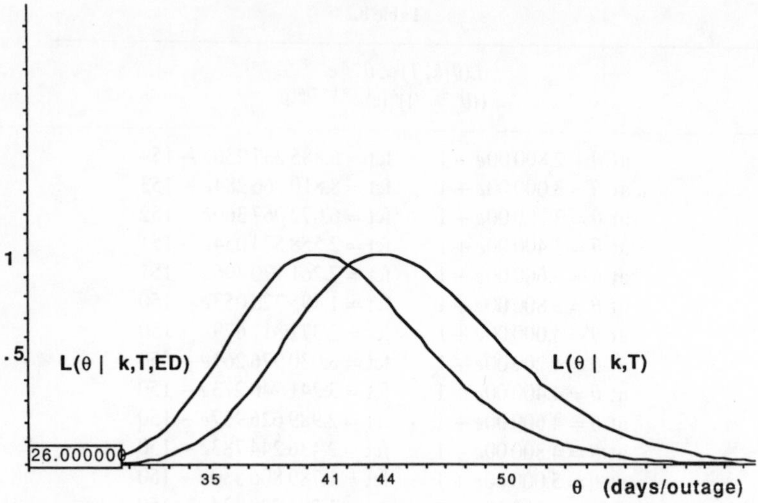

Figure 8.10

Above, both likelihoods are plotted. To avoid computational problems, we divided each likelihood by its maximum value. This procedure makes the maximum value of the integrated likelihood equal to 1 but does not change its shape. Therefore, both functions can be compared in the same graph (see Figure 8.10).

As may be easily seen, the integrated likelihood that considers the experimental design gives more weight to smaller values of θ. In this instance, the sampling rule was indeed informative.

8.6 DISCUSSION

8.6.1 Discussion by Kjell A. Doksum

The Barlow, Irony and Shor (BIS) chapter presents an interesting analysis of a study where the likelihood and statistical inference is influenced by the experimental design and in particular by a certain missing data mechanism. They use influence diagrams to help explain how the design influences the statistical inference. Their chapter inspired me to consider an extension and a variation of their setup. First we give the extension.

BIS experiments with a predictor variable

If we are trying to estimate the expected excess life over 30 days of a power plant we may obtain a better estimate by incorporating information in a covariate such as $x =$ the age of the power plant. For the study described in the BIS chapter, the basic parameters with the covariate x incorporated are λ_x and θ_x, where for a power plant of age x, λ_x is the expected number of outages of at least 30 days and θ_x is the expected duration in excess over 30 days of outages of at least 30 days. For the ith power plant, we observe the age X_i, the number K_i of recorded outages of at least 30 days and the excess duration $Y_{i1}, \ldots, Y_{iK_i}$ over 30 days of these outages. For the ith power plant, we also have the unobservable $N_i =$ number of outages of at least 30 days.

We assume that, given age, the power plants are independent, and given that plant i has $K_i = k_i$ outages, the excess duration times $Y_{i1}, \ldots, Y_{ik_i}$ are independent. For the ith power plant, given $X_i = x_i$, we make the same distributional assumptions on N_i, K_i and $Y_{i1}, \ldots, Y_{ik_i}$ as in the BIS chapter. Let $T_i = \sum_{j=1}^{K_i} Y_{ij}$, then the likelihood is proportional to

$$\left[\prod_{i=1}^{r} \left(\frac{\lambda_{x_i}}{\theta_{x_i}}\right)^{K_i}\right] \exp\left\{-\sum_{i=1}^{r} [h_i \lambda_{x_i} c(\theta_{x_i}) - (T_i/\theta_{x_i})]\right\} \tag{8.1}$$

where r is the number of power plants in the study, h_i the number of operating hours of experience for plant i during the study, and $c(\theta_{x_i})$ the probability that for a power plant of age x_i, an occurred excess over 30 days will be detected; i.e. from the BIS chapter,

$$c(\theta_{x_i}) = 0.68 + [1 - \exp(-30/\theta_{x_i})]0.023\theta_{x_i}.$$

At this point, we consider two parametrizations

$$\lambda_{x_i} = \alpha_1 + \alpha_2 x_i, \qquad \theta_{x_i} = \beta_1 + \beta_2 x_i \tag{8.2}$$

and

$$\lambda_{x_i} = \exp\{\alpha_1 + \alpha_2 x_i\}, \qquad \theta_{x_i} = \exp\{\beta_1 + \beta_2 x_i\}. \tag{8.3}$$

The second parametrization has the advantage that λ_{x_i} and θ_{x_i} are positive for parameters $\alpha_1, \alpha_2, \beta_1, \beta_2$ in R, while for the first parametrization restrictions are needed on the range of the parameter values. The most appropriate parametrization, which may be different from equation (8.2) and (8.3), could be determined by plotting nonparametric estimates, such as nearest neighbor estimates, of λ_x and θ_x as functions of x.

Finally, if a parametrization such as (8.2) or (8.3) involving four parameters $\alpha_1, \alpha_2, \beta_1$ and β_2 is decided on, then the likelihood of $\alpha_1, \alpha_2, \beta_1, \beta_2$ can be obtained from equation (8.1). Now this likelihood can be used to produce estimates of the parameters which then in turn will yield an estimate of the expected excess θ_x over 30 days of the downtime of a power plant of age x.

Next a variation of the model in the BIS paper will be considered.

Stratification based on the dependent variable

It is pointed out in the BIS chapter that one of the key properties of the experimental design they consider is that smaller outages tend to be missed. This is similar to 'dependent variable stratified sampling' where the chance that a dependent variable is observed depends on its size. An analysis similar to the BIS analysis can be carried out for such studies. In particular we will consider the variable probability sampling scheme described by, among others, Rubin (1976) and Jewell (1985). In this scheme we have s strata for the dependent variable Y defined for given constants $K_1,\ldots,K_s$ by

$$S_1=\{y:y<K_1\},\quad S_2=\{y:k_1\leqslant Y<K_2\},\ldots,S_{s-1}=\{y:K_{s-1}\leqslant y<K_s\},\quad S_s=\{y:y\geqslant K_s\}.$$

The sampling is carried out such that $\Pr(Y\in S_i)=p_i$, where $p_1,\ldots,p_s$ are given, known numbers.

Given the value $(x_1,\ldots,x_p)=\mathbf{x}$ of the covariate vector $\mathbf{X}=(X_1,\ldots,X_p)$, Y satisfies the linear model

$$Y=\mathbf{x}\boldsymbol{\beta}+e \tag{8.4}$$

where $\boldsymbol{\beta}$ is a column vector of regression parameters and e are independent with continuous distribution function F_0. We observe independent vectors $(\mathbf{X}_1,Y_1),\ldots,(\mathbf{X}_n,Y_n)$ and assume that given $\mathbf{X}_i=\mathbf{x}_i$, Y_i satisfies model (8.4). Now, using Bayes' theorem, the likelihood of $\boldsymbol{\beta}$ is

$$L(\boldsymbol{\beta}|\mathbf{x},\mathbf{y},ED)=\prod_{i=1}^{n}\prod_{j=1}^{s}\left[\frac{p_jf_0(y_i-\mathbf{x}_i\boldsymbol{\beta})}{\sum_{j=1}^{s}p_j\Pr(Y_i\in S_j|\mathbf{x}_i,\boldsymbol{\beta})}\right]^{I[y_i\in S_j]} \tag{8.5}$$

where $I[\ ]$ denotes the indicator function and

$$\Pr(Y_i\in S_j|\mathbf{x}_i,\boldsymbol{\beta})=F_0(K_j-\mathbf{x}_i\boldsymbol{\beta})-F_0(K_{j-1}-\mathbf{x}_i\boldsymbol{\beta}).$$

Example. Suppose T is a failure time whose distribution depends on a vector of explanatory variables $\mathbf{X}=(X_1,\ldots,X_p)$. Suppose that, given $\mathbf{X}=\mathbf{x}$, T has the Weibull distribution

$$\Pr(T\leqslant|\mathbf{x},\boldsymbol{\beta})=1-\exp\{-(\theta_{\mathbf{x}}t)^{\gamma}\},\qquad \gamma>0 \tag{8.6}$$

where $\theta_{\mathbf{x}}=\exp\{\mathbf{x}\boldsymbol{\beta}\}$, and $\boldsymbol{\beta}=(\beta_1,\ldots,\beta_p)$ is a vector of regression parameters. Suppose $(T_1,\mathbf{X}_1),\ldots,(T_n,\mathbf{X}_n)$ are independent vectors obtained using variable probability sampling based on the strata $I_1,\ldots,I_s$, and that T_i given $\mathbf{X}_i$ satisfies model (8.6). Let $Y_i=\log T_i$ and $S_j=\log I_j$, where the log of an interval is the interval of logs of the endpoints. Now the likelihood of $\boldsymbol{\beta}$ is given by equation (8.5) with $F_0(y)$ equal to the extreme value distribution

$$F_0(y)=\exp\{-\exp(\gamma y)\}.$$

If γ is known, the resulting likelihood can be used to produce an estimate of $\boldsymbol{\beta}$. If γ is unknown, equation (8.5) can be used as a joint likelihood for $\boldsymbol{\beta}$ and γ and both $\boldsymbol{\beta}$ and γ can be estimated from the data.

Jewell (1985) considers the semiparametric case where F_0, rather than being extreme value, is an unknown distribution function. The present Weibull model-based approach will produce better estimates of $\boldsymbol{\beta}$ for small to moderate sample sizes where variance swamps bias.

8.6.2 Reply

Kjell Doksum has suggested some thoughtful and interesting modelling alternatives to our model. He has made a variety of suggestions—both Bayesian and sample theoretic. The principle of parsimony has not been invoked.

Herman Rubin questioned our use of the Poisson model at the conference. The model was both convenient and based on our uncertainty concerning the power plant outage rate in our very large population of plants. Another analyst might use a different model, but there is no mathematically justified 'right' model. This was a posterior not a prior Bayesian analysis.

ACKNOWLEDGEMENTS

This research was partially supported by the US Army Research Office under Contract No. DAAG29-85-K-0208 with the University of California, and by CNPq, Brasilia, Brazil under Contract No. 20.2926/85-MA.

REFERENCES

Barlow, R. E. and Proschan, F. (1988) Life distribution models and incomplete data. In P. R. Krishnaiah and C. R. Rao (eds) *Handbook of Statistics*, vol. 7, Elsevier, Amsterdam, pp. 225–249.

Barlow, R. E. and Campo, R. (1975) Total time on test processes and applications to failure data analysis, *Reliability and Fault Tree Analysis*, SIAM, Philadelphia, Pa, pp. 451–81.

Howard, R. A. and Matheson, J. E. (1981) Influence diagrams. In R. A. Howard and J. E. Matheson (eds) (1984) *The Principles and Applications of Decision Analysis*, vol II, Strategic Decisions Group, Menlo Park, Calif.

Jewell, N. P. (1985) Least squares regression with data arising from stratified samples of the dependent variable, *Biometrika*, **72**, 11–21.

Raiffa, H. and Schlaifer, R. (1961) *Applied Statistical Decision Theory*, MIT Press, Cambridge, Mass.

Roberts, H. V. (1967) Informative stopping rules and inference about population size, *J. Am. Stat. Assoc.*, **62**, 763–75.

Rubin, D. B. (1976) Inference and missing data, *Biometrika*, **63**, 581–92.

APPENDIX I

Derivation of $p(\mathbf{y}|k, \theta, ED)$ and $C(\theta)$

Let y denote the excess over 30 days of a forced outage greater than 30 days. Let $f(y|\theta)$ be the conditional density of y given that the corresponding forced outage has exceeded 30 days. In this case, θ is the unknown parameter related to the duration of such outages. Let us compute the probability of observing y given θ and the experimental design (ED):

$$\Pr(\text{observe } y|\theta, ED)\,\mathrm{d}y = \Pr(\text{observe } y|y \text{ had occurred}, \theta, ED) \times \Pr(y \text{ had occurred}|\theta, ED)\,\mathrm{d}y.$$

The corresponding density can be written as

$$p(y|\theta, ED)\,\mathrm{d}y = \Pr(\text{observe } y|y \text{ had occurred}, \theta, ED) \times f(y|\theta)\,\mathrm{d}y.$$

Now our problem is to find Pr(observe $y|y$ had occurred, θ, ED). Let $\Delta = 30$ days. The calendar period considered in this case, p, consists of $m = 17$ whole quarters. Therefore, an outage of length u will not be observed if its start falls in one of the m intervals each of length $2\Delta - u$. For instance, an outage of length 50 days will be missed if its start falls between the 60th and the 70th day of any one of the 17 whole quarters. That is, if its start falls in any of the 17 intervals of length 10 days that begin in the 61st day of each quarter.

Thus, when the excess is $y = u - \Delta$, the probability that an actual outage will be missed is

$$\begin{cases} \dfrac{m(\Delta - y)}{p} & \text{when } y < \Delta \\ 0 & \text{otherwise.} \end{cases}$$

If the density of y given θ is $f(y|\theta)$, given the experimental design, observed y will have density

$$p(y|\theta, ED) = \left\{ f(y|\theta) \times \left[1 - \frac{m(\Delta - y)^+}{p} \right] \right\} \times [C(\theta)]^{-1}$$

with $C(\theta)$ equal to the integral of the expression within the outer brackets. In other words, $C(\theta)$ is the normalizing constant, i.e. it is the probability that an occurred excess over 30 days will not be missed.

Suppose that n downtimes exceeding 30 days had occurred. Then the probability of observing k such downtimes given n, θ, ED, will be

$$p(k|n,\theta,ED) = \binom{n}{k} C(\theta)^k [1 - C(\theta)]^{n-k}.$$

Now we need to choose a parametric model for $f(y|\theta)$. To do this, the available data were plotted according to a procedure explained in Barlow and Campo (1975), see Appendix III. The plot suggests:

$$f(y|\theta) = \frac{1}{\theta}\exp\left(\frac{-y}{\theta}\right).$$

Here θ is the mean of excess (over 30 days) of downtimes exceeding 30 days.

Let $\mathbf{y}$ be the vector of observed downtimes exceeding 30 days. Given that we have observed k such downtimes, θ and ED, the density of $\mathbf{y}$ will be

$$p(\mathbf{y}|k,\theta,ED) = \prod_{i=1}^{k} \frac{f(y_i|\theta)\left[1 - \frac{m(\Delta - y_i)^+}{p}\right]}{C(\theta)}$$

or

$$p(\mathbf{y}|k,\theta,ED) = [\theta \cdot C(\theta)]^{-k} \exp\left(-\frac{T}{\theta}\right) \prod_{i=1}^{k}\left[1 - \frac{m(\Delta - y_i)^+}{p}\right],$$

where $T = \sum_1^k y_i$ (sum of recorded excesses) and

$$C(\theta) = 1 - \frac{m\Delta}{p} + [1 - \exp(-\Delta/\theta)]\frac{m\theta}{p}.$$

APPENDIX II

Table AII.1 Fossil fuel electrical power units 575 MW and larger

Date	Unit	Downtime duration (hours)
Quarter 1 1976		
2/18/76	Amos Unit 1	1412
1/30/76	H. L. Bowen Unit 1	1018
2/22/76	Kincaid No. 2	1660
2/07/76	Ninemile Point No. 4	1294
Quarter 2 1976		
4/01/76	H. L. Bowen Unit 1	4390
4/01/76	Cardinal Unit 2	792
5/17/76	Monroe No. 1	781
4/20/76	W. H. Sammis No. 6	733
Quarter 3 1976		
7/20/76	Bowline Point Unit 1	1469
8/08/76	Kincaid No. 2	1125
Quarter 4 1976		
10/11/76	H. L. Bowen Unit 2	797
12/20/76	Ninemile Point No. 5	2755
Quarter 1 1977		
3/07/77	Amos Unit 2	2925
1/18/77	Chalk Point Unit 3	806
3/21/77	Cliffside Unit 5	996
2/28/77	Gorgas Unit 10	720
2/05/77	Mohave Unit 2	1161
2/14/77	Ninemile Point No. 4	2548
1/03/77	W. H. Sammis No. 6	4432
Quarter 2 1977		
4/30/77	Astoria Project	3913
4/08/77	Baxter Wilson Unit 2	940
6/24/77	H. L. Bowen Unit 1	1053
5/27/77	Bowline Point Unit 2	1631
4/05/77	Oswego Unit 5	1035

Table AII.1 (*Contd.*)

Date	Unit	Downtime duration (hours)
Quarter 3 1977		
8/10/77	Belews Creek Unit 1	773
8/08/77	Chalk Point Unit 3	915
9/30/77	Chalk Point Unit 3	846
7/07/77	Sherburne Unit 1	1521
Quarter 4 1977		
11/06/77	Amos Unit 2	850
11/09/77	Baldwin Unit 2	792
11/30/77	Cumberland Unit 1	766
11/23/77	Kincaid No. 1	1928
11/15/77	La Cygne Unit 1	961
10/04/77	W. H. Sammis No. 7	1257
Quarter 1 1978		
2/24/78	Cumberland Unit 1	851
1/30/78	Harrison Unit 2	3625
2/01/78	Mohave Unit 1	3528
3/10/78	Ninemile Point No. 5	2216
1/09/78	W. H. Sammis No. 7	3109
Quarter 2 1978		
5/05/78	Gaston Steam Plant Unit 5	864
5/15/78	Marshall No. 3	1408
5/19/78	Ninemile Point No. 4	2958
5/03/78	Tradinghouse Creek Unit 2	2188
Quarter 3 1978		
7/01/78	Centralia Unit 1	1559
7/16/78	Ninemile Point No. 4	3557
9/29/78	Ormond Beach Unit 2	776
Quarter 4 1978		
11/06/78	Keystone No. 1	768
10/03/78	Oswego Unit 5	1247

Table AII.1 (*Contd.*)

Date	Unit	Downtime duration (hours)
Quarter 1 1979		
2/17/79	Conesville Unit 4	2411
1/01/79	Hatfield No. 1	4167
1/01/79	Hatfield No. 1	3320
1/01/79	Mt Storm No. 1	2159
2/03/79	Paradise No. 1	3424
3/30/79	Ravenswood No. 3	2903
Quarter 2 1979		
4/01/79	Baxter Wilson Unit 2	1672
6/28/79	Harrison Unit 2	3858
4/01/79	La Cygne Unit 2	1287
5/27/79	Mohave Unit 1	792
Quarter 3 1979		
7/26/79	Astoria Project	1034
8/06/79	Harrison Unit 1	5116
8/30/79	Hudson No. 2	761
8/20/79	Keystone No. 1	1003
7/08/79	La Cygne Unit 1	920
7/23/79	Ormond Beach Unit 2	799
9/22/79	Pittsburg Unit 7	1573
9/29/79	W. F. Wyman Unit 4	1815
Quarter 4 1979		
10/01/79	Centralia Unit 2	1278
11/22/79	Mt Storm Unit 3	2586
Quarter 1 1980		
1/12/80	Hatfield No. 1	1287
Quarter 2 1980		
5/05/80	Belews Creek Unit 2	1360
5/14/80	Kincaid No. 1	903
5/04/80	W. H. Sammis No. 7	1286

APPENDIX III

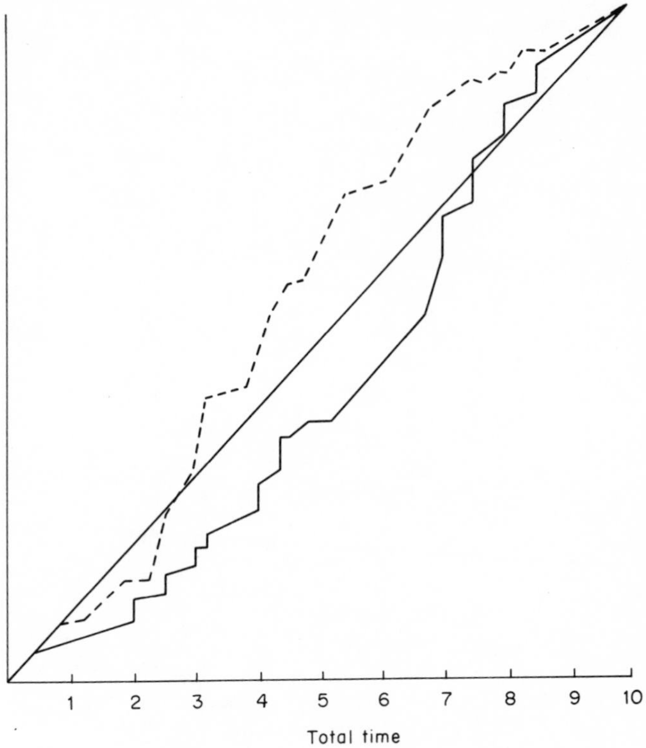

Figure AIII.1 Total time on tests plots

CHAPTER 9

Real Time Influence Diagrams for Monitoring and Controlling Mechanical Systems

A. M. Agogino and K. Ramamurthi, *University of California, Berkeley, USA*

ABSTRACT

This chapter describes an architecture for real time influence diagrams for monitoring and controlling mechanical systems. The IDES (influence diagram based expert system) performs probabilistic inference and expected value decision making in integrating dynamic sensor readings, statistical data and subjective expertise in real time. The IDES 'greedy' decision algorithm is presented along with a complexity analysis for the same. An application to milling machine monitoring and control is described in which influence diagrams are used to represent and evaluate temporal signal data in real time. The IDES fusion of acoustic, force and current sensors achieves effective prediction and control performance with relatively simple signal processing. Research issues concerning sensor-based real time influence diagrams are discussed.

9.1 INTRODUCTION

In-process diagnostic and monitoring systems require reasoning in real time about failure and process states based on information from prior experience and current sensor readings. Often in complex sensor-based systems, the interpretation of sensor readings and the prediction of process states are indirect and non-deterministic. In real time systems, there is always a trade-off between the quality of the signal and information processing and the response rates. The *diagnostician's problem* is to identify likely process states from the observable

Influence Diagrams, Belief Nets and Decision Analysis
Edited by R. M. Oliver and J. Q. Smith

sensor readings. In the unattended milling machine problem described in this chapter, this corresponds to determining the state of the cutting process ('normal', 'worn tool', 'broken tool' or 'tool chatter') given the observables (current, dynamometer and acoustic sensor readings). The *controller's problem* is to efficiently invoke an optimal course of action given these observables and process goals. In the milling machine application, the supervisory controller must respond to the sensory information in a way that prevents tool, workpiece or machine damage.

In this chapter, we use influence diagrams for integrating dynamic sensor readings, statistical data and subjective expertise for solving both the inference and control problems in computer-integrated applications requiring real time performance. The influence diagram based expert system (IDES) provides both an influence diagram representational scheme and an efficient computational architecture for performing probabilistic inference and expected value decision making. An extended version of the milling machine application can be found in Agogino *et al.* (1988a). Integration with 'analogical reasoning' schemes is described in Agogino *et al.* (1988b).

9.2 SENSOR-BASED REAL TIME EXPERT SYSTEMS

In spite of recent progress in signal processing and computer intelligence, human beings possess a unique ability for processing complex types of information and making inferences. Numerical algorithms for control have difficulty in incorporating the experiential knowledge that is effectively used by expert diagnosticians and equipment operators (Wright and Bourne, 1988). On the other hand, our abilities are limited. Our response rates may be unacceptably slow for real time applications and we can be overwhelmed by large amounts of sensory input (Kao *et al.*, 1987).

Although there have been a number of new products in the commercial market, current expert systems tools, hardware vehicles, and theoretical methodologies are inadequate or limited for many mechanical diagnostic and manufacturing applications in which explicit representation of uncertainty is a requirement. Laffey *et al.* (1988) provide an excellent review of real time knowledge-based systems in which they claim that none of the systems which were studied would be considered real-time monitoring or control applications. They define the following characteristics as challenges to 'traditional knowledge-based problem-solving techniques': (1) nonmonotonic or time varying aspects of the data base; (2) need for continuous operation; (3) asynchronous event sequences; (4) need for direct integration of sensor data; (5) uncertain or missing data; (6) high (fast) performance; (7) temporal reasoning; (8) ability to change focus of attention; (9) guaranteed response times; and (10) integration with procedural components.

9.2.1 Management of uncertainty

In this chapter we focus on the problems involved with uncertain or missing data. Rule-based expert systems using binary logic have trouble accounting for imperfections in the measuring instruments. Applications based on simple binary rules are prone to making false-positive and false-negative predictions. Spurious sensor readings and sensor degradation over time have been reported as major problems in rule-based systems (Fox, 1983; Laffey *et al.*, 1988; Rege and Agogino, 1986a).

Certainty factors have been used as a means to incorporate probabilistic knowledge in rule-based expert systems; the rules being weighted by a belief factor. In the medical diagnostic system MYCIN, the knowledge is stored in rules of the form 'If E_i then D_j' where E_i represents evidence that supports disease hypothesis D_j. The certainty factor $CF(D_j, E_i)$ associated with each rule ranges between -1 and 1. Positive certainty factors quantify relative (not absolute) support for the hypothesis and negative numbers quantify the relative support against the hypothesis.

Conditional independence is assumed in the combining rules that give the joint factor when multiple pieces of evidence support one disease hypothesis. Heckerman and Horvitz (1987) have shown that the MYCIN function for combining evidence is analogous to Bayes's formula assuming that each knowledge rule is conditionally independent (assuming a probabilistic interpretation of certainty factors). Thus the benefits of approaches using a simple weighting of rules must be compared against the loss in theoretical rigor and robustness (Cheeseman, 1985). All too often, joint and conditional dependencies are critical aspects of diagnostic reasoning. Probabilistic influences more often than not are highly interdependent.

There is growing support in the artificial intelligence research community for the use of influence diagrams (probabilistic influence diagrams are sometimes called Bayes' networks or belief networks) for representing uncertain knowledge in building expert systems (Heckerman and Horvitz, 1987; Henrion, 1987; Henrion and Cooley, 1987; Holtzman, 1985; Horvitz, 1987; Horvitz *et al.*, 1986, 1988; Pearl, 1986). Influence diagrams provide an intuitive graphical framework for representing complex interdependencies and rules for combining evidence that are based on rigorous probability theory. The IDES developed at the University of California at Berkeley (Agogino and Rege, 1987) is an implementation of influence diagram technology with efficient algorithms designed for real time applications.

9.2.2 Representing conditional independence with influence diagrams

Influence diagrams provide an attractive graphical scheme for explicitly codifying conditional independence between critical probabilistic variables. The salient information in the diagram is, in fact, not which variables influence each other, but rather, which ones do not influence each other given the conditioning information. Thus we define influence by its dual concept 'lack of influence'.

Definition 1. Consider two variables x and y. Here y does not influence x given C if $\Pr(x/y, C) = \Pr(x/C)$, where C represents the state of information and other conditioning variables. The reverse implication is also true for all cases except the degenerate case where $\Pr(y/C) = 0$. □

In graph-theoretic terms, an influence diagram can be represented as a directed graph $G = (V, A)$ where V is a set of n nodes and A is the set of directed arcs joining these nodes. Given n state variables $x_1, x_2, \ldots, x_n$ in the problem being modeled, there are $n!$ expansions of the associated joint probability distribution. An influence diagram represents one expansion of the joint distribution which can be manipulated to realize other expansions through application of Bayesian probability theory. An expert may find one expansion more natural than others in codifying and conditioning subjective probabilities. Yet another expansion may be computationally more efficient in solving the diagram in order to answer a specific inference question. In manipulating the diagram during a solution sequence, any intermediate diagram must be 'consistent' to the original in the following sense. (Interested readers are referred to Smith, 1989, for a rigorous axiomatic development of this concept, using implication.)

Definition 2. An influence diagram $G' = (V', A')$ is said to be consistent with another influence diagram $G = (V, A)$ if and only if the set of explicit independencies $I_{G'}$ associated with G' can be derived from the set I_G associated with G where V' is a subset of V (Rege and Agogino, 1986b, 1988). □

Thus, for diagram G' to remain consistent with diagram G, there should not be any assertion about conditional independence between variables in G' which cannot be concluded from the information in the original diagram G. Consequently, the terms in the expansion represented by the new diagram should be computable from the terms in G. (Note, however, that it is possible for the information on conditional independence in G' to be a proper subset of that in G.)

The concept of consistency provides us with an important tool to compare the information on conditional independence asserted by two experts, say 'A'

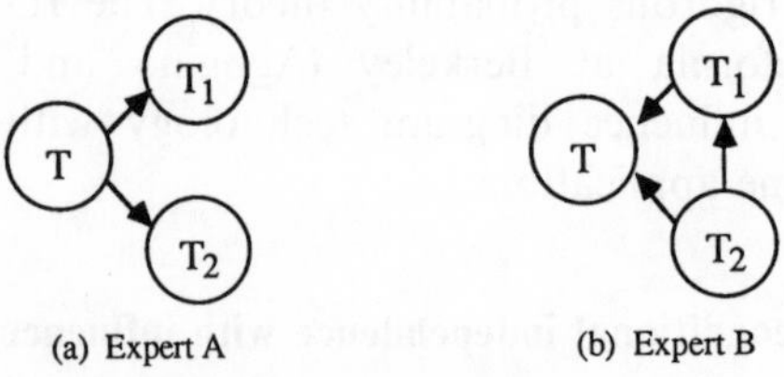

Figure 9.1 Topologically consistent world views: (b) is consistent with (a) (Rege and Agogino, 1988).

and 'B', who provide different influence diagram models of the same problem domain. Consider the two influence diagrams shown in Figure 9.1. The first, Figure 9.1(a), shows that expert A believes that the variable T influences the variables T_1 and T_2 which are conditionally independent of each other given T. Thus in the view of expert A the information flow, so to speak, is from T to T_1 and T_2. On the other hand expert B views the problem domain as in Figure 9.1(b) implying that he is more comfortable with inferring the probability distribution for T given T_1 and T_2. Further, he asserts that in his view of the world T_1 and T_2 are not independent. If we treat the variables T_1 and T_2 as 'observables' (e.g. readings of thermocouples for the same process) and T as a state variable that is not directly observable (the temperature of that process) we notice the cognitive differences in the thought processes of the two experts. Expert A sees the temperature as influencing the observable thermocouple readings T_1 and T_2; in other words, he knows the accuracy of each thermocouple given the process temperature. Expert B, on the other hand, feels more comfortable inferring the process temperature from the two thermocouple readings. In fact, he can make an inference about the reading on one thermocouple given the reading of the other (hence the arc from T_2 to T_1). According to Definition 2, expert B has a view topologically consistent with that of expert A, but his experiential knowledge may differ. Expert A might be a scientist who sees a causal mapping from process temperature to thermocouple readings (causal reasoning). Expert B may be a plant operator who has considerable experience in inferring the true process temperature from thermocouple readings (diagnostic reasoning).

Shachter and Heckerman (1987) note that a 'causal' representation is often less complex in that it captures more information concerning conditional independence (as is the case in Figure 9.1a), although this may be in the direction opposite of the intended usage. Tversky and Kahnemann (1980) also observe that the probability of consequence given cause (causal reasoning) is psychologically more readily available than the probability of cause given consequence (diagnostic reasoning). However, they caution that their experimental studies indicate that influences perceived as causal may exhibit a confidence bias unjustified by the data. The advantage of modeling with influence diagrams is that the model can be created in the direction that the expert has experienced the world, but can be transformed directly in order to answer a specific query. We go on to argue, that for real time applications, some of these transformations can be performed off-line so as to minimize the computational cost required in actual operation.

9.3 UNATTENDED MILLING MACHINE PROBLEMS

Unattended manufacturing has all of the attributes that challenge conventional rule-based expert system as outlined by Laffey *et al.* (1988) with particular emphasis on: multiple sensors that require some level of signal processing; imperfect or uncertain mapping between sensor readings and critical states; complex interdependencies; and real time response needs. The unattended milling machine problem will be used as an example to demonstrate the strength of the IDES approach under these challenging conditions. A brief review of previous work with current in-process sensor technologies applied to milling applications will be provided followed by an introduction to the theory of influence diagrams and their application to the unattended milling machine problem.

Various sensors and signal processing techniques have been investigated for application to unattended machining. Current approaches to in-process monitoring include dimension sensors (Tlusty and Andrews, 1983), motor current sensors (Stein *et al.*,1984; Stein and Shin, 1985), force dynamometers (Altintas *et al.*, 1985; Lan and Naerheim, 1985), accelerometers (Braun *et al.*, 1987) and acoustic emission sensors (Dornfeld, 1984, 1985; Diei and Dornfeld, 1985a, b).

These studies reflect the progress made in the last decade towards developing unattended computer-integrated manufacturing systems. However, they also reveal practical limitations that restrict the full utilization of these sensor technologies today. Agogino *et al.* (1988a) summarize:

1. There are no simple deterministic models relating the state of the tool chip interface to sensor readings. All of the models have varying degrees of uncertainty associated with them. How should machines be programmed to respond to this 'noisy' information?
2. No single sensor is clearly preferred over the others. Each has specific advantages and disadvantages depending on the operation and the environmental conditions. How can diverse sensor readings be 'fused' in order to make intelligent inference and control decisions?
3. Most of the approaches to date use a single sensor for use in techniques that require extensive signal processing and are only applicable to specific machining conditions. How can these techniques be generalized to a broader class of operations without adding more burden to the already heavy computational requirements in signal processing?
4. Manufacturing researchers, engineers, and operators possess both first-principle and experiential knowledge concerning the interpretation of this sensory data. How can this subjective and often qualitative information be utilized in intelligent automated manufacturing systems?

Agogino *et al.* (1988a) propose influence diagrams as a framework for integrating operator's expertise, first-principle knowledge and experimental data for the wide range of sensors possible for intelligent in-process monitoring and control. The use of multiple sensors reduces the sensitivity of the system to any particular sensor's drawbacks, requiring less precision than needed with a single sensor and thus potentially requiring less sophisticated signal processing. The non-deterministic nature of the inference problem and 'noisy' sensor data is handled by operations with Bayesian probability. In the IDES implementation, features are extracted from the raw sensor data and used by the expert system to answer specific queries and send control signals out to the machine operator or controller.

9.4 SENSOR-BASED INFLUENCE DIAGRAMS

The diagnostician's and controller's problem have been proposed by Rege and Agogino (1986a) and Agogino *et al.* (1988a) as a means of describing sensor-based diagnostic inference and control decision making in real time applications.

9.4.1 The diagnostician's problem

Consider a system with the set of state variables denoted by $X = \{x_1, x_2 \ldots x_n\}$. Let S be a set of sensors or observables within the system and F be a set of potential failure nodes, such that

$$X \supset S = \{s_1, s_2 \ldots s_n\} \quad \text{and} \quad X \supset F = \{f_1, f_2 \ldots f_n\}.$$

The diagnostician's problem is to assess the likelihood of failures for various combinations of hypothesized states given some combination of sensor readings. In mathematical terms the diagnostician's problem can be stated as determining $\Pr(F/S)$. The single-valued diagnostician's problem is simply to identify the most likely failure event in the set F given the sensor readings S (i.e. to identify 'i' such that $\Pr(f_i/S) \geqslant \Pr(f_j/S) \forall j, j \neq i$).

Classification of State Nodes

For applications in diagnostic reasoning and control three categories of state nodes are useful (Figure 9.2).

Sensor nodes. A sensor node represents sensor measurements that can be directly observed by the operator or controlling system. It might also represent a physical state of the system that is immediately obvious to the operator's senses, such as sight, hearing or smell. Often features from the raw data are used as sensor nodes in the influence diagram model. In our milling machine example the

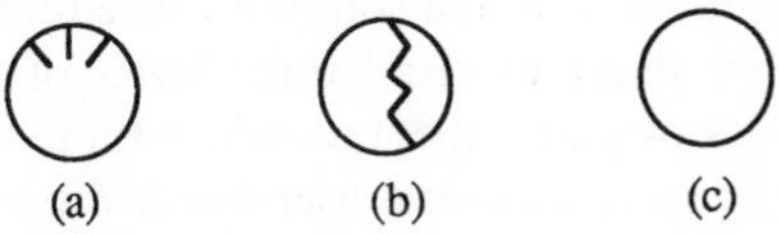

Figure 9.2 Types of state nodes: (a) sensor, (b) failure and (c) intermediate.

sensor nodes are features from current, dynamometer, and acoustic emission measurements, e.g. RMS, ΔRMS.

Failure nodes. As the name implies, a failure node represents states of a physical component in the system which may be the cause or symptom of, or contribute to, the initiation of the diagnostic search. Since a sensor itself may fail, some or all of the sensor nodes may belong to this class. Possible failure events for the milling machine failure node are 'machine tool chatter', 'cutting tool wear' and 'cutting tool break'.

Intermediate nodes. These are useful nodes for modeling but are not failure or sensor nodes (i.e. they will not be goal or conditioning nodes in a diagnostic query). They generally represent intangibles in the problem which cannot 'fail' or which are not measured directly by any sensor in the system. They are useful in modeling complex mechanical and manufacturing systems and in providing a structure for updating system parameters with human and statistical knowledge obtained over time.

9.4.2 IDES implementation

Heckerman and Horvitz (1987) convincingly argue that certain classes of dependencies cannot be modeled by modular rules in a natural or efficient manner unless strong forms of conditional independence are assumed. They and others argue that influence diagrams allow the expert to explicitly model dependencies and independencies in a compact and intuitive graphical representation and are thus an ideal representation for probabilistic expert system. Influence diagrams capture not only 'heuristic' or 'shallow' knowledge but also attempt to encode explicitly the structure of the physical model in the mind of the expert. Further, the sensor-based influence diagram architecture (Rege and Agogino, 1986a) explicitly allows the possibility of aberrations and failures of sensors themselves. The IDES (Agogino and Rege, 1987) implementation was developed at the University of California at Berkeley. It was written in the compiled language C, making it transportable to a large number of mainframe and microcomputer systems with data acquisition and real time capabilities.

The IDES architecture is highly modular in form. Constructed on the expert system paradigm requiring separation of control and data, the domain knowledge base is separated from the solution procedure. The knowledge base consists of the influence diagram at all three levels of data abstraction. The control strategy consists of graph-theoretic algorithms that create a sequence of transformations in order to answer a user-directed probabilistic inference or control query. The sequence of transformations is then sent to the data acquisition system and numerical processor for further numerical computation.

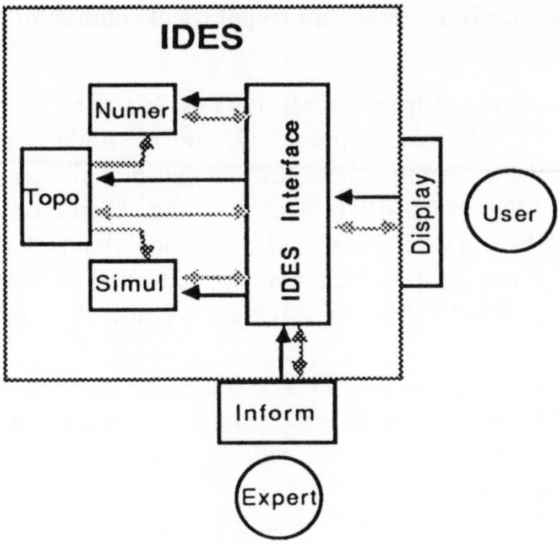

Figure 9.3 IDES architecture

This separation of the knowledge base from the control strategy provides a flexibility in the architecture that allows the user to change the structure and the parameters of the model in real time.

The IDES architecture consists of separate modules for knowledge base development, user interaction, control and coordination of the modules (Figure 9.3). The Inform architecture is designed for user-directed knowledge acquisition of the influence diagram model (Moore and Agogino, 1987). GraphIDES is a color graphics MS/DOS™ implementation of this architecture (Lambert and Agogino, 1987; Ramamurthi, 1988). IDES Interface is the module which co-ordinates all of the modules and is the interface with the user. Topo and Numer are the modules for control of the solution at the topological (or symbolic) and numerical levels respectively (Bronstein, 1987). There are several search algorithms employed by Topo; the default 'greedy' search algorithm is described in Rege and Agogino (1986b, 1988) along with a complexity analysis for probabilistic inference. The IDES 'greedy' decision algorithm will be described in this chapter. Simul is a simulation module for testing the efficiency of solution algorithms (Bronstein, 1987). Using Simul, one could perform sensitivity analysis at the design stage and suitably trade off between accuracy of reasoning and the computational time to obtain timely decisions in real time systems. During real time processing Simul can be used to estimate the cost of competing algorithms in order to pick the one that is most efficient for a specific query (Rege and Agogino, 1988). In the schematic of the IDES architecture given in Figure 9.3 control lines are solid and data lines are dotted.

9.4.3 IDES greedy decision algorithm (topological solution of controller's problem)

Among the many kinds of transformations possible, the IDES is based on the following three: (1) state node removal, (2) control node removal and (3) arc reversal. For the theory behind these transformations, the reader is directed to the references (Miller *et al.*, 1976; Howard and Matheson, 1984; Olmsted, 1984; Shachter, 1985, 1986; Rege and Agogino, 1988). The IDES 'greedy' implementation of these topological transformations and their computational complexity are described below for solving the controller's problem.

Input. An influence diagram $G=(V,A)$, the value node x_v, the sequentially ordered control/decision nodes $d_J; v \notin J$, state nodes of known values $x_K; j, v \notin K$ and relevant state nodes of unknown values $x_U; k, j, v \notin U$.

Output. An influence diagram with a single node, the value node x_v, and an array D of d_{J*}, the optimal decision sequence. Let $M' = \{1, 2, \ldots, n\} \backslash v$ be the nodes to be removed (all nodes but the value node for a decision problem). Let M be the 'current' version of M'. Initially $M = M'$.

1. *Barren node removal.* Remove all nodes in M which have no successors (barren nodes). Update the successor sets for nodes which directly preceded the barren nodes. If M is empty, then end, else go to step 2.
2. *Known sensor node reduction.* Propagate values of known sensor nodes in K to their non-decision node successors and remove all successor arcs. Remove the known sensor nodes with no predecessors. Eliminate any disjointed nodes that are made barren, by identifying the nodes in the graph that are unconnected to the value node. If M is empty, then end, else go to step 3.
3. *State node removal.* Remove all state nodes in M which have only one non-decision node successor each. The removal is done in order of number of predecessors—with the node having the least number of predecessors removed first. Update successors, predecessors, etc. Repeat till all nodes in M with only one non-decision node successor are exhausted. If M is empty, then end, else go to step 4.
4. *Control node removal.* Remove the control/decision node in J which directly precedes the value node if all other conditioning predecessors of the value node are also informational predecessors of the control node. Remove all informational arcs to the absorbed control node and remove any of these nodes that are consequently disjointed from the graph and made barren, by identifying the unconnected elements of the graph. If M is empty, then end, else if any control node is removed, then go to step 3, else go to step 5.
5. *Arc reversal.* Select the state node in M with the least number of non-decision node successors. Let this node be x.

   Pick a state node successor of x (excluding x_v), check if there is more than one path from x to that successor. If yes then reject it, else say y is that successor. Reverse the arc $(x \rightarrow y)$ with concomitant updating of predecessor, successor sets, etc. Check if x has only one non-decision node successor—if so go to step 2. Otherwise go back to.

Complexity Analysis

The following nomenclature is used in analyzing the complexity of this algorithm.

A is an $n \times n$ adjacency matrix that represents the digraph. If there exists an arc from i to j, then the (i, j) element of **A** is '1', else '0';
K is the set of all known sensor nodes and let the size of $K, |K| = f$;
J is the set of all decision nodes and let its size, $|J| = d$;
S is the set of all informational predecessors of all the decision nodes and $|S| = e$. It should be noted here that $S \supset K$.

We use an array **exist[1 × n]** to keep track of whether a node exists in the diagram or has been removed already ($exist[i] = 1$ if the ith node exists, else zero) and another array **removable[1 × n]** to indicate whether a node can be removed or not from the diagram ($removable[i] = 1$, if the ith node can be removed, else zero). Two more arrays are used to store information about sensor nodes and the decision nodes. **Known-node[1 × n]** array indicates which nodes are sensor nodes ($known\text{-}node[i] = 1$ if ith node is a sensor node, else zero) and similarly **decision-node[1 × n]** array indicates which nodes are decision nodes. We maintain one other array **successors[1 × n]** in which $successors[i]$ gives the number of direct successors of x_i at any stage of the algorithm.

Initialization. We search through the matrix **A** and count the number of successors for each node and initialize the 'successors' array: this takes $O(n^2)$ number of operations.

Barren node removal. For each node $x_i, i \neq v$, we check if $exist[i] = 1$ (otherwise the node has already been removed) and $removable[i] = 1$ (verify that the node can be removed). If so, we access $successors[i]$ and check if its value is 0 (i.e. whether it is barren). If true, we remove the node by making $exist[i] = 0$. We also make all the entries in the ith column of the matrix **A** which are 1 into 0. During this process, for each entry $A(m, i)$ which is 1, we reduce the value of $successors[m]$ by 1. This takes $O(n)$ steps. Since there can be at most $(n - 1)$ barren nodes, the total time for removal of barren nodes is $O(n^2)$ steps.

Known sensor node reduction. For each node x_i, we check if $exist[i] = 1$ ($i \in M$), $removable[i] = 1$ and $known\text{-}node[i] = 1$ ($i \in K$). If all are true, then check if it has any predecessors. If no predecessors, then delete the sensor node and since it does not have any predecessors, the 'successors' array need not be updated. Hence, deleting a sensor node takes only a constant number of steps and there could be at most f sensor nodes with no predecessors. Hence this would take $O(f)$ steps. On the other hand, if any sensor node has predecessors, then only the arcs to its successors are removed. There could be at most $(n-1)$ successors and hence there are $O(n)$ steps. Repeating this for each known node (there can be at most f known nodes), we get $O(fn)$ algorithm. Hence considering both cases, with and without predecessors, this could take $O(fn)$ steps in the worst case situation.

Due to removal of arcs from the sensor node or the removal of sensor node itself, many nodes could have become disjointed from the graph and thereby made barren. Hence the components of the undirected graph connected directly or remotely to the value node are identified by a **reconnect** routine. For this purpose, an array **connected[i]** is used to mark the connected elements, both predecessors and successors, of any node in the graph (*connected*$[i]=1$ if ith node is connected to the graph, else zero). One other array **visited[i]** is used to check if all the connected nodes are visited in the graph (*visited*$[i]=1$ if ith node is connected and also visited, else zero). To begin with, the nodes connected to the value node are marked 'connected' and the value node is marked 'visited'. Now, each connected node is successively visited, thereby sequentially searching for all the connected components of the graph. The search ends when the counter in the iterative 'while' loop exceeds the total number of nodes in the graph. To find the connected nodes (predecessors and successors) of any node in the graph, it takes $O(n)$ steps, and there can be at most n connected nodes. Hence the 'reconnect' routine takes $O(n^2)$ steps to find the connected components of the graph.

State node removal. For each node x_i, we check if $i \in M, i \neq v$, and $i \notin J, K$. If true, then access *successors*$[i]$ and if this is equal to 1, then check if the successor does not belong to J. If so, then remove the node x_i with update of the 'successors' array. Each removal takes $O(n)$ steps as explained in the case of barren node removal and there can be at most $\{n-(e+d)\}$ nodes removed in this stage: so it takes $O(n\{n-(e+d)\})$ steps.

Control node removal. For each node x_i, we check if $i \in M, i \in J$ and x_i is a direct predecessor of value node x_v. If all the three conditions are satisfied, then we check if any predecessor of x_v is also a predecessor of x_i. While true, we continue checking till all the predecessors of x_v are exhausted. At the worst case, the maximum number of checks possible is e, and hence this consumes $O(e)$ steps. If all the predecessors of x_v are also informational predecessors of x_i, then remove x_i with update of the 'successors' array. Each removal alone takes $O(n)$ steps as before and hence the checking and removal, both take $O(n+e)$ steps.

In this step, at most d control nodes can be removed and hence this takes $O(d\{n+e\})$ steps.

If any control node is removed, as explained earlier, all the informational arcs to the absorbed control node are deleted. Consequently, due to the deletion of arcs, many nodes could have become disjointed from the graph and thereby made barren. Like in the previous case, here too, we use the 'reconnect' routine to search for the connected components of the graph. This, as explained earlier, takes $O(n^2)$ steps.

Arc reversal. For each node x_i, we check if $i \in M, i \neq v, i \notin J$, and $i \notin S$. If true, then we check if all of its successors do not belong to J. If all the conditions are met, then we initialize an array of those that satisfy the above conditions and let us call this **workable[1 × n]** (*workable*$[i] = 1$ if all conditions are satisfied for the ith node, else zero). To check if any of the successors of x_i belong to J, it takes $O(n)$ steps, and we do this for all the n nodes. Hence this takes $O(n^2)$ steps. Among the nodes in this array, i.e. *workable*$[i] = 1$, we select one with the lowest value of *successors*$[i]$. Since, at most there could be $\{n-(e+d+1)\}$ nodes in the array workable, this takes $O(n-(e+d))$ steps.

For this node (say x), we pick a successor (say y): *reverse*$(x \rightarrow y)$ and then check if a cycle is formed. This is done as follows: we change the (x, y)th element to 0 and the (y, x)th element to 1. We then use any standard cycle check algorithm $(O(n^2))$ to determine if the diagram so obtained has a cycle. If so we reject arc $(x \rightarrow y)$ and go on to another successor of x. If the diagram does not have a cycle, we update the predecessors/successors as follows: scan the xth column of **A** for predecessors of x. For each row z in this column which is 1, we check element (z, y)—if it is 0, we make it 1 and increment *successors*$[z]$ by 1. We do the same with the yth column. Updating the predecessors takes $O(n)$ steps. In the worst case we would have to perform the cycle check for each successor of node x till we find an arc which can be reversed.

As explained earlier, the node x has the least number of successors among the nodes in 'workable' array. For complexity analysis, we would like to know the maximum number of successors x can have at worst. Now, consider the subgraph on the nodes in the 'workable' array. Since the graph is acyclic, the subgraph should have at least one node without any successors in 'workable' array. However, it can have successors outside the subgraph. Based on the conditions on which 'workable' has been formed, the maximum number of successors the leaf node in subgraph can have is $(e + 1)$, i.e. only the informational predecessors of decision nodes and the value node, x_v. Hence, the maximum number of successors the node x can have is $(e + 1)$. Hence at most e cycle checks need to be performed before an arc from x to y can be reversed. At the next stage $(e - 1)$ cycle checks would have to be done and so on till all but one arc from x is reversed, whence x can be removed. Thus the removal of single node in 'workable' requires at most $O(e^2)$ cycle checks and therefore $O(n^2e^2)$ computations. Finally, $\{n-(e+d)\}$ nodes will be removed and therefore the complexity of this phase is $O(e^2(n-[e+d])n^2)$.

It should be noted here that if there are no informational nodes, i.e. $e=0$, as indicated by the above algorithm, there would be no arc reversals. The complexity will be dominated by state node removal of $O(n^2)$. When there are no decision nodes, the controller's problem is analogous to an inference problem with the value node as the goal node. The result obtained here with $d=0$, i.e. no arc reversals, is same as the result of the complexity analysis of the inference problem with no conditioning nodes (Rege and Agogino, 1988). Also, if we treat the informational predecessors and the decision nodes as analogous to conditioning nodes (c) in the inference problem, we arrive at the same order of complexity as given in Rege and Agogino (1988) which is $O(c^2(n-c)n^2)$, where $c=e+d+1$.

The convergence of this algorithm can be proved by suitably modifying the method outlined in Shachter (1986). From step 1 to step 4, the diagram can be reduced successively if a possibility for removal of a node exists. If no node is removed in steps 1 to 4, then in step 5, by the well-ordered principle of acyclic directed graphs, it is always possible to find one node whose outgoing arcs can be reversed until it has only one successor. This node would be ultimately removed in either step 2 or step 3, when the algorithm repeats itself. Hence, as explained above, all the state nodes and known nodes with no decision node as successor would be removed eventually. At this stage, as explained in Shachter (1986), there would be at least one decision node which could be absorbed into the value node. Eventually all the control nodes are removed in the above manner and the remaining relevant state and known nodes are absorbed into the value node. Therefore, for a finite number of nodes, the algorithm converges in finite time, since at each cycle of the algorithm at least one node is removed.

Example. Sensor-based inference (Agogino *et al.*, 1988a; Rege and Agogino, 1986a). Suppose we know that a failure node 'F' influences an intermediate state variable 'I' that can only be observed through sensor node 'S' (Figure 9.4a). At a numerical level, this implies that we have $\Pr(F)$, $\Pr(I/F)$ and $\Pr(S/I)$. Sensor-based inference is estimating the likelihood of a failure from a specific value of the sensor reading, i.e. $\Pr(F/S)$. This can be accomplished by sequential application of the rules for arc reversal and node absorption as shown in Figures 9.4(b) and (c).

Each topological operation in Figure 9.4 corresponds to the following

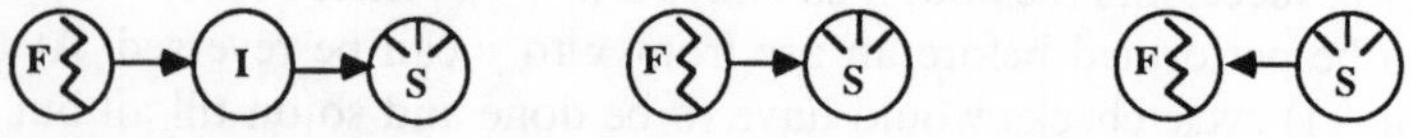

(a) Original model (b) Absorb I into S (c) Reverse Arc from F to S

Figure 9.4 Sensor-based inference with goal: $Pr(F/S)$.

functional evaluations that begin with the known quantities $\{\Pr(F), \Pr(I/F)$ and $\Pr(S/I)\}$ and end with the goal, $\Pr(F/S)$.

Figure 9.4(b): $$\Pr(S|F) = \sum_{\Omega_I} \Pr(S|I)\Pr(I|F) = \sum_{\Omega_I} \Pr(S, I|F) \tag{9.1}$$

Figure 9.4(c): $$\Pr(S) = \sum_{\Omega_F} \Pr(S|F)\Pr(F) = \sum_{\Omega_F} \Pr(S, F) \tag{9.2}$$

$$\Pr(F|S) = \frac{\Pr(S|F)\Pr(F)}{\Pr(S)}. \tag{9.3}$$

Example. Sensor-based control (Agogino *et al.*, 1988a; Ramamurthi and Agogino, 1988). In a controller's problem, a timely and suitable decision from an array of available control strategies needs to be made based on the sensor readings (Figure 9.5a) such that the system is directed towards an optimal state at any instant of time. The cost function in this optimization problem is the cost of control and it depends on the failure state of the system and the controller's decision. The set of decisions available to the controller are represented by decision node D and the cost function is described by the value node V. As in the example in Figure 9.4, after the node absorption and arc reversal transformations, Figure 9.5(a) is reduced to Figure 9.5(c). At this juncture, an optimal decision corresponding to the known state of the sensor is chosen, thereby minimizing the cost of control.

It should be noted here that in both class of problems, inference and control, the causal representation (arcs into sensor nodes as shown in Figure 9.5) of the system results in arc reversals between sensor nodes and the intermediate nodes. Also, the readily available knowledge about the state of the system from the sensory information cannot be utilized until the arcs into the sensor nodes are reversed. Hence, the delay in propagation of the sensory information to the intermediate nodes about the state of the system and also the increasing number of arc reversals can result in an excessive computational burden for the quantitative solution of the influence diagram. In a real time situation, this computational delay could adversely affect the controller's performance. However, a diagnostic representation (arcs out of sensor nodes) would eliminate arc reversals between sensor nodes and intermediate nodes. Further, the known

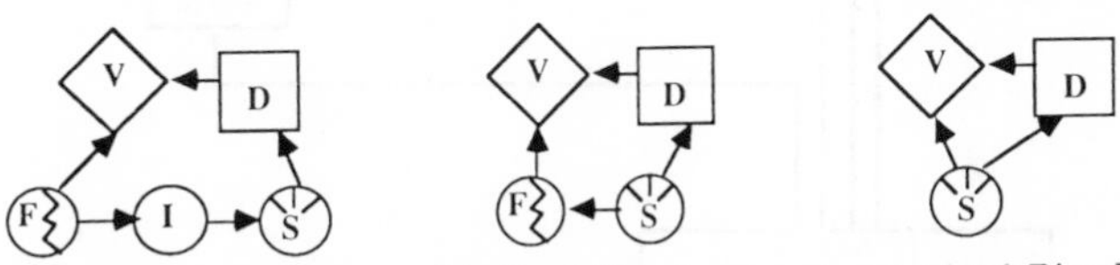

(a) Original model (b) Absorb I into S and reverse arc from F to S (c) Absorb F into V

Figure 9.5 Sensor-based control.

states of the sensor nodes could be propagated into its successors, thereby reducing the conditioning space of the intermediate nodes. Hence, in real time application of influence diagrams, like the following milling machine example, the response time and the computational efficiency could be significantly improved by representing the system as a diagnostic model. In a situation where the experts view the system as a causal model, a faster response could be arrived at, perhaps, through preprocessing of the casual model and implementing an equivalent diagnostic representation in real time applications.

9.5 MILLING MACHINE MONITORING AND CONTROL

The development of unattended machining requires automating the human operator's function in monitoring and making appropriate adjustments in the

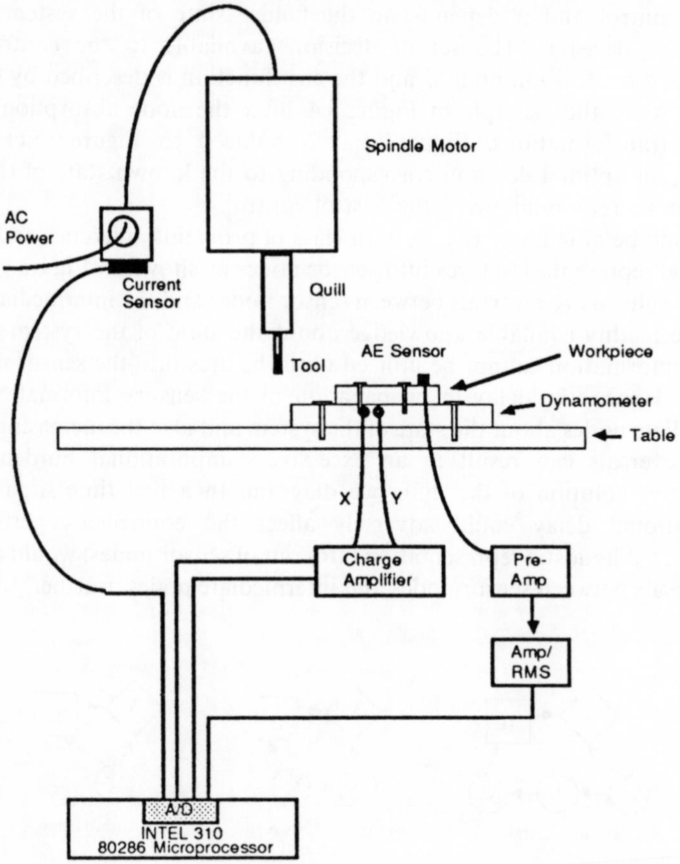

Figure 9.6 Schematic of the milling machine set up (Agogino *et al.*, 1988a).

state of the machine. As a demonstration of its potential in intelligent machining, IDES was applied to a numerically controlled (NC) upright Bridgeport milling machine. As shown schematically in Figure 9.6, the amplified sensor outputs of acoustic emission, dynamometer and a.c. current sensor were fed into an Intel 310 computer based on an Intel 80286 microprocessor equipped with an analog to digital (A/D) converter as an add-on board. Data were acquired from four channels, one each for the AE signal, dynamometer X and Y directions, and the current sensor. Data were sampled every 1 ms from all four channels, in approximately 750 μs, over the cutting parameters and machining conditions defined by this study (Agogino *et al.*, 1988a).

In order to develop a 'real time simulation' that did not compete with daily users for time on the milling machine the data were uploaded from the Intel 310 to a DEC Microvax™ computer. The simulation reads the actual sensor data from files on the Microvax™ as if it were reading them in real time. Feature extraction from the multiple sensor data is performed over sets of 100 sensor readings, which correspond to the values of sensor nodes in the influence diagram. IDES gives a diagnosis of the current state of the machine tool and also recommends a decision to the operator or controller based on these sensor readings. The simulation repeats itself after every 100 ms time interval in which 400 new data points (one reading every milliseconds from four channels) and 400 previous data points (for temporal trending) are analyzed in order to arrive at the optimal control decision.

9.5.1 Feature identification and extraction

Feature identification involves identifying important signal features and their interrelationship with the unobservables of interest. Feature extraction denotes the mapping of the numerical sensor readings to symbolic values of the sensor nodes. Nii *et al.* (1982) refer to this as the 'signal-to-symbol' problem in building expert systems.

In the milling machine example, because some of the features extracted are temporal in nature, a time history of the sensor output is required. For example, to extract the frequency content of the acoustic emission signal over each 100 ms time interval, a Fourier transform of the signal based on its time history of 100 data points was performed. In general the length of the time interval used will be limited by the response time required by the application. In controlling a milling machine a fast response time is desirable. We have found the 100 ms time interval to give satisfactory results; however, on some machines faster responses may be preferable for detecting tool breakage (Agogino *et al.*, 1988a).

A brief summary of the development of the milling machine influence diagram model follows. The interested reader is referred to Agogino *et al.* (1988a) for more detail.

Acoustic emission

Acoustic emission signals were obtained from an Endevco 920A piezoelectric transducer mounted directly on the workpiece. The RMS signal was fed through an A/D converter in the Intel 310 computer for data acquisition. Acoustic emission can be an important predictive measurement for detecting chatter, the self-excited vibration that occurs when a machine tool exceeds its stability limit. If detected, chatter can usually be corrected by reducing the depth of cut or feed rate. If left undetected, chatter can cause tool, workpiece or machine damage. When chatter occurs, the AE signal increases dramatically in amplitude. This trend can be observed by comparing Figure 9.7 with Figure 9.8: Figure 9.7 shows the AE RMS signal for entry of the cutting tool into the workpiece with

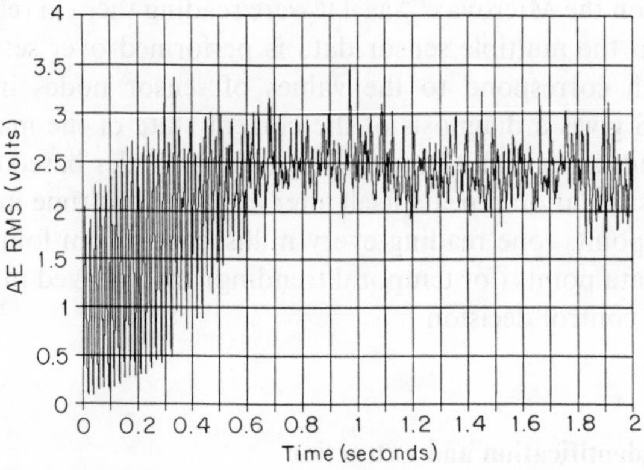

Figure 9.7 AE vs time for entry cut without chatter (Agogino *et al.*, 1988a).

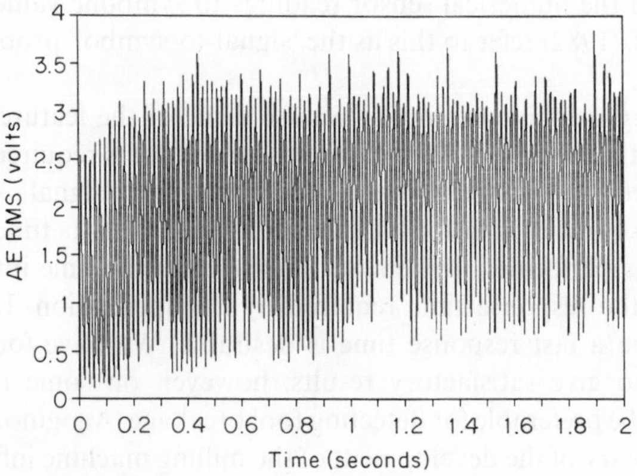

Figure 9.8 AE vs time for entry cut with chatter (Agogino *et al.*, 1988a).

no chatter, whereas Figure 9.8 shows an entry cut with chatter. The large initial amplitude in both cases is due to the entry of the tool into the workpiece. Once the tool is completely in the workpiece this large amplitude dies down when chatter is not present. When chatter is present, however, this high amplitude is maintained. Thus the AE magnitude and change in magnitude are good indicators of chatter. Further, the frequency content of the signal can be used to strengthen the hypothesis of whether or not chatter is present.

These predictive indicators of chatter in the AE signal (AE magnitude, change in AE magnitude and frequency content) were used as nodes in the related influence diagram model shown in Figure 9.9. The magnitude of the signal was defined as the difference between maximum and minimum values and the change in magnitude defined from the previous 100 ms time interval. At the numerical level, the frequency node was represented by binary values: 'on' only if the energy levels of frequencies above 100 Hz were above characteristic levels for the machine and 'off' otherwise.

The AE signal can also be useful in detecting tool wear (the average AE signal increases as the tool wears) and in detecting tool breakage (the AE RMS typically exhibits a high amplitude peak at the moment of tool fracture which is followed by a sharp drop in signal amplitude to a level below that of normal machining.) Thus the mean RMS value can be used to detect tool wear as modeled in Figure 9.10(a) ('ΔMEAN' represents the change in the average AE signal from the previous time interval to the present one.) In the influence diagram model for tool breakage in Figure 9.10(b) the peak AE RMS is compared to a threshold level. The 'PEAK' node is a binary node that is set to 'yes' if the maximum AE RMS exceeds the threshold and 'no' otherwise.

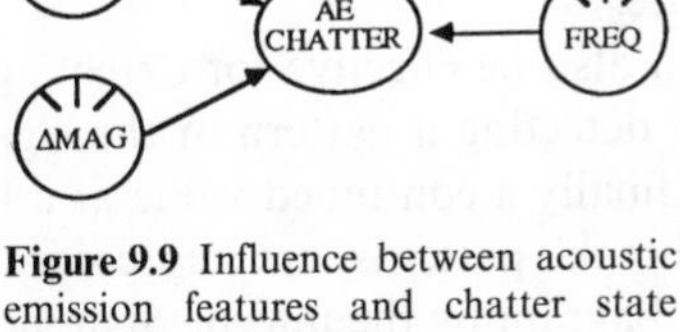

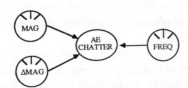

Figure 9.9 Influence between acoustic emission features and chatter state (Agogino *et al.*, 1988a).

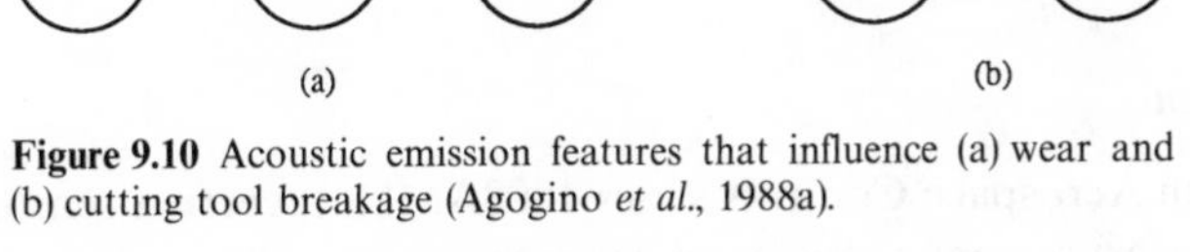

Figure 9.10 Acoustic emission features that influence (a) wear and (b) cutting tool breakage (Agogino *et al.*, 1988a).

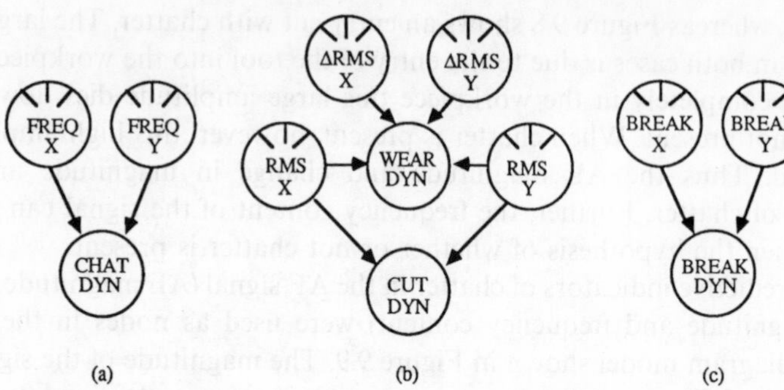

Figure 9.11 Features of the dynamometer signal (Agogino *et al.*, 1988a).

Force sensing

A Kistler 9257A dynamometer (piezoelectric transducer) was used to measure the force in the milling plane of the workpiece. As shown in Figure 9.6, the dynamometer was bolted to the milling machine table and the workpiece bolted onto the dynamometer.

Chatter can also be detected by the frequency content of the dynamometer as modeled in Figure 9.11(a). Because an increase in tool wear results in an increase in the cutting force, the magnitude and the change in the cutting force RMS along both *X* and *Y* directions are useful in predicting worsening tool wear. The Kistler dynamometer is extremely sensitive to any variation in the force applied to the workpiece, and thus the increases in the forces during entry and exit of the multi-toothed milling tool must also be considered in interpreting the RMS signal from the dynamometer. As shown in the influence diagram in Figure 9.11(b), the magnitude and change in RMS of the dynamometer signal are used to detect the wear state, in conjunction with the use of the RMS to determine whether the cut (shown as the intermediate node labeled 'CUT DYN') is in an entry or exit state.

The dynamometer can also be effective for detecting cutting tool breakage. This is accomplished by detecting a pattern in the signal showing a large rise followed by a drop and finally a continued value at a level above the previous average value. These 'break' patterns are represented in the dynamometer portion of the influence diagram by the intermediate nodes labeled 'BREAK X' and 'BREAK Y' in Figure 9.11(c). This pattern recognition technique for the IDES implementation has proven to be both effective and fast, requiring much less signal processing than techniques using time series analysis and linear discriminant functions.

Motor current

An American Aerospace Controls series 1003AMl a.c. current sensor (based on induction measurements) was used to measure spindle motor current. The current drawn by the motor is directly proportional to the torque exerted by the motor and hence is proportional to the cutting force. Because of the sharp changes in the motor load during the tool entry and exit, the current sensor is an excellent detector of the cut state. This is extremely valuable because these transitions are difficult to distinguish with the other sensors.

9.5.2 Sensor fusion

The 30 node composite influence diagram shown in Figure 9.12 was obtained by combining the influence diagram modules in Figures 9.9–9.11 and adding the effect of the spindle motor current sensor node. The event space of each

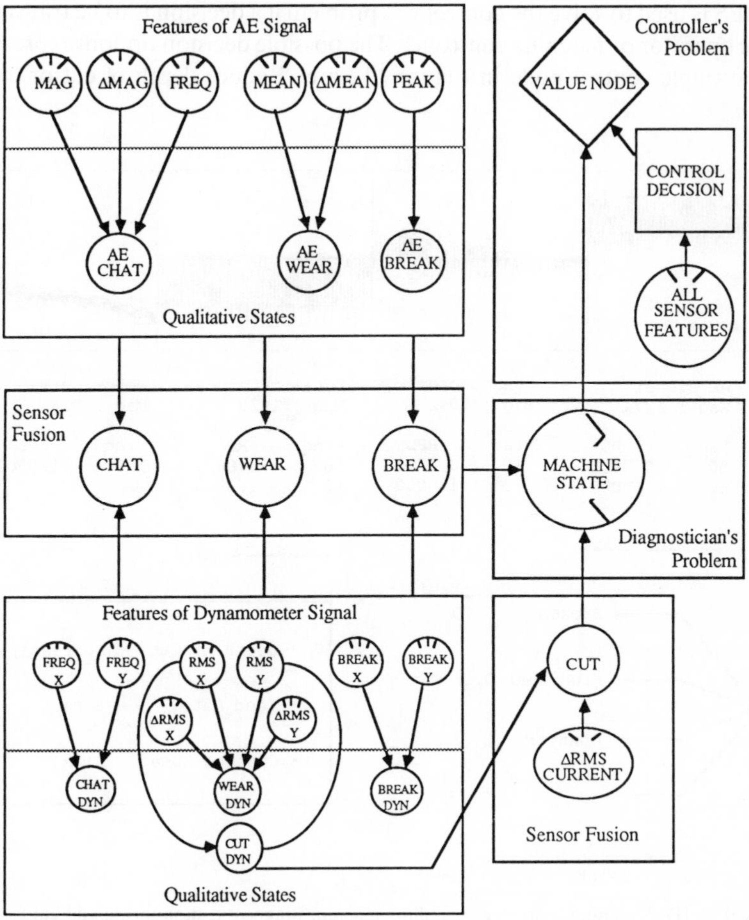

Figure 9.12 Composite influence diagram for milling machine problem (Agogino *et al.*, 1988a).

node ranges from two to four events. The goal of the diagnostician's problem is to infer the machine state from the observable sensor readings. The single failure node in Figure 9.12 represents the fusion of information concerning the machine state from the qualitative features of the AE dynamometer and current sensors. This node can take on values corresponding to 'chatter', 'worn tool', 'broken tool' and 'OK', which denotes a normal machine state. The IDES inference engine uses Bayesian probability to propagate the implications of the sensor readings to the likelihood estimates for each possible state of the milling machine. Conditional independence is only assumed where there are no arcs drawn in the influence diagram in accordance with both expert opinion and the experimental data. Arcs between sensor nodes, for this application, are not drawn under the assumption that all diagnostic queries will be conditioned on these nodes.

IDES is used to solve the controller's problem if a decision is to be transmitted to the operator or machine controller. The possible decision options represented by the single control node in Figure 9.12 are: 'Reduce depth of cut or cutting

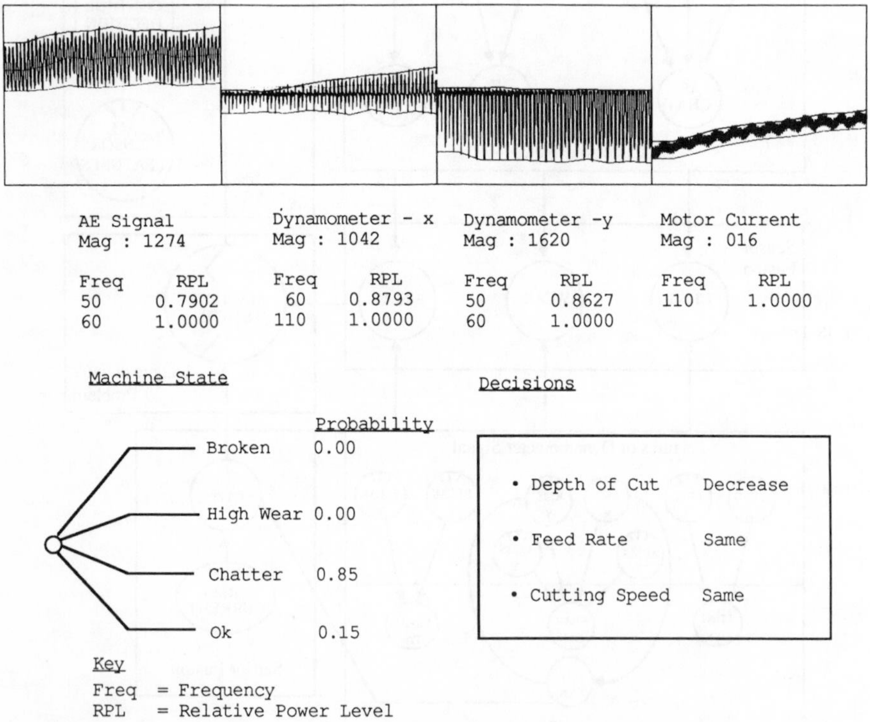

Figure 9.13 IDES computer display for milling machine diagnostic, monitoring and control. Freq: frequency, RPL: relative power level (Agogino *et al.*, 1988a).

speed or feed rate', 'Retract tool (and call operator)' or 'No change'. IDES uses dynamic programming to optimize a parametrized utility function over the control options and assessed uncertainty of the possible failure states.

The results are presented on a computer screen in both graphical and text form for the benefit of the machine operator or the floor supervisor (Figure 9.13). Plots of each signal (AE, dynamometer, and motor current) are displayed in 100 ms windows. Under each plot are the magnitudes of the two most dominant frequencies and their relative energy levels (energy of the dominant frequency normalized to one). The probability distribution of the machine state and the recommended control decisions are highlighted below the signal plots.

9.6 DEVELOPMENT AND TUNING

The topology of the influence diagram (identification of relevant features and dependencies) was obtained from dialogues with experts and from a study of literature in the area. The probability distributions encoded at the numerical level of this model were based on subjective estimates that were tuned for the specific machine tested. No explicit physical models were used to represent the cutting process; only statistical data and experimental knowledge derived from the experts in the form of subjective probability distributions were used in arriving at the optimal decisions. The experts found it natural to model the problem such that arcs were drawn from sensor features to failure states which, as explained earlier, is an ideal computational structure for diagnostic reasoning. Due to this type of diagnostic representation, only a few topological transformations on the influence diagram were needed in order to optimize the structure for efficient computation in solving the controller's problem.

One advantage of influence diagrams is that they can be used to answer arbitrary queries concerning the system states. Suppose one of the sensors fails or is removed? IDES need not condition its recommendations on that sensor. Consider the elimination of the dynamometer sensor, which is intrusive and is not very practical for use in most day-to-day machining operations. Even in this case, without the dynamometer, IDES made reliable predictions, though as expected, the uncertainty factor was wider than when the dynamometer was included.

9.7 RESULTS

IDES made correct diagnostic predictions in sixteen cases and optimal control decisions in nineteen cases in twenty-two trial cuts. The event with the maximum probability was compared to the state actually observed in the cut in each of these cases (e.g. Table 9.1). In the cases in which the most likely failure event did not correspond to the actual event (e.g. Table 9.2), the strength of the prediction was not strong (probabilities close to 50 percent) and thus justified

Table 9.1 Detection of chatter with control decision to 'reduce depth of cut' (trial cut no. D4D; depth of cut, 60 1/1000 in; feed rate, 15 in/min; cutting speed, 1600 rpm; no. samples, 3000; description, 'end exit/chatter'; time interval no. 2 (Agogino *et al.*, 1988a).

Probability	OK	Chatter	Worn	Broken	Decision
With dynamometer	0.15	0.85	0.00	0.00	Reduce depth of cut
Without dynamometer	0.00	0.47	0.38	0.15	Reduce depth of cut

Table 9.2 Ambiguous inference and conservative decision to 'reduce depth of cut' (trial cut no. D3D; depth of cut, 30 1/1000 in; feed rate, 15 in/min; cutting speed, 1600 rpm; samples, 3000; description, 'normal'; time interval no. 3 (Agogino *et al.*, 1988a).

Probability	OK	Chatter	Worn	Broken	Decision
With dynamometer	0.47	0.53	0.00	0.00	Reduce depth of cut

the display of the entire distribution of events, not just the one with highest probability. In the nineteen optimal control cases the decision of the system corresponded to that which a human operator of the machine would recommend. The results for a typical cut with and without the dynamometer sensor are shown in Table 9.1.

In the three examples where the decision was suboptimal the error was on the 'safe' side in all cases. Table 9.2 demonstrates one of these cases. The system predicts nearly equal probabilities for 'chatter' and the 'OK' state resulting in an ambiguous inference (although the most likely event is 'chatter' but with relatively weak support). IDES responds in a robust fashion, recognizing that the data are noisy and the uncertainty level is high. The decision made is to reduce the depth of cut because of the strong possibility that the tool is chattering. The actual state was 'OK' (no chatter, tool breakage or worn tool). The ideal decision should have been to make no change in the machine settings. Thus the actual decision recommended is conservative. This reflects the risk aversion in the utility function used in this example (Agogino *et al.*, 1988a). The parametrized utility function can always be modified to adjust the level of risk aversion and relative costs.

9.8 RESEARCH EXTENSIONS

The signal processing for this application was performed entirely in software taking approximately 100 ms on a DEC Microvax™ II. Solution of the diagnostician's problem or controller's problem takes another 100 ms each. Because it is desirable to have much smaller cycle times when implementing the system in a feedback control loop to the machine tool, several strategies are under investigation. Although the milling machine influence diagram was optimized off-line for a useful set of inference queries, some of the decision queries could be compiled out into set-points on a machine controller or by rules in software; e.g. *If RMS of acoustic emission is above a set-point then the tool is broken and retract the tool.* Hardware solutions under consideration include: (1) implementing part of the system on a microchip; (2) using add-on boards for signal processing in hardware; and (3) using array processing for the numerical computation (Agogino *et al.*, 1988a).

The application of IDES to the milling machine monitoring and control problem has stimulated research and development in the following areas:

1. Application of the work in reasoning by analogy and induction to optimize the features used in the influence diagram model and improve the probability estimates;
2. Automatic generation of the influence diagram by inductive learning from a given set of examples (Russell, 1986; Russell *et al.*, 1988);
3. Integrating the influence diagram model with neutral network schemes (Agogino, 1988);
4. Use of the Simul module to estimate the time-computational costs of alternate algorithms (Rege and Agogino, 1988) or models (Horvitz *et al.*, 1986) and make meta-control decisions in choosing the appropriate algorithm or model to use, trading off accuracy and response time.

Research on the use of IDES for supervisory control of robotic manipulators has led to development of dynamic influence diagrams that represent the time varying attributes of an evolving system (Ramamurthi and Agogino, 1988). Using the dynamic variables and their derivatives that define the motion of a manipulator arm, the performance of the adaptive controller of the arm is modeled in a real time influence diagram paradigm. The suitability of IDES as a supervisory expert controller to predict the incipient instability under unpredictable external disturbances and take corrective action in real time has been investigated in this study with positive results. The real time expert system, with the information from sensors, dynamically optimizes the cost of control and as a result chooses between robust auxiliary controller and the non-robust adaptive controller depending on inferences made from the observable variables.

ACKNOWLEDGEMENTS

The authors would like to thank Dave A. Dornfeld and Sabbir S. Rangwala of the RAMP (Robotics, Automation, Manufacturing and Production) Group in the Department of Mechanical Engineering at UC Berkeley for their expert advice and assistance in developing the milling machine application. Kenneth M. Schneider developed the original milling machine model using the IDES software and Sampath Srinivas extended the application. Much of this chapter is derived from Agogino *et al.* (1988a, b). Further details of the milling machine application can be found in Agogino *et al.* (1988a). Machine learning extensions using reasoning by analogy are described in Agogino *et al.* (1988b) and are primarily the efforts of Ramanathan Guha and Stuart Russell. The authors also acknowledge Ashutosh Rege for his theoretical contributions concerning sensor-based inference using influence diagrams. The influence diagram work was funded, in part, through NSF grant DMC-8451622 and the State of California's Project MICRO (Microelectronic Research Opportunities) Program with matching funds from the General Electric Company and the IBM Corporation.

9.9 DISCUSSION

9.9.1 Discussion by Jack Breese

This chapter describes one of the first efforts to use influence diagrams for modeling and controlling a mechanical device. It presents an architecture for real time reasoning with influence diagrams, develops an innovative algorithm for decision-theoretic inference, and provides some insights into knowledge engineering with influence diagrams for mechanical devices. While the IDES system convincingly demonstrates that diagnostic and control reasoning using influence diagrams at time scales appropriate for unattended milling machine operations are feasible, it is less clear what motivates the use of influence diagrams as the real time delivery vehicle for this knowledge. An alternative approach would be to use influence diagrams as tools for designing diagnostic and control rules off-line, and subsequently implemented and delivered in the most efficient manner possible (for example as a set of rules). There are a number of possible justifications for the explicit use of influence diagrams in real time, some of which were hinted at in the chapter, all of which require further research.

An influence diagram is a more flexible and general representation than a set of hard-coded rules. An ID can be used to answer arbitrary probabilistic queries, make decision recommendations and be used as a tool to explore a problem structure. A series of diagrams can encode different probability and value models for different part types in the milling machine application. An interesting issue is expression of risk aversion for different parts in a manner consistent with higher level controls and objectives. The system must be able to flexibly adapt to changing situations and parts.

Learning is a key element of intelligent behavior. Data collection can be used to update probability distributions, but are there ways for a real time system to learn about the structure of the models it is using? In many instances there is a fundamental trade-off between achieving objectives on the one hand and gathering information about the properties of a system on the other.

A final topic is control of reasoning—a real time system must reason about its own problem-solving activities as well as the objective level task at hand. Using an influence diagram is expressive and flexible, but also computationally expensive. An architecture for real time monitoring and control should be able to (1) demonstrate a means of developing approximate solutions in limited amounts of time and (2) devise a method for deciding how much computation and reasoning to put into a particular task. A complete answer is of little use if it is not available in a timely fashion.

The chapter by Agogino is an ambitious step towards development of real time systems for mechanical devices. Though in principle capable of closed loop supervisory control of a milling machine, the system has been designed and demonstrated primarily as an advisor to a human operator. Additional benefits of an influence diagram approach versus more conventional methods for real time control will be clearer as the system is applied in more autonomous situations and addresses more fully the issues outlined above.

9.9.2 Reply

The discussant has provided an excellent response to his own perceptive questions concerning the use of influence diagrams (ID's) as real time delivery vehicles. I agree with all of his points and admit that many of them are the subjects of ongoing research projects. I will expand on some of his comments below.

I agree that once the query is specified in advance, for either a decision or inference problem, the solution sequence can be optimized off-line and compiled into a simple set of rules or a look-up table. For example, I believe that this *off-line optimization* approach would be superior to real time ID processing for the 'tool breakage' failure mode decision. An appropriate rule might be:

> IF the acoustic AE RMS amplitude is above a critical threshold level.
> THEN cutting tool is broken, retract the cutting tool and call the operator.

On the other hand, if an event occurs which was not originally modeled and optimized off-line (e.g., failure of one of the sensors) the structure of a decision query may change. The failed sensors may no longer be informational predecessors of the decision nodes. Through self-inspection or via an external query from the operator, the real time ID system could be used to explore this possibility. The ability to respond to unplanned events and make spontaneous adaptations is a mark of an intelligent system: one that can adapt and reconfigure itself to respond to unforeseen events. This adaptive ability is even more pronounced for systems that can respond to a general inference query. In IDES, the influence diagram and ID solver provide both the data structure and inference procedure for answering arbitrary queries on the system in real time, without requiring the preprocessing and storage overhead required to solve and store every possible query, as is done with the 'undirected graph completion' methods.

I also agree that machine learning is an essential attribute of an intelligent real time system. I and my students have performed some preliminary work in off-line machine learning as discussed briefly in the 'Research Extensions' section of this paper (analogical reasoning, inducing influence diagrams and neural networks) and as described respectively in references by Russell *et al.* (1988), Agogino *et al.* (1988b) and Agogino (1988).

The IDES Simul module is our approach to meta-level reasoning about the system's own problem solving activities. The Simul module can be used to estimate the time-computational costs of alternate algorithms (Rege and Agogino, 1988) or choices of models (Horvitz *et al.*, 1986). Although Simul's features were not utilized in the milling machine example, it is in theory capable of estimating the costs of competing algorithms and models in order to trade off accuracy and response time consistent with the system's objectives and risk factors.

REFERENCES

Agogino, A. M. (ed.) (1988) *Research Summaries of the Berkeley Expert Systems Technology Laboratory*, Summer 1988, Dept of Mechanical Engineering, University of California, Berkeley, Calif.

Agogino, A. M., Srinivas, S. and Schneider, K. (1988a) Multiple sensor expert system for diagnostic reasoning, monitoring and control of mechanical systems, *Mechanical Systems and Signal Processing*, **2**(2), 165–85.

Agogino, A. M., Guha, R. and Russell, S. (1988b) Sensor fusion using influence diagrams and reasoning by analogy: application to milling machine monitoring and control, *Artificial Intelligence in Engineering: Diagnostics and Learning*, Computational Mechanics Publications, Southampton, pp. 333–57.

Agogino, A. M. and Rege, A. (1987) IDES: influence diagram based expert system, *Mathematical Modelling*, **8**, 227–33.

Altintas, Y., Yellowsley, I. and Tlusty, J. (1985) The detection of tool breakage in milling, *Sensors and Controls in Manufacturing*, ASME-PED, **18**, 41–8. (*Note*: The name of Y. Altintas was misspelled as Y. Attanis in the publication.)

Braun, S., Rotberg, J. and Lentz, E. (1987) Signal processing for single tooth milling monitoring, *Mechanical Systems and Signal Processing*, **1**(2), 185–96.

Bronstein, M. (1987) Documentation for IDES: Influence Diagram based Expert System, Berkeley Expert Systems Technology Lab., Dept of Mechanical Engineering, University of California, Berkeley.

Cheeseman, P. (1985) In defense of probability, *Proceedings of the Ninth International Joint Conference on Artificial Intelligence*, AAAI, **2**, 1002–9.

Diei, E. N. and Dornfeld, D. A. (1985a) Acoustic emission from the face milling process—the effect of process variables, *Sensors and Controls in Manufacturing*, ASME-PED, **18**, 75–84.

Diei, E. N. and Dornfeld, D. A. (1985b) A model of tool fracture generated acoustic emission during machining, *Sensors and Controls in Manufacturing*, ASME-PED, **18**, 33–9.

Dornfeld, D. A. (1984) The role of acoustic emission in manufacturing process monitoring, *Proceedings of the Sympsium on Sensor Technology on Untended Manufacturing*, SME.

Dornfeld, D. A. (1985) Acoustic emission monitoring and analysis of manufacturing processes, *Proceedings 12th NSF Conference on Production Research and Technology*, SME.

Fox, M. S. (1983) Techniques for sensor-based diagnosis, *Proceedings of the Eighth International Joint Conference on Artificial Intelligence*, AAAI, **1**, 158–63.

Heckerman, D. E. and Horvitz, E. J. (1987) On the expressiveness of rule-based systems for reasoning with uncertainty, *Proceedings of the Sixth National Conference on Artificial Intelligence*, AAAI, **1**, 121–6.

Henrion, M. (1987) Practical issues in constructing a Bayes belief network, *Proceedings of the Third Workshop on Uncertainty in Artificial Intelligence*, AAAI, pp. 132–9.

Henrion, M., and Cooley, D. R. (1987) An experimental comparison of knowledge engineering for expert systems and for decision analysis, *Proceedings of the Sixth National Conference on Artificial Intelligence*, AAAI, **2**, 471–6.

Holtzman, S. (1985) Intelligent decision systems, Ph.D. dissertation, Engineering-Economic Systems, Stanford University. (Reprinted by Strategic Decisions Group, Menlo Park, Calif.)

Horvitz, E. J. (1987) Reasoning about beliefs and actions under computational resource constraints, *Proceedings of the Third AAAI Workshop on Uncertainty in Artificial Intelligence*, p. 412.

Horvitz, E. J., Breese, J. S. and Henrion, M. (1988) Decision theory in expert systems and artificial intelligence, *International Journal of Approximate Reasoning*, **2**, 247–302.

Horvitz, E. J., Heckerman, D. E. and Langlotz, C. P. (1986) A framework for comparing alternative formalisms for plausible reasoning, *Proceedings of the Fifth National Conference on Artificial Intelligence*, AAAI, **1**, 210–14.

Howard, R. A. and Matheson, J. E. (1984) Influence diagrams, *The Principles and Applications of Decision Analysis*, vol. 2, Strategic Decisions Group, Menlo Park, Calif.

Kao, S., Laffey, T., Schmidt, J., Read, J. and Dunham, L. (1987) Real-time analysis of telemetry data, *Proceedings of ESIG—Third Annual Expert Systems in Government Conference*, Washington, D.C., 19–23 Oct.

Laffey, T. J., Cox, P. A., Schmidt, J. L., Kao, S. M. and Read, Y. (1988) Real-time knowledge-based systems, *AI Magazine*, **9**(1), 27–45.

Lambert, M. and Agogino, A. M. (1987) A graphical interface to an influence diagram based expert system, *Proc. of the Second International Joint Conference on Human–Computer Interaction*, p. 324.

Lan, M. S. and Naerheim, Y. (1985) In process detection of tool breakage in milling, *Sensors and Controls in Manufacturing*, ASME-PED, **18**, 49–56.

Miller, A. C., Merkhofer, M. M., Howard, R. A., Matheson, J. E. and Rice, T. R. (1976) *Development of Automated Aids for Decision Analysis*, Final Technical Report, DARPA No. 2742, SRI International, Menlo Park, Calif.

Moore, E. A. and Agogino, A. M. (1987) INFORM: an architecture for expert-directed knowledge acquisition, *International Journal of Man–Machine Studies*, **26**(2), 213–30. (Also in *Knowledge Acquisition Tools for Expert Systems*, Academic Press, 1988, vol. 2, pp. 227–44.

Nii, H. P., Feigenbaum, E. A., Anton, J. J. and Rockmore, A. J. (1982) Signal-to-symbol transformation: HASP/SIAP case study, *AI Magazine*, **3**, No. 2, 23–35.

Olmsted, S. M. (1984) On representing and solving decision problems, Ph.D. dissertation, Engineering-Economic Systems Department, Standford University.

Pearl, J. (1986) On the logic of probabilistic dependencies, *Proceedings of the Fifth National Conference on Artificial Intelligence*, AAAI, **1**, 339–43.

Ramamurthi, K. (1988) User's manual for GraphIDES: Graphical Influence Diagram based Expert System, Berkeley Expert Systems Technology Lab., Dept of Mechanical Engineering, University of California, Berkeley.

Ramamurthi, K. and Agogino, A. M. (1988) Real time expert system for fault tolerant supervisory control, *Proceedings of the 1988 ASME International Computers in Engineering Conference*, **2**, 333–40.

Rege, A. and Agogino, A. M. (1986a) Sensor-integrated expert system for manufacturing and process diagnostics, *Knowledge-based Expert Systems for Manufacturing*, ASME-PED, **24**, 67–83.

Rege, A. and Agogino, A. M. (1986b) Representing and solving probabilistic inference problems in expert systems, *Proceedings of the ICS-86, International Computer Symposium*, IEEE, **3**, 1685–91.

Rege, A. and Agogino, A. M. (1988) Topological framework for representing and solving probabilistic inference problems in expert systems, *IEEE Transactions on Systems, Man and Cybernetics*, **18**(3), 402–14.

Russell, S. J. (1986) Analogical and inductive reasoning, Ph.D. dissertation, Computer Science Department, Standford University.

Russell, S., Srinivas, S. and Agogino, A. M. (1988) Inducing influence diagrams from examples, Working Paper 88-0201-0, Berkeley Expert Systems Technology Lab., Dept of Mechanical Engineering, University of California, Berkeley.

Shachter, R. D. (1985) Intelligent probabilistic inference, *Proceedings of the AAAI Workshop on Uncertainty and Probability in Artificial Intelligence*, 237–44.

Shachter, R. D. (1986) Evaluating influence diagrams, *Operations Research*, **36**(6), 871–82.

Shachter, R. D. and Heckerman, D. E. (1987) A backwards view of assessment, *Artificial Intelligence Magazine*, Fall.

Smith, J. (1989, to appear) Influence diagrams for statistical modelling, *Annals of Statistics*.

Stein, J. L. and Shin, K. C. (1985) Current monitoring of field controlled DC spindle drives, *Sensors and Controls for Automated Manufacturing and Robotics*, ASME-PED, **18**, 57–66.

Stein, J. L., Colvin, D., Clever, G. and Wang, C. H. (1984) Current monitoring on DC servo machine tool feed drives, *Sensors and Controls for Automated Manufacturing and Robotics*, ASME-PED, **18**, 45–63.

Tlusty, J. and Andrews, G. C. (1983) A critical review of sensors for unmanned machining, *Annals of the CIRP*, **32**(2), 563–72.

Tversky, A. and Kahnemann, D. (1980) Causal schema in judgements under uncertainty. In M. Fishbein (ed.) *Progress in Social Psychology*, Lowrence Ernbaum, Hillsdale, NJ.

Wright, P. K. and Bourne, D. A. (1988) *Manufacturing Intelligence*, Addison-Wesley, Reading, Mass.

CHAPTER 10

Multi-target Tracking using Influence Diagram Models

C. Robert Kenley and Thomas R. Casaletto, *Lockheed Missiles & Space Company, Inc., Sunnyvale, USA*

ABSTRACT

Tracking of many targets in a dense environment is one of the greatest technical challenges facing strategic defense systems. Previous efforts to solve this problem have modeled each track as a statistically independent entity, which generates excess false tracks. In the midcourse phase of strategic defense, closely spaced objects deployed in Keplerian trajectories from the same post-boost vehicle (PBV) are more similar than they are different. This is because over 99 percent of the kinetic energy of the object comes from the trajectory of the PBV, and the remainder comes from small amounts of energy used to eject individual objects. This commonality between objects is modeled naturally using a Gaussian influence diagram, which captures correlated motion of the objects in a rigid pattern relative to the PBV trajectory and independent motion from individual object deployment characteristics. Empirical studies of this new technique show superior performance of the influence diagram model over the independent entity model.

10.1 INTRODUCTION

Strategic defense surveillance systems may have to acquire and track large numbers of ballistic objects deployed exoatmospherically from post-boost vehicles (PBVs). These objects may be deployed in closely spaced clusters to reduce the ability of the defense to accurately identify and track reentry vehicles (RVs) that must be intercepted. Current multi-target tracking algorithms model the objects being tracked as statistically independent. One such algorithm is the multiple hypothesis tracker, which processes measurements when they are

Influence Diagrams, Belief Nets and Dicision Analysis
Edited by R. M. Oliver and J. Q. Smith

received and maintains all feasible assignments of a frame of track measurements, deleting tracks when they no longer receive measurements updates. In regions of high target density, the independent tracks formed using this method do not have the required accuracy to keep the number of hypotheses from growing exponentially. This chapter applies a Gaussian influence diagram (Shachter and Kenley, 1989; Kenley, 1986) to model and analyze the dynamic probabilistic system of a cluster of objects. It uses relatively little information about RV cluster dynamics to effectively identify tracks and associate measurements to tracks on a frame-to-frame basis.

This chapter presents a simple two-object, linear-motion model in Section 10.2 to demonstrate the basic concepts of cluster tracking, and Section 10.3 provides parametric sensitivity analysis of the simple model. Section 10.4 presents empirical results from a study of frame-to-frame association of line-of-sight measurements on clusters of objects following Keplerian trajectories, and compares performance of the independent object multiple hypothesis tracker with the influence diagram cluster tracker.

10.2 TWO-OBJECT, LINEAR-MOTION CLUSTER MODEL

First, we will construct a simple model to demonstrate the basic concepts of cluster tracking using Gaussian influence diagrams. In the systems we are studying, the velocity of a single object is comprised of two components, $\bar{v}$ the velocity of the centroid and Δv the difference between the velocity of the object and the centroid. For exoatmospheric objects, the centroid velocity is near $6\,\mathrm{km\,s^{-1}}$ and the difference between the individual object and centroid velocities is on the order of meters per second. The standard deviation of the centroid velocity among cases studied was $300\,\mathrm{m\,s^{-1}}$ and the standard deviation of Δv varied from 2 to $200\,\mathrm{m\,s^{-1}}$.

A Gaussian influence diagram representation of the relationship between the individual velocities, the centroid velocity, and individual Δv's is shown in Figure 10.1. The labels next to each node represent the conditional variance of the variable represented by the node, given the values of its predecessors in the network, and the labels on the arcs represent the linear sensitivity of the variable with respect to a unit change in the predecessor variable from which the arc is constructed. This linear sensitivity arc coefficient is traditionally known as the coefficient of partial regression of the successor on the predecessor. The double circle nodes represent variables that are deterministic, linear functions of their predecessors, and hence, have conditional variances of zero. Applying the node reduction rules from Shachter and Kenley (1989), we can remove $\Delta v_1, \Delta v_2$, and $\bar{v}$ from the diagram, resulting in the two-node velocity model in Figure 10.2. After reduction of these variables from the model, the velocities of the two objects are seen to be correlated in proportion to the ratio of the variance of the centroid velocity to the sum of the variances of the centroid velocity and

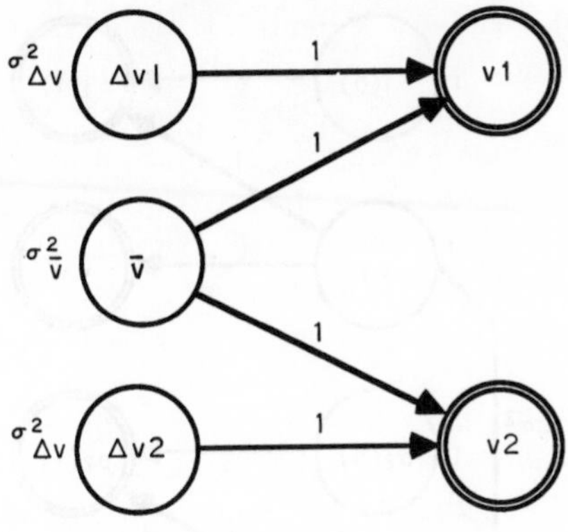

Figure 10.1 Centroid–Δv model

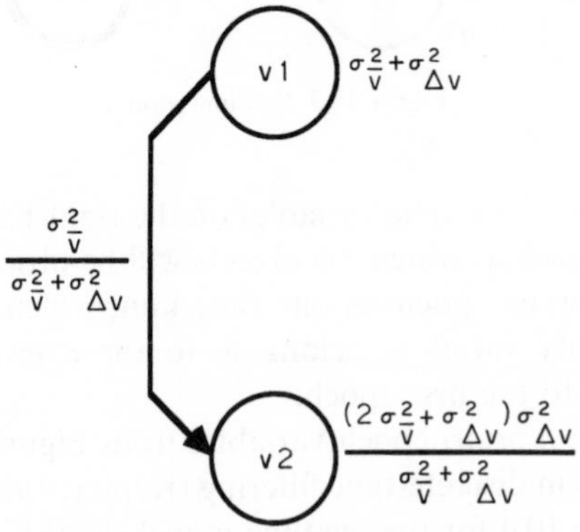

Figure 10.2 Correlated velocity model.

Δv. In the RV tracking problem, this ratio is very near unity, which indicates that the motion of the objects follows a nearly rigid pattern dominated by the velocity of the centroid caused by the relatively large amount of energy inherited from the PBV by all the objects in the cluster.

To reduce the complexity of symbol manipulation for parametric sensitivity analysis, we define a model with velocity uncertainties constrained so that $\sigma_{\bar{v}}^2 + \sigma_{\Delta v}^2 = 1$ (velocity unit)^2, and $\sigma_{\bar{v}}^2 \equiv \sigma^2$. The prior state of information on the state vectors is represented by the epoch zero variables on the left-hand side of Figure 10.3. Velocity uncertainties are derived by substituting the constraints into the appropriate expressions from Figure 10.2, and the baseline standard deviation of the positions of the two objects is one distance unit. The velocities of the two objects are uncorrelated and have unit variance if the variance of the centroid $\sigma^2 = 0$. On the other hand, the second velocity is identically equal

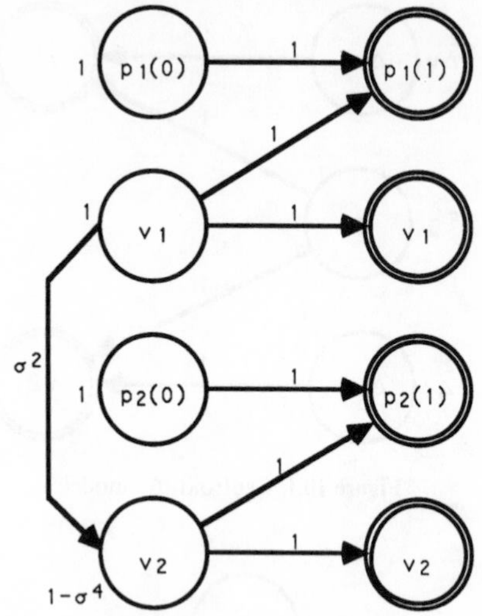

Figure 10.3 Baseline model.

to the first velocity if $\sigma^2 = 1$. The variables on the right-hand side of Figure 10.3 represent the first epoch at which the objects will be observed. For the baseline model, the time between epochs is one time unit, which is shown by ones on the arcs from velocity variables belonging to the *a priori* epoch to position variables belonging to the first epoch.

Elimination of the *a priori* epoch variables from Figure 10.3 is equivalent to the time update step in discrete-time filtering (Kenley, 1986), the result of which are shown in Figure 10.4 for the baseline model. At this point, all state vector components are correlated, depending on the value of σ^2. If $\sigma^2 = 0$, the two objects' state vectors are independent, identically distributed random vectors. If $\sigma^2 = 1$, the second velocity is a deterministic function of the first velocity as before, and the second position has unity conditional variance and arc coefficient one from the first velocity.

Up to this point, we have modeled our prior information about cluster dynamics and demonstrated the process of performing a time update. In Figure 10.5, the likelihood model for data collection is added to the diagram. In the baseline model, the measurements z_1 and z_2 are observations of the positions of the respective objects, which is indicated by arc coefficients of unity from the respective position nodes. The measurements are noisy, and we have assumed that the conditional variance of the measurement error is unity. The next step is the key to measurement assignment and update processing using the influence diagram model. The influence diagram version of Bayes's rule is performed by reversal of the arcs from the prior nodes to the likelihood nodes in Figure 10.5. The result of this sequence of scalar reversals is shown in

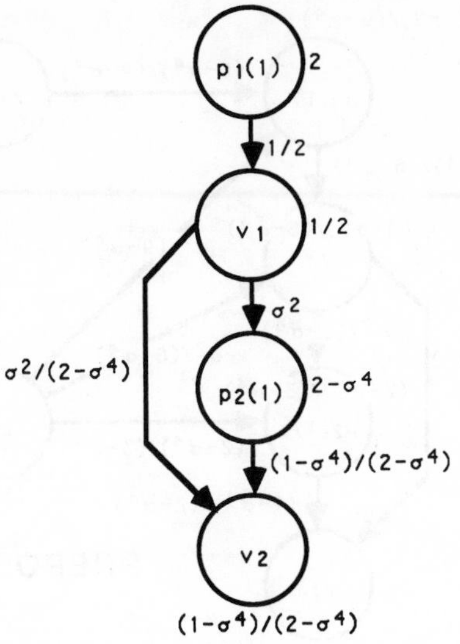

Figure 10.4 Baseline model after first time update.

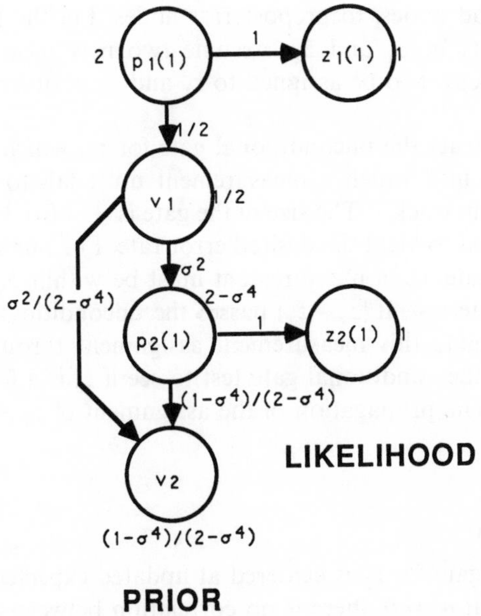

Figure 10.5 Prior–likelihood model.

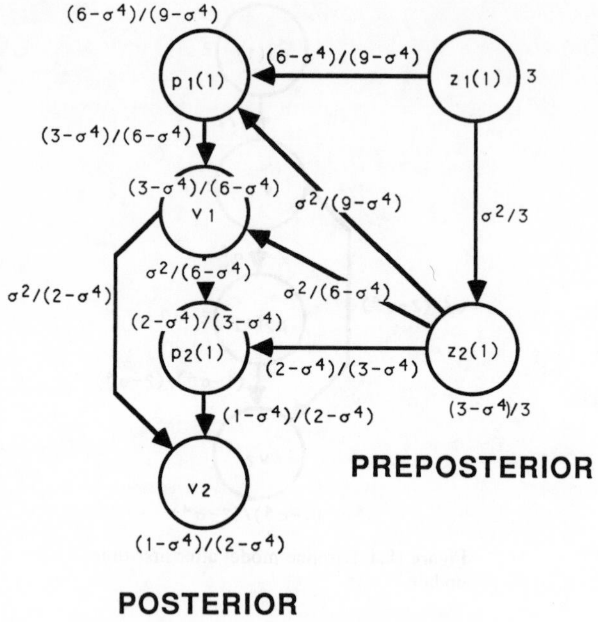

Figure 10.6 Posterior–preposterior model.

Figure 10.6, where the prior nodes have been transformed to posterior nodes; and the likelihood nodes, to preposterior nodes. Let the prior expectation of the measurements be $\hat{z}_1$ and $\hat{z}_2$. Assume two new measurements z_a and z_b arrive from the sensor to be assigned to z_1 and z_2 in order to update the state vectors.

First, we construct the unconditional gate for z_1, which is a credible region about the mean into which a measurement must fall to be considered as a measurement from track 1. The size of the gate is $\pm$ a fixed number of standard deviations selected to yield the desired error rate. For our example, assume the gate is a ± 3-σ gate, i.e. a measurement must be within $3\sqrt{3}$ distance units of $\hat{z}_1$ to pass the gate test. If $|z_a - \hat{z}_1|$ passes the unconditional gate test, the next step is to propagate this measurement assignment through the preposterior model to set up the conditional gate test to see if z_b is a feasible measurement to assign to z_2. The propagation of the assignment of z_a is as follows:

1. $\Delta_1 \leftarrow z_a - \hat{z}_1$;
2. $\hat{z}_1 \leftarrow \hat{z}_1 + \Delta_1$;
3. $\hat{z}_2 \leftarrow \hat{z}_2 + \sigma^2 \Delta_1/3$.

The conditional gate for z_2 is centered at updated expected value of z_2 shown in step 3 above. If $\sigma^2 = 0$, there is no correlation between the objects, and the gate is centered on the original expected value. If $\sigma^2 = 1$, the gate is shifted by one-third of the deviation of z_a from the prior mean of z_1. The width of the conditional gate is a fixed number times the *conditional* standard deviation of z_2, which is $[(3-\sigma^4)/3]^{1/2}$. For the independent object case when $\sigma^2=0$, the conditional standard deviation is 1, but for the most strongly correlated case when $\sigma^2=1$, it is $\sqrt{(2/3)} \approx 0.816$, which is a narrower gate. This shifting of the expected value and tightening of the gate using the influence diagram for correlated tracks is the key to producing fewer false assignments. Also, a hypothesis that a particular permutation of the measurements is associated with the fixed sequence of predicted measurements is tested by considering each scalar element of the permutation in order until an element fails the conditional gate test. At the point of failure, no further hypotheses that are extensions of the failed hypothesis are considered, saving a large amount of computation that would be wasted to determine that a doomed hypothesis is going to fail. For example, consider a case with five objects and a new frame of five measurements. Let the measurements be indexed by $\{a,b,c,d,e\}$ and the tracks by $\{1,2,3,4,5\}$. When testing the hypothesis that the permutation $\{a,b,c,d,e\}$ is the correct measurement assignment to the sequence $\{1,2,3,4,5\}$, the assignment of measurement 'a' with track 1 may pass the unconditional gate test, but the conditional assignment that measurements 'b' is from track 2, given measurement 'a' is assigned to track 1, may not pass the conditional gate test.

At this point, no further testing of the permutation $\{a,b,c,d,e\}$ is required, and the permutations $\{a,b,c,e,d\}$, $\{a,b,d,c,e\}$, $\{a,b,d,e,c\}$, $\{a,b,e,c,d\}$, and $\{a,b,e,d,c\}$ need no further testing either. If all the assignments in the permutation pass the successive conditional gate tests, it is equivalent to the permutation passing an unconditional χ^2 test with five degrees of freedom. If the χ^2 test is used to test association hypotheses, it is necessary to test all $5!=120$ permutations to maintain a given error rate, but the influence diagram structure can be used to achieve the same error rate with much less computation. In the strategic defense midcourse problem where there can be dozens of objects within a cluster, the computational savings become significant.

Once the measurement assignment is made, the arc coefficients from the preposterior nodes to the posterior nodes are used to update the state vectors. This provides additional computational efficiency over standard tracking techniques, where two sets of matrix inversions are performed, one for gate construction and one for measurement updates. For $\sigma^2=0$, the updates of the first and second object state vectors are independent, because the arc coefficients become zero from z_2 to p_1 and v_1, and from p_1 and v_2 to p_2 and v_2. For $\sigma^2=0$, the posterior variance of p_1 is $\frac{2}{3} \approx 0.667$, but for $\sigma^2=1$, it is $\frac{5}{8}=0.625$. Thus, the tracking accuracy of correlated objects is better than the accuracy of independent tracks. This leads to smaller *unconditional* gates for correlating the first measurement when using influence diagram cluster tracking, which improves performance with respect to generating fewer excess tracks and restricting the depth of search when testing an incorrect permutation.

10.3 SENSITIVITY ANALYSIS OF THE TWO-OBJECT, LINEAR-MOTION MODEL

This section presents parametric sensitivity of the baseline two-object model with respect to the *a priori* variance of the centroid velocity and the frame time between observations. The sensitivities demonstrate regions where the influence diagram model provides superior results to the independent entity model.

Figure 10.7 shows sensitivity of key parameters that determine association gate sizes with respect to the variance of the centroid velocity. Figure 10.7(a) shows sensitivity of the unconditional gate for the first through sixth sample. These early samples are important, because this is the region where objects are most closely spaced and have a high potential for producing multiple association hypotheses that cannot be rejected. For the unconditional gate, the influence diagram offers strongly increasing returns when the centroid variance is 80 percent or more of the total velocity variance. The correlation coefficient between the two measurements is shown in Figure 10.7(b). For centroid variances greater than 95 percent, the correlation between measurements is larger for samples two and three than for the first sample. This larger correlation produces a stronger shift of the conditional gate when the measurement associated to z_1 is not centered at the expected value. This represents correction for an underlying bias of the velocity when the centroid variance is large enough to allow significant deviation of the centroid velocity from its expected value. The allowable correction for bias begins to decrease after four samples are collected, and state vector estimates become more reliable. As the number of samples increase, the correlation between z_1 and z_2 approaches zero, which suggests a strategy that uses the cluster tracker to initiate tracking on a closely spaced cluster until the correlation has converged to a level that allows tracking of the objects as independent entities. The conditional gate variance

$$\sigma^2(z_2|z_1) = \sigma^2(z_1)[1 - \rho^2(z_1, z_2)]$$

is shown in Figure 10.7(c), which shows stronger sensitivity to centroid variance than the unconditional gate. The influence diagram sequential association procedure described in Section 10.3 uses a parallelogram acceptance region as shown in Figure 10.8. The area of the parallelogram is equal to the product of the conditional and unconditional gate widths and is plotted in Figure 10.7(d). The area combines all the information in Figure 10.7(a)–(c) in a single figure of merit to determine the region where influence diagram cluster association may produce better results than the independent entity model, which is represented by $\sigma^2 = 0$ in the sensitivity plots. The gate area parameter shows strong sensitivity for samples two and three, which may be justify designing an influence diagram tracker when the centroid variance is a low as 40 percent of the total variance.

Figure 10.9 shows sensitivity with respect to the frame time between observations when the centroid velocity variance is assume to be 95 percent,

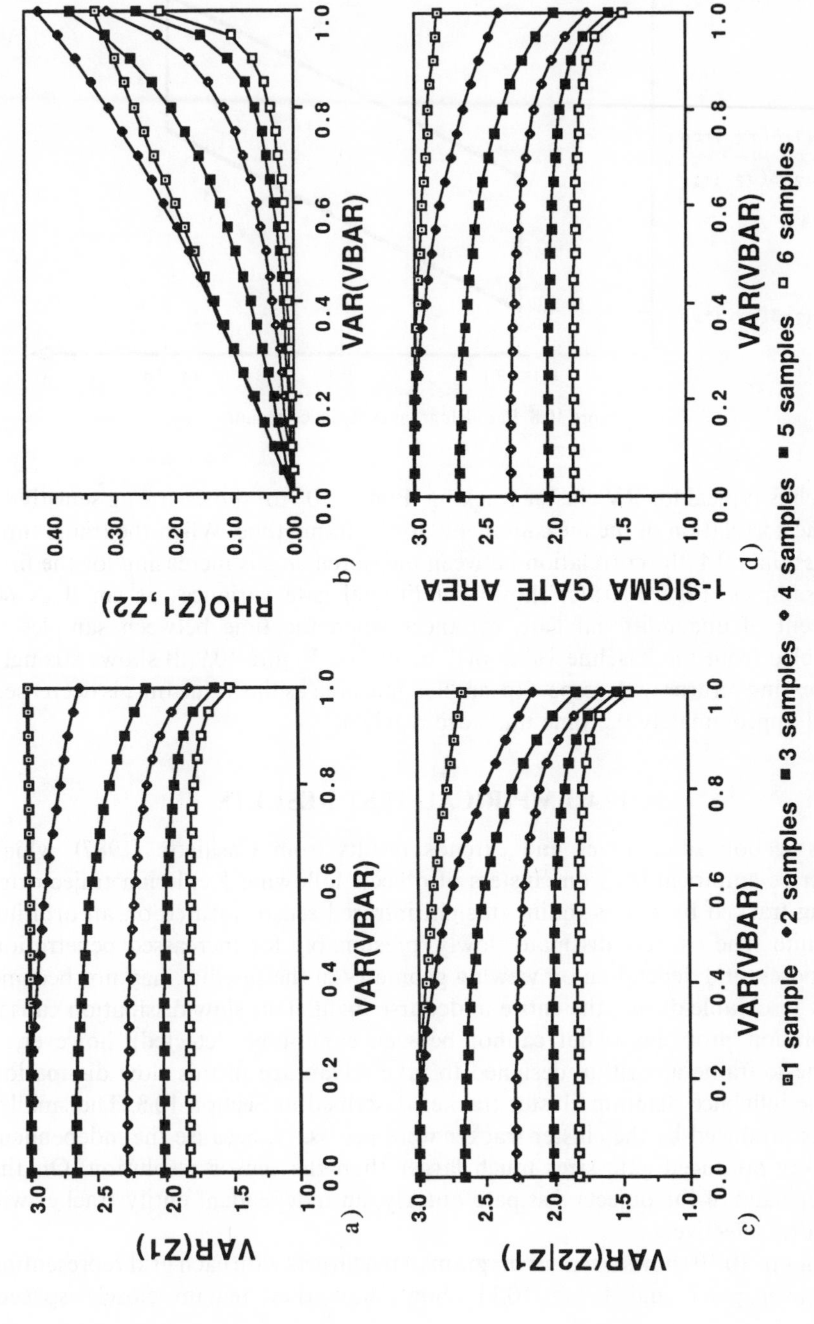

Figure 10.7 Sensitivity analysis with respect to centroid variance.

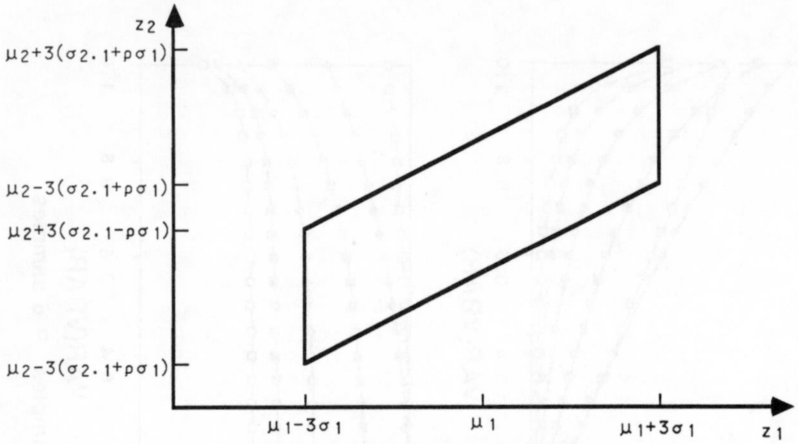

Figure 10.8 Parallelogram acceptance region.

which is typical for RV cluster tracking. Figure 10.9(b) shows strong sensitivity of the correlation of the measurements to the frame time. When the frame time is less than 0.4, the correlation between measurements is increasing for the first six samples. Figure 10.9(c) shows conditional gate variances as small as 60 percent of unconditional gate variances when the time between samples is doubled from the baseline value of 1 time unit. Figure 10.9(d) shows strongly increasing returns with respect to smaller gate area as the frame time is decreased until approximately 0.2 time units are reached.

10.4 EMPIRICAL TEST RESULTS

This section summarizes and extends results from Casaletto (1987), which describe empirical tests on clusters of objects following Keplerian trajectories being tracked by a passive line-of-sight infrared sensor located on an orbiting satellite. The clusters dissipate slowly, presumably for increassed penetration purposes, and depending on viewing geometry of the satellite may not become fully resolvable during the entire midcourse flight. This slow dissipation causes resolution problems (what cannot be seen cannot be detected); however, a frame-to-frame algorithm designed to take advantage of this slow dissipation is the influence diagram cluster tracker described in Section 10.3. The smaller gates produced by the cluster tracker were necessary, because the independent tracker produced gate sizes much larger than the sensor resolution. On the other hand, if the objects dissipate quickly, an independent entity tracker will perform effectively.

Figure 10.10 shows a scatter diagram of ten targets with each grid representing a sensor pixel, and Figure 10.11 shows how they remain closely spaced.

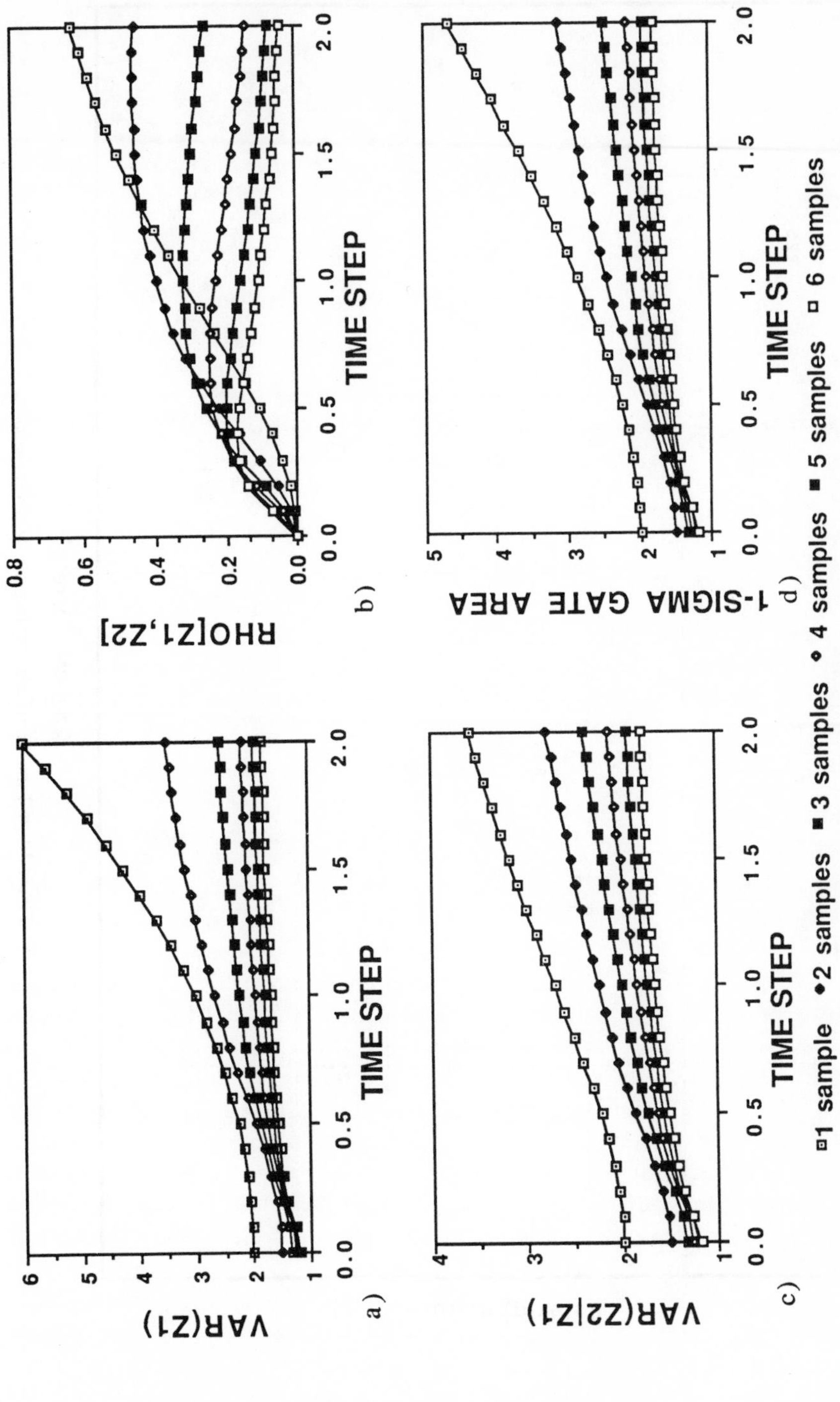

Figure 10.9 Sensitivity analysis with respect to time.

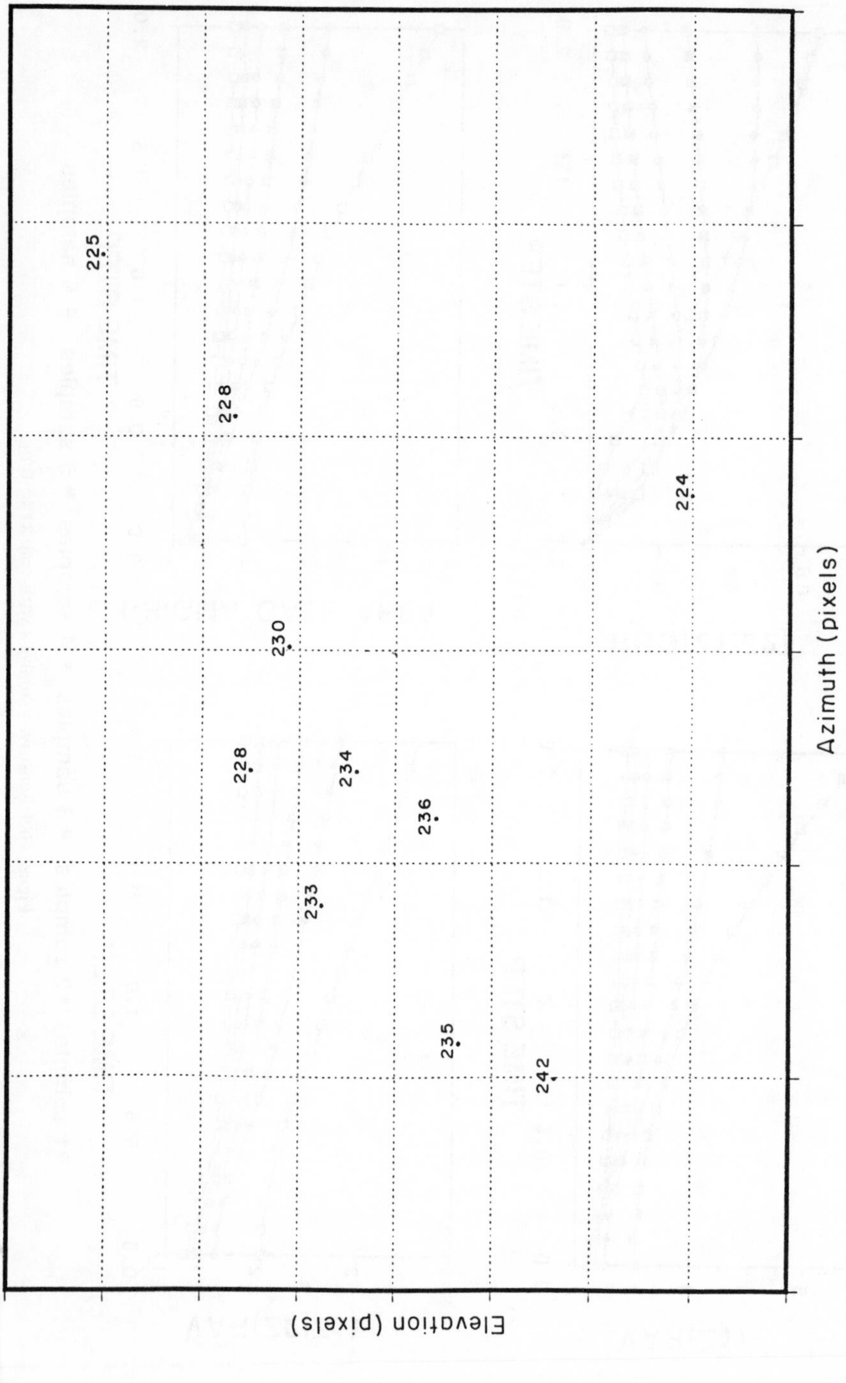

Figure 10.10 First-frame scatter diagram of ten targets.

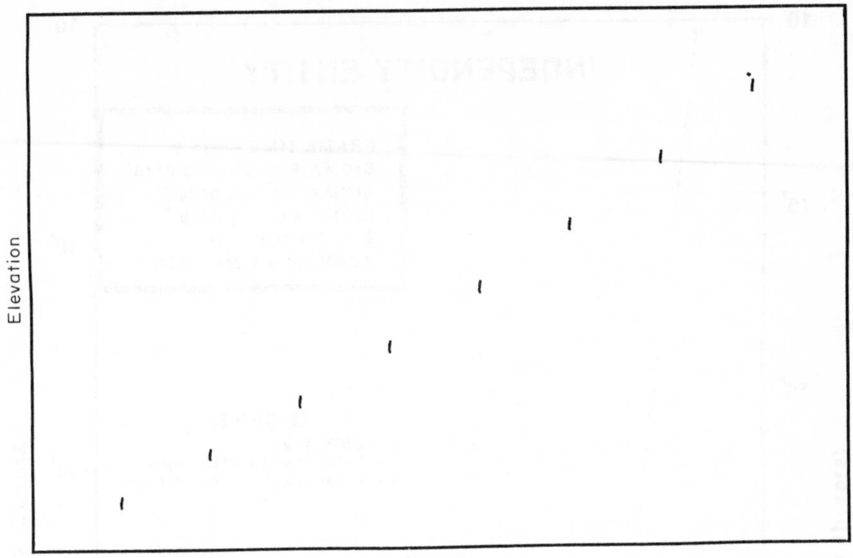

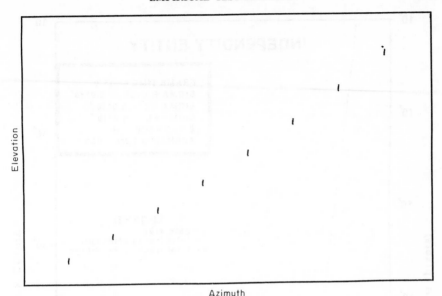

Figure 10.11 Multiple-frame scatter diagram of ten targets.

Figure 10.12 compares the influence diagram cluster tracker and the individual target tracker for this cluster using a 15 s frame time. The cluster of ten targets is tracked in each case, and the track gate size and number of tracks generated are used as measures of algorithm effectiveness. In Figure 10.12(a), the individual target tracker must have a large gate in frame two, because the velocity vector of each track is not well known . This causes many gates to overlap in frame two, resulting in a number of excess tracks. By frame three, the gates have decreased a sizeable amount, but remain large enough to cause excess tracks. The merge logic used in the algorithm reduces the number of excess tracks somewhat, but not entirely. Note the oscillatory nature of the total number of tracks after splitting and merging. Even in steady state, the gata area remains about 5 pixels indicating that two contacts within 2 pixels of each other will have overlapping gates. Figure 10.12(b) has two gate size curves, one for the unconditional gate and one for the first conditional gate. The unconditional gate size is approximately the same size as the gate for the individual target tracker, but the unconditional gate reaches a steady state size of approximately 1.5×1.5 pixels. *Note that there are no excess tracks in this run.* This is a result of using the information that these contacts are slowly dissipating due to the large amount of energy they receive in common from the PBV.

Figure 10.13 shows algorithm results with a 30 s frame time and a 13-object cluster. Similar to the 15 s frame time results, the individual target tracker shows many excess tracks, a steady state gate size of 6 pixels, and oscillatory behavior

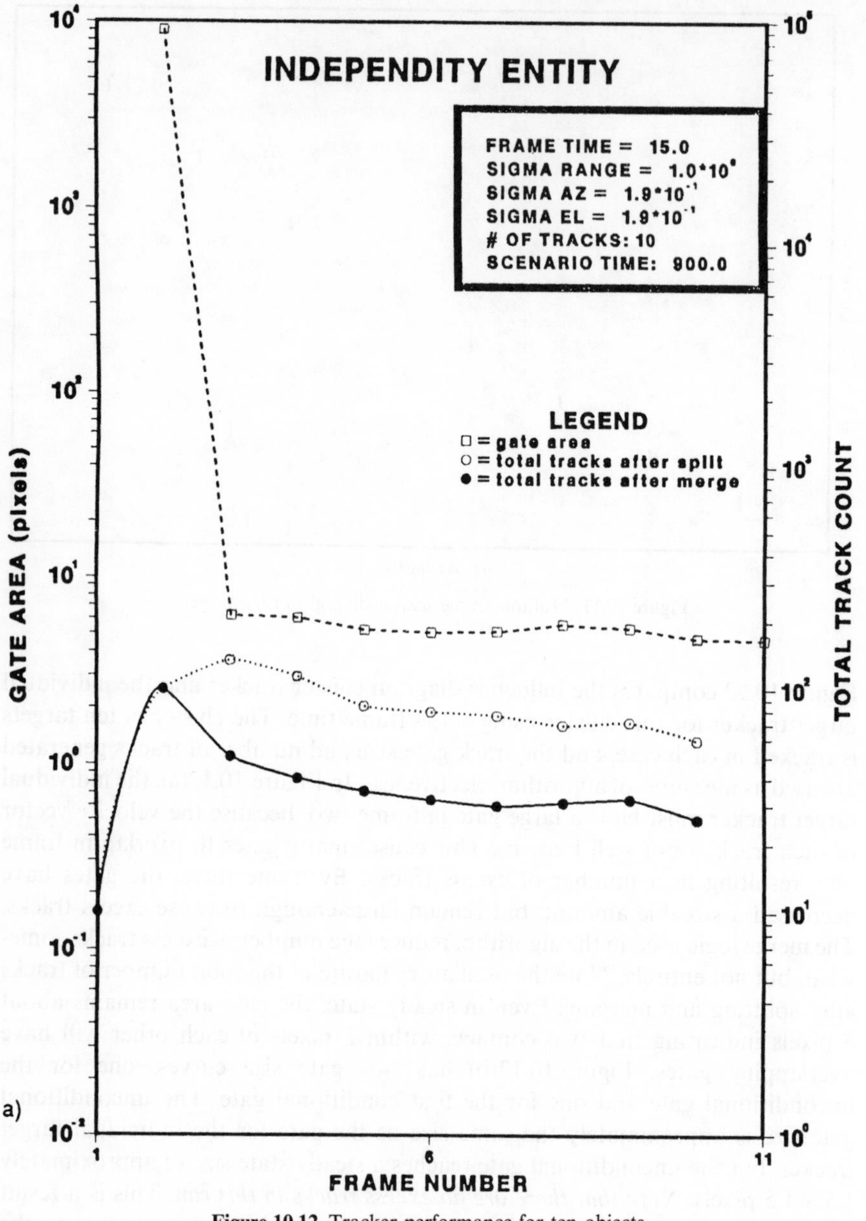

Figure 10.12 Tracker performance for ten objects.

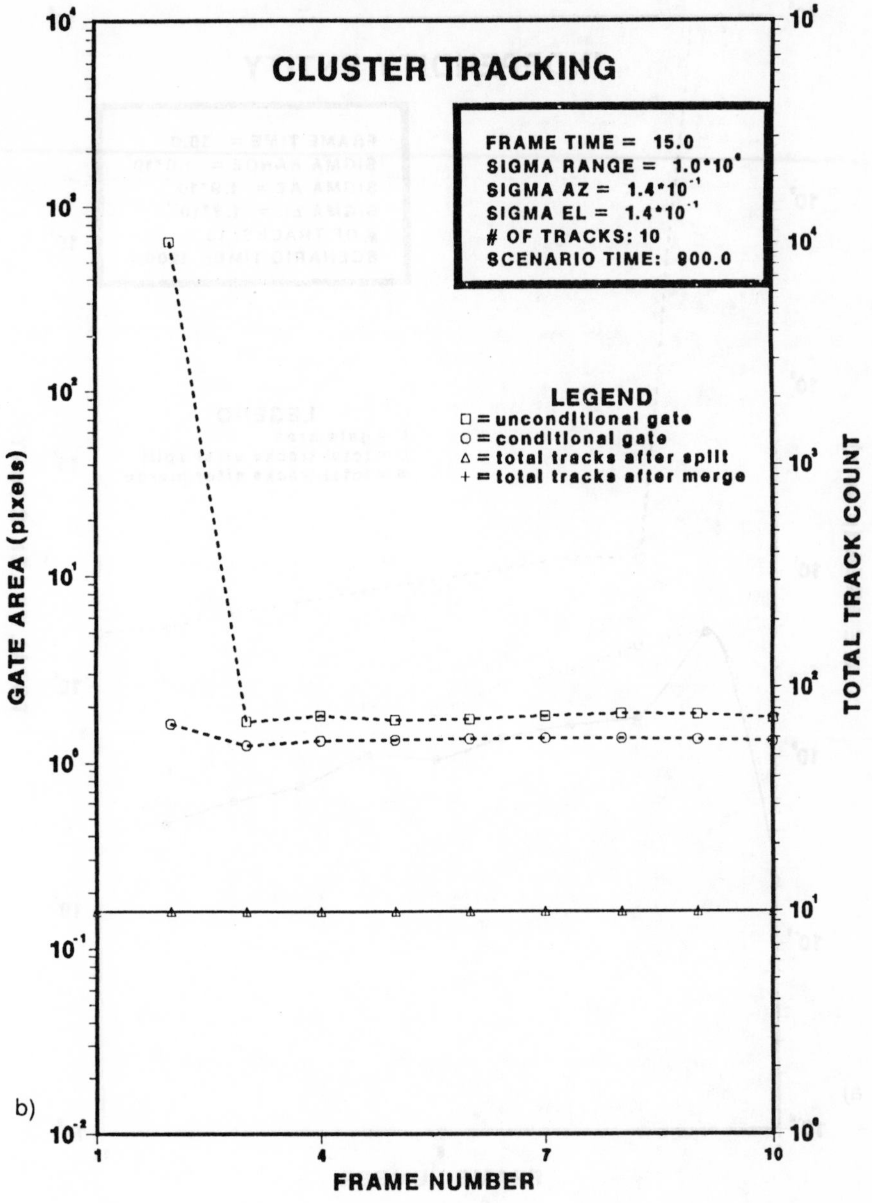
CLUSTER TRACKING
FRAME TIME = 15.0
SIGMA RANGE = 1.0*10⁰
SIGMA AZ = 1.4*10⁻¹
SIGMA EL = 1.4*10⁻¹
OF TRACKS: 10
SCENARIO TIME: 900.0
LEGEND
□ = unconditional gate
○ = conditional gate
△ = total tracks after split
+ = total tracks after merge
GATE AREA (pixels)
TOTAL TRACK COUNT
FRAME NUMBER
b)

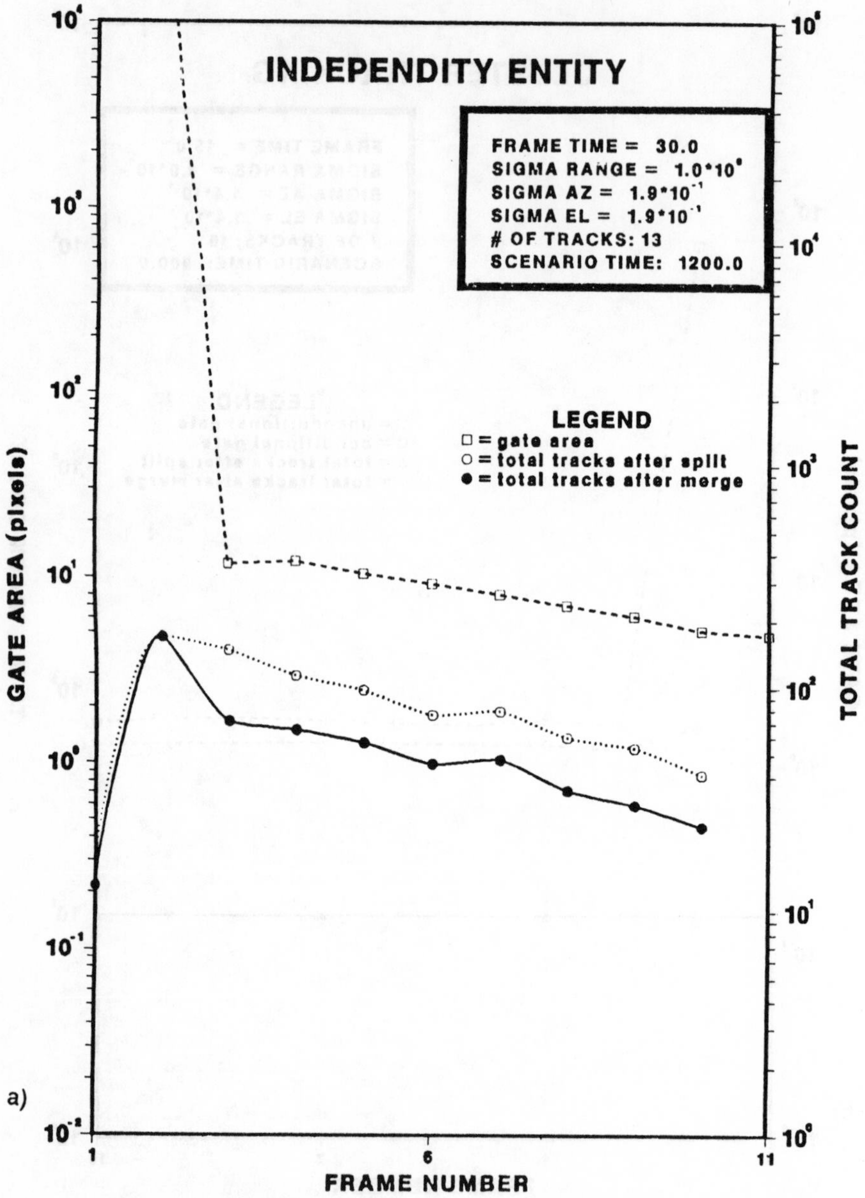

Figure 10.13 Tracker performance for thirteen objects.

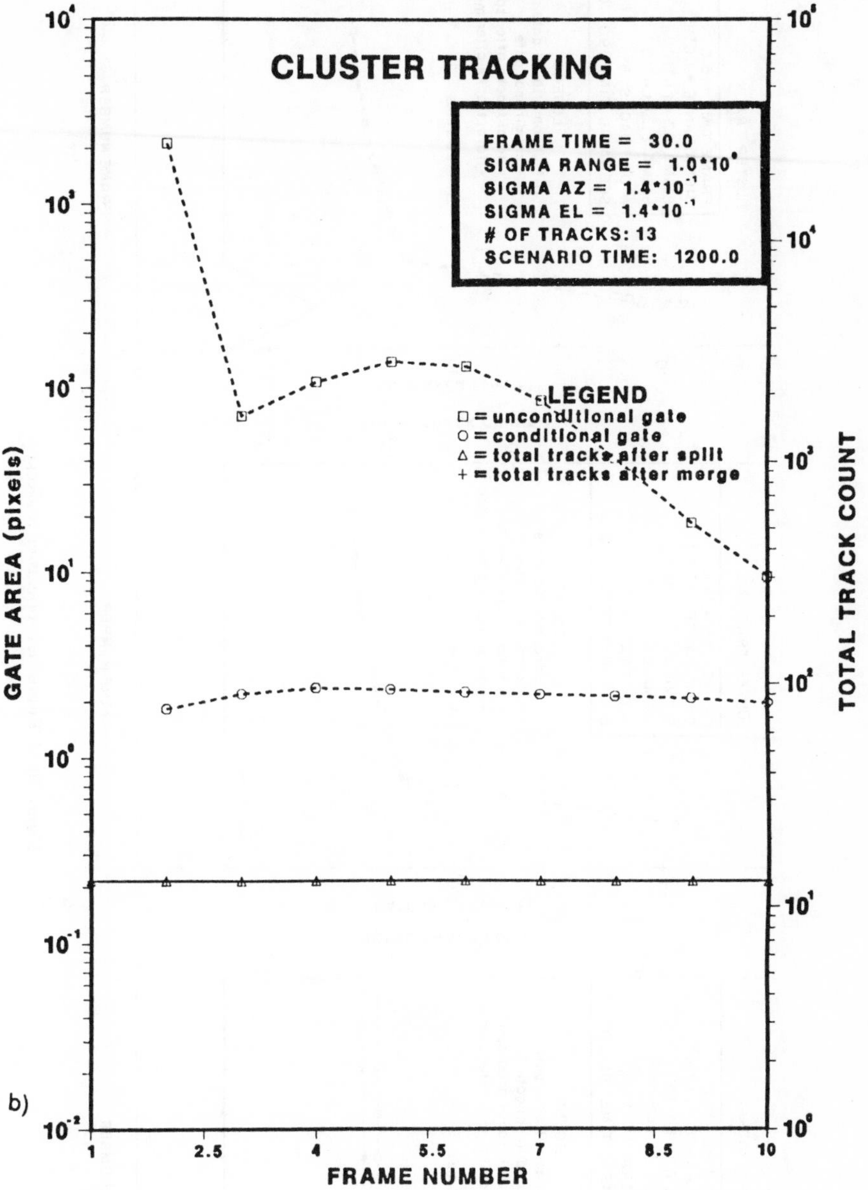
CLUSTER TRACKING
FRAME TIME = 30.0
SIGMA RANGE = 1.0*10^0
SIGMA AZ = 1.4*10^-1
SIGMA EL = 1.4*10^-1
OF TRACKS: 13
SCENARIO TIME: 1200.0
LEGEND
□ = unconditional gate
○ = conditional gate
△ = total tracks after split
+ = total tracks after merge
GATE AREA (pixels)
TOTAL TRACK COUNT
FRAME NUMBER

b)

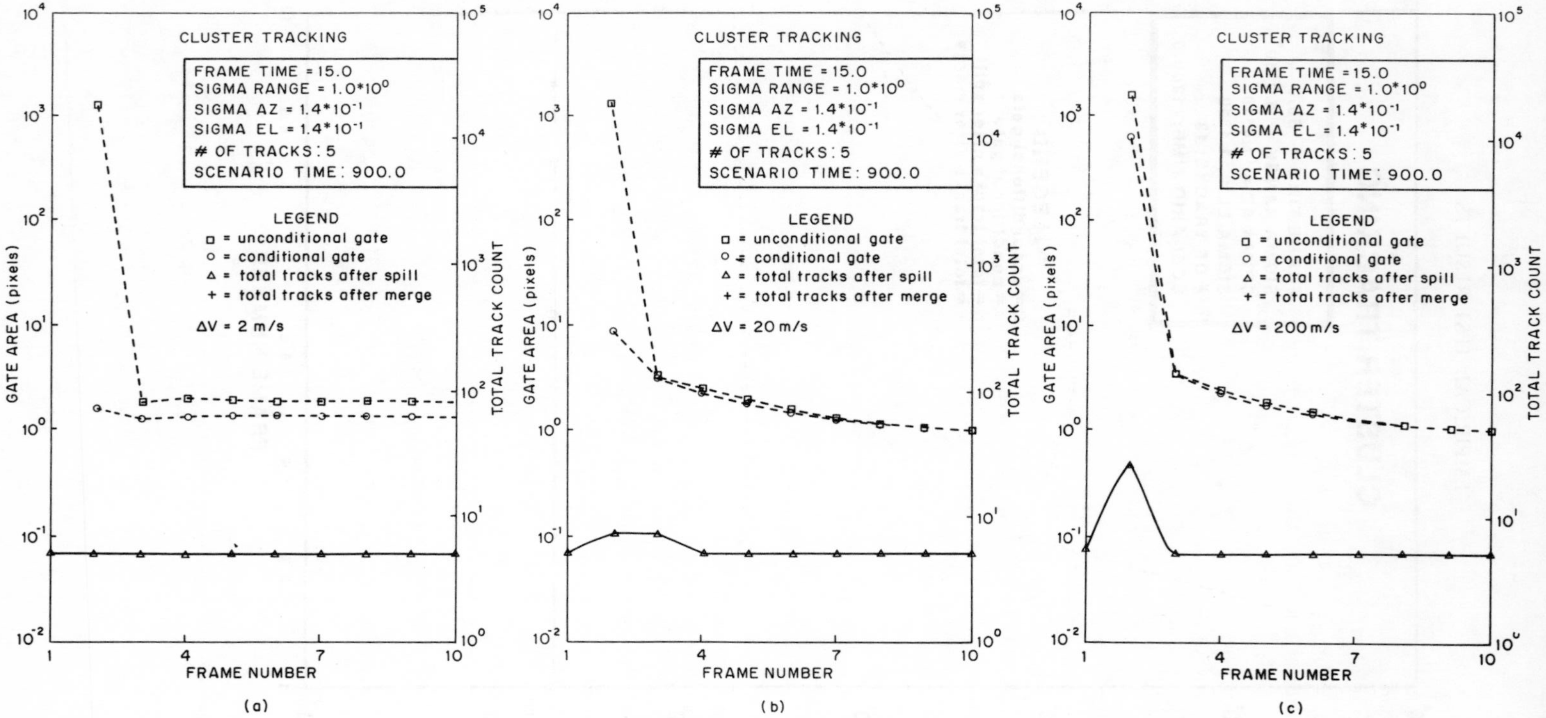

Figure 10.14 Robustness of cluster tracking.

of the splitting and merging processes. *Again, the cluster tracker produces no excess tracks.* The cluster tracker unconditional gate is again near the size of the individual target tracker gate. The conditional gate is approximately 1.7 pixels. After 300 s of tracking with the cluster tracker, the unconditional and conditional gates are the same size, indicating that the tracks can be tracked as individual objects from this point in time.

An important area for investigation is robustness of the cluster tracker with respect to assumptions about the a priori variance of Δv. Results of a sensitivity analysis are shown in Figure 10.14, where the *a priori* 3-σ value for Δv is assumed to be 2, 20, and 200 m s^{-1}. In the case analyzed, three times the empirical standard deviation of Δv is close to 2 m s^{-1}. When the model uses a 3-σ value that is one or two orders of magnitude too large, the cluster tracker has no excess tracks after the fourth frame of data. Although the *a priori* uncertainty regarding the variance is incorrect, the correlated velocity structure allows the model to learn the underlying values well enough to sort out the true picture.

10.5 CONCLUSIONS

Typically, the art of probabilistic modeling has focused on defining a model that exploits stochastic independence without sacrificing representativeness to the point of being inadequate to solve the problem at hand. This has led to many elegant academic solutions in search of a problem, and frustrated data systems engineers with no tools to solve their complex real-world problems. In this chapter, we have demonstrated that the Gaussian influence diagram tool can be used to exploit stochastic dependence to solve a difficult real-world data systems engineering problem that was not solved using state-of-the art models based on stochastic independence.

The empirical results for tracking of RV clusters are summarized in Figure 10.15, which has the abscissa equal to the diameter of the uncertainty

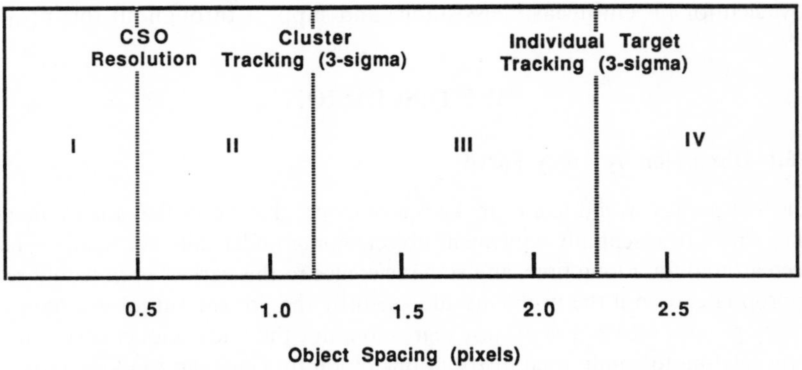

Figure 10.15 Performance regions for different tracking approaches.

gate for measurement samples. For a 15 s frame time, the cluster tracker drives the steady-state conditional gate diameter to 1.2 pixels; and the individual target tracker drives its gate to 2.2 pixels. Objects separated by distances in region I of Figure 10.15 cannot be resolved by the sensor. Objects separated by distances in region II are resolvable, but are too close to guarantee that the cluster tracker will not generate excess tracks. Objects separated by distances in region III can be tracked effectively by the cluster tracker, but will not be tracked by the individual target tracker. Finally, objects separated by distances in region IV can be handled by the individual target tracker.

The algorithms developed for the empirical studies presented in this chapter assumed that the N tracks seen in frame one would all reappear in frame two to be associated with the previous frame of data. If the object spacing is near the resolution limit of the sensor, objects can disappear from one frame to the next when two or more resolved objects from the first frame fall within one resolution cell on the next frame. Another possibility is that new objects are deployed within the cluster between frames. An algorithm has been developed to handle such cases (Casaletto, 1988), but empirical studies have not been completed to test the algorithm.

Originally, the concept of applying the Gaussian influence diagram to frame-to-frame association arose from application of the diagram for system performance analysis of a pattern recognition algorithm. The cluster tracker is a variant of that analysis, and there is great potential for applying the Gaussian influence diagram to other pattern recognition problems, such as robotic vision systems and image processing systems. Also, the tracking results can be applied and extended to track naval and land-based battle groups.

ACKNOWLEDGEMENT

This work was supported by Independent Research and Development funds at Lockheed Missiles & Space Company, Inc. Our gratitude goes to Dr Mike Kovacich for his enthusiastic assistance and support throughout this project.

10.6 DISCUSSION

10.6.1 Discussion by Terry Speed

The first point I would like to make is a bibliographical one. Gaussian influence diagrams—or essentially equivalent objects going under different names—have been around for a long time, and some reference to this earlier work would seem appropriate, even if the problems addressed by the present author are novel. In a very general seense it is all just regression: not the linear model sort, but the form relating to conditional distributions of jointly Gaussian random variables. The ideas here go back at least to Yule (1907), if no earlier. Much of the literature on path analysis in the social sciences, see e.g. Duncan (1975) rests on this same body of knowledge. More recently, Wermuth (1980) connected the models of path analysis with those of econometrics, the link being the covariance selection models introduced in Dempster (1972). Wermuth's paper also contains an idea closely related to the notion of d-separation discussed by Pearl at this conference, namely the concept of *reducible zero pattern* (see Wermuth, 1980, Definition 1, p. 965). I make this observation because it has seemed to me that many of the people at this conference are not aware of highly relevant work in areas other than their own.

With this said, let me turn to the contents of the present chapter. It struck me that the framework and the problem addressed were very close to those one meets in linear systems theory, and that it might be helpful to keep things as close to this theory as possible, highlighting the difference where appropriate. The state of our system at time t is simply the array $(\mathbf{p}(t), \mathbf{v}(t)) = ((p_i(t), v_i(t)): i = 1, \ldots, n)$ of positions and velocities of the n objects, and these states seem to evolve according to the usual laws of mechanics. The observations are the true positions, corrupted by errors, and in a possibly scrambled form. Thus we observe not $\mathbf{z}(t) = (z_j(t): i = 1, \ldots, n)$ but $z_{\sigma_t}(t) = (z_{\sigma_t(i)}(t): i = 1, \ldots, n)$ where σ_t is a (presumably random) permutation of the labels $\{1, \ldots, n\}$.

What then is the problem the chapter is addressing? It would seem to be the following variant upon the standard one from linear systems theory: calculate

$$E\{\mathbf{p}(t), \mathbf{v}(t))|\mathbf{z}_{\sigma_t}(t)\}$$

and

$$\mathrm{Cov}\{(\mathbf{p}(t), \mathbf{v}(t))|\mathbf{z}_{\sigma_t}(t)\}$$

assuming a joint Gaussian distribution for the random vectors $\mathbf{p}(t), \mathbf{v}(t)$ and $\mathbf{z}(t)$ and some distribution for the permutation σ_t. At least this is how a systems theorist or a Bayesian might see things. The author does not, however, but proceeds without any probabilistic model for σ_t, preferring to use some combination of Gaussian influence diagrams, with its Bayes-like reversing of arrows, and a sequence of tests of hypotheses aimed at deducing σ_t.

Some comments would seem to be in order: not too many, though, because I have no desire to contribute in even an indirect way to the solution of a 'Star Wars' problem. First, what have influence diagrams go to do with this problem and its solution? In my view the answer seems to be nothing at all! It is no surprise to me that writing $v_i(t) = \bar{v}(t) + \Delta v_i(t), i = 1, \ldots, n$ and assuming that the deviations $\Delta v_i(t)$ are uncorrelated gives better results than simply assuming the $V_j(t)$ are uncorrelated: this improvement has nothing to do with influence diagrams. Indeed if it were not for the random permutation σ_t, the desired calculations could be elegantly carried out by a very simple form of the Kalman filter. Second, why is it necessary to abandon the Bayesian paradigm and perform a sequence of tests which in effect estimate σ_t? And even if this is done, the sequence of increasingly more stringent conditional tests seems to owe nothing to influence diagrams. I would not have though it difficult to devise a distribution and a time evaluation for σ_t which is no less complex computationally than the approach of the present chapter.

Summarizing, this chapter raises a problem which is a variant of a very simple one from linear systems theory: the labels of the observed variables get scrambled. A solution is offered which somewhat unconvincingly utilizes the formalism of Gaussian influence diagrams while the real issue—a probability model for this scrambling process—is ignored. Modeling this process and making use of Bayes' theorem would seem to be the way to make progress on this sort of problem.

10.6.2 Reply

First, we want to thank Dr Speed for his comments on our chapter. We appreciate that he devoted his time and talent to reviewing our chapter, even though in doing so, he has donated some of his time to the 'Star Wars' effort, which certainly is not his favorite charity.

We agree with the comments regarding previous work in the field of regression models and their relationship to Gaussian influence diagrams. Indeed, the reference to Yule (1907) also appears in the cited work of Kenley (1986) and Shachter and Kenley (1989). Yule's theory motivated the definition of the parameters for the Gaussian influence in those papers, and all that remained was to derive the detailed algorithms to update the parameters when applying the standard influence diagram operations of removal and reversal.

The restatement of our problem is correct. The comments regarding the requirement for a Bayesian to have a probabilistic model for σ_t are duly noted, but we also point out that complete definition of an optimal decision rule requires that a loss function must be provided to the pure Bayesian. Being good engineers, we did take shortcuts to get on with providing a solution to a technical problem that must be solved in 'real time'. Although we did not follow the pure Bayesian approach, we would like to point out that one initial formulation of the problem that failed miserably was a very non-Bayesian analysis of variance model with $\sum \Delta v_j = 0$.

It was no surprise to Dr Speed that our correlated model for $v_j(t)$ provides better results than assuming that the velocities are uncorrelated, but as happens with many problems of this type, hindsight indeed does make the results seem obvious, even though quite a few bright people have been struggling to find an acceptable solution over several years. The definition of progress is very relative. Dr Speed suggests that investigating the stochastic process for σ_t and applying Bayes' theorem would lead to progress on this sort of problem. Our results show the shortcuts we took did not affect the system's capability to determine the correct permutation rapidly without having to process all $n!$ permutations, as would be required by the pure Bayesian approach suggested by Dr Speed. This type of performance meets our definition of progress.

One final comment is in order. The experience of the authors and other practitioners of influence diagram modeling is that the power of the approach is in formulating the problem to be solved and not getting lost in the technical details of manipulating equations and proving theorems to find a solution to a real-world system design or decision problem. Unfortunately, one only comes to appreciate this feature of influence diagrams by rolling up his or her sleeves and performing probabilistic engineering by drawing and manipulating influence diagrams. This intutive, creative process is difficult to convey through the written word, but it is our hope that Dr Speed and others roll up their sleeves with us and discover the power of the influence diagram.

REFERENCES

Casaletto, T. R. (1987) *Multiple Target Tracking Using Cluster Association*, SSTS Engineering Memorandum DSE-MAK-001, Lockheed Missiles & Space Company, November.

Casaletto, T. R. (1988) *$N \rightarrow M$ Cluster Association*, SSTS Engineering Memorandum DSE-MAK-005, Lockheed Missiles & Space Company, February.

Dempster, A. (1972) Covariance selection, *Biometrics*, **28**, 157–75.

Duncan, O. D. (1975) *Introduction to Structured Equation Models*, Academic Press, New York.

Kenley, C. R. (1986) Influence diagram models with continuous variables, Ph.D. dissertation, Department of Engineering-Economic Systems, Stanford University, Stanford, Calif. June.

Shachter, R. D. and Kenley, C. R. (1989a) Gaussian influence diagrams, *Management Science*, **35**, 527–50.

Wermuth, N. (1980) Linear recursive equations, covariance selection and path analysis, *J. Amer. Statist. Assoc.*, **75**, 963–72.

Yule, G. U. (1907) On the theory of correlation for any number of variables, treated by a new system of notation. *Proc. Roy. Soc.*, **A79**, 182–93. See also *Statistical Papers of G Udny Yule*, M. G. Kendall (ed.), Griffin, 1971.

CHAPTER 11

A Socio-technical Approach to Assessing Human Reliability

Lawrence D. Phillips and Patrick Humphreys *London School of Economics, UK*, David Embrey, *Human Reliability Associates, UK* and Douglas L. Selby, *Oak Ridge National Laboratory, USA*

ABSTRACT

This chapter presents a new approach for determining human error rates in complex technical systems. The approach consists of a technical component, an influence diagram, and a social component in which assessments required by the influence diagram are generated by groups of experts working together in groups. Iterative use of a ten-step procedure, described in the chapter, facilitates the generation of a model that is acceptable to the group. The approach was tested in the field by applying it to events concerning pressurized thermal shock for two nuclear power stations in the United States. Critiques on the last day indicated general satisfaction with the approach, which compares well with established technologies for human reliability assessment.

11.1 BACKGROUND

This chapter presents a new approach to assessing human reliability in complex technical systems, and reports its use in the study of pressurized thermal shock events for two nuclear power stations in the United States. There are both social and technical components to the approach, so to keep this in mind, and to provide an identifying acronymn, we are calling the methodology 'socio-technical assessment of human reliability', or STAHR, for short.

It is important to emphasize that STAHR does not provide the definitive technical fix to a problem on which a great deal of effort has already been

Influence Diagrams, Belief Nets and Decision Analysis
Edited by R. M. Oliver and J. Q. Smith

spent. It does, however, provide regulators and risk assessors with a methodology that has certain advantages and disadvantages compared to existing approaches. How useful it proves to be in practice has only begun to be determined, but work to date indicates that additional research on STAHR is warranted.

A key feature of this approach is that it draws on two fields of study: decision theory and group processes. Decision theory provides the form of a model which allows the desired error rates to be determined, while group processes provide the input data through the group interaction of experts who are knowledgeable about the factors influencing the event whose error rate is being assessed. The different perspectives of these experts, if managed effectively by the group, can lead to informed, useful inputs to the model. Thus, the validity of any error rates that are produced by the model depends not only on the technical model itself, but also on the social processes that help to generate model inputs.

The impetus for the STAHR approach began in 1982 at an Oak Ridge meeting addressing methods for assessing human reliablity in pressurized thermal shock studies. One of us (Phillips) introduced influence diagram technology (Howard and Matheson, 1980) as a potentially easier modeling tool than event trees or fault trees. The main advantage of influence diagrams from a technical perspective is that they capitalize on the independence between events and model only dependencies; that is, the influence diagram organizes the dependencies as a system of conditional probabilities.

By the early spring of 1983, the Decision Analysis Unit at the London School of Economics together with Human Reliability Associates in Lancashire, England, had developed the technology to the point where it could be tested. A field study, sponsored by Oak Ridge National Laboratories, was carried out in late May 1983 at Hartford, Connecticut, to address operator actions associated with thermal shock events that could occur at the Calvert Cliffs Unit 1 nuclear power station. Participants and the sponsor considered that STAHR was a useful approach which provided valid quantifications of operator's actions, but that further refinements were needed. After several changes and modifications to the approach, another field exercise was held in November–December near Raleigh, North Carolina for similar events at the H. B. Robinson plant.

No further development of the STAHR approach has occurred since 1984, though it has been presented at several conferences and workshops, and compared to other methodologies for assessing human reliability (OECD, 1985). It has fared well in these comparative studies, even though less than $20 000 has been spent in developing STAHR.

This chapter presents the state of the STAHR approach following the two field studies. First we explain the approach to modeling that is applied in STAHR. Next, we present the influence diagram that emerged from the field studies, and explain the ten-step procedure that was followed in applying the approach. We then talk about group processes, which must be managed effectively for successful application of STAHR. The approach is then illustrated with extracts from the field studies, and we conclude with some overall observations.

11.2 INTRODUCTION

Influence diagrams were developed in the mid-1970s at the Stanford Research Institute (Miller *et al.*, 1976), then applied and further developed at Decisions and Designs, Inc. (Selvidge, 1976) for intelligence analysis, all without a single paper being published in a professional journal. Howard and Matheson (1980) extended the theory and showed that any decision tree can be represented as an influence diagram, but not all influence diagrams can be turned into decision trees unless certain logical transformations are performed on the linkages between nodes in the diagram.

We presume at this point that the reader is acquainted with influence diagram technology or has read the tutorial on influence diagrams presented in the introduction to this book. The STAHR influence diagram is simpler than those presented in the introduction because it consists only of event nodes: a target event (that a human operator will successfully perform a particular action), and related events that literally can influence the target event. The influence diagram provides the technical means for organizing the conditional probability assessments that are required for calculating the unconditional probability of the target event.

It is, of course, important to recognize that no probability is ever unconditional. All events shown on the influence diagram occur within some context, and it is this context that establishes conditioning events that are not usually shown in the notation on the influence diagram. Thus, in applying this technology, it is important to establish at the start of every assessment procedure what these common conditioning events are.

11.2.1 Requisite models

Where does the influence diagram come from in STAHR, and how are the necessary assessments obtained? Mainly through human judgments obtained from experts working in groups. That is the 'socio' component of the STAHR approach. The theory behind this component was developed by Phillips (1984) who has also provided illustrative case studies (Phillips, 1982, 1989). The key idea is that groups of experts are brought together to work in an iterative and consultative fashion to create a requisite model of the key issues that concern the group. A judgmental model is considered requisite if it is sufficient in form and content to resolve the key issues, or to solve the problem at hand.

A requisite model is developed by consulting the key players, people who have information, judgments and experience relevant to the key issues. The process of creating a model is iterative, with current model results being shown to the key players who can then compare the current results with their own holistic judgements. Any sense of discrepancy is explored, with two possible results: intuition and judgment may be found lacking or wrong, or the model itself may be inadequate or incorrect.

Thus, the process of creating a requisite model uses the sense of unease felt by the key players about current model results, and this sense of unease is used to develop the model further and to generate new intuitions about the issues. When the sense of unease has gone, and no new intuitions emerge, then the model is considered requisite. The aim of requisite modeling is to help the key players toward a shared understanding of the problem, thus enabling decision makers to act, to create a new reality.

A requisite model is neither optimal nor normative, is rarely descriptive, and is at best conditionally prescriptive. A requisite model is about a shared social reality, the current collective understanding of the key players. Any one individual will have a more detailed understanding of some key issues, so the model is a simplified representation of each individual's perspective. By combining many perspectives in one model, new metaperspectives emerge.

Requisite models are appropriate when there is a substantial judgmental element that must be made explicit in order to resolve some key issues. Because judgment, intuition and expertise are important ingredients of requisite models, there can be no external reality that can serve as a single criterion against which optimality could be judged. Thus, requisite models are not optimal models. Nor are they normative in the sense that they describe the behavior of idealized, consistent decision makers; that claim would be too strong for requisite models. Nor can they be considered as descriptive models in the sense that they describe the behavior of actual people. Requisite models are stronger than that; they serve as guides to action though they may not themselves model alternative courses of action. Requisite models are not satisficing models, either; on the contrary, a requisite model attempts to overcome limitations on human processing of information due to bounded rationality.

11.2.2 Modeling human error rates

Requisite modeling seems ideally suited for the determination of human error rates in complex technical systems. The human operator in a complex system cannot, for the purpose of determining error rates, be treated as an unreliable machine component. In determining error rates for machines, two fundamental assumptions are made. First, that all machines of a particular type are identical as far as error rates are concerned. Second, that all machines of a particular type will be operating within environmental bounds over which the error rate remains unchanged. Neither of these assumptions is true for the human operator.

Each person is different from the next, and not even requiring certain standards of training and competence can ensure that other factors, such as those affecting morale and motivation, will not have overriding effects on the error rates. Third, environmental factors can have a substantial impact on human error rates. The same operator may perform differently if he is moved to a new plant, even though it may be of the same design, if, for example, teams function differently in the two plants. In short, people are different, and the environments they operate in are different, both from plant to plant and, within a plant, from time to time. Human error rates are not, then, unconditional figures that can be assigned to particular events. Rather, they are numbers that are conditional on individuals, and on the social and physical environment in which they are operating.

The effective assessment of error rates should take these conditioning influences into account. Technically, the STAHR approach does this by using the influence diagram to display the conditioning influences, and by using the educated assessments of experts to provide judgments that can take account of the uniqueness of the influences for a particular plant.

11.2.3 Validity of assessments

This raises the question of whether experts can provide assessments that are valid. Our view is that given the right circumstances people can provide precise, reliable and accurate assessments. This viewpoint is elaborated in Phillips (1987), but some authorities believe that bias is a pervading element in probability assessment (Fischhoff and MacGregor, 1982). Unfortunately, virtually none of the research that leads to the observation of bias in probability assessments has been conducted under circumstances (such as those explained in Staël von Holstein and Matheson, 1979) that would facilitate good assessments.

In addition to following those procedures, two additional factors can contribute to obtaining good probability assessments. One is the structure of the relationships of events whose probabilities are being assessed; good structure promotes good assessments. If the influence diagram is meaningful to the assessors, then when new probabilities that are implied by the structure of relationships are calculated and checked against the assessors' intuitions, these comparisons will provide a meaningful basis for subsequent revisions and improvements in the network of assessments. Since the validity of a single probability assessment has no meaning, it is important to create sets of related probabilities, where the meaningfulness of the probabilities is given by the structure of their relationships.

The other factor that contributes to good assessments is the use of groups of experts. If the group processes are managed well, then it is possible for 'many heads to be beter than one'. Each participant contributes his or her perspective on the issues such that when final assessments are made, all participants are better informed than if they had made individual assessments without the benefit of group discussion.

The success of the STAHR approach depends, in part, on the presence of a group facilitator who is acquainted with the literature on probability assessment, is knowledgeable about techniques that facilitate good assessments and is experienced in working with groups of people. Research has not yet determined how critical this role is, but we are confident that the necessary expertise and skills can be acquired with reasonable effort by potential group facilitators.

In summary, there is nothing in the research literature to suggest that people are incapable of making good assessments. Weathermen in several countries do it now; on all occasions when they say there is a 60 percent chance of rain tomorrow, it rains on 60 percent of the following days. Thus, weather forecasters are said to be 'well calibrated'; the STAHR approach tries to arrange for circumstances that will promote 'well-calibrated' probability assessments. However, calibrating the very low probabilities that emerge from the STAHR approach, or indeed any other approach, is technically difficult because of the low error rates implied. There are simply too few opportunities to determine whether the weatherman's low probabilities of rain in the desert are realistic.

11.3 THE STAHR INFLUENCE DIAGRAM

The influence diagram for operator actions in managing pressurized thermal shock events in nuclear power stations is shown in Figure 11.1. The initial version of this model was developed by discussions with human factors specialists, probabilistic risk analysts, nuclear engineers and other experts. In particular, a two-hour session with a group of experienced operators from nuclear power stations identified most of the key influencing factors that appear in the current model.

The current version of the model has benefited from the observations of over two dozen participants in the two field studies, and from the experience of using it. We do not yet know whether this influence diagram is generic in the sense that it can handle all events in which operators are expected to take actions. Possibly parts of the diagram are generic and others need to be developed to fit the specific situation. At any rate, the STAHR approach is sufficiently flexible that modifications to Figure 11.1 can be made to suit the circumstances, or, indeed, entirely different influence diagrams can be drawn.

11.3.1 The influences

The top node in Figure 11.1 represents the target event. For example, if an alarm in the control room signals that some malfunction has occurred and the operator attempts to correct the malfunction by following established procedures, the target event might be that the operator correctly performs a specified step in the procedures.

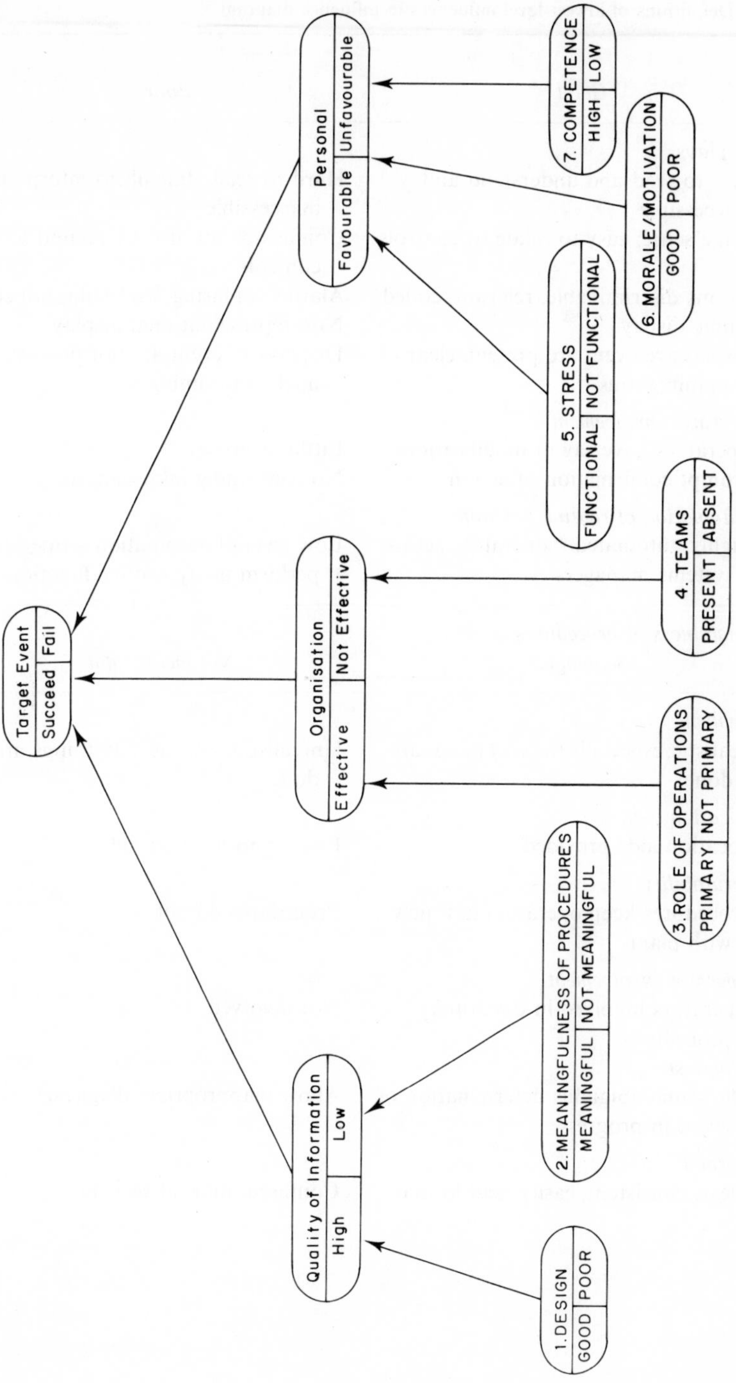

Figure 11.1 The STAHR influence diagram for pressurized thermal shock events.

Table 11.1 Definitions of lowest-level influences in influence diagram

1. *Design*	
Good	*Poor*
(i) *Displays*	
Easy to read and understand and accessible	Hard to read, difficult to interpret; inaccessible
Make sense; easy to relate to controls	Confusing; not directly related to controls
Alarms discriminable, relevant, coded	Alarms confusing, irrelevant, not coded
Mimic dispay	Non-representational display
Displays re event are present, clear, unambiguous	Displays re event are not present, unclear or ambiguous
(ii) *Operator involvement*	
Operators have say in modifications	Little or no say
Prompt confirmation of action	No confirming information
(iii) *Automation of routine functions*	
Highly automated—operators act as systems managers	Low level of automation—operators perform many routine functions

2. *Meaningfulness of procedures*	
Meaningful	*Not meaningful*
(i) *Realism*	
Realistic; especially the way things are done	Unrealistic; not the way things are done
(ii) *Location aids*	
Location aids provided	Few or no location aids
(iii) *Scrutability*	
Procedures keep operators in touch with plant	Procedures do not
(iv) *Operator involvement*	
Operators involved in developing procedures	Not involved
(v) *Diagnostic*	
Allow unambiguous determination of event in progress	Allow inappropriate diagnosis
(vi) *Format*	
Clear, consistent, easily read format	Confused, difficult to read

3. *Role of operations*

Primary	*Not primary*
(i) *Accountability*	
All other functions report to operations supervisor	Only operations staff report to operations supervisor
(ii) *Relationship to maintenance and other functions*	
Good relations	Antagonism
(iii) *Paperwork*	
About right	Excessive
(iv) *Operator involvement*	
Operators have a say in how the place is run	No say

4. *Teams*

Present	*Absent*
(i) *Shifts*	
Allow teams to stay together	Prohibit team formation
(ii) *Roles*	
Well-defined accountabilities. Scope for exercising judgement	Poorly defined accountabilities, or rigid job descriptions so little scope for exercising judgment
(iii) *Training*	
Team members train together	Team members not trained together

5. *Stress*

Helpful	*Level not helpful*
(i) *Shifts*	
No jet lag	'Permanent jet lag'
(ii) *Time available*	
Adequate	Too little
(iii) *Operating objectives*	
No conflict	Conflict
(iv) *Transient related stress*	
Appropriate	Understressed or overstressed

Table 11.1 (*Contd.*)

6. *Morale/motivation*	
Good	*Poor*
(i) *Status of operators*	
Treated as professionals	Treated as laborers
(ii) *Career structure*	
Operators can find best level in organisation	Peter principle operates (i.e. over-promoted)
(iii) *Physical/mental well being*	
Operators physically and mentally capable of performing job	Job performance adversely affected by physical and/or mental impairment

7. *Competence*	
High	*Low*
(i) *Training*	
Operators generally well trained in emergency procedures	Poorly trained in emergency procedures
(ii) *Certification*	
Peer review is used	No peer review
(iii) *Performance feedback*	
Operators given periodic feedback on performance	No feedback
(iv) *Experience*	
Operators experienced in dealing with target event	Operators inexperienced

The influence diagram shows three major influences on the target event. One is the quality of information available to the operator, a second is the extent to which the organization of the nuclear power station contributes to getting the work done effectively, and the third influence is the impact of personal and psychological factors pertaining to the operators themselves. That is, the effective performance of the target event depends on the physical environment, the social environment and personal factors.

Each of these three major factors is itself influenced by other factors. The quality of information is largely a matter of good design of the control room and of the presence of meaningful procedures. The organization is effective if the operations department has a primary role at the power station, and if the organization at the power station allows the formation of teams. Personal factors will contribute to effective performance of the target event if the level of stress experienced by the operators is helpful, if morale and motivation of the operators is good, and if the operators are highly competent.

These seven 'bottom-level' influences actually describe the power station, its organization and its operators. Each of these seven influences is defined in some detail in Table 11.1. For example, design is a matter of displays, operator involvement and automation of routine functions. The design of a particular power station would be judged good if all the descriptors in the column headed 'good' were characteristic of the station, and would be judged 'poor' if the right column descriptions were true. Of course, most power stations fall between these extremes; they are mixtures of good and poor, or of degrees of 'goodness' and 'poorness'. To facilitate subsequent assessment, it may be helpful to define the collection of all the poor descriptions as characterizing a 'minimally acceptable' or 'barely licensable' power station, with all the 'good' descriptions representing a 'maximally feasible' station.

11.3.2 The STAHR procedure

Using the influence diagram is a matter of applying the following ten steps (but not necessarily in this order):

1. Describe all relevant conditioning events.
2. Define the target event.
3. Choose a middle-level event and assess the weight of evidence for each of the bottom influences leading into this middle event.
4. Assess the weight of evidence for this middle-level influence conditional on the bottom-level influences.
5. Repeat steps 3 and 4 for the remaining middle and bottom influences.
6. Assess probabilities of the target event conditional on the middle-level influences.
7. Calculate the unconditional probability of the target event and the unconditional weight of evidence of the middle influences.
8. Compare these results to the holistic judgments of the assessors. Revise the assessments as necessary to reduce the discrepancy between holistic judgments and model results.
9. Iterate through the above steps as necessary until the assessors have finished refining their judgments.
10. Do sensitivity analyses on any remaining group disagreements. Report either point estimates if disagreements are of no consequence, or ranges if disagreements are substantial.

In step 1, participants describe the general setting in which the target event might occur as well as all conditions leading up to the target event. Assessors are reminded that this description and statement of initial conditions form a context for their subsequent assessments, and that these assessments are conditional on this context.

In the second step, the target event is defined in such a way that its occurrence or non-occurrence is capable, at least theoretically, of confirmation without additional information. Thus, 'rain tomorrow' is a poorly defined event, whereas 'less than 0.1 mm of precipitation in a rain gauge located at weather station x' is a well-defined event.

In carrying out step 3, the assessors might begin by focusing attention on the left-most middle node, 'quality of information', and assess weights of evidence for the bottom influences 'design' and 'procedures'. This is done with reference to the definitions of those bottom influences. For example, with respect to the design influence, the group of assessors must decide whether on balance the design of the particular power station is more similar to the descriptions under the 'good' column or the 'poor' column. The assessors may find it helpful to imagine a continuous dimension between 'good' and 'poor' and then to determine where on the dimension this particular power station lies with respect to the event in question. In short, the assessors are judging numbers that reflect relative weight of evidence as between the poles of the design influence. The weight of evidence would also be judged for the next bottom node, 'meaningfulness of procedures', only here six different factors, from 'realism' to 'format', must be taken into account in making the judgment.

The weights of evidence placed on the poles of each dimension are assigned as numbers that sum to 1. Thus, letting w_1 represent the weight of evidence on display being good and w_2 representing the weight of evidence on procedures being meaningful, then the assessments for these two bottom nodes can be represented as follows:

	Good	Poor
DESIGN	w_1	$1 - w_1$

	Meaningful	Not meaningful
PROCEDURES	w_2	$1 - w_2$

Step 4 requires the assessment of probabilities for the quality of information, a middle-level influence, conditional on the lower-level influences. The poles of the two bottom-level influences combine to make four different combinations, good design and meaningful procedures, good design and not-meaningful procedures, poor design and meaningful procedures, and poor design and not-meaningful procedures. Each of these four combinations describes a hypothetical power station of the sort under consideration, and these hypothetical stations are kept in mind by the assessors when they determine the weights of evidence for the quality of information. This can be set out as follows:

If DESIGN and PROCEDURES		*then* QUALITY OF INFORMATION *is* High	Low	JOINT WEIGHTS
Good	Meaningful	w_3	$1-w_3$	w_1w_2
Good	Not meaningful	w_4	$1-w_4$	$w_1(1-w_2)$
Poor	Meaningful	w_5	$1-w_5$	$(1-w_1)w_2$
Poor	Not meaningful	w_6	$1-w_6$	$(1-w_1)(1-w_2)$

For example, w_3 is the weight of evidence that the quality of information is high, given that design is good and procedures are meaningful. here 'high' quality of information does not mean an ideally perfect power station; instead, it is meant to represent a power station where both design and procedures are of a high, yet practically realistic standard. Also, 'low' quality of information does not mean some abysmally bad standard but rather a standard that is minimally licensable. The assessments w_3 through w_6 capture possible interactions between design and procedures. This is a key feature of the influence diagram technology, and experience to date suggests that it is an important feature for human reliability assessment. For example, in some power stations good design in the control room may compensate to some extent for procedures that are not very meaningful, whereas if design were poor, the additional burden of procedures that are not meaningful could be very serious indeed.

The weights are assessed in such a way that they are assumed to follow the probability calculus. That is, because the weights are multipliers, any assignment of zero will have the effect of 'zeroing out' any combination of influences to which the zero weight is applied. To see this more clearly, consider the four hypothetical stations described above at step 4 (good design and meaningful procedures, good design and not meaningful procedures, etc.). The weights associated with these four combinations of two influences can be obtained by multiplying the weights of evidence assessed in step 3; these products are shown as JOINT WEIGHTS in the table above. Note that the product rule for probabilities applies, so if any individual weight is zero, then the product is also zero.

To calculate the unconditional probability for the quality of information being high, it is necessary to multiply these four joint weights by the weights w_3 through w_6 and then to add these four products. That is,

$$w(\text{high}) = w_3w_1w_2 + w_4w_1(1-w_2) + w_5(1-w_1)w_2 + w_6(1-w_1)(1-w_2).$$

Note that this calculation makes use of both the product and addition laws of probability. It is the repeated application of these two laws that allows unconditional weights at higher nodes to be determined. The unconditional weights determined for the quality of information serve as weights on the rows of probabilities in the matrix for the target-level event, and the types of calculations just illustrated are repeated to obtain the unconditional probabilities for the target event.

Returning now to the ten-step procedure, step 5 requires that steps 3 and 4 be repeated for the rest of the middle- and bottom-level influences. Thus, weights of evidence are assessed for the role of operations and for teams, then a matrix of conditional probabilities is assessed for the organizational influence conditional on the lower-level influences. The same procedure is then followed in making the necessary assessments for personal factors.

Step 6 requires, for the first time, assessments of probabilities. However, these probabilities are for the target event conditional on the middle-level influences. In a sense, what is being assessed is conditional error rates, that is, assessors are giving their judgments about what error rates would be under the assumption of particular patterns of influences. Obviously, if data were available to provide these rates, that would help to establish the validity of the STAHR results. However, such data were not available for the two field studies, so it was necessary to rely on expert judgment.

Since the quality of information can be either high or low, the organization can be either effective or not, and personal factors can be favorable or unfavorable, there are eight possible combinations of these influences. A separate error rate associated with the target event is assessed for each of those eight combinations.

This is not a particularly easy job for assessors because they must keep in mind three different influences as well as their possible interactions. Favorable personal factors, for example, may well save the day even if the organization is not effective, and may even compensate to some extent for low-quality information. Insofar as the middle-level influences interact, this stage in the assessment process is important for it allows assessors to express the effect on error rates of these interactions.

The task can be made easier by first considering the most favorable combination of influences which would imply the lowest error rate, then the most unfavorable combination yielding the worst error rate. Then a series of paired comparisons between various combinations, holding two influences constant while looking only at the effect of the third, will result in a coherent set of assessments.

We found that our assessors were helped by providing them with displays of a probability-odds scale. The scale itself showed three cycles of log-odds (four cycles would have been better) starting at odds of 1:1. The left side of the scale was calibrated in probabilities, with the right side displaying odds in the form

x:1 (chances, as in 1 out of 100, might have been even more undestandable to our groups).

Step 7 is best carried out by a computer which can apply the multiplication and addition laws of probability to determine the unconditional probability of the target event as well as the next-lower influences. Several such programs are described in this volume.

In step 8, the unconditional probabilities and weights of evidence for the middle-level influences are given to the assessors who then compare these results to their own holistic judgments. Discrepancies are usually discussed in the group and revisions made as necessary to any assessment. It is often useful to give feedback about the unconditional weights of evidence for middle-level influences immediately after the middle-level matrix has been assessed.

Step 9 indicates that iteration through the first eight steps may occur as individual assessors share their perceptions of the problem with each other, develop new intuitions about the problem and revise their assessments. Eventually, when the sense of unease created by discrepancies between current model results and holistic judgments disappear, and when no new intuitions arise about the issues, model development is at an end and the model can be considered requisite.

Since individual experts may still disagree about certain assessments, it is worthwhile as the tenth step to do sensitivity analyses to determine the extent to which these disagreements influence the unconditional probability of the target event. An easy, but not entirely satisfactory, way to do this is to put in first all those assessments that would lead to the lowest probability for the target event and see what its unconditional value is, then put in all assessments that would lead to the largest probability, thus determining a range of possible results. The difficulty with this is that no individual in the group is likely to believe all of the most pessimistic or all of the most optimistic assessments, so the range established by this approach is unduly large. It should not be difficult, however, to develop easy and effective procedures for establishing realistic ranges for the probability of the target event, ranges that accommodate the actual variation of opinion in the group.

This has been only a very brief description of the stages that appear to be necessary for applying the influence diagram technology. The steps are certainly not intended as a rigid procedure to be followed without deviation. Instead, they should be thought of as an agenda that will guide the work of the group.

11.4 GROUP PROCESSES

So far, little has been said about the group processes that form the second component to the STAHR approach. A key assumption here is that groups can generate better assessments than any individual. Particularly for human reliability assessment in complex systems, there is unlikely to be any single individual with an unbiased perspective on the issues. Although each individual may be biased in his or her view, the other side of the coin is that each person has something worthwhile to contribute to the overall assessment. It is within the context of the group that different perspectives on the issues can most effectively be revealed and shared with others so that the group can take as its main function the generation of assessments that accommodate these different perspectives.

To ensure that all perspectives on the problem are fairly represented, it is important to establish a climate in which information is seen as a neutral commodity to be shared by all regardless of status or investment in the problem. To help create this climate it is important to establish the role of group consultant. This individual needs to be conversant with the technical aspects of influence diagrams and with probability assessment, and needs a working knowledge of group processes (Low and Bridger, 1979).

The group consultant should be seen by the group as an impartial facilitator of the work of the group, as someone who is only providing structure to help the group think about the problem. Although the group facilitator needs some minimal acquaintance with the principles of nuclear power generation and with the key components of the plant itself, it is probably not desirable that he or she be a specialist in nuclear power, otherwise it might be more difficult to maintain a neutral, task-oriented climate in the group. Thus, a major role for the group facilitator is not to tell people what to think about the issues but how to think about them.

The other major role for the facilitator is to attend to the group processes and intervene to help the group maintain its task orientation. The group can easily become distracted from its main task because viewpoints in the group will often diverge. The cognitive maps that a design engineer and a reactor operator have of the same system will in many respects be quite different, yet each will at times insist on the validity of his or her particular viewpoint. The group consultant must help to legitimize each of these viewpoints and to explore them in generating useful assessments.

To a certain extent, adversarial processes may operate in these groups. Operators will openly criticize certain aspects of design, and design engineers may well be contemptuous of procedures that they deem to be unnecessary if only people would operate the system properly. Trainers of operators may be somewhat skeptical of the optimistic 'can-do' attitude of the operators, for they have often seen things go badly wrong in simulator exercises. Operators may feel that anyone who has not had 'hands-on' experience in the real control room rather than a simulator is out of date at best and simply out of touch at worst. Unless the facilitator manages the groups processes effectively, minor squabbles can easily turn into major confrontations that seriously divert the group from its effective work.

This discussion is not meant to imply that the group should be composed so as to reduce adversarial processes. On the contrary, an underlying assumption of the STAHR approach is that diversity of viewpoint is needed if good assessments are to be generated. Differences are to be confronted openly in the group, and taken seriously regardless of the status of the holder of the viewpoint. Thus, diversity of viewpoint is a key criterion in composing groups. At the very least, these roles should be represented: group consultant, technical moderator to help direct the discussion on technical issues, trainer of nuclear power station operators, reliability and systems analyst, thermohydraulics engineer and procedures specialist, one or two other engineers with specialized knowledge of the power station and of the target events, probabilistic risk analyst, and, of course, reactor operators.

11.5 THE FIELD TESTS

To convey a sense of how STAHR operates in practice, we report here a representative exercise that draws on the two field tests conducted so far. Results given here are typical of those obtained in the field tests, but should not be ascribed to either of the two actual reactors.

11.5.1 Prior preparation

Both field tests confirmed the importance of preparing well for the group sessions. Participants must be chosen so that all prespectives on the operation of the nuclear power plant are represented in the group, and to ensure that the necessary expertise will be available. Target events must be selected and defined well. The letter inviting experts to participate should state the objectives of the exercise, list the target events that will be considered, indicate what preparation is expected of participants, provide an appended summary description of the STAHR approach and encourage participants to clear their diaries so they can give their full attention throughout the meeting without interruptions.

11.5.2 Introduction

The first session began with a brief introduction to the project and a statement of the goal, which was to use this new technology in carrying out human reliability analyses of selected events associated with pressurized thermal shock incidents. Participants than introduced each other, and the background to the current project was explained. This was followed by a brief description of the role of human judgment in risk assessments, with particular emphasis on the view of probability as an expression of a degree of belief. The conditions under which good calibration of probability assessments could be expected were also described. The group then adjourned for a tour of the simulator; this gave all participants a better idea of the layout and configuration of the control room in the actual plant.

After the tour, the group reconvened and the list of target events to be considered during the week was discussed. Although each event was different, in every case an operator was expected to perform successfully some mitigating action.

11.5.3 Practice session

The group chose one of the target events to be the subject of a practice session. Four initial conditions, which define the conditioning events, were agreed:

1. The target event occurs near the end of the core refueling cycle.
2. The reactor is at hot 0 percent (532°F).
3. The atmospheric dump valve (ADV) is open.
4. The main feedwater system is in bypass mode.

Because the originally stated target event was not well defined, participants redefined it as follows:

> Operator will recognize that ADV is open and will isolate ADV line within 30 minutes.

To ensure that all members of the group were reasonably familiar with the technical operation of the system, engineers knowledgeable about the reactor described the main steam header and also the main feed valve and bypass valve of the main feedwater system. The group was then introduced to influence diagrams and their relationship to event trees. An influence diagram similar to that shown in Figure 11.1 was presented, together with definitions of the bottom-level influences. Considerable discussion of the influences followed, with the result that the definitions of some influences were slightly changed and extended. Table 11.1 gives the end result of changes that resulted from both field studies. (The original influence diagram showed a link from the team influence to the stress influence, but this was found to be unworkable during a practice session, so the link was dropped.)

Most of the practice session was spent in discussions that helped to generate the assessments for the target event. It was apparent that the group did not find it particularly easy to make these assessments, and considerable disagreement about the appropriate numbers characterized the discussions. Eventually, however, consensus judgments emerged, and the unconditional probability of the operator successfully completing the target action was determined to be 0.937. However, because this was the first effort of the group, the figure was not taken very seriously. Introduction, training and practice session were completed on the first day.

Later in the week, after participants had gained experience in using STAHR, this event was reevaluated. The resulting probabilities for success and failure were 0.964 and 0.036 respectively, failure being attributed almost entirely to personal factors.

11.5.4 Application of STAHR

Over the next three days the group applied the STAHR methodology to twelve remaining target events. The following session includes most of the elements that might be found in any particular session.

The following initial conditions were given to the group:

1. The steam-line break consists of a 1 ft^2 hole.
2. The reactor is at full power.
3. The break is outside the containment vessel.

A definition of the target event was at first rather elusive. Starting with the operator recognizing that a steam-line break has occurred, the group considered several intermediate actions before arriving at the following:

> Operator throttles charging pumps after primary pressure reaches high-pressure safety injection (HPSI) head.

It was agreed that this event would determine whether or not the operator would successfully control the repressurization.

In arriving at their assessment, the group followed the ten-step procedure outlined above. Discussion of the input assessments took about four hours, with considerable diagreement expressed for over one-half of the assessments. Finally, a set of assessments was agreed upon as a base case, and this yielded a probability of success for the target event of 0.974. When the contentious assessments were replaced by the most pessimistic values, the target success probability dropped to 0.867. When they were replaced by the most optimistic assessments, the success probability rose to 0.992.

In fairness, it should be said that no individual in the group believed all of the pessimistic or all of the optimistic assessments that went into these two sensitivity analyses. Thus, the range of success probability from 0.867 to 0.992 considerably exceeds the range that would have been obtained if each individual's assessments had been tried in the influence diagram. Looked at differently, the range of the failure rate, 0.008–0.133, is little more than 15 to 1, which is considerably less that the factors of 100 or even 1000 that occasionally characterize the uncertainty in failure rates obtained by other methods.

The 0.026 failure frequency (1 – 0.974) was attributed both to personal factors and to the quality of information available to the operator (control room design and procedures). The quality of information was considered to be the factor which could be improved most easily. Specifically, the importanc of the operator action could be better defined in the procedures and a P/T CRT plot with the acceptable range of operation marked would greatly improve the quality of information.

Additional sensitivity analyses were performed to see what the effect would be of improving the procedures and control room design. This was simulated in the influence diagram by moving the weights of evidence to 100 on both these bottom-level influences. When that was done, the probability of success rose to 0.986. A minimally licensable plant was also simulated by assigning 0 to both design and procedures, with the resulting probability of success dropping to 0.880.

If all the bottom-level influences are scored at 0, then the probability of success is 0.546. This suggests that the operator in a barely adequate plant still has a better than 50 percent chance of performing this particular target event successfully. Similarly, in the maximally feasible plant, characterized by a score of 100 on all the bottom-level influences, the probability of success moves to 0.992.

The group next considered perturbations on the steam-line break scenario, though the target event was always to control repressurization. Interestingly, although the perturbations led to changes in several assessments, some lower and some higher, these tended to balance each other with the net result that the 0.974 probability of success and 0.026 probability of failure were maintained.

The group then considered what would happen in a loss of cooling accident (LOCA) rather than a steam-line break. In this case, the success and failure probabilities were evaluated to be 0.968 and 0.032 respectively. The increased failure rate was due almost exclusively to the perception that the information in the LOCA procedures associated with performing this action was less informative than that found in the procedures for steam-line breaks.

In similar fashion the remining target events were modeled, and a critique of the STAHR approach was held at the end of the last day.

11.6 DISCUSSION

At least one day is needed to train participants in the STAHR approach, and building confidence and familiarity takes another day. Then, new target events take no more than two hours for the complete set of assessments, and variations on already completed target events take less than one hour to complete. (A total of 31 assessments are required for the influence diagram shown in Figure 11.1.)

During the critique held on the last day, many participants wanted to compare STAHR with the THERP (Technique for Human Error Rate Prediction) approach of Swain and Guttman (1984). One person said that THERP was easy to understand and the results could be communicated easily, but the STAHR approach was more comprehensive. Another felt that the STAHR approach has the disadvantage that it requires experts to participate, whereas

THERP does not. However, another participant insisted that the successful application of THERP does require experts. Another person commented that the STAHR approach requires participants to go back to first principles, whereas the handbook flavor of THERP gives only approximate results. Participants felt that a major advantage of the STAHR approach was its ability to capture interactions of factors.

Discussion about the adequacy of the influence diagram suggested that most of the influence diagram was generic and so could be considered for any event, but that improvement was needed to the design influence so as to separate out general control factors from target-specific factors. Groups in both field studies agreed that the group processes had worked well; there was good interaction, everyone participated, and status differences did not intrude.

Some participants felt that it would have been useful to have mock-ups of the control room available in the conference room. All participants agreed that a free flow of information was essential to the work of the group, and for this reason felt that observers should not be alowed, not even Nuclear Regulatory Commission personnel or power station management representatives.

Finally, several people said they were persuaded that the numbers generated by the STAHR approach were believable and they were eager to try the process again on different events. One person commented that he did not like the error rates that were emerging from the assessment, but that they were believable.

11.7 CONCLUSION

The two field tests have demonstrated that the STAHR approach has potential for assessing human reliability in complex technical system. By concentrating on influences rather than individual events, it avoids the bottom-up detail of event trees and fault trees with their attendant bushiness in attempting to cover all event-sequences that could possibly happen, and the high costs of applying those approaches for complex systems.

Unlike handbook-based approaches, STAHR cannot be applied simply by following some established steps in a procedure. Rather, considerable flexibility is built into STAHR, and participants are forced to think about fundamental aspects of the operation of the system. Because of this, probabilities for target events cannot be developed quickly. In addition, heavy demands are placed on the limited time of many experts, so the application of STAHR is not inexpensive.

At this stage, STAHR is not a full developed technology; further development is needed of the procedures and techniques, and additional field studies will help to evolve the approach. Eventually, when the benefits of using an improved STAHR approach are weighed against its costs, and then compared to other approaches, it may well take its place alongside other technologies that facilitate human reliability assessment.

11.8 DISCUSSION

11.8.1 Discussion by Ward Edwards

Human beings embedded in complex, important technologically sophisticated man–machine systems are at least as likely to fail as are the machine components of such systems. Systematic analysis of the probability of system failure produced by operator failure is called human reliability assessment. The natural tool for such assessments is one form or another of a fault tree. Fault trees intended for human reliability assessment differ from those appropriate to other kinds of failures in two ways. First, frequentistic or test data relevant to the needed conditional probabilities are never available. This only means that judgments must be made, which is routine for fault trees in any case. Second, the variables that enter into the fault tree are quite different in kind from those with which we are familiar. In this chapter, for example, they include not only design of information-providing displays but also organizational characteristics, stress, presence or absence of teamwork, and morale. We should not be surprised; such topics are obviously relevant to human error. But embedding them in a probabilistic structure, in this case an influence diagram, and judging appropriate probabilities are unfamiliar activities.

The chapter I am discussing presents an eleven-node influence diagram representation intended to be somewhat generic for operator errors in nuclear power plants. This diagram, like any other influence diagram, is a way of conceptualizing a probabilistic problem in order to facilitate assessments. It was developed in the course of work with nuclear power experts, and was tried out in a four-day meeting in 1983. The diagram was developed as a result of the group interactive process so characteristic of Phillips's work; Phillips was the facilitator. As usual, the diagram was modified in the course of its use. As the chapter makes clear in its disussion of requisite modeling, nothing is fixed or prespecified in group processes of the kind known as decision conferencing, of which this was an example. (However, tools of thought such as Bayes' theorem, multi-attribute utility measurement, maximization of expected utility, and Pareto analysis turn up in the output of decision conferences with such high probability that absence of all of them would be an amazing event indeed.)

The bulk of the chapter is devoted to an illustrative example taken from that application. I do not believe other examples of this particular approach to human reliability assessment exist, though I do not understand why not.

A technical problem recurs and recurs in the example. The target event, failure of a human operator to respond appropriately to an indication of a malfunction, is easily thought of as having a probability. But some of the intervening quantities, assessed as though they were probabilities and assumed to be just that in the influence diagram formalism, are somewhat less easy to interpret. What is the probability that personal factors are favorable, given that morale and motivation are good, competence is high, and operators are experienced? Such intermediate variables as personal factors are called 'influences' in the chapter, perhaps to indicate some concern about thinking of them as events having probabilities.

There seems to me to be less to this problem than meets the eye. Dichotomization of a continuum is familiar. We might have little difficulty with a fault tree that treated pressure as high, medium, or low—with each term carefully defined. We do not have an explicit operational definition of personal factors (or of quality of information provided to operators, or of effectiveness of the organization). But we do have fairly clear operational definitions of the bottom-level variables. More important, we have a structure of assessed conditional probabilities. What seems crucial, and is much facilitated by decision conferencing procedures, is that the entire structure, including lablels, input numbers, and output numbers, including its probabilities of human error, hang together and make sense to those who generated it. This is what Phillips means by a 'requisite model'. And in this sense, the definition of requisiteness is quite clear: a model is requisite if and only if none of those involved in generating it are dissatisfied with it or wish to change it.

Should one believe the output of such models? The answer obviously depends on what 'believe' means. Normally, and in this example, the output is one on which those who generated the model are willing to act. That is clearly an important way for a model to be believeable. In contexts like nuclear power plant operator reliability, another kind of believeability would consist of having a procedure capable of producing numbers that would be convincing to those who had not been present during its execution. It is far from clear that this procedure would meet such a test. It is also far from clear that any presently available alternatives to it would do much better. The test of effectiveness has a 'for-this-day-and-train-only' character to it. That is quite likely to be bothersome to those who like to place firm credence in models. It is less bothersome to those, like me, who see the alternative to requisite modeling as being no explicit procedure or test at all.

While this would not be my choice as a way of introducing novices to influence diagrams, it is an important application. It would be interesting to compare its results with those of other procedures for human reliability assessment. The procedure is sufficiently efficient so that one could ask the group involved to consider discrepancies between different approaches. In this field, at present, no 'gold standard' of methodological validity exists. Exploration of inconsistencies among methods with an eye to understanding and perhaps resolving them is the best one can hope for outside of the psychological laboratory. One could also imagine laboratory approaches to exploration of the idea. Unfortunately, they would probably be too contrived to be persuasive.

REFERENCES

Fischhoff, B. and MacGregor, D. (1982) Subjective confidence in forecasts, *Journal of Forecasting*, **1**(2), 155–72.

Howard, R. and Matheson, J. (1981) *Influence Diagrams*. In R. A. Howard and J. E.

Matheson (eds) (1984) *The Principles and Applications of Decision Analysis*, vol. II, Strategic Decisions Group, Menlo Park, Calif.

Low, K. B. and Bridger, H. (1979) Small group work in relation to managment development. In B. Babington-Smith and B. A. Farrell (eds) *Learning in Small Groups: A Study of Five Methods*, Pergamon Press, London, pp 83–102.

Miller, A. C., Merkhofer, M. M. and Howard, R. A. (1976) *Development of Automated Aids for Decision Analysis*, Stanford Research Institute, Menlo Park, Calif.

OECD (1985) *Expert Judgment of Human Reliability*, Restricted CNSI Report No. 88, Nuclear Energy Agency, Paris.

Phillips, L. D. (1982) Requisite decision modelling: a case study, *The Journal of the Operational Research Society*, **33**(4), 303–11.

Phillips, L. D. (1984) A theory of requisite decision models, *Acta Psychologica*, **56**, 29–48.

Phillips, L. D. (1987) On the adequacy of judgmental forecasts. In G. Wright and P. Ayton (eds) *Judgmental Forecasting*, Wiley, Chichester, pp. 11–30.

Phillips, L. D. (1989) Requisite decision modelling for technological projects. In C. Vlek and G. Cvetkovich (eds) *Social Decision Methodology for Technological Projects*, Kluwer Academic Publishers, Dordrecht, pp 95–110.

Selvidge, J. (1976) *Rapid Screening of Decision Options*, Technical Report 76–12, Decisions and Designs, McLean, Va.

Stäel von Holstein, C-A. S. and Matheson, J. (1979) *A Manual for Encoding Probability Distrubutions*, SRI Project 7028, SRI International, Menlo Park, Calif.

Swain, A. D. and Guttman, H. E. (1984) *Handbook of Human Reliability Analysis with Emphasis on Nuclear Power Plant Applications*, NUREG/CR-1278, US Nuclear Regulatory Commission, Washington, DC.

CHAPTER 12

Bayesian Updating of Event Tree Parameters to Predict High Risk Incidents

Robert M. Oliver and H. J. Yang *University of California, Berkeley, USA*

ABSTRACT

In this chapter the authors use chance influence diagrams to describe event trees used in safety analyses of low-probability high-risk incidents. Event trees are graphical representations that embed the roads to failure in a logical way; they lay out the various accident sequences of different risk levels and enumerate which subsystem failures are included in these sequences. This chapter shows how the branch parameters (chance of subsystem failure) used in the event tree models can be updated by a Bayesian method based on the observed counts of certain well-defined subsets of accident sequences. The influence diagram affords us a simple updating procedure based on arc reversals in a specially structured subgraph. This subgraph has a single unobservable branch parameter shared by two or more branches in the tree using only the data on accident sequence counts that use the subsystems (branches) of interest. The chapter concludes with a numerical example which shows how information contained in low severity incidents can be used to improve the prediction of the most severe incidents which, typically, are so rare that they have not yet occurred.

12.1 INTRODUCTION AND BACKGROUND

The event trees that are used to characterize accident sequences in safety analyses of the nuclear power industry typically have a very large number of branches and therefore a very large number of parameters that determine the conditional probability of a particular accident sequence. Using expert opinion and engineering knowledge, these parameters are usually estimated from fault trees whose top event yields the desired probability of branch failure. As far as we know there is no explicit, formal updating scheme in which historical data on

Influence Diagrams, Belief Nets and Decision Analysis
Edited by R. M. Oliver and J. Q. Smith

particular accident sequences are then used to obtain the posterior distribution of these branch parameters. This chapter proposes a model and a Bayesian procedure which accomplishes the updating in real time as counts of subsystem failures become available through testing or operational experience.

Several models and analysis techniques have been proposed to describe incident escalation in the nuclear power industry. Their degree of merit will depend on how well they represent the assumed mechanism of interaction, what parameters they require, what insights they yield, and, ultimately, on how accurate they are at prediction. Groer (1984) and Islam and Lindgren (1986) analyzed the Three-Mile-Island-type partial core-melts; Lewis (1984a, b) has been more concerned with the analysis of complete core-melts. Chow and Oliver (1988) have obtained posterior distributions of failure rates and recalculated some of the earlier predictions for median time to next incident using more recent data. More recently, Chow, *et al.* (1988) have proposed models which describe how incidents of low severity can escalate to incidents of higher severity and thereby provide a method whereby information on low-severity incidents can be used to sharpen the prediction for high-severity ones.

The prediction of risk in large-scale high-risk, low-probability incidents must be looked at in terms of general engineering knowledge and operational experience. Incidents may be the result of a chain of unlikely events, mistakes, and misjudgements. The important point to keep in mind is that lower and higher severity incidents and different accident sequences are interlinked and that one type may contain information helpful in predicting the other.

12.2 SAFETY STUDIES, EVENT TREES AND SEVERITY CLASSIFICATION

To study the progression of accident sequences in an organized and logical way safety analyses use event trees to trace how various malfunctions or subsystem failures influence the accidents and the risks. The drawing of an event tree is very much the choice of the model-builder and the type of engineering subsystems one is dealing with. Figure 12.1 shows an event tree that is typical of many accident sequences in a nuclear power plant; the tree starts with an initiating event, corresponding to a relatively low-severity incident. As a result of other failures the precursor may escalate into a more severe incident. Although we have selected an example from the nuclear industry the construction of event trees, identification of accident sequences and estimation of branch parameters is a major part of most safety studies including systems for detecting and avoiding ship collisions, systems used to avoid mid-air collisions and many others. Each branch in the tree represents the operation or failure of a particular subsystem which is identified in the column heading; if the branch drops we have a failure of that subsystem, if the branch rises the subsystem is available or operates successfully. The number on each falling branch is the chance that the subsystem

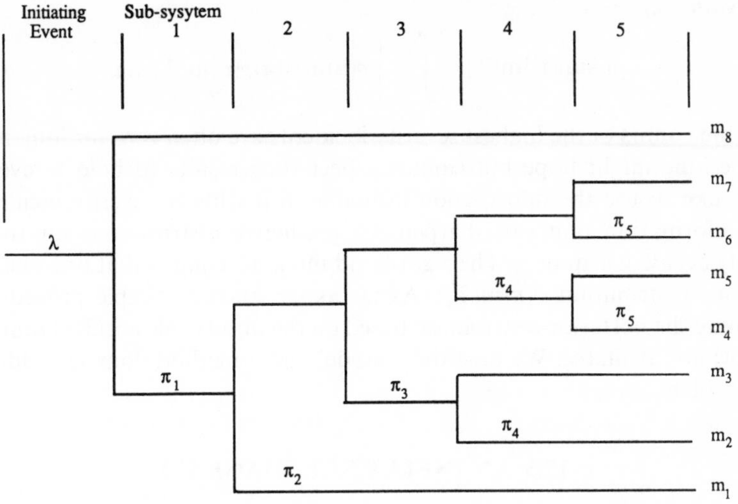

Figure 12.1 A representative event tree for power plant accidents.

at the head of the column will fail; we refer to it as the branch parameter. In general the relationship between accident sequences, ending states and risk severity may be very complex. Over a long period of time $(0, T)$ (in the case of nuclear reactors the relevant period is usually measured in reactor-years of operating experience) subsystem failures and successes will result in counts of accident sequences $m_1 = m_1(T)$, $m_2 = m_2(T), \ldots, m_k = m_k(T)$, where the index $k = 1, 2, \ldots, K$ enumerates the different accident sequences in the event tree. Although an ordering is not required, severity levels are often ordered so that $m_1(T)$ represents the count of most severe incidents, $m_K(T)$ represents the least severe incident. If subsystem 1 succeeds, the initiating events do not escalate to more severe incidents, regardless of whether downstream subsystems do or do not fail. On the other hand, the failure of subsystem 1 escalates the initiating events to various levels of severity depending on the failure status of following subsystems. An important feature of the event tree is that certain subsystems can influence more than one accident sequence in which case we have a 'shared' banch parameter. For example, the probability of failure of subsystem 4, π_4, influences two branches and the second, fourth and fifth accident sequences. The conditional parameter π_5, influences two branches and the fourth and sixth accident sequences, whereas subsystems 1, 2 and 3 affect only one branch apiece.

In general, if $\mathbf{m}(T)$ is the observable (vector) count of accident sequences at various risk levels, if $\Delta\mathbf{m}(\tau) = \mathbf{m}(T + \tau) - \mathbf{m}(T)$ is the count growth in the interval of time $(T, T + \tau)$, and if $\boldsymbol{\pi}$ is the unobservable vector of branch parameters, we are interested in predicting the future growth of incidents based on past observations,

$$p(\Delta\mathbf{m})(\tau)|\mathbf{m}(T) = \int \cdots \int p(\Delta\mathbf{m}(\tau)|\boldsymbol{\pi})p(\boldsymbol{\pi}|\mathbf{m}(T))\mathrm{d}\boldsymbol{\pi}.$$

Since the counts of the highest severity incidents are often zero for long periods of time (one might hope but cannot expect such results to hold forever) one would like to use the information available in the lower severity incidents to yield information that can sharpen the predictive distributions for the high severity accident sequnces. The right-hand integral requires that one obtain the posterior distribution $p(\boldsymbol{\pi}|\mathbf{m}(T))$. As far as we know a simple procedure for obtaining the posterior distribution based on the observable accident sequences, $\mathbf{m}(T)$, is not available. We describe a model and a methodology for addressing this problem.

12.3 AN INFLUENCE DIAGRAM

In the influence diagrams of this chapter nodes denote random variables and directed arcs denote the fact that a conditional probability of the node at the arrowhead need only be assessed in terms of its direct predecessor nodes. We follow a convention used earlier: unshaded nodes containing Greek letters denote unobservable parameters that influence the observable counts of events which are denoted by shaded nodes. For further discussion of the structure of influence diagrams one should refer to articles by Howard and Matheson (1984), Shachter (1987) and Smith (1987). We mention, without proof, that the absorption of a node corresponds to taking expectations with respect to that node conditional on its predecessor nodes; reversal of an arc, when allowed, corresponds to the use of Bayes' rule. Deterministic nodes (denoted by double circles) are known exactly given the values of the predecessor nodes since they are simply the sum of counts of subsets of accident sequences passing through the branch of a particular subtree. The influence diagram in Figure 12.2 and the event tree in Figure 12.1 share the same branch parameters. At each (non-deterministic) node, except for the initiating events node, there are always two predecessor nodes, one of which is the unobservable branch parameter with the appropriate subscript and the other is an observable node of counts of accident sequences. Figure 12.2 also recognizes the eight different accident sequences which add up to the total count, $m_1 + \cdots + m_8$ in the right-hand column of Figure 12.1. Obviously, the accident sequence is affected by the failures or successes of downstream subsystems. Also, a single subsystem that fails may have a very different effect depending on whether the accident sequence being considered has progressed through the top rather than the bottom of the tree. For example, the failure of subsystem 4 has a very different effect depending on whether subsystem 3 does or does not fail. In the former case we may obtain a very severe incident.

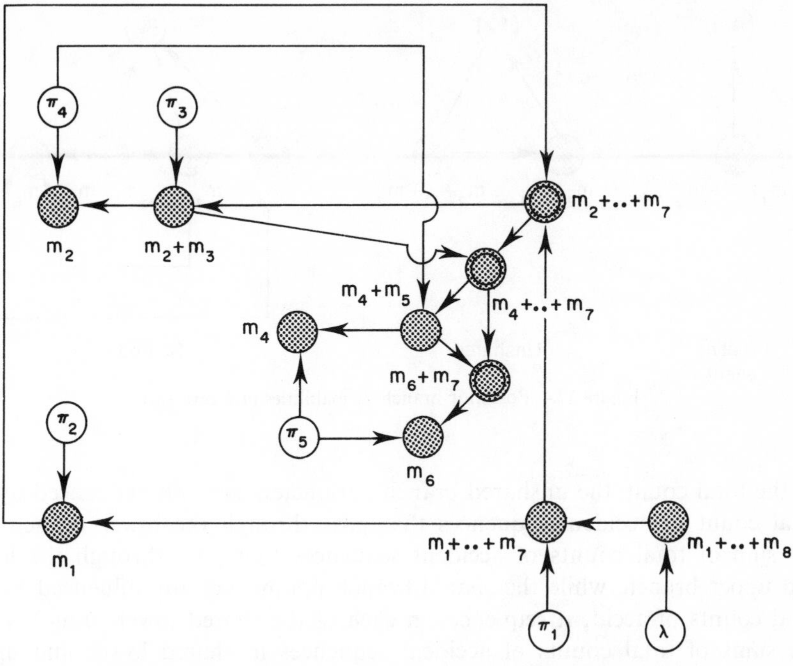

Figure 12.2 An influence diagram for the event tree in Figure 12.1.

On inspection of the influence diagram in Figure 12.2 it can be seen that there is a special structure for each of the random nodes influenced by an unobservable parameter, π_i, and a random node corresponding to a count of accident sequences. For example, in Figure 12.2, only $m_6 + m_7$ and π_5 influence m_6. Similarly $m_4 + m_5$ and the (same) π_5 influence m_4. These influences are also shown in Figure 12.3 where we display the subgraphs that only include direct successors of π's or any other observable nodes which provide us with information needed to update the π's. By arc reversal we obtain the results in Figure 12.4. The posterior distribution of initiating events is only influenced

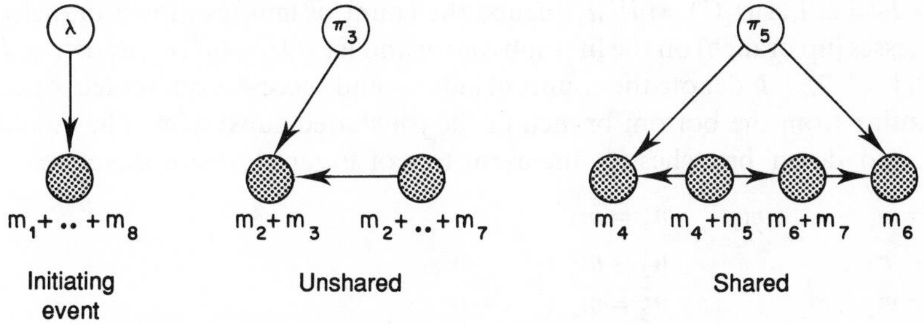

Figure 12.3 Nodes influenced by branch parameters and counts.

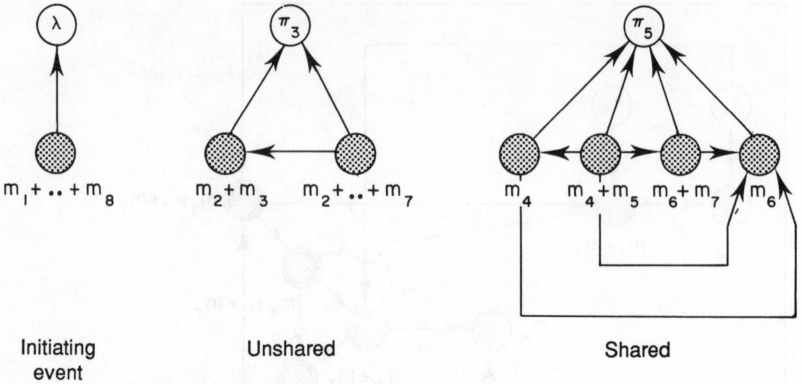

Figure 12.4 Posterior branch probabilities (arc reversal).

by the total count, the unshared branch parameters are only influenced by the total count of accident sequences that pass through the lower branch and the sum of total counts of accident sequences that pass through the lower and upper branch, while the shared branch parameters are influenced by the total counts of accident sequences in each of the shared lower branches and the sums of total counts of accident sequences in shared lower and upper branches.

12.4 SEPARABLE UPDATING THEOREM

Define π_j, $(j = 1, 2, \ldots, J)$ as the chance of jth subsystem failure and m_k, $(k = 1, 2, \ldots, K)$ to be the count of the kth accident sequence and d_j^i, u_j^i, are the counts of event at 'down' and 'up' branches that correspond to the jth subsystem in the ith $(i = 1, 2, \ldots I_j)$ shared branch from the bottom. Obviously, I_j is the total number of shared branches for the jth subsystem. In Figure 12.1, for example, there are five subsystems and eight accident sequences so that $J = 5$, $K = 8$. Subsystems four and five are shared by two branches and the first, second and third subsystems influence only one branch so that $I_1 = I_2 = I_3 = 1$ and $I_4 = I_5 = 2$. Let $D_1(1) = (d_1^1, u_1^1)$ denoe the count of failures (down branch) and successes (up branch) on the first subsystem and let $D_j(i) = (d_j^1, u_j^1, d_j^2, u_j^2 \ldots, d_j^i, u_j^i)$ with $i = 1, 2, \ldots I_j$ denote the counts of failures and successes on 'shared' branches counting from the bottom branch of the jth shared subsystem. The counts on 'up' and 'down' branches for the event tree of Figure 12.1 are therefore:

$$
\begin{array}{llll}
d_1^1 = m_1 + \cdots + m_7 & u_1^1 = m_8 & & \\
d_2^1 = m_1 & u_2^1 = m_2 + \cdots + m_7 & & \\
d_3^1 = m_2 + m_3 & u_3^1 = m_4 + \cdots + m_7 & & \\
d_4^1 = m_2 & u_4^1 = m_3 & d_4^2 = m_4 + m_5 & u_4^2 = m_6 + m_7 \\
d_5^1 = m_4 & u_5^1 = m_5 & d_5^2 = m_6 & u_5^2 = m_7
\end{array}
$$

and the data sets that we will use to update the posterior distributions of branch parameters are

$$\begin{aligned}
D_1(1) &= (d_1^1, u_1^1) = (m_1 + \cdots + m_7, m_8)\\
D_2(1) &= (d_2^1, u_2^1) = (m_1, m_2 + \cdots + m_7)\\
D_3(1) &= (d_3^1, u_3^1) = (m_2 + m_3, m_4 + \cdots + m_7)\\
D_4(2) &= (d_4^1, u_4^1, d_4^2, u_4^2) = (m_2, m_3, m_4 + m_5, m_6 + m_7)\\
D_5(2) &= (d_5^1, u_5^1, d_5^2, u_5^2) = (m_4, m_5, m_6, m_7).
\end{aligned}$$

Note that each d or u denotes the count of accident sequences that pass through down or up branches. The set of counts in $D_4(2)$, for example, is not the same as the set $(m_2, m_3, m_4, m_5, m_6, m_7)$. Here, $D_j = D_j(I_j)$ represents a set of counts which are sufficient to update π_j so that $p(\pi_j|D) = p(\pi_j|D_j)$. Finally, let $D = \{m_1, m_2, \ldots, m_K\}$ denote the set of all observable counts and $m = \sum_k m_k$ denote the total number of initiating events. Under the assumption that branch parameters are *a priori* independent, the posterior distribution of parameters is given by the

12.4.1 Separable updating theorem for branch probabilities

$$\begin{aligned}
p(\lambda, \pi|D) &= p(\lambda, \pi_1, \ldots, \pi_J|m_1, \ldots, m_K)\\
&= p(\lambda|m)p(\pi_1|D_1(1))p(\pi_2|D_2(I_2))\cdots p(\pi_J|D_J(I_J))
\end{aligned}$$

with each factor on the right-hand side obtained by the recursions (C_{ij} are normalizing constants)

$$\begin{aligned}
p(\pi_j|D_j(i)) &= C_{ij}p(\pi_j|D_j(i-1))p(d_j^i|\pi_j, d_j^i + u_j^i) \qquad i = 1, 2, \ldots, I_j\\
p(\pi_j|D_j(1)) &= C_{1j}p(\pi_j)p(d_j^1|\pi_j, d_j^1 + u_j^1) \qquad j = 1, 2, \ldots, J.
\end{aligned}$$

A proof is obtained by considering the subgraph of the influence diagram associated with each subsystem. Figure 12.5 shows two different cases of failures influenced by the parameter π_j. In the first case the branch parameter is not shared: since there are no predecessor nodes for the parameter π_j and since $d_j + u_j$ but not d_j, may have predecessor nodes, the posterior distribution of π_j is proportional to the product of likelihood and the prior, $p(d_j|\pi_j, d_j + u_j)p(\pi_j)$

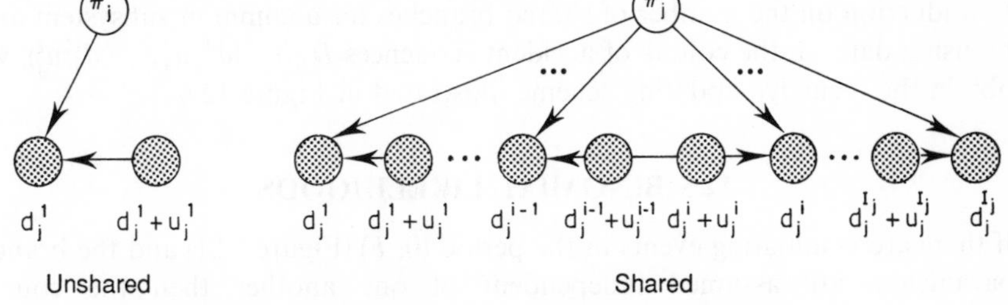

Figure 12.5 Counts influenced by a single branch parameter.

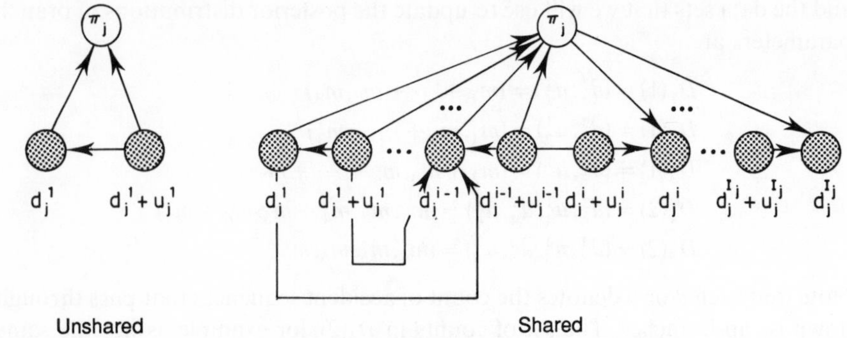

Figure 12.6 Inferring π from counts.

and independent of all other incident counts and all other branch parameters. Arc reversal is shown in the left-hand side of Figure 12.6 for the unshared branch. Now consider the case where more than one branch in the event tree shares a common subsystem. The counts must obviously be dependent because they sum to a number less than or equal to the total number of initiating accidents; we therefore show an undirected arc between nodes $d_j^{i-1} + u_j^{i-1}$ and $d_j^i + u_j^i$ since our results do not depend on the direction of the arrow. The other arcs connecting $d_j^i + u_j^i$ and $d_j^l + u_j^l$, $i \neq l$, are omitted; we point out that no arcs enter nodes having 'down' branch counts. Inference on the parameter π_j is slightly more complicated because of the effects of shared branches. Updates must now depend not only on the priors of π_j but also upon the likelihoods of d_j^i gven π_j and $d_j^i + u_j^i$ with $i = 1, 2, \ldots, I_j$. It would appear from Figure 12.4 that the posterior distribution for π_j is, in general, a complicated function of the observables in those cases where there are shared branch parameters. Fortunately there is considerable simplification in that updating can be done recursively. First,

$$p(\pi_j | d_j^1, u_j^1) = C_{1j} p(\pi_j) p(d_j^1 | \pi_j, d_j^1 + u_j^1).$$

With the additional data d_j^2, u_j^2, we obtain

$$\begin{aligned} p(\pi_j | d_j^1, u_j^1, d_j^2, u_j^2) &= C p(\pi_j) p(d_j^1 | \pi_j, d_j^i + u_j^i) p(d_j^2 | \pi_j, d_j^2 + u_j^2) \\ &= C_{2j} p(\pi_j | D_j(1)) P(d_j^2 | \pi_j, d_j^2 + u_j^2). \end{aligned}$$

By induction on the number of shared branches for a common subsystem and by using data on the counts of accident sequences $D_j(i) = (d_j^1, u_j^1, \ldots, d_j^i, u_j^i)$, we obtain the recursive updating scheme illustrated in Figure 12.6.

12.5 BINOMIAL LIKELIHOODS

If there are m initiating events in the period $(0, T)$ (Figure 12.1) and the branch parameters are assumed independent of one another then the counts

$\mathbf{m}(T) = \{m_1, \ldots, m_8\}$ of the eight different accident sequences with $m_1 + \cdots + m_8 = m$ can be assumed to have the multinomial distribution, i.e.

$$(m_1, m_2, \ldots, m_8 | m, \pi_1, \pi_2, \ldots, \pi_5) \sim \mathrm{MN}(m, \pi_1\pi_2, \pi_1(1-\pi_2)\pi_3\pi_4, \ldots, (1-\pi_1)).$$

Marginally, the count of any single accident sequence is therefore binomial. For example, the distribution of the count of the second accident sequence conditional on m is independent of π_5,

$$(m_2 | m, \pi_1, \pi_2, \ldots, \pi_5) \sim \mathrm{Bin}(m, \pi_1(1-\pi_2)\pi_3\pi_4).$$

When the likelihood of d_j^i is assumed to be a binomial distribution conditional on π_j and $d_j^i + u_j^i$, one obtains a simple result for the updating process. Define $K_j(d)$ and $K_j(u)$, respectively, as the sets of accident sequences that pass through the down and up branches using subsystem j; we also use the notation $K_j = K_j(d) \cup K_j(u)$ to denote the totality of all accident sequences that are influenced by the availability or failure of subsystem j. There is a close connection between the 'up' and 'down' branch counts that we have used in previous sections of the paper and the counts obtained by adding up accident sequences over these sets. We denote the sums of accident sequences over the different sets by

$$T_j = \sum_{i=1}^{I_j} d_j^i = \sum_{k \in K_j(d)} m_k, \qquad S_j = \sum_{i=1}^{I_j} u_j^i = \sum_{k \in K_j(u)} m_k.$$

The set of all accident sequences is represented by $K = \bigcup_j K_j$ since every sequence must use at least one subsystem. Using this notation it is easy to see that the influence diagrams in Figures 12.5 and 12.6 can be further simplified as the two counts S_j and T_j carry sufficient information to update π_j. The proof is based on the special structure of the binomial likelihood which yields the result that if π_j is common to several branches then the posterior distribution of π_j conditional on data D is

$$p(\pi_j | D) = p(\pi_j | D_j(I_j)) = Cp(\pi_j)\pi_j^{T_j}(1-\pi_j)^{S_j}$$

with total counts S_j and T_j defined as before. Thus in the case of binomial likelihoods the inference problem is considerably simplified as shown in Figure 12.7.

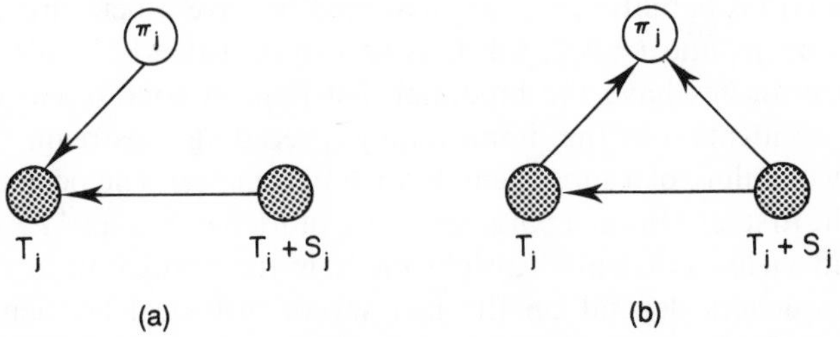

Figure 12.7 Inference with binomial likelihoods. (a) Before reversal. (b) After arc reversal.

12.6 BETA PRIORS

When the prior distribution of π_j is assumed to be a beta distribution with parameters given by (α_j, β_j), and the likelihood of d^i_j is binomial conditional on π_j and $d^i_j + u^i_j$, the posterior distributions of the branch parameters are also beta distributions with new parameters reflecting the sum of all accident sequences which pass through 'up' and 'down' branches sharing a common subsystem.

$$(\pi_j | D) \sim \mathrm{Be}\left(\alpha + \sum_{k \in K_j(d)} m_k, \beta + \sum_{k \in K_j(u)} m_k \right).$$

The above is obtained by using the fact that if d^i_j conditional on π_j and $d^i_j + u^i_j$ is binomial for $i = 1, \ldots, I_j$, then the sum of all accident sequences passing through down branches sharing subsystem j is also binomial conditional on the same π_j and the sum of all accident sequences passing through subsystem j:

$$\sum_{k \in K_j(d)} m_k \left| \sum_{k \in K_j} m_k, \pi_j \sim \mathrm{Bin}\left(\sum_{k \in K_j} m_k, \pi_j \right) \right..$$

With a beta prior the updated postrior has an α term which is the sum of all relevant branch failures influenced by the shared parameter $\boldsymbol{\pi}$, while the β term includes the sum of all upper branch successes.

12.7 NUMERICAL EXAMPLE

We illustrate with a numerical example based on the event tree in Figure 12.1. The growth of accident counts over time is shown in Table 12.1. Notice that the large majority of accident sequences are included in the top path of the tree which, in our example, corresponds to the least severe sequence, that over long periods of time the counts for the most severe sequences, m_1 and m_2, are zero and that the rate of initiating events corresponds to something less than one per time period (per reactor year if the example were to apply to reactor safety). The purpose of choosing the high rate of initiating events is to dramatize the effect of very small counts in severe accident sequences against a background of very large counts of less severe accident sequences. The prior distributions of conditional probabilities, π_i's, are assumed to have a beta distribution with parameters α_i, β_i, and $i = 1, 2, 3, 4, 5$, as shown in Table 12.2. We assume that counts on branches have the binomial distribution conditional on the total count of accidents on the immediately preceding upstream branch and probability of failure of a subsystem (branch parameter). The posterior expected values of the first four branch parameters are plotted in Figure 12.8 as a function of time and yields substantial insight on how the prediction of future severe accident sequences depend on the less severe historical accident sequences. The sharing of information in the branch parameters is an essential ingredient

Table 12.1 Counts of accident sequences over time

T	$m_8(T)$	$m_7(T)$	$m_6(T)$	$m_5(T)$	$m_4(T)$	$m_3(T)$	$m_2(T)$	$m_1(T)$
100	80	1	0	0	0	0	0	0
200	130	1	1	0	0	0	0	0
300	250	2	1	1	0	0	0	0
400	360	2	2	1	0	0	0	0
500	390	2	2	1	0	0	0	0
600	500	3	2	1	1	0	0	0
700	580	3	2	2	1	0	0	0
800	650	3	2	2	1	0	0	0
900	750	3	2	2	2	0	0	0
1000	810	4	2	3	2	1	0	0
1100	890	4	2	3	2	1	0	0
1200	960	4	3	3	2	1	0	0
1300	1050	4	3	3	2	1	0	0
1400	1170	4	3	3	2	1	0	0
1500	1250	5	3	3	2	1	0	0

Table 12.2 Parameters of beta priors for branch probabilities

	Subsystem 1	Subsystem 2	Subsystem 3	Subsystem 4	Subsystem 5
α_i	0.001	0.001	0.1	0.2	3
β_i	0.1	0.05	1	2	9

of this change. We observe no sequences in the bottom path of Figure 12.1 since $m_1(T) = 0$; as a result the expected value of π_2 decreases over time and results in an increasing expected time until the first accident sequence of that type. Although the count of the second sequence from the bottom is also zero, π_4 is shared by a branch in the fourth and fifth accident sequences that lead to less severe incidents and for which we have several incident counts. As a result the expected value of π_4 increases over time; this growth contrasts sharply with the effect that zero counts of the first accident sequence count have on the posterior distribution of π_2. One can trace the effects of changes in counts of different sequences upon the posterior expectation of π_3 which has an approximately ninefold increase at $T = 1000$. Although we do not include a graph for the posterior expectation of π_5 it fluctuates around values close to the prior expectation of about one-third. As we get more counts over time, the posterior distributions of the branch parameters often becomes sharper due to

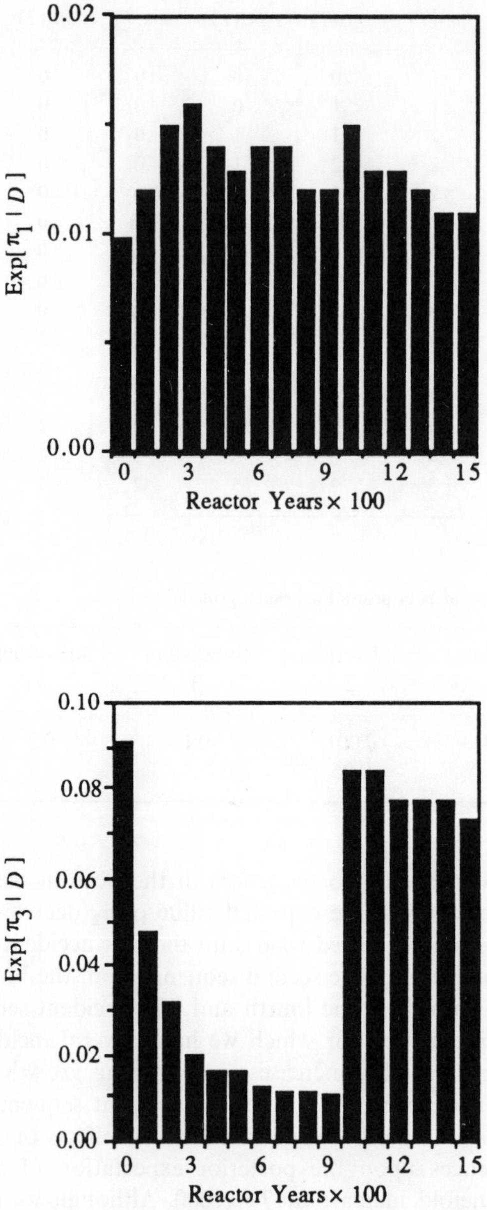

Figure 12.8 Posterior expectation of branch parameters versus time.

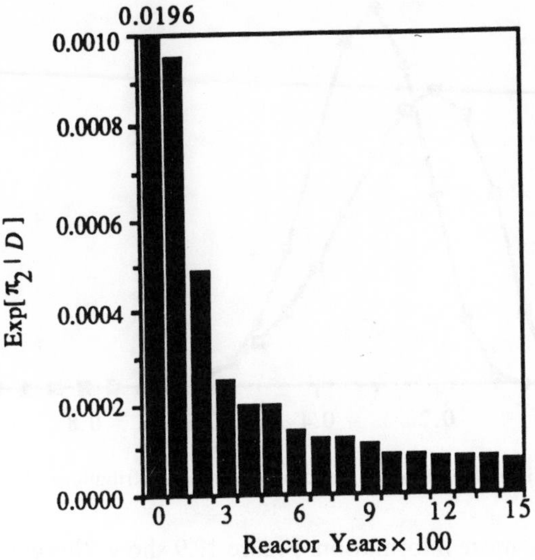

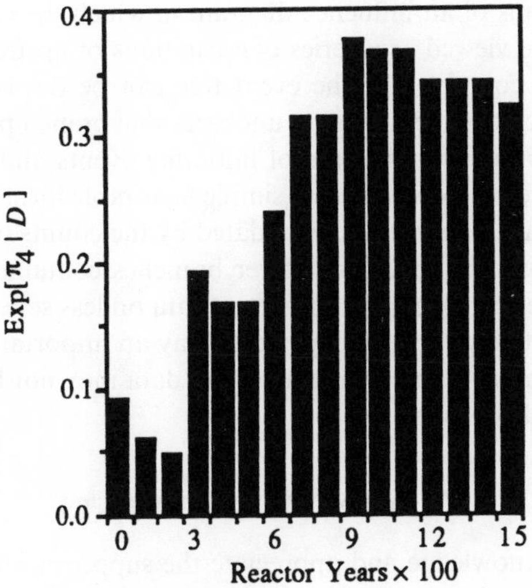

Figure 12.8 (*Contd.*)

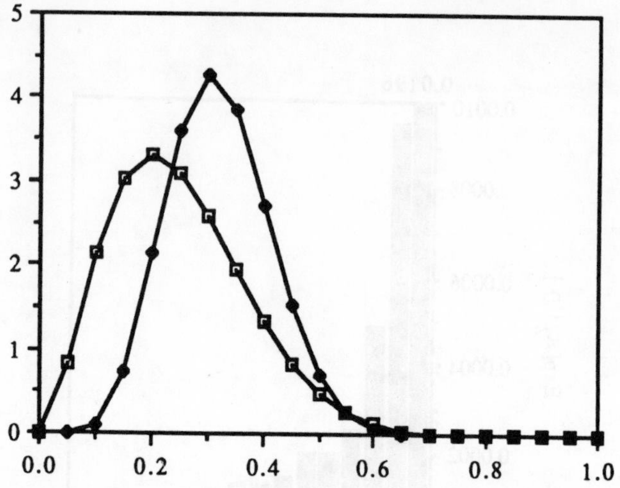

Figure 12.9 Prior (□) and posterior (◇) distributions of π_5.

the availability of more information. Figure 12.9 shows this effect by comparing the prior and posterior distribution of π_5 at $T = 1500$.

12.8 CONCLUSION

In this chapter we have described event trees for low-probability accident sequences in terms of an influence diagram in which the count of downstream sequences can be viewed as a series of escalations of upstream incident counts. All information contained in the event tree can be displayed in an influence diagram that includes the priors on unobservable branch parameters. The joint posterior distribution for the rate of initiating events and the probabilities of subsystem failure are expressed in a simple factorable form in which each factor corresponds to a subsystem and is updated by the counts of accident sequences that pass through the upper and lower branches containing the parameter of interest. A numerical example shows how data on less severe incidents, that are obtained in relatively short time periods, play an important role in sharpening predictions for more severe incidents where data may not become available for very long periods of time.

ACKNOWLEDGEMENTS

The authors acknowledge and appreciate the support of the Office of Nuclear Regulatory Research in the US Nuclear Regulatory Commission through their Grant NRC-04-87-104, the National Science Foundation and the Lawrence Livermore Laboratory.

12.9 DISCUSSION

12.9.1 Discussion by Nora Smiriga

This chapter is an excellent illustration of the power of influence diagrams. In my opinion, it is much easier to see how to update the distributions of π_j's by using influence diagrams than by using event trees. In addition, the authors were aware of the importance of good notation. By introducing d^i_j and u^i_j, one can describe the influence diagram in terms of λ, the π_j, the d^i_j, and the u^i_j; and the arc reversals can easily be carried out. The authors clearly show how to generalize the problem when branch parameters are shared by several subsystems. The chapter does not require the ordering of the accident sequence $m_1, \ldots, m_k$; in practice it might also be difficult to order these in terms of severity levels. It is also possible that some power plants might have some m_i which are indistinguishable in their effects, making it difficult and costly to diagnose the failure of a subsystem. Finally, the assumed independence of the prior distributions does not guarantee the independence of the posterior distributions. However, in this case one can see from the influence diagram—after performing the arc reversals—that $p(\lambda, \pi | D)$ will be the product of the posterior distributions.

The authors' work is clearly valuable to NRC, because it shows how one can easily update the distributions of the π_i's using test data.

12.9.2 Discussion by A. P. Dempster

The point that Oliver and co-authors have been making in this and several cited papers is that data from low-severity accidents should be jointly modelled with data from high-severity accidents when estimation of low probabilities of the high-severity types are at issue. The point is valid and important, but needs to be pushed further.

The artificial example presented by Oliver and Yang supposes 1264 occurrences of a certain initiating event in 100 reactors followed for 15 years. Among the 1264 events only 14 followed the relatively dangerous track of failure of subsystem 1, and among the 14 none followed the supposed highly dangerous track of failure of subsystem 2. Hence, the occurrence rate of initiating events at about 0.85 per reactor year is quite accurately determined, as is the predicted subsequent failure rate of subsystem 1 at around 10^{-2}. But the data are obviously quite uninformative about the critical further failure rate of subsystem 2. Almost every statistician will point out that a failure rate could be as high as 20 percent and still a string of 14 successive nonfailures will occur with chance at least 4.4 percent, which is barely significant at the 0.05 level. Even the most fervent Bayesian must accept this as evidence that the prior distribution of π_2 has a large influence on the posterior and hence on any judgement about the probability of the high-severity accident.

The particular beta prior with $\alpha_2 = 0.001$ and $\beta_2 = 0.05$ shown in Table 12.2 could surely not be accepted easily. The implied prior probability of failure is about 2 percent, which might be justifiable, but notice the drastic effect of a single sample observation as shown in the upper right panel of figure 12.8, whereby a single nonfailure changes the 2 percent by a factor of 20 to about 0.1 percent. Note that a single failure would have had an even more drastic effect of raising the odds of failure on the next trial to about 95 percent chance of failure. Clearly, no regulator could place credence in a statistical method that places so much weight on the outcome of a single trial.

The correct lesson of the example, it seems to me, is that the risk analyst needs to find other data or other arguments to come up with a credible prior for π_2. I believe this means 'borrowing strength' from knowledge of failure rates of systems 'like' subsystem 2 in some way to be described and argued convincingly. For the Bayesian, this means a much more difficult task of prior assessment than envisioned in Table 12.2, because it implied that the prior failure rates in some large class of systems must be modelled as dependent in some meaningful way. The resulting data analysis would go much beyond that envisaged by simple influence diagrams.

12.9.3 Discussion by Ross D. Shachter

This note provides a simpler proof for the separable updating theorem for branch probabilities. The authors develop this theorem to provide a general and tractable mechanism for the computation of posterior probabilities on the basis of evidence about event trees. The proof presented in the chapter is fairly hard to understand because it appears to depend on a judicious application of Bayes' theorem and a fortuitous ordering of the events from the tree. In fact, their result is considerably more robust than it might seem from the chapter.

First, consider an influence diagram, shown in Figure 12.10(a), which corresponds to the event tree in their chapter. This diagram is topologically equivalent to the influence diagram displayed in Figure 12.2, but the nodes are reorganized into three columns: the uncertain parameters on the left; the 'down' evidence in the middle, and the 'up' evidence on the right. For any fork in the event tree, if the probability of going 'down' is given by parameter π, then the evidence consists of the number of times we observe 'down', D, and the number of times we observe 'up', U. We can equivalently use the number of visits to the fork, V, as evidence, in which case $U = V - D$ is a deterministic function. Therefore all of the 'up' nodes in the third column are deterministic functions of their predecessors.

When we evaluate the information in the influence diagram to compute posterior distributions for the parameters in the first column, the evidence in the second and third columns has already been observed. The 'likelihood principle' states that we should take advantage of whatever that evidence reveals about the parameters, but we should not otherwise care about the prior distribution for our observations. Thus, if we treat the evidence for each fork as the pair '$U; D$' (or alternatively '$U; V$') we do not need to keep track of the *a priori* dependence among the observations. Figure 12.10(b) shows the influence diagram after applying the likelihood principle.

Once we have constructed the second influence diagram, the proof of the theorem is straightforward. We assume that each of the parameters is *a priori* independent. We also assume that we have evidence for each fork and that there is exactly one parameter describing the behavior at a given fork. It is therefore clear from the diagram shown in Figure 12.10(b) that each of the parameters will be independent a posteriori. Furthermore, the posterior distribution for each parameter can be obtained through repeated applications of Bayes' theorem, or a single application with a likelihood formed from the product of the associated likelihoods. As in the chapter, there is no need to make any assumptions about the nature of the prior distributions or the likelihood functions.

12.9.4 Reply to A. P. Dempster

The authors agree with almost everything the discussant says, including the final paragraph, which effectively states that the risk analyst must come up with credible priors for parameters. The main criticism seem to be that since a single failure has an 'incredible' effect in the case of the parameter π_2, one must find other ways to assess credible priors. Unfortunately, there is little evidence or expert opinion in some cases so that about the best one can do is make a guess, use a large variance for the prior and let experience and experimental data help out.

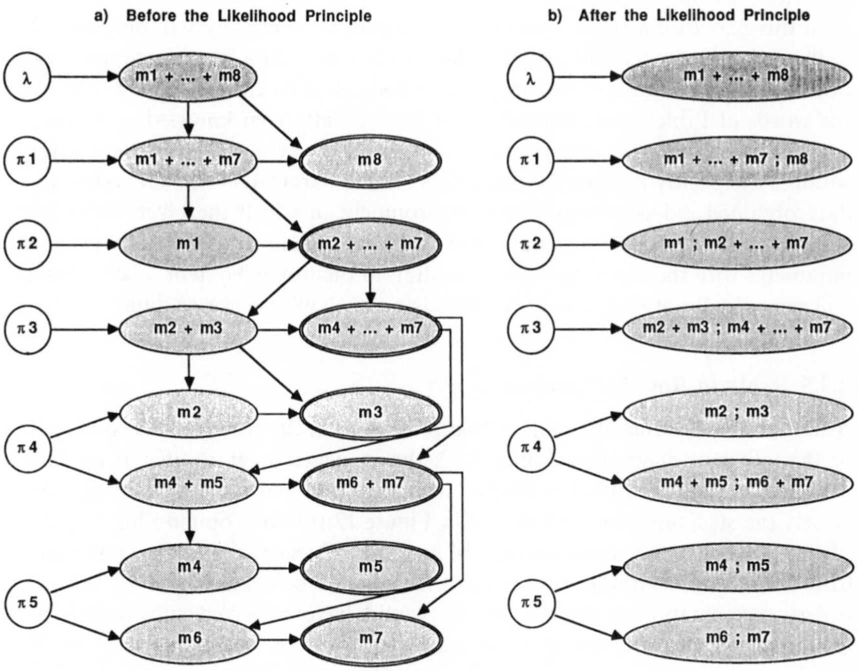

Figure 12.10

Table 12.3 **Comparisons of prior and posterior π_2**

	Prior				Posterior (1500 RYs)			
	α_2	β_2	Mean	Variance	α'_2	β'_2	Mean	Variance
Old π_2	0.001	0.05	0.0196	0.0183	0.001	14.05	7.117×10^{-5}	4.728×10^{-6}
New π_2	1	50	0.0196	3.697×10^{-4}	1	64	0.0154	2.330×10^{-4}

The prior distribution of π_2 that we used is an example of a situation where its expected value is believed to be around 2 percent but insufficient engineering knowledge is available; we therefore assigned a relatively large variance. Had we had better knowledge or experience we would have selected a sharper prior but we cannot force such good fortune. By way of response and illustration we now compare expectation and variance of π_2 when we assume $\alpha_2 = 1$ and $\beta_2 = 50$ rather than 0.001 and 0.05 (the original paper). These new parameter values lead to the same prior expectation for π_2 but to very much smaller prior variance, about a factor of 200 smaller; thus, we are very certain about the prior $E[\pi_2]$. (see Table 12.3.)

In this new case a single obsrvation of failure or nonfailure will only slightly reduce the expected value and the effect of the new data on the variance will not be very great. But this only emphasizes what all of us know to be the case—in the words of Professor Dempster—'borrow strength from knowledge of failure rates of subsystems like subsystem 2'. One natural way is to extend the conversation by explicitly requiring that π_2 be a shared parameter with subsystem test data obtained independently of the environment in which the given event tree is being analyzed. If such test data is available π_2 should be modeled as a shared parameter with the event tree corresponding to the test subsystem. The question still remains: What does one do if there is no such experience or data?

12.9.5 Reply to Ross D. Shachter

Although the influence diagram in Figure 12.10(a) in the comments is equivalent to the influence diagram in Figure 12.2, the influence diagram offered by Ross Shachter in Figure 12.10(a) is simpler to understand and resembles much more closely the structure of the event tree in Figure 12.1; in our opinion his diagram offers readers a better understanding of how the influence diagram is constructed from the event tree. Except for the fact that it would be unfair to the discussant's orderly presentation of a critique, we would have preferred substituting his diagram for our own. His diagram also closely resemble the one used by N. Smiriga in her oral presentation at the conference.

Based on the likelihood principle Shachter then offers a simplified diagram in which updating of all shared parameters can be visualized in terms of two nodes.

In fact one can go even further in that all observables (on which updating is based) can be merged into a single node as illustrated in Figure 12.11 for parameter π_5. Note, however, that relevant quantities within the node are separated by semicolons and there are as many elements in the data vector of the single node as there are predecessor nodes for the shared parameter (see Figures 12.5 and 12.6). It is certainly true that a merged node is a more compact way of representing the data in an influence diagram but if one is interested in completing the dirty details of calculating posteriors one must keep separate the elements of the data vector—it matters not whether this is done with a vector of counts, $D_j(I_j)$, as in the original paper, with semicolons in Figure 12.11, or with two nodes in Figure 12.10(b).

If u is the count on a down branch and v is the number of visits, it is true that we only need the likelihood $p(u, v|\pi)$ and prior $p(\pi)$ to get the posterior distribution $p(\pi|u, v)$, but to do this we need access to the conditional distribution $p(u|v, \pi)$ and $p(\pi)$.

Figure 12.12(a) is implied by all the diagrams to its right but the likelihood principle for cases (b), (c) and (d) in Figure 12.12 are given by

$$p(\pi|D) = Cp(\underline{u|\pi})p(\underline{v|\pi})p(\underline{\pi})$$
$$p(\pi|D) = Cp(v|u)p(\underline{u|\pi})p(\underline{\pi}) = Cp(\underline{u|\pi})p(\underline{\pi})$$
$$p(\pi|D) = Cp(\underline{u|v, \pi})p(v)p(\underline{\pi}) = Cp(\underline{u|v, \pi})p(\underline{\pi})$$

each of which yields a very different posterior distribution but only one of which is correct for our problem. We have underlined the factors that matter in the joint density, and for ease of notation have used C throughout for the normalizing constants even though they differ in each case. In the case of a four-node diagram with four data elements and one shared parameter, $D = (u, v, x, y)$, where u and x correspond to the number of downs and v and y are the number of visits. Use of the likelihood principle in Figure 12.10(b) leads to a factorization for the

Figure 12.11 Merging all sufficient statistics into one node.

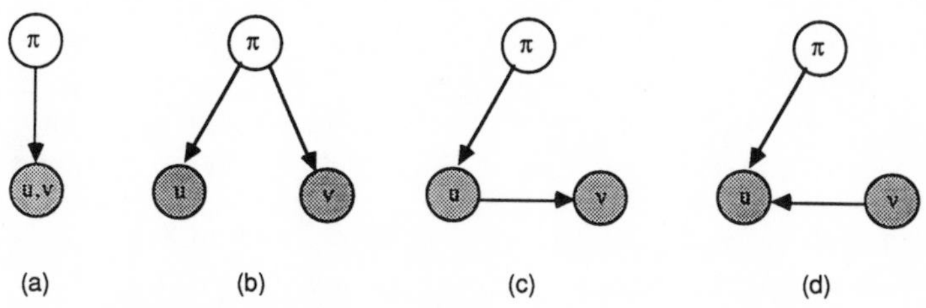

Figure 12.12 Examples of some IDs whose posterior is $p(\pi|D)$.

posterior represented by

$$p(\pi \mid D) = Cp(u, v \mid \pi)p(x, y \mid \pi)p(\pi).$$

Again, this factorization is correct but misleading in that the one we must actually use to obtain the posterior is not the one suggested but

$$p(\pi \mid D) = C\,p(x \mid y, \pi)\,\underline{p(u \mid y)}\,p(u \mid v, \pi)\,\underline{p(y)}\,p(\pi) = Cp(x \mid y, \pi)p(u \mid v, \pi)p(\pi).$$

The likelihood principle shows that the number of factors is one more than the number of down branches.

We very much appreciate and thank Nora Smiriga, A. P. Dempster and Ross D. Shachter for their thoughtful remarks.

REFERENCES

Chow, T. C. and Oliver, R. M. (1988) Predicting nuclear incidents, *Jour. of Forcasting*, **7**, 49–61.

Chow, T. C., Oliver, R. M. and Vignaux, G. A. (1988) *A Bayesian Escalation Model to Predict Nuclear Incidents and Risk*, Report ORC 87-3, University of California at Berkeley. To appear in *Operations Research*.

Groer, P. G. (1984) Bayesian estimates for the rate of Three Mile Island type releases. In R. A. Waller and V. T. Covello (*eds*), *Low Probability High-Consequence Risk Analysis*, New York, pp 127–136.

Howard, R. A. and Matheson, J. E. (1984) Influence diagrams. In R. A. Howard and J. E. Matheson (eds) *Reading on the Principles and Applications of decision Analysis*, vol. II, Strategic Decisions Group, Menlo Park, Calif., pp 721–762.

Islam, S. and Lindgren, K. (1986) How many reactor incidents will there be?, *Nature*, **322**, 691–2.

Lewis, H. W. (1984a) Bayesian estimation of core-melt probability, *Nuclear Science and Engineering*, **86**, 111–12.

Shachter, Ross D. (1987) Evaluating influence diagrams, *Operations Research*, **34** (26), 871–82.

Smith, J. W. (1987) *Diagrams of Influence in Statistical Models*, Research Report 99 Department of Statistics, University of Warwick, Coventry, UK

CHAPTER 13

Using Influence Diagrams in Multiattribute Utility Analysis—Improving Effectiveness through Improving Communication

Miley W. Merkhofer, *Applied Decision Analysis, Inc., Menlo Park, CA, USA*

ABSTRACT

Many decision analysts routinely use influence diagrams to structure problems for decision analysis. The diagrams have two major characteristics that motivate their use for this purpose. First, they provide an extremely effective vehicle for communication between the analysts and those with expertise relevant to the problem-under study. Second, once constructed, the diagrams can be straightforwardly quantified and solved to provide probabilities and evaluations of decision strategies. This chapter addresses the first of these advantages. Examples are used to describe and illustrate methods that have been found effective for using influence diagrams to identify and organize important factors into models useful for decision analysis.

13.1 INTRODUCTION

Early discussions of influence diagrams in the decision analysis literature described the methodology as a 'frontend' for automated decision aids (Miller *et al.*, 1976). It is easy to understand the reason for this perspective. The diagrams are intuitively understandable because the formal definition of influence coincides well with the intuitive notion possessed by most people. At the same time, the diagrams capture information sufficient for normative analysis in a compact, precise representation that can be conveniently stored in computer memory. The computer can also manipulate the diagrams using well-specified rules to generate problem representations that are more convenient for analysis. Not surprisingly, most of the research related to influence diagrams has concentrated on automating the transition from influence diagram formulation

Influence Diagrams, Belief Nets and decision Analysis
Edited by R. M. Oliver and J. Q. Smith

to analysis (e.g. Miller *et al.*, 1976; Korsan and Matheson, 1978; Merkhofer and Leaf, 1981; Howard and Matheson, 1981; Olmsted, 1983; Shachter, 1986, 1988).

Influence diagrams provide a benefit beyond serving as a frontend for automated decision aids. Many of the same characteristics that make them useful for communicating with computers also make them effective for communications among people. In short, influence diagrams are effective tools for formulating, structuring, and summarizing a problem. This chapter describes the use of influence diagrams not merely as a device for communicating with computers, but rather as a means for facilitating communication among all parties to an analysis.

One area in which incluence diagrams have been found to be particularly effective for problem formulation, structuring, and communication is in applications of multiattribute utility analysis (MUA) to health and environmental decisions. By incorporating influence diagrams into the MUA process, it is possible to simplify some of the more difficult steps of the analysis. In addition, the credibility of MUA applications has been enhanced by the ability of influence diagrams to clearly and concisely document underlying assumptions.

13.2 MULTIATTRIBUTE UTILITY ANALYSIS OF HEALTH AND ENVIRONMENTAL PROBLEMS

MUA is well suited to the demands of health and environmental decision problems (Keeney and Nair, 1977; Hobbs, 1979; Keeney, 1980; Merkhofer and Keeney, 1987). It permits evaluating options against multiple objectives. Competing objectives, such as 'maximize public health' and 'minimize economic costs', are common in health and environmental problems. MUA also embodies a strong and appealing logic. The MUA logic is useful because health and safety choices are often controversial and are frequently challenged in court.

The five basic steps of MUA are:

1. Identifying decision objectives;
2. Establishing attributes for measuring the degree to which objectives are achieved;
3. Quantifying uncertainty over attributes;
4. Calculating decision strategies that maximize expected utility and performing sensitivity analysis.

Conducting these steps can be difficult, especially in the case of health and environmental decisions. One source of difficulty is the complexity of health and environmental problems. The number of variables and their interrelationships can tax even the most capable and computationally well-equipped analysts. Another source of difficulty is the many participants to such decisions. Each must accept the analysis if it is to have credibility. This requires ensuring that experts contribute according to their areas of expertise and that policymakers retain the responsibility of providing policy judgments. Everyone must understand the analysis. In addition, policymakers must feel confident of their ability to explain the analysis to those who exercise oversight authority, including their superiors, Congress, and the public.

Influence diagrams can be used to facilitate the MUA process. In particular, the ability of an influence diagram to aid communication and understanding can be used effectively within three of the MUA steps—identification of objectives, establishment of attributes, and quantification of uncertainty over attributes. In addition, the ability of influence diagrams to serve as a basis for automating analysis can be used to facilitate the last step—calculation of optimal decision strategies and sensitivity analysis. Examples of how this can be accomplished are provided below.

13.3 INFLUENCE DIAGRAMS

An influence diagram is a network consisting of nodes and directed arcs with no cycles. In nontechnical terms, this means that an influence diagram is a graphical representation of a problem consisting of simple geometric shapes (e.g. circles and squares) connected by arrows such that there are no paths following the arrows that lead in cycles. The geometric shapes, or nodes, represent the uncertain variables and decisions relevant to the problem. The directed arcs, or arrows, represent probabilistic dependencies and information flows.

There are three types of nodes in influence diagrams: chance nodes, decision nodes, and value nodes. According to convention, chance nodes have a circular shape and decision nodes have a square or rectangular shape. The shape of value nodes is not so well standardized. Depending on the analyst, they might be circular, hexagonal, or a rectangle with rounded corners.

The directed arcs connecting the nodes are one of two types: informational arcs (which point into decision nodes) or conditional arcs (which point into chance and value nodes). Informational arcs represent information known to the decision maker at the time the decision must be made. Conditional arcs represent probabilistic dependence—the probability distribution for the successor node (the node pointed to by the arrow) is conditionally dependent on the predecessor node (the node from which the arrow originates). The most natural source of dependence is a cause–effect relationship. However, conditional arcs need not represent causality.

Although the individual elements of influence diagrams are simple, the diagrams can be used to represent highly complex problems. For example, Figure 13.1 illustrates an influence diagram constructed to help structure an

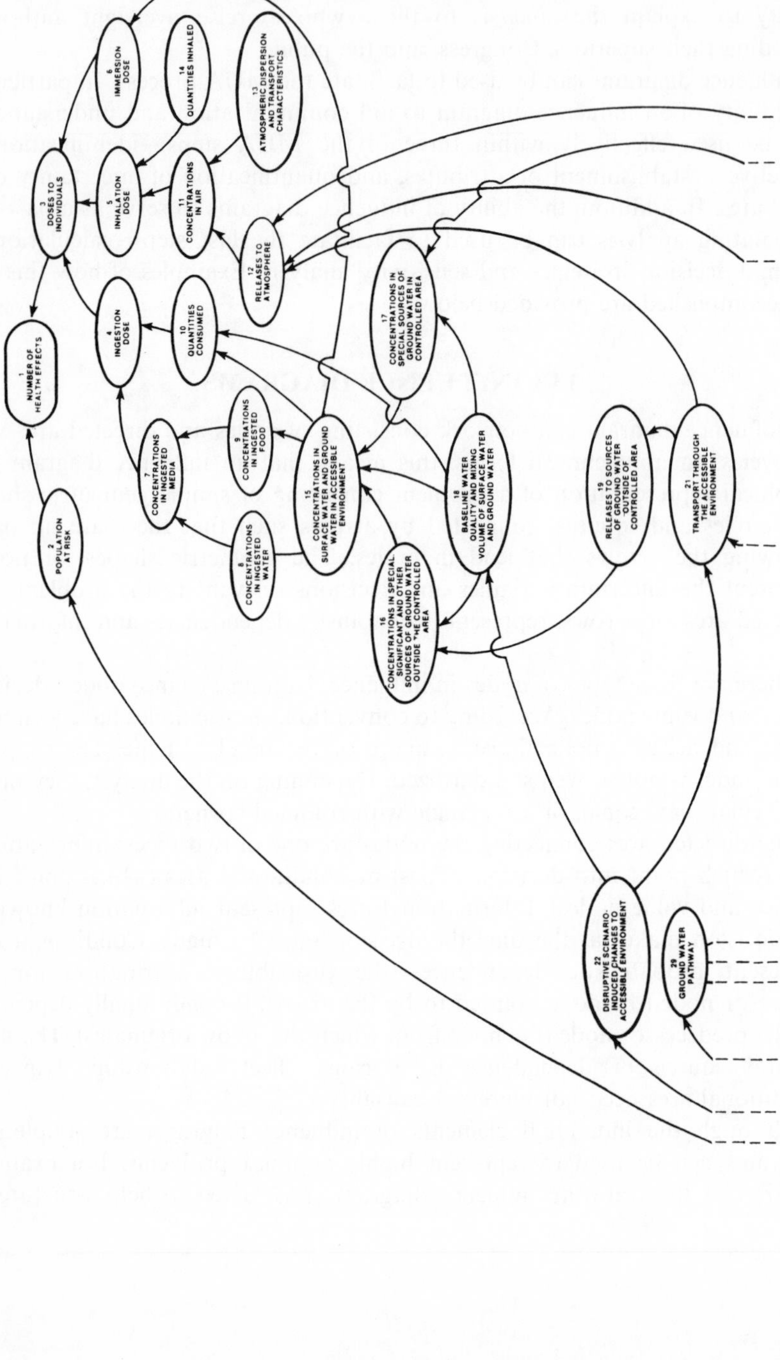
1 NUMBER OF HEALTH EFFECTS
2 POPULATION AT RISK
3 DOSES TO INDIVIDUALS
4 INGESTION DOSE
5 INHALATION DOSE
6 IMMERSION DOSE
7 CONCENTRATIONS IN INGESTED MEDIA
8 CONCENTRATIONS IN INGESTED WATER
9 CONCENTRATIONS IN INGESTED FOOD
10 QUANTITIES CONSUMED
11 CONCENTRATIONS IN AIR
12 RELEASES TO ATMOSPHERE
13 ATMOSPHERIC DISPERSION AND TRANSPORT CHARACTERISTICS
14 QUANTITIES INHALED
15 CONCENTRATIONS IN SURFACE WATER AND GROUND WATER IN ACCESSIBLE ENVIRONMENT
16 CONCENTRATIONS IN SPECIAL SIGNIFICANT AND OTHER SOURCES OF GROUND WATER OUTSIDE THE CONTROLLED AREA
17 CONCENTRATIONS IN SPECIAL SOURCES OF GROUND WATER IN CONTROLLED AREA
18 BASELINE WATER QUALITY AND MIXING VOLUME OF SURFACE WATER AND GROUND WATER
19 RELEASES TO SOURCES OF GROUND WATER OUTSIDE OF CONTROLLED AREA
21 TRANSPORT THROUGH THE ACCESSIBLE ENVIRONMENT
22 DISRUPTIVE SCENARIO-INDUCED CHANGES TO ACCESSIBLE ENVIRONMENT
29 GROUND WATER PATHWAY

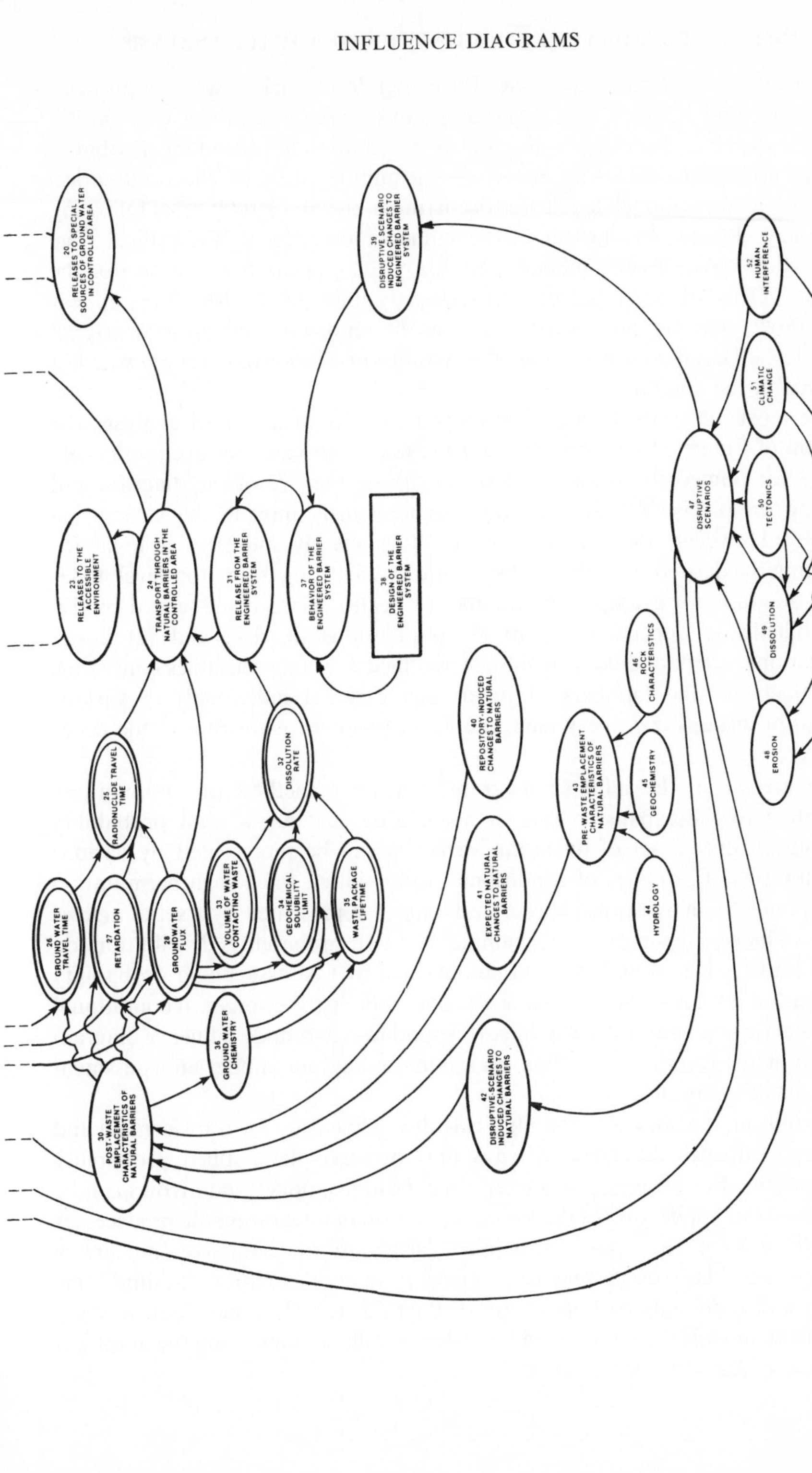

Figure 13.1 Influence diagram constructed for an MUA of potential sites for a nuclear waste repository. Diagram indicates the factors influencing the numbers of postclosure health effects in 10 000 years.

MUA evaluation of alternative sites for a high-level nuclear waste repository (Merkhofer and Keeney, 1987). The diagram represents a model wherein the specified variables determine a probability distribution for one of the attributes, or value nodes, selected for the analysis—the number of health effects (fatalities) attributable to radionuclide releases occurring in the first 10 000 years following repository closure. As illustrated, the influence diagram is hierarchical. The factors that most directly influence health effects appear near the top of the diagram. Detailed, basic factors that indirectly influence health effects appear lower down. The decision of where to site the repository influences nearly all of the basic factors and (to reduce the number of connecting arrows) was not included in the diagram.

Like most influence diagrams developed to help structure an analysis, the diagram of Figure 13.1 is complex, and the uncertain variables are too loosely defined for immediate formal analysis. (Limiting the size of the diagram and insisting on precise definitions would unnecessarily impede the structuring process.) However, the hierarchical structure clearly indicates a logic for obtaining the desired probability distribution. According to the top-level nodes in the diagram, for example, the number of health effects is influenced by (i.e. probabilistically dependent upon) (1) populations at risk, and (2) doses. Populations at risk and doses would be specified as vector variables consisting, respectively, of the numbers of potentially exposed individuals in various population groups (e.g. the young, the aged, pregnant women) and the doses each receive.

The arrows in the influnce diagram indicate a method of computation. According to the expansion rule of probability calculus, a joint probability distribution over a set of uncertain variables can be represented by various permutations of products of conditional probabilities. Each such permutation corresponds to an assessment order and may be represented by a set of arrows in an influence diagram. For example, a joint probability distribution on uncertain variables A and B could be obtained by first assessing a probability distribution on A and then assessing a probability distribution on B conditional on A. This assessment order would correspond to a two-node influence diagram with an arrow from A to B. Thus, an influence diagram implies an assessment order for its solution.

In addition, the laws of probability calculus define rules for transforming and reducing influence diagrams in ways that preserve the implied probability distributions. For example, under certain conditions, nodes and arrows can be eliminated (an application of the logical operation of integrating the predecessor variable out of the joint probability distribution). Similarly, an arrow between two chance nodes may be reversed. (Usually this requires adding some arrows and represents an application of Bayes' rule.) Thus, new, but logically equivalent, models may be defined by systematically manipulating the graphical structure of the influence diagram.

13.4 BENEFITS OF USING INFLUENCE DIAGRAMS

Constructing an influence diagram, such as that shown in Figure 13.1, serves several purposes:

- First, it helps ensure a systematic, balanced exploration of issues. The hierarchical structure promotes a 'top-down' approach to modeling. Attention is focused initially on key variables and their relationships, and only later on the details. Because the diagram is generated backwards from the value attributes, the identified variables tend to fan away uniformly, resulting in a more balanced model (Owen, 1978). Thus, the diagram tends to counter the natural tendency towards giving excessive attention to some areas at the expense of others.
- A second benefit of constructing an influence diagram is that it encourages communication among participants. The process forces experts to specify explicitly their views as to policy and intervening variables. It enables participants to efficiently share their thinking with colleagues.
- A third benefit occurs once the influence diagram has been constructed. The diagram provides a picture of the problem that is relatively easily examined, explored, and, if appropriate, changed. People often have difficulty perceiving the dependencies among choices, uncertainties, and outcomes. The influence diagram provides a useful overview. Interested parties can step back and see the problem in a global way that often stimulates fresh thinking.
- A fourth benefit occurs because the picture provided by the influence diagram can be used as a 'knowledge map'. In many ways, this knowledge map is similar to a road map (Smithin, 1980). A road map shows how towns are linked to one another, and illustrates possible routes between them. Some areas covered by the map are more familiar than others, yet a close study may reveal routes that might otherwise be overlooked. Other areas are less well known and the map acts as a guide through them. In the same way, influence diagrams bring problem considerations together and show how they are related. It can be used to follow routes or chains of arguments. In some cases the diagram may remind users of familiar ideas; in others it can be used to explore new considerations and suggest possibilities for action.
- Finally, a major benefit comes when it is time to analyze the model represented by the influence diagram. Because an influence diagram is logically precise, there is no need to transform the diagram to a form appropriate for analysis. The same graphical representation that is so effective for modeling is also effective for analysis. Because transformations of influence diagrams correspond to applications of probability calculus, graphical techniques can be used to transform the diagrams to more convenient forms without getting bogged down in mathematical details.

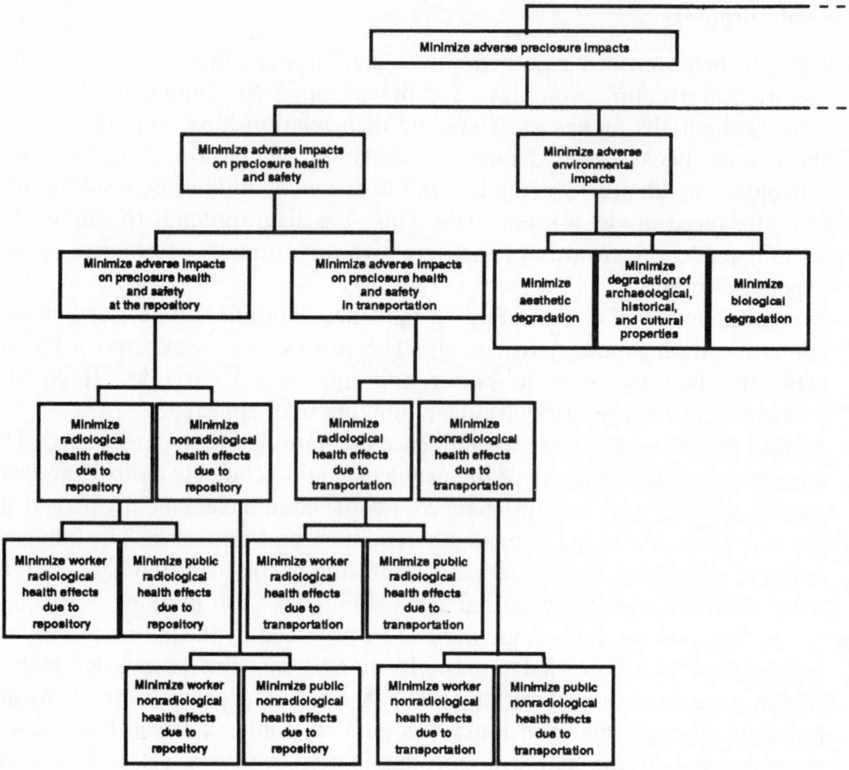

Figure 13.2 Objectives hierarchy constructed for an MUA of potential sites for a nuclear waste repository.

13.5 USING INFLUENCE DIAGRAMS TO FACILITATE THE IDENTIFICATION OF DECISION OBJECTIVES

The fundamental principle underlying MUA is that the desirability of an alternative is determined by the extent to which it achieves decision objectives. Thus, identifying and structuring objectives is the first step in an MUA. Decision analysts typically develop the objectives by interviewing decision makers (and, perhaps, decision stakeholders).

The process of developing objectives involves two stages (Keeney, 1980). First, candidate objectives are listed. For analyses conducted for government agencies, statements of objectives can often be found in enabling legislation, agency guidelines, and other documentation. The objectives are typically of two distinct types: fundamental objectives and means objectives. Fundamental

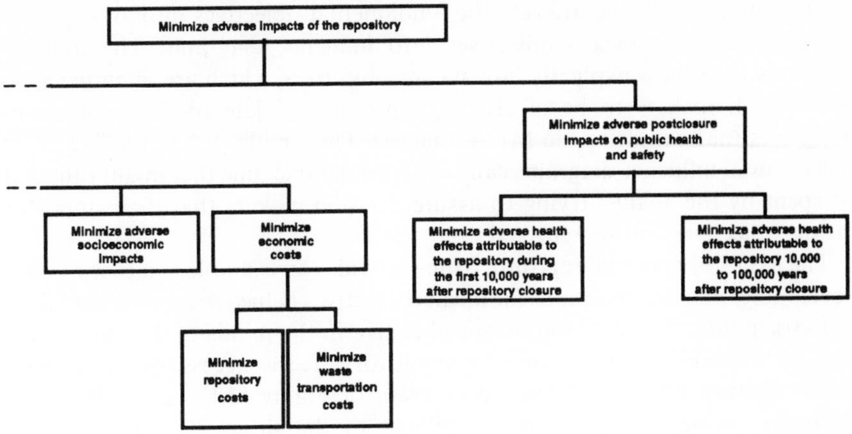

objectives specify the essential reason for interest in the decision problem. The objective of minimizing health effects is a fundamental objective. Means objectives are important for their influence on the degree to which fundamental objectives are met. An example of a means objective is the objective of minimizing radionuclide releases from a nuclear waste repository.

The second stage of the process of developing objectives consists of organizing objectives into a hierarchy. Typically, only fundamental objectives appear in the objectives hierarchy*. Figure 13.2 shows the hierarchy of fundamental objectives developed for the MUA of potential repository sites. The hierarchy defines general objectives (like maximizing health and safety) in terms of more specific, lower-level objectives (such as minimizing the incidences of specific types of injuries, sicknesses, and fatalities). The objectives hierarchy is recommended because its construction helps ensure that no 'holes' (missing objectives) occur in the analysis and helps eliminate situations where double counting might result (because holes and redundancies are more easily identified from the hierarchy).

The fact that both means objectives and fundamental objectives are generated in the first stage of the process, but only fundamental objectives appear in the objectives hierarchy, creates a problem for the analyst. The various participants in the decision are typically focused on means objectives. Using only the fundamental objectives from the generated list can create the impression that the analysis will not be sensitive to the issues of greatest concern to decision makers.

*Fundamental objectives are used because it is often difficult to define means objectives that satisfy necessary independence requirements, the relationships among means objectives tend to be complex, and it is more difficult to obtain agreement on means objectives from diverse groups.

The solution is to incorporate the fundamental objectives into an objectives hierarchy and the means objectives into influence diagrams. The influence diagrams then show explicitly how means objectives, which are often perceived by decision makers to be of crucial importance, relate to the fundamental objectives that will drive the MUA analysis. The simple step of placing means objectives in influence diagrams can save considerable time that might otherwise be spent by the analyst trying to assure decision makers that their important means objectives will be reflected.

The MUA of potential repository sites provides an example of a case in which developing influence diagrams for means objectives helped secure the confidence of participants. The most important objective in the minds of decision makers was to minimize releases from the repository to the accessible environment (node number 23 in the influence diagram of Figure 13.1). (This objective is expressed in nearly all legislation related to the siting of a nuclear waste repository.) Prior to the development of the influence diagram, the response of some of the managers who reviewed the objectives hierarchy (Figure 13.2) was that the analysis would be vulnerable to legal challenges because it did not reflect the requirement of minimizing radionuclide releases. Constructing the influence diagram was crucial to securing the confidence of decision makers that the MUA analysis could be demonstrated to be legally sound.

13.6 USING INFLUENCE DIAGRAMS TO ESTABLISH ATTRIBUTE MEASUREMENT SCALES

According to the MUA approach, a means must be found for measuring the degree to which alternatives achieve objectives. This is accomplished by defining an attribute measurement scale for each lowest-level objective in the objectives hierarchy.

There are two types of attribute scales—natural scales and constructed scales (Keeney, 1986). Natural scales are established scales that enjoy common usage and interpretation. For example, cost, measured in millions of dollars, is a natural scale for measuring performance against the objective of minimizing economic costs. Constructed scales are scales developed specifically for the problem at hand. Constructed scales often consist of verbal descriptions of distinct levels of impact.

Although decision analysts prefer the use of natural scales, many of the objectives underlying health and environmental problems have no natural scales. Thus, constructed scales must be developed. Developing the constructed scales, however, can be difficult. Difficulty arises because attributes for health and the environment are usually complex, and it is hard to identify the factors or dimensions that are most important to measuring their achievement.

Influence diagrams may be used to systematically develop constructed scales. There are three steps. In the first step, an influence diagram is constructed to identify the factors that influence the degree to which a complex objective is achieved. As an illustration, one of the objectives shown in Figure 13.2 is minimizing degradation to archaeological, historical, and cultural properties (such properties might be damaged, for example, during the construction of the repository or the roads leading to it). Figure 13.3 shows the diagram developed to identify the factors judged to influence the degree of degradation that would occur. The diagram was constructed using a top-down approach in which lower-level factors were identified by asking specialists to describe the measurements or studies they would undertake to assess the higher-level factors.

In the second step, the factors having the greatest influence at each level of the diagram are identified. A factor might be judged to have greater influence because its range of variability is greater than the others, or because even a small change in that factor can lead to a substantial change in the assessment of the higher-level factor. Factors having the greatest influence may be identified with double ovals. For example, the double ovals in Figure 13.3 indicate the factors judged to have the greatest influence on the level of degradation to archaeological, historical, and cultural properties. In this example, the specialists felt that the most important factors were the number of properties impacted, their significance, their amenability to mitigation, and the magnitude of the effects.The factor judged most significant to determining the magnitude of the effects was the location of the properties and proximity to areas affected by repository construction.

The third step in the process of developing a constructed scale is to describe several alternative degrees of impact in terms of the factors identified in the influence diagram with the double ovals. Table 13.1 shows the attribute scale developed from the influence diagram of Figure 13.3. The scale defines six alternative levels of impact in terms of the considerations identified with the double ovals. In the instructions for using this or similar scales, it must be pointed out that the combination of outcomes specified for any given score do not necessarily represent the only way that the corresponding score could be obtained. For example, although a score of two is defined in terms of impacts to two properties of major significance or ten properties of minor significance, the score could be justified in the case of other combinations of numbers of affected major and minor properties, provided that the net impact, on balance, would be judged as comparable.

13.7 USING INFLUENCE DIAGRAMS TO FACILITATE QUANTIFYING UNCERTAINTY OVER ATTRIBUTES

The third step in the MUA process is to quantify uncertainty over the various attributes selected for the analysis. In the case of health and environmental applications, existing cause–effect models would seem helpful for conducting this step. For example, models have been developed to describe releases of

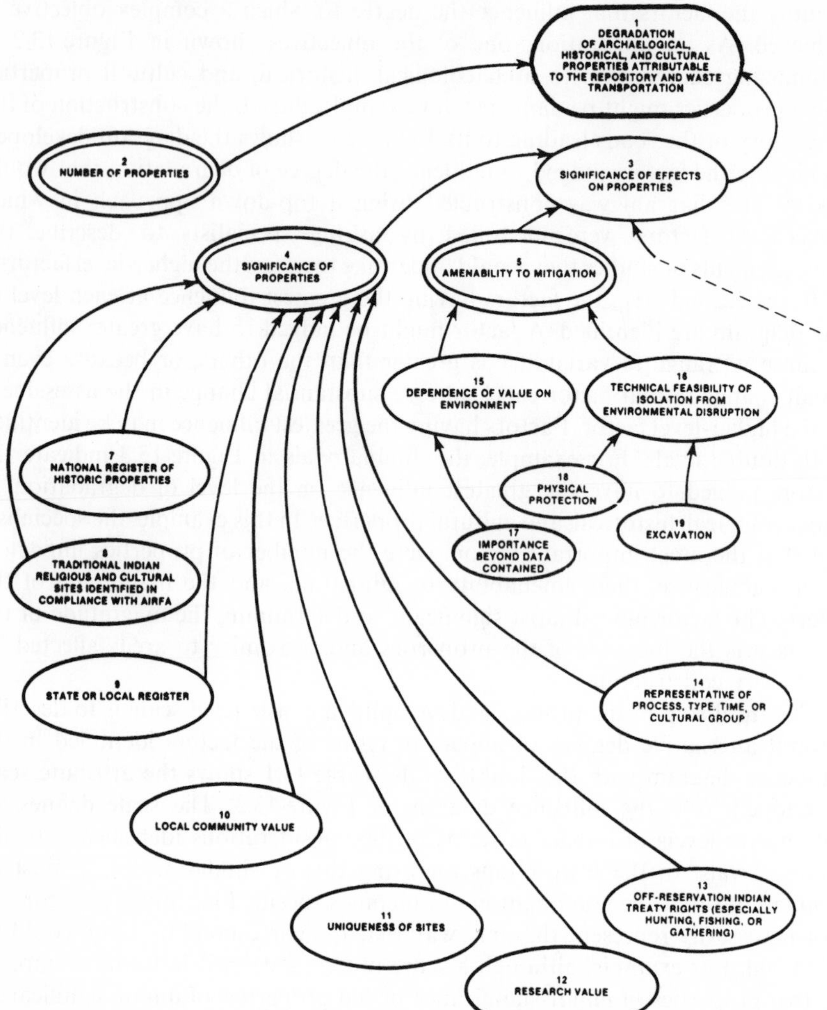

Figure 13.3 Another influence diagram constructed for MUA of potential repository sites. The diagram indicates factors that influence the degradation of archaeological, historical, and cultural properties.

hazardous substances or dangerous forms of energy due to failures of technological systems, the movement and transformations of substances released into the environment, the exposures of people or the things they value to hazards, and the consequence of such exposures (Covello and Merkhofer, 1990). The problem with using cause–effect models is that the models are often inadequate and biased. For example, in the case of repository siting, complex models have

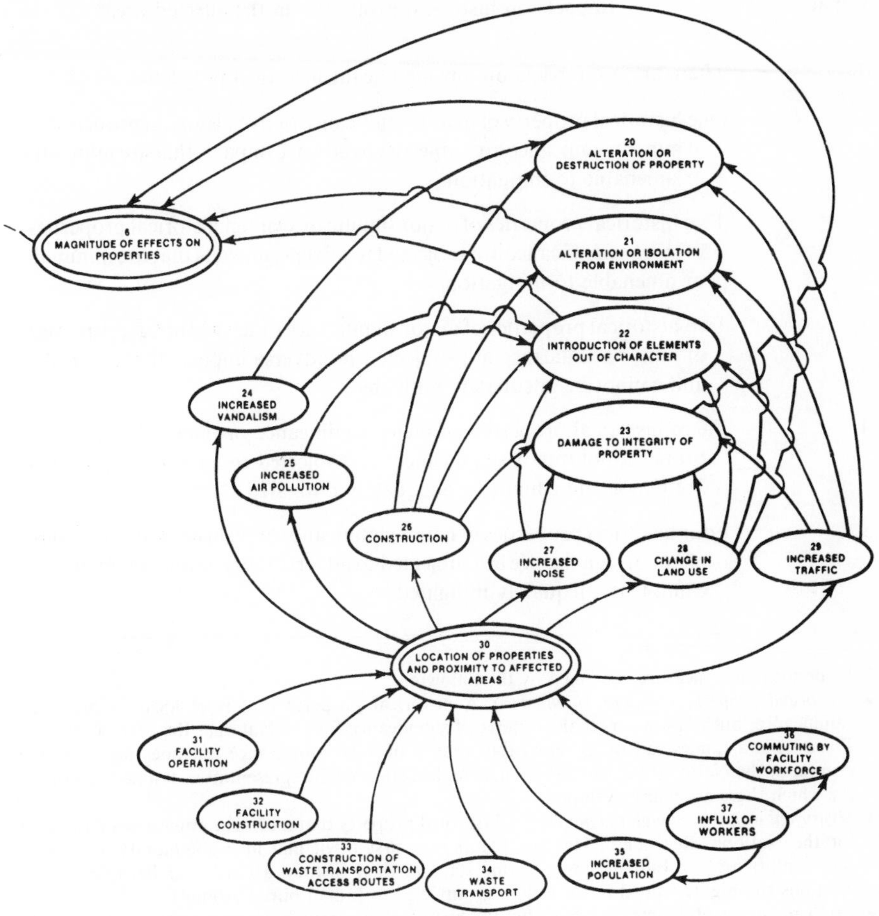

Figure 13.3 (*Contd.*)

been developed for estimating the release of radionuclides and their transport through groundwater. The models, however, are incomplete in that they fail to account adequately for the effects of heat generated by the waste. The models are also biased in that they adopt conservative assumptions that 'err on the side of protection of public health'. Thus, although complex models are typically available, it is often difficult to use them directly in MUA.

Influence diagrams provide an alternative, often more useful, basis for quantifying uncertainty over attributes. There are several ways in which the diagrams might be used. For example, influence diagrams in which attributes

Table 13.1 Constructed scale for measuring degradation to archaeological, historical, and cultural properties from the repository and waste transportation

Impact level	Impacts on historical properties in the affected area[a]
0	There are no impacts on any significant historical properties
1	One historical property of major significance or five historical properties of minor significance are subjected to adverse impacts that are minimal or amenable to mitigation
2	Two historical properties of major significance or ten historical properties of minor significance are subjected to adverse impacts that are minimal or amenable to mitigation
3	Two historical properties of major significance or ten historical properties of minor significance are subjected to adverse impacts that are major and cannot be adequately mitigated
4	There historical properties of major significance or fifteen historical properties of minor significance are subjected to adverse impacts that are major and cannot be adequately mitigated
5	Four historical properties of major significance or 20 historical properties of minor significance are subjected to adverse impacts that are major and cannot be adequately mitigated

[a]The performance measure is defined by the following:

- *Historical property of minor significance*: A historical property that is of local or restricted significance, but does not meet the criteria of significance for the National Register of Historic Places (e.g. a homestead or miner's cabin that is of local importance but does not meet the criteria of the National Register; an archaeological site that is representative of a period of time for which there are many examples).
- *Historical property of major signifance*: A historical property that meets the criteria of significance for the National Register of Historic Places (e.g. first town hall in a community; cave sites representative of an Indian people at one stage of their history; a Civil War battlefield) or a religious site highly valued by an Indian group (e.g. an Indian burial ground).
- *Minimal impacts*: Impacts that may alter the historical property, but will not change its integrity or its significance.
- *Major impacts*: Impacts that change the integrity or the significance of the historical property.
- *Amenable to mitigation*: The character of the historical property is such that it is possible to mitigate adverse impacts, reducing major impacts to minor or eliminating adverse impacts (e.g. impacts on an archaeological site that is significant because of the data it contains can be mitigated by excavating and analyzing those data; subsurface sites located within the controlled area may be protected under agreements made to guarantee that they will not be disturbed; a historical site can be adequately protected from vandals by erecting physical barriers).
- *Not amenable to mitigation*: The character of the historical property is such that impacts cannot be adequately mitigated because the value depends on the relationship of the historical property to its environment (e.g. a historical property of religious significance; a historical property that has value beyond the data contained; an archaeological site that is too complex for adequate excavation given state-of-the-art techniques).

are top-level value nodes may be constructed and solved to obtain the desired probability distributions. If there are no influences across the diagrams and there are no downstream decisions (decision nodes with informational arcs pointing into them), then the diagrams are independent of one another. Provided that appropriate conditional probability distributions are available for all predecessor variables, the diagrams may be solved individually to obtain probability distributions (conditioned on decisions) for their top-level attributes. Probability encoding (e.g. Spetzler and Staël von Holstein, 1975; Merkhofer, 1987) may be used to obtain the necessary inputs to the diagrams, and available cause–effect models may be used indirectly in the process by ensuring that the experts whose probabilities are encoded exercise the models and understand their limitations prior to providing their probability estimates.

In practice, however, complex influence diagrams such as that of Figure 13.1 are rarely solved. The difficulty is that assessing conditional probability distributions is extremely time-consuming. Thus, it is usually necessary to grossly simplify most influence diagrams to make probability assessment practical. Here again, the diagrams provide a useful benefit. Helpful simplifications, such as eliminating variables and reversing the order of conditioning; may be accomplished through the systematic rules for graphically manipulating influence diagrams. Thus, alternative simplifications may be easily explored to assess the demands they would place on probability encoding. When all available assessment orders seem equally difficult, distinct but logically equivalent assessments may be developed to provide redundancy. Added confidence can then be obtained by iteratively adjusting the various assessments to obtain consistency.

Even if the top-level attribute in the influence diagram is assessed directly, the influence diagram is still useful. The influence diagram provides a way of applying and more clearly documenting the logic underlying probability judgments. Specifically, the ability of an influence diagram to serve as a knowledge map provides two benefits. The first benefit comes when the diagram is used within the formal process of probability encoding. The influence diagram serves as a reminder of the numerous factors that bear on the uncertainty and the nature of their influences. In the MUA evaluation of potential repository sites, for example, the influence diagrams of Figures 13.1 and 13.3 played key roles in the assessment of possible levels of impact. Figure 13.1 was used to facilitate a group probability encoding exercise in which eleven experts debated repository performance and ultimately reached consensus over the probabilities of various disruptive events and the conditional probabilities of radionuclide releases.

The second benefit comes when judgmental assessments are explained and defended. The influence diagram shows explicitly the major factors that were considered. By pointing to a factor in the diagram used to support probability assessment, an expert can make a good case that that factor was, in fact, considered. Conversely, if reviewers identify an important consideration that does not appear in the associated influence diagram, a good case can be made that the expert did not consider, or did not significantly weight, that consideration when arriving at the professional judgment.

13.8 USING INFLUENCE DIAGRAMS TO COMPUTE EXPECTED UTILITIES AND PERFORM SENSITIVITY ANALYSIS

The fact that influence diagrams may be analyzed and solved to obtain an optimal decision strategy makes them useful for conducting the final step of an MUA. If the influence diagrams constructed for the various attributes in the analysis are combined to produce an influence diagram for the total decision problem, the combined model can be solved to obtain the MUA results.

The combined influence diagram representing the MUA may be obtained by combining the various influence diagrams constructed for each attribute through the objectives hierarchy. For example, the influence diagrams of Figures 13.1 and 13.3 would be appended, respectively, to the corresponding Figure 13.2 objectives, namely, 'minimize postclosure health effects during the first 10 000 years' and 'minimize degradation of archaeological, historical, and cultural properties'. Other influence diagrams would be similarly appended to their corresponding objectives. The various segments of the diagram must then be reviewed to identify duplications, influences among factors, and information flows between the various segments of the combined diagram. Duplicated variables may be combined and arrows added to represent influences and information flows. The lines connecting the objectives must be replaced by arrows, of course, to correspond to influence diagram notation and to indicate that the degree of achievement of lower-level objectives influences the degree of achievement of the corresponding higher-level objectives. Lastly, the diagram must be simplfied, using the rules for the graphic transformation of influence diagrams, to obtain a model with realistic probability distribution input requirements.

The combined influence diagram has a special structure that simplifies its solution. The objectives in the original objectives hierarchy are now value nodes that show explicitly the separable nature of preferences for decision outcomes. These nodes are deterministic because the multiattribute utility function, developed in step 4 of the MUA process, specifies an equation for computing utility, a measure of value that may be disaggregated to each of the higher-level objectives. Specifically, the nodes will be either summations or multiplications. (The reason for this is that the multiattribute utility function is, in nearly all practical applications, linear, multiplicative, or a combination of linear and multiplicative elements. Objectives are deliberately selected to meet independence conditions that imply such forms.) Tatman (1985) has specified the rules for reducing influence diagrams containing summation and multiplication nodes and provides an algorithm that exploits the separable nature of MUA influence diagrams. Thus, the solution sequence can be automated.

13.9 CONCLUSIONS AND SUMMARY

Influence diagrams are being in used in creative ways to improve communication and thereby facilitate decision analysis. In the case of MUA, influence diagrams can be used in several steps of the process. The diagrams can be used to facilitate the identification of objectives, the development of constructed scales for measuring the degree to which objectives might be achieved, and the development and defense of attribute assessments.

As with most versatile tools, the effective use of influence diagrams requires considerable skill. Common pitfalls include a misconceived desire to draw arrows from every chance node to every decision node (since all such factors are important and, therefore, 'influence' decisions); an incorrect view that small diagrams are trivial and, accordingly, a tendency to make the models too large; and confusions and difficulties associated with time sequencing. Nevertheless, when used skillfully, influence diagrams can be enormously effective. They provide a compact, easily interpreted, and computationally powerful means for analysis. Among other applications, they have been found to be exceedingly helpful for the analysis of health and environmental problems.

ACKNOWLEDGEMENTS

Preparation of this paper was partially supported by NSF Grant No. SES-8606906. The opinions expressed by the author do not necessarily represent the views of the National Science Foundation.

13.10 DISCUSSION

13.10.1 Discussion by G. A. Vignaux

This chapter links two techniques that are interesting to us as practitioners with an application area that is important to us as citizens. The first technique, influence diagrams (IDs), is used to develop models, elicit information, and communicate structure and assumptions to clients. The second, multiattribute methods, particularly multiattribute utility analysis (MAU), is used to evaluate the complicated multidimensional results of decision alternatives.

The application area is the study of really complex problems and, particularly, of what is ultimately of concern to us all, decisions involving health and the environment.

Elicitation and communication

The chapter describes the use of IDs to formulate, communicate, and expose problem structure during several phases of the MUA solution process.

Those of us who came to IDs late and particularly those brought up on a strict diet of Raiffa, Lindley, and simple OR decision applications, tend to see IDs as a method to supplant the decision tree; being more powerful, more intellectually satisfying, and neater to draw. In this chapter, their use as a calculation framework is hinted at rather than explained; it is clear that the prime value of IDs, as the author sees them, is in the communications role.

In this role, they exhibit some degree of failure in this chapter. I found it very hard to read some of the node lables and numbers in the more complicated diagrams. Since they are referred to in the text, it would add to the value of the chapter if the diagrams could be made clearer or if parts of them could be reproduced on a larger scale.

One question that is proper to raise in this context is whether IDs are the best of the several available tools for communicating model structure. For example, if the number of people already using a particular method is a significant factor in the decision to adopt it (and for the computer scientists I should only need to mention the continuing existence of FORTRAN as an illustration of this effect) the Ishikawa (1976) (fishbone) diagram would have an overwhelming advantage over IDs at least in the number of users, if not in theoretical or graphical elegance. There is, of course, naturally a close topological correspondence between the two types of diagrams.

One particular advantage of IDs in the possibility of extension from communication to calculation, though their use in this way appears not to be so important in MUA applications. Even as a vehicle for calculation the ID is not unique. It is clear also that there is a psychological difference between the two uses of IDs which may be worth noting. In elicitation and communication one is naturally bound to concentrate on the dependencies demonstrated in the diagram, on the association between one variable and another, on the arcs that are drawn between nodes. In contrast, when engaged in calculation, the powerful constraints are embedded in the independencies, where arcs are missing between nodes. In making the transition from one use to the other the importance of this difference may easily be overlooked and the need for careful checking of independencies may be forgotten.

The value of documenting assumptions and structures during the development of complicated applications is critical. This is important not only as a tool for the development team themselves and a vehicle for explanation to others involved but also for those who must evaluate the study later. Gass (1971, 1983) proposes that the development team formally construct a modeling audit trail to record the history of changes in assumptions and model structures in detail sufficient for later evaluators to follow. The ID, used as the author proposes, would be a valuable, perhaps ideal, part of this trail.

Multi-attribute utility analysis

The author has made a very satisfying link between IDs and MUA. He is careful to distinguish where he has found IDs to be useful in the MUA process, in model development, in organizing means (but not fundamental) objectives, in establishing constructed attribute scales, in quantifying uncertainty over attributes, and in the determination of expected utilities and sensitivity analysis, and is careful not to claim them as a universal tool.

The author comments that traditional cause–effect models are generally 'inadequate and biased' and that the authors have found them 'difficult to use... directly in MUA'. This is a damning indictment of standard physical modeling methods, which, if supported, should cause concern to all the technical community. These models embed much more sophisticated scientific experience and technical knowledge than can be represented in a simple ID, or even a multiattribute model of some type. As I understand the author, the problem is that they often fail to include factors that turn out to be important to a particular analysis or that they are oriented towards a particular aspect such as 'protection of public health'. I agree that their technical brilliance is often misdirected towards the more physically tractable aspects of the problem. Surely, though, the observed inadequacy and bias is not an inherent property of these models, sufficient to sentence them to be thrown out completely, as suggested. Instead, the process of their development should be improved by better analysis—perhaps using IDs to decrease the chance of ignoring important factors. Then we may find an imbalance between the preciseness of the physical effects for which there is a good theoretical basis and the fuzziness of the softer socioeconomic effects that would now be included. One could argue then, I concede, that the whole model should be changed to balance the degrees of fuzziness in different parts and that the preciseness of the physical components might be quite inappropriate.

The last phase, where a number IDs may have to be combined to construct a 'super-ID', must be immensely difficult and subject to error. We collect vast amounts of data of different types together with complicated interconnections, and the need for documentation must be associated with the need to store, maintain, and keep track of it all. I wonder if the authors, or anyone else, have considered using standard relational database techniques as an aid to this process. Surely there may be some value in a combination of IDs with database methods to structure and manipulate the data. In particular we could use a modified database system to (a) store several associated IDs together with their data, and (b) automatically combine many IDs, as described in Section 13.9, sorting out the duplication of variables and their interconnections.

Health and environmental problems

After a moment's reflection, one can recognize the existence of intergroup difficulties common in any large-scale problem such as the health and environmental problems described by Merkhofer. If ever there is to be a resolution, even if not a solution, to such problems, it is clear that each of the 'players' or 'player-groups' associated with the dispute should feel that the factors that concern them have been recognized in the analysis. Though we cannot claim, I think, that we have the final answer to such problems in MUA or in IDs, or in other methods, at least we may be developing tools that help us to communicate a little better with those with whom we disagree.

REFERENCES

Covello, V. T. and Merkhofer, M. W. (1990) (forthcoming), *Risk Assessment: Methods and Approaches for Quantifying Health and Environmental Risks*, Plenum Press, New York.

Gass, S. I. (1977) Evaluation of complex models, *Comput & Ops. Res.*, **4**, 27–35.

Gass, S. I. (1983) Decision-aiding models: validation, assessment and related issues for policy analysis, *Operations Research*, **31**, 603–31.

Hobbs, B. F. (1979) *Analytical Multiobjective Decision Methods for Power Plant Siting: A Review of Theory and Applications*, Division of Regional Studies, Brookhaven National Laboratory, Upton, New York (August).

Howard, R. A. and Matheson J. E. (1984) Influence diagrams, In R. A. Howard and J. E. Matheson (eds) *Readings on The Principles and Applications of Decision Analysis*, vol. II, Strategic Decisions Groups, Menlo Park, Calif. pp 721–762.

Ishikawa, K. (1976) *Guide to Quality Control*, Asian Productivity Organization, Tokyo.

Keeney, R. L. (1980) *Siting Energy Facilities*, Academic Press, New York.

Keeney, R. L. (1986) *Identifying and Structuring Values*, Systems Science Department, Institute of Safety and Systems Management, University of Southern California, Los Angeles (December).

Keeney, R. L. and Nair K. (1977) Selecting nuclear power plant sites in the Pacific Northwest using decision analysis. In D. E. Bell, R. L. Keeney, and H. Raiffa (eds) *Conflicting Objectives in Decisions*, Wiley Interscience, New York, pp 298–322.

Korsan, R. J. and Matheson, J. E. (1978) *Pilot Automated Influence Diagram Decision Aid*, SRI International, Menlo Park, Calif.

Merkhofer, M. W. (1987) Quantifying judgmental uncertainty: methodology, experiences, and insights, *IEEE Transactions on Systems, Man, and Cybernetics*, SMC, **17**(5) 741–52 (September/October).

Merkhofer, M. W. and Keeney, R. L. (1987) A multiattribute utility analysis of alternative sites for the disposal of nuclear waste, *Risk Analysis*, **7**(2) 173–94 (June).

Merkhofer, M. W. and Leaf E. B. (1981) *A Computer-Aided Decision Structuring Process*, Final Report prepared for Operational Decision Aids Project, ONR Contract No. N00014-80-C-0237, SRI International, Menlo Park, Calif. (June).

Miller, A. C. III, Merkhofer, M. W., Howard, R. A., Matheson, J. E. and Rice, T. R. (1976) *Development of Automated Aids for Decision Analysis*, Research report to Defense Advanced Research Projects Agency by SRI International, Menlo Park, Calif. (May).

Olmsted, S. M. (1983) On representing and solving decision problems, Ph.D. thesis, EES Department, Stanford University, Stanford, Calif.

Owen, D. L. (1978) The use of influence diagrams in structuring complex decision problems, presented at the Second Lawrence Symposium on Systems and Decision Sciences. In R. A. Howard and J. E. Matheson (eds) *Readings on The Principles and Applications of Decision Analysis*, vol II, Strategic Decision Group, Menlo Park, Calif; pp 763–771.

Shachter, R. (1986) Evaluating influence diagrams, *Operations Research*, **34**(6), 871–82.

Shachter, R. (1988) Probabilistic inference and influence diagrams, *Operations Research*, **36** (4), 589–604.

Smithin, T. (1980) Maps of the mind: new pathways to decision-making, *Business Horizons*, (December), 24–28.

Spetzler, C. S. and Staël von Holstein, C.-A. (1975) Probability encoding in decision analysis, *Management Science*, **22** 340–58.

Tatman, J. A. (1985) Decision processes in influence diagrams: formulation and analysis, Ph.D. dissertation, EES Department, Stanford University, Stanford, Calif. (December).

PART IV

Problems and Applications: Medical

CHAPTER 14

An Influence Diagram Approach to Medical Technology Assessment

Ross D. Shachter, *Stanford University, USA*
David M. Eddy and Victor Hasselblad, *Duke University, USA*

ABSTRACT

As medical technologies become more complex and more expensive, society is faced with increasingly difficult resource allocation decisions throughout the health care system. Although there is a clear need for quantitative tools to compare the effects of different technologies on health outcomes, there are many problems with applying classical statistical techniques to the analysis of health technologies. These include a shortage of direct evidence on the health outcomes of interest and differing qualities of experimental design. The experimental evidence must also be corrected for factors such as patient demographics, dosage, or the quality of health care available.

Many of these problems can be overcome through a Bayesian approach, the 'confidence profile method', which computes a posterior probability distribution for the effect of each technology on the health outcomes of interest. In this chapter, we develop an influence diagram representation for the concepts underlying the confidence profile method. The influence diagram structure allows the model formulation to be general but simple to understand. At the same time, it leads to approximation methods for evaluating the medical technology assessment. This approach is illustrated with an analysis of the effect of a thrombolytic agent (tissue-type plasminogen activator) on one-year survival from heart attacks.

14.1 INTRODUCTION

As medical technologies become more complex and more expensive, society is faced with increasingly difficult choices. These dilemmas extend throughout the

Influence Diagrams, Belief Nets and Decision Analysis
Edited by R. M. Oliver and J. Q. Smith

health care system from physicians selecting tests and treatments for individual patients, to third-party payors who must approve procedures for reimbursement. Traditionally, these issues have been resolved through a subjective, qualitative process. However, as resources become more limited while the number of available choices grows rapidly, there is a clear need for quantitative tools to compare the effects of different technologies on health outcomes.

There are many problems with applying classical statistical techniques to the analysis of health technologies. For most technologies, there is little direct evidence of the effect of the technology on the health outcomes of interest. The quality of the experimental design can vary greatly from randomized controlled trials (RCTs) to retrospective studies and even anecdotes. Evidence from even the best designed experiments must be adjusted to account for the applicability to a particular assessment, correcting for factors such as patient demographics, dosage, or the quality of health care available. Finally, much of the evidence relates to intermediate outcomes (such as cholesterol level) rather than a health outcome of interest (such as survival).

Many of these problems can be overcome through a Bayesian approach to combining the evidence from empirical studies and the subjective expert judgments needed to complete the analysis. This approach is incorporated into the 'confidence profile method' for evaluating the evidence from different forms of empirical evidence and accounting for the biases and applicability of each piece of evidence (Eddy, 1989). The adjusted evidence can be combined, along with explicit subjective assessments, to yield posterior probability distributions for the effect of each technology on the health outcomes of interest. Numerical integration software has been developed to analyze these problems in a user-friendly environment on the personal computer (Hasselblad, 1987).

The confidence profile method and the software have been successfully applied to a wide variety of medical technology assessments, including the efficacy of thrombolytic agents in the treatment of heart attacks (Eddy, 1986), the value of mammography screening (Eddy *et al.*, 1988), the efficacy of antabuse (disulfiram) in the treatment of alcoholism (Critchfield and Eddy, 1987), and the cost-effectiveness of neonatal screening for maple syrup urine disease (Hasselblad and Critchfield, 1987).

In this chapter, we develop an influence diagram representation for the confidence profile method. The influence diagram unifies and simplifies the concepts of confidence profiles. It introduces a discipline in which the basic, underlying parameters and their relationships are defined before experimental evidence is introduced. The influence diagram structure allows the model to be fully general and yet fundamentally simple. Assumptions of prior distributions, dependence, and experimental design are explicit. As a result, adjustments for biases and errors can be stated clearly in a modular fashion.

The influence diagram representation leads to several methods for evaluating the medical technology assessment once it has been formulated: a linear approximation method (Shachter, 1988b), Monte Carlo integration, and posterior mode analysis. We illustrate the confidence profile method with an analysis of the effect of a thrombolytic agent (tissue-type plasminogen activator) on one-year survival from heart attacks.

In Section 14.2, we introduce the fundamental definitions and concepts of the confidence profile method. Section 14.3 presents the influence diagram model used to relate the population parameters to the experimental evidence. Section 14.4 formulates the example of the effect of thrombolytic agents on survival after heart attack. Section 14.5 considers the factors that can affect the applicability of experimental evidence to the parameters of interest and how to adjust for them. A brief discussion of three solution methods is presented in Section 14.6, along with the solution to the example from Section 14.4. Finally, Section 14.7 contains some conclusions and suggestions for further research.

14.2 THE MEDICAL TECHNOLOGY ASSESSMENT PROBLEM

The goal of a technology assessment is to choose among courses of action to be taken, such as the choice of treatment for a particular disorder or the administration (or abandonment) of a screening test. There is considerable uncertainty about the consequences of different courses of action. First, the underlying biological processes and the effects of interventions are inherently stochastic, so there is no infallible prediction in an individual case. Second, there is additional uncertainty arising from the limited knowledge of these processes, particularly in the case of new and emerging technologies.

In general, the results of the assessment should be the expected utility for some decision maker of pursuing different policies (as a function of the health outcomes), so that the decision maker can choose the optimal policy for action. In this chapter, we consider a significant subset of that problem, which focuses on the probabilistic analysis. We will not be explicitly considering a utility function or the choices among policies, but limiting our scope to estimating the posterior distributions for the health outcomes based on choices among health technologies. In this way the analysis could be applied by many different decision makers, who might not be able to agree on the choice of a particular utility function.

Although the method presented here can incorporate any experimental evidence considered relevant, it is nevertheless based on a number of subjective estimates. These arise from the choice of model and the choice of prior distributions for the uncertainties in the model. Since the results are in the form of posterior probability distributions, prior distribution estimates are unavoidable and the judgment of the modeler plays a key role in the analysis. Also, much of the information available for the analysis is in the form of expert knowledge. Our goal is for these subjective judgments to be explicit and clearly documented.

In order to build a model, we need some precise definitions of medical concepts. The following definitions form the basis for the confidence profile method.

A *healthy technology* is used broadly throughout this chapter to include any intervention that might affect a health outcome. Examples include health education, diagnostic tests, treatments, pain control, and psychotherapy.

A *health outcome* is an outcome of a disease or injury that people can experience and care about. Examples of health outcomes are life and death, pain, disfigurement, disability, and anxiety.

Uncertain quantities of interest are represented as (population) parameters in the mathematical model. These can be probabilities, rates, or simply numeric values. Examples are the probability that a patient will survive a year after a heart attack, the rate at which alcoholics will resume drinking after treatment, and the number of days they will consume alcohol in the year following treatment.

The goal of the analysis in this chapter is to obtain a posterior distribution for the change or *effect* on the health outcomes of one health technology (the *paradigm treatment*) as contrasted with an alternative technology (the *paradigm control*). Both technologies are examined in the context of the *paradigm circumstances of interest*, with respect to the provider, population, disease, and setting. The behavior of each technology is described by parameters in the model, θ_t for the paradigm treatment and θ_c for the paradigm control. For dichotomous health outcomes these parameters would be probabilities. The effect in this case is most often expressed as either the difference of the probabilities, $\delta = \theta_t - \theta_c$, the relative risk (their ratio), $\rho = \theta_t/\theta_c$, or the odds-ratio, $\varepsilon = [\theta_t/(1 - \theta_t)][\theta_c/(1 - \theta_c)]^{-1}$. For example, if the chance of the health outcome (e.g. death within one year) without the technology is 0.8, and the chance of the health outcome with the technology is 0.4, the different effect measures are $\delta = -0.4, \rho = 0.5$, and $\varepsilon = 0.375$. All three measures indicate a significant effect from the technology in reducing the probability of death. For continuous-valued health outcomes (e.g. weight or IQ), the effect is usually measured as the difference or ratio in the magnitude of the health outcomes.

Our knowledge about parameters is expressed in terms of a prior probability distribution and experimental evidence obtained from experts and the medical literature. For example, to assess the effect of changing dietary cholesterol on the chance of a heart attack, we might construct the influence diagram shown in Figure 14.1(a). We would need to assess our prior conditional distribution for a heart attack given diet. We would also want to incorporate any medical evidence available on their connection.

In examining the literature, we would discover that while there are no studies directly linking diet to heart attacks, there is considerable evidence about an indirect connection involving the serum cholesterol level, as shown in the diagram in Figure 14.1(b). Serum cholesterol is an *intermediate outcome*, a marker of biological changes that might indicate or affect the probability or

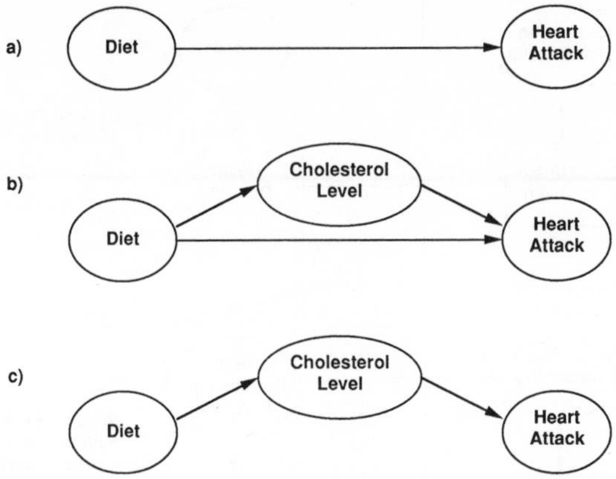

Figure 14.1 Intermediate outcomes.

magnitude of the health outcome. Other examples are blood pressure and the reperfusion of a coronary artery after treatment for a heart attack. Some intermediate outcomes are also health outcomes, since people care about them and they provide information about other health outcomes. An example is obesity.

To learn about the effect of diet on heart attacks, we would now have to gather data on the effect of diet (the technology) on serum cholesterol, and the combined effects of diet and serum cholesterol on heart attacks (the health outcome). It is convenient to make a Markov assumption when it is justified, as shown in the diagram in Figure 14.1(c), namely that once we know the serum cholesterol level, diet provides no additional information about heart attacks. In that case, when gathering evidence for factors affecting heart attacks, we would only need to consider serum cholesterol levels.

When evaluating a diagnostic or screening technology, where the purpose of the technology is to provide information, it is important to consider the *followup actions*, the treatment decisions made on the basis of that information. The general influence diagram for this problem is drawn in Figure 14.2(a). A screening test provides imperfect information about an underlying disorder to the physician who must choose a treatment which affects the health outcome. In the case of glaucoma shown in Figure 14.2(b), we might use tonometry to detect high intraocular pressure so that a treatment can prevent blindness. Modeling the doctor's decision (and patient's compliance) as uncertain, we obtain the diagram in Figure 14.2(c). The efficacy of screening clearly depends on three factors: the ability of the screening test (tonometry) to detect the disorder (the intermediate outcome, high intraocular pressure); the doctor's decision to treat on the basis of the test results (the followup action); and the effect of the treatment on the chance of blindness (the health outcome).

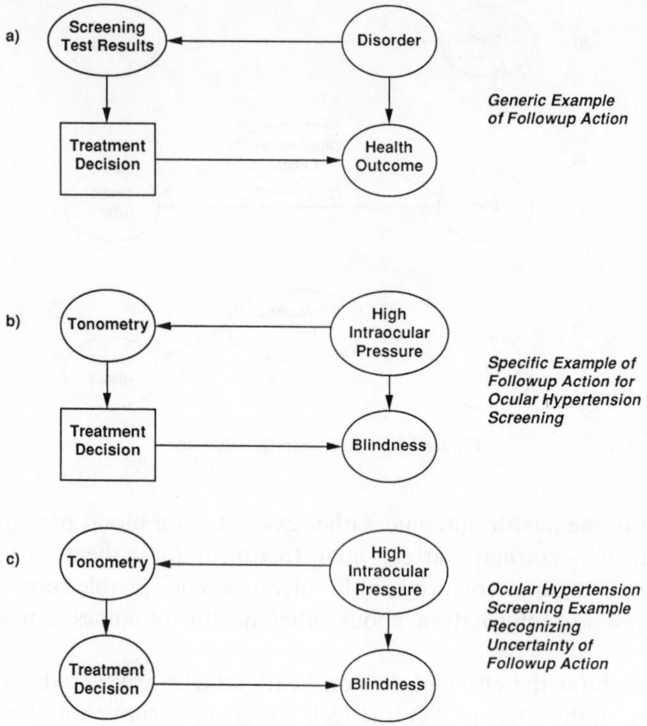

Figure 14.2 Analysis of screening efficacy.

14.3 THE INFLUENCE DIAGRAM MODEL

Although the ultimate problem of a medical technology assessment can be described by an influence diagram with decisions as to which treatments, if any, to recommend, and for which classes of patients, we are concentrating our attention in this chapter on the analysis of experimental data. In the influence diagrams we construct there are two kinds of variables, corresponding to the observed experimental evidence and the unobservable parameters. The parameters are uncertain quantities, represented by clear oval nodes in the diagram, and the observed evidence variables are represented by shaded oval nodes in the diagram. The influence diagram model we construct is applicable to many problems beyond medical technology assessment. (For examples, see Oliver and Yang, 1989, and Datta and Oliver, 1989.) It is similar in many respects to the models of Kiiveri *et al.* (1984), Pearl (1986) and Spiegelhalter (1987), in which directed graphs are used to represent model structure and evidence propagation.

14.3.1 Parameter variables

Each parameter in the model corresponds to a random variable. We distinguish between two sorts of parameters, basic and deterministic. A prior distribution, $P\{\beta\}$, must be specified for the basic parameters, $\beta = (\beta_1, \ldots, \beta_b)$. The deterministic parameters, $\theta = (\theta_1, \ldots, \theta_m)$, on the other hand, are specified as a deterministic function, $\theta(\beta)$, of the basic parameters. (Although the deterministic parameters can be explicitly defined as functions of both basic and deterministic parameters, they are implicitly functions of the basic parameters alone.) As a convention, we will assume that the basic parameters are always included as the first b deterministic parameters:

$$\theta_i(\beta) = \beta_i \qquad \text{for } i = 1, \ldots, b.$$

It therefore follows that $m \geqslant b$, with equality holding only when $\theta(\beta) = \beta$.

The dependence assumptions made during assessment are indicated in an influence diagram, such as the one shown in Figure 14.3. Note that the deterministic parameters are random variables, since they are functions of random variables. Nonetheless $\mathrm{Var}\{\theta(\beta) | \beta\} = \mathbf{0}$, because all of the uncertainty in the parameters arises from the basic parameters.

In our experience in medical technology assessment, a handful of deterministic functions have provided considerable modeling power. (Remember that they can be nested, since a deterministic parameter can depend on a combination of basic and deterministic parameters.) The trivial case is the constant function

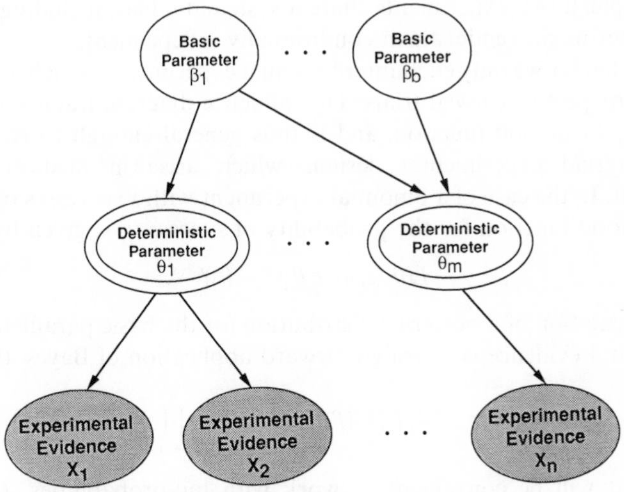

Figure 14.3 Different types of variables in the model.

which defines a constant parameter and depends on no other parameters. The simplest non-trivial function is a transformation, in which there is a single conditioning variable. The most useful transformations are the logarithmic, $\ln(x)$, exponential, e^x, odds, $x/(1-x)$, inverse odds, $x/(1+x)$, and scaling $c_1 + (c_2 - c_1)x$. The only other (primitive) functions we have needed in our analyses have been linear and quadratic forms. Examples of the uses of these deterministic functions in technology assessments are given in Sections 14.4 and 14.5.

14.3.2 Experimental evidence

The observed experimental evidence, $X = (X_1, \ldots, X_n)$, is represented in the influence diagram by shaded nodes, conditioned by their related parameters. As a convention, only deterministic parameters can condition experimental evidence nodes, as shown in Figure 14.3. (Recall that all of the basic parameters are also included as deterministic parameters.) For each evidence variable X_j there is a likelihood function $L_j(\theta | X_j)$. We assume that the different samples of experimental evidence are conditionally independent given the parameters:

$$L\{\theta | X\} = \prod_{j=1}^{n} L_j\{\theta | X_j\}$$

and

$$L\{\beta | X\} = \prod_{j=1}^{n} L_j\{\theta(\beta) | X_j\}.$$

This is often made possible through the introduction of additional parameters. For example, if two experiments share a systematic bias, including that bias as a parameter might render them conditionally independent.

In this chapter we only encounter binomial experiments, which describe single arms of prospective clinical trials. The influence diagram framework allows us to use any likelihood function, and is thus general enough to represent most of the myriad experimental designs which arise in medical technology assessment. In the case of a binomial experiment with s successes out of n trials, the likelihood function for the probability of success θ_j is given by

$$L(\theta_j | s, n) \propto \theta_j^s (1 - \theta_j)^{n-s}.$$

The calculation of a posterior distribution for the basic parameters given the experimental evidence is a straightforward application of Bayes' theorem:

$$P\{\beta | X\} \propto P\{\beta\} L\{(\theta(\beta) | X\} = P\{\beta\} \prod_{j=1}^{n} L_j\{\theta(\beta) | X_j\}.$$

At times it will be convenient to work with log-probabilities, *LP*, and log-likelihoods, *LL*, for which Bayes' theorem takes the corresponding form:

$$LP\{\beta|X\} = \ln P\{\beta|X\} = k + LP\{\beta\} + LL\{\theta(\beta)|X\}$$
$$= k + LP\{\beta\} + \sum_{j=1}^{n} LL_j\{\theta(\beta)|X_j\}$$

where k is simply a normalizing constant.

14.3.3 Additional assumptions

The influence diagram model presented above provides a complete framework for the representation of the relationships among the parameters of interest and the experimental evidence. Although the model is theoretically straightforward to analyze once it has been constructed, the practical computation might prove intractable. Often, additional assumptions are made to allow for easier assessment or more efficient computation. Note that none of these assumptions are required in our model.

A1. Single parameter evidence

In many models, the likelihood function for each piece of evidence relates (directly) to only one deterministic parameter:

$$L_j\{\theta|X_j\} = f(\theta_k).$$

When this is satisfied, the evidence indexes $\{1, \ldots, n\}$ can be partitioned into sets $N_1, \ldots, N_m$, such that the likelihood can be written as

$$L\{\theta|X\} = \prod_{i=1}^{m} \Pi_{j \in N_i} L\{\theta_i|X_j\}.$$

A2. Basic parameter independence (a priori)

It is often convenient to assume that, a priori, the basic parameters are independent. This can be a reasonable assumption when a noninformative prior is being used in the analysis and the intent is for the data to dominate in the posterior distributions. Explicitly stated, this condition means that

$$P\{\beta\} = \prod_{i=1}^{b} P\{\beta_i\}.$$

A3. Only basic parameters

This is the situation in which $m = b$ and $\theta(\beta) = \beta$, that is, the only deterministic parameters are the basic ones.

Although each of these additional assumptions is reasonable, in concert they are deceptively strong, since they imply posterior independence of the parameters as in Oliver and Yang (1989):

$$P\{\beta|X\} \propto \prod_{i=1}^{b} P\{\beta_i\}[\Pi_{j\in N_i} L_j\{\beta_i|X_j\}] \propto \prod_{i=1}^{b} P\{\beta_i|X\}.$$

In most of our models, assumption A2 has been satisfied, and often A1 as well. As a rule, we have posterior dependence of our parameters as in Datta and Oliver (1989).

14.4 FORMULATION EXAMPLE

The influence diagram model presented in Section 14.3 provides the framework to model most medical technology assessments. In this section, this is demonstrated for the assessment of mortality risk with different treatments following a heart attack.

14.4.1 The problem

Tissue-type plasminogen activator (t-PA) is one of several thrombolytic agents used to dissolve (lyse) blood clots (thrombi) in coronary arteries after heart attacks, with the intention of restoring blood flow through the coronary artery (reperfusion), and thereby increasing the chance of survival. Estimating the effect of t-PA on survival is complicated by the fact that there is no single randomized controlled trial that compares the effect of t-PA with conventional care (CC) or any other thrombolytic agents on long-term (one-year) survival. The available studies of t-PA (shown in Table 14.1) involve intermediate outcomes (e.g.

Table 14.1 Evidence for t-PA analysis

Study	Treatment	Outcome	Successes	Trials
TIMI Study Group	t-PA	Reperfusion	78	118
(1985)	IVSK	Reperfusion	44	122
Collen *et al.*	t-PA	Reperfusion	25	33
(1984)	Placebo	Reperfusion	1	14
Yusuf *et al.*	IVSK	Mortality	412	2672
(1985)	Placebo/ conven. care	Mortality	501	2612
Kennedy *et al.*	TA (ICKS)	Mortality|Reperfusion	5	93
(1985)	TA (ICSK)	Mortality|No Reperfusion	6	41
	Conven. care	Mortality|Reperfusion	0	14
	Conven. care	Mortality|No Reperfusion	17	102
	Conven. care	Reperfusion	14	116

perfusion and reperfusion), short-term outcomes (in-hospital mortality), and different controls (placebo, CC, and intravenous streptokinase (IVSK)). In addition to the studies described in Table 14.1, a large number of studies have examined other thrombolytic agents—IVSK, intracoronary streptokinase (ICSK), and urokinase (UK) (Yusuf *et al.*, 1985). While they do not provide direct evidence about t-PA, they contain information to compare the various controls used in studies involving t-PA.

An assessment of t-PA has been performed using the confidence profile method (Eddy, 1986, Eddy, 1989) and analyzed with the confidence profile software (Hasselblad, 1987). A couple of features of the problem make it complicated. First, the studies compare the technology with different controls. For example, Collen *et al.* (1984) compared t-PA with conventional care, TIMI Study Group (1985) compared t-PA with IVSK, Kennedy *et al.* (1985) compared ICSK with conventional care, and other randomized trials have compared IVSK with conventional care, and ICSK with conventional care. Second, the studies contain information about different intermediate or health outcomes. For example, TIMI Study Group (1985) provides information about reperfusion, Yusuf *et al.* (1985) provide information on actual one-year mortality, and Kennedy *et al.* (1985) contains information on both.

Both multiple controls and intermediate outcomes are the rule rather than the exception in technology assessment. They can increase the dependence and hence the dimensions in the analysis, making evaluation of results more complicated. Nonetheless, they are handled in a straightforward manner with the influence diagram model, so the formulation of the problem stays fairly simple.

14.4.2 Chains

The parameters of interest in the t-PA assessment are the probabilities of one-year mortality given three different technologies: t-PA, IVSK, and conventional care, and their differences. A simple influence diagram for the treatment decision in which our goal is to minimize the probability of one-year mortality is drawn in on Figure 14.4(a). The necessary data to construct the diagram are, of course, the probability of mortality for each of the three technologies, so we could build our model with those three probabilities as basic variables. However, some of the data which we want to include involve an intermediate event, reperfusion (the return of blood flow to the affected heart muscle). Therefore, we construct the more detailed influence diagram shown in Figure 14.4(b). In fact, we could obtain the diagram in Figure 14.4(a) from the one in Figure 14.4(b) by performing a node removal operation on the 'reperfusion' node (Shachter, 1988a).

The data needed for the diagram in Figure 14.4(b) are the probabilities of reperfusion given each technology, 'pRep|T' = Pr{Reperfusion | Technology T},

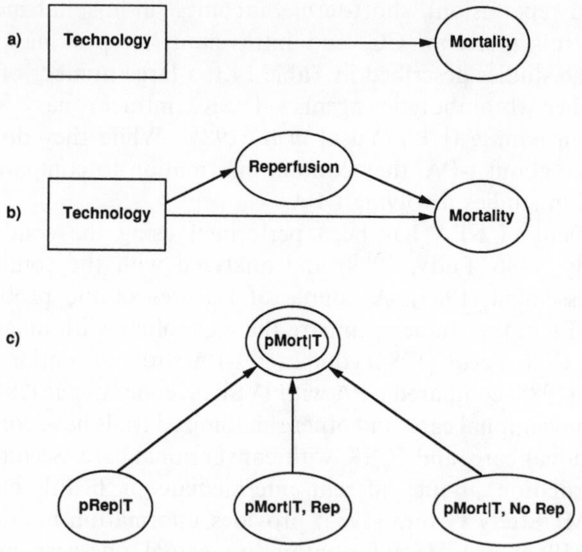

Figure 14.4 Example of chaining.

and probabilities of mortality given each technology and whether reperfusion has occurred, 'pMort|T, Rep' = Pr{Mortality|Technology T, Reperfusion} and 'pMort|T, No Rep' = Pr{Mortality|Technology T, No Reperfusion}. From those data we could compute the data for the diagram in Figure 14.4(a), using conditioning, just as it is done in the node removal operation:

$$\text{pMort}|\text{T} = \text{pRep}|\text{T}\cdot\text{pMort}|\text{T, Rep} + (1 - \text{pRep}|\text{T})\cdot\text{pMort}|\text{T, No Rep}.$$

These parameters are shown in the influence diagram in Figure 14.4(c). There are three basic parameters, 'pRep|T', 'pMort|T, Rep', and 'pMort|T, No Rep', and there is a single deterministic parameter, 'pMort|T', which is a deterministic function (a quadratic form) of the other three. This is the process we call chaining.

If the Markov assumption is applied to this chain (the probability of mortality given whether reperfusion has occurred does not depend on the technology), then we would need fewer basic parameters to represent the problem. In the analysis of the t-PA problem, such a simplifying assumption was made (Eddy, 1986). It was assumed that the probability of mortality given whether reperfusion has occurred is the same for all thrombolytic agents (TA), namely 'pMort|TA, Rep' and 'pMort|TA, No Rep'.

14.4.3 Completing the formulation

The influence diagram for the t-PA assessment is shown in Figure 14.5. There are seven basic parameters, 'pMort|CC, Rep', 'pMort|CC, No Rep', 'pMort|TA,

Rep', 'pMort|TA, No Rep', 'pRep|CC', 'pRep|t-PA', and 'pRep|IVSK'. The basic parameters are assumed to be independent, *a priori*, assumption A2 from Section 14.3, and each basic parameter is assigned a beta (0.5, 0.5) prior distribution. (Although these assumptions are arbitrary and difficult to justify, the intent is to set up the problem so that the data can dominate the results. If we had a more informative choice for the prior distribution $P\{\beta\}$ we would use it, although we should be careful not to use any of the information from the evidence included in forming those prior distributions. The sensitivity of the conclusions to this choice of prior distribution is examined later in this section.) There are four deterministic parameters shown in Figure 14.5: 'pMort|CC', 'pMort|t-PA', and 'pMort|IVSK' are computed using the chaining formula, and 'pMort|t-PA – pMort|CC' has a difference as its deterministic function. We also create two other difference deterministic parameters to compare IVSK with CC and t-PA, but they are omitted from in the figure to improve readability.

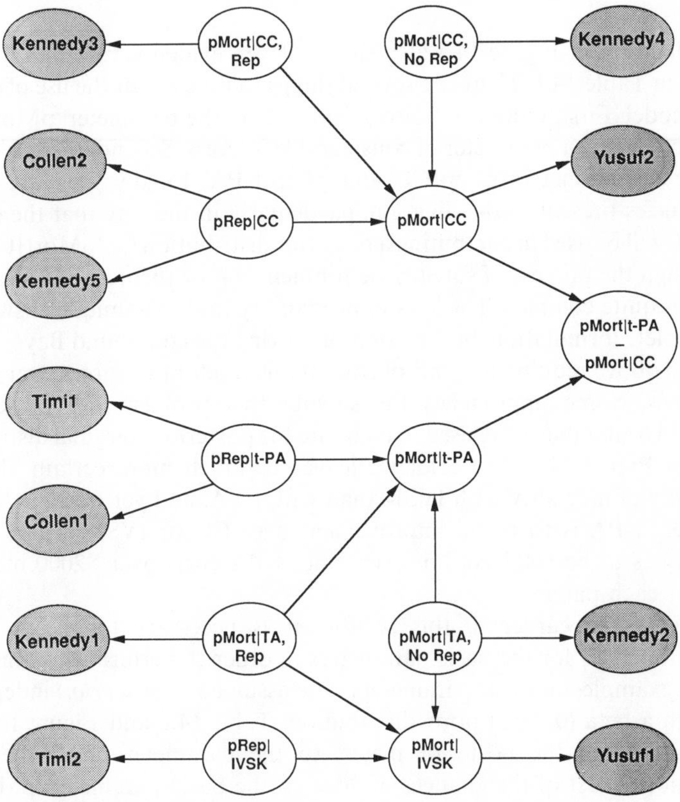

Figure 14.5 Use of tissue-type plasminogen activator (t-PA) immediately following a myocardial infarction. Effect on one-year mortality.

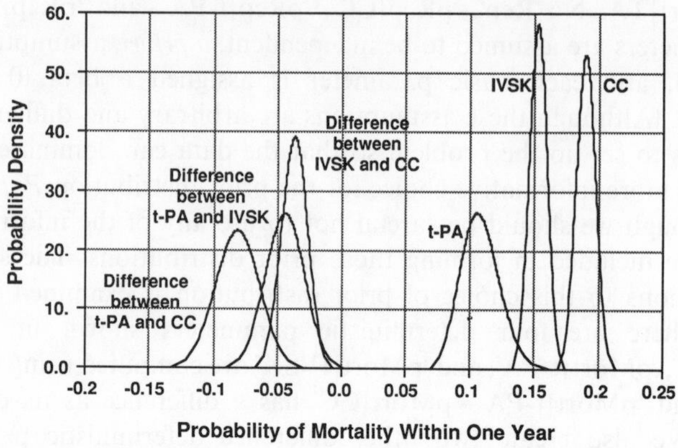

Figure 14.6 Comparison of tissue-type plasminogen activator relative to intravenous streptokinase and conventional care for treatment of myocardial infarction.

The diagram in Figure 14.5 also shows eleven evidence variables, based on the data in Table 14.1. There are several things to note about the use of evidence in this model. First, there is no direct evidence on the parameter 'pMort|t-PA'. This is, of course, a parameter of considerable interest. Second, there is multiple evidence for parameters 'pRep|CC' and 'pRep|t-PA'. Finally, there are complex dependencies present in the diagram, particularly in the way that the evidence for IVSK will be used in computing a posterior distribution for 'pMort|t-PA'.

Although the process of solving the problem is now theoretically straightforward, it is quite complex. The laws of probability lead to a unique answer given this complete formulation, but the steps involving the compound Bayes formula presented above require integrals of continuous random variables over multiple dimensions. Some approaches for solving this problem are presented in Section 14.6, and they were used to compute the posterior marginal distributions shown in Figure 14.6. The evidence leaves us much more certain about the probability of mortality with IVSK than with t-PA, and yet we can clearly see how likely t-PA is to be an improvement over CC or IVSK. There are still policy issues to be resolved, however, since t-PA costs over $2000 more than IVSK for each patient.

As mentioned earlier in this section, it is necessary to specify a prior distribution $P\{\beta\}$ for the basic parameters in order to perform this analysis. In the t-PA example, the basic parameters were assumed to be *a priori* independent, each with a beta (0.5, 0.5) prior distribution. Table 14.2 and Figure 14.7 show the sensitivity of the posterior means to the parameter of the beta prior distribution. Most of the changes are due to the basic parameter 'pMort|CC, Rep', for which there are fourteen cases of evidence and no deaths. It is not surprising that the posterior for this parameter would be so sensitive to the

Table 14.2 Sensitivity of the posterior means to the basic parameter priors

	Beta parameters for prior distributions				
	(0.05, 0.05)	(0.1, 0.1)	(0.5, 0.5)	(1, 1)	(5, 5)
pRep|CC	0.111	0.111	0.115	0.119	0.143
pRep|t-PA	0.682	0.682	0.681	0.680	0.671
pRep|IVSK	0.355	0.355	0.357	0.359	0.373
pMort|CC, Rep	0.005	0.009	0.042	0.075	0.203
pMort|CC, No Rep	0.213	0.213	0.209	0.206	0.190
pMort|TA, Rep	0.059	0.059	0.063	0.067	0.093
pMort|TA, No Rep	0.205	0.205	0.203	0.202	0.192
pMort|CC	0.190	0.190	0.190	0.190	0.192
pMort|t-PA	0.105	0.106	0.108	0.110	0.126
pMort|IVSK	0.153	0.153	0.153	0.153	0.155
pMort|t-PA − pMort|CC	−0.085	−0.084	−0.082	−0.080	−0.067
pMort|IVSK − pMort|CC	−0.037	−0.037	−0.037	−0.037	−0.037
pMort|t-PA − pMort|IVSK	−0.048	−0.047	−0.046	−0.043	−0.029

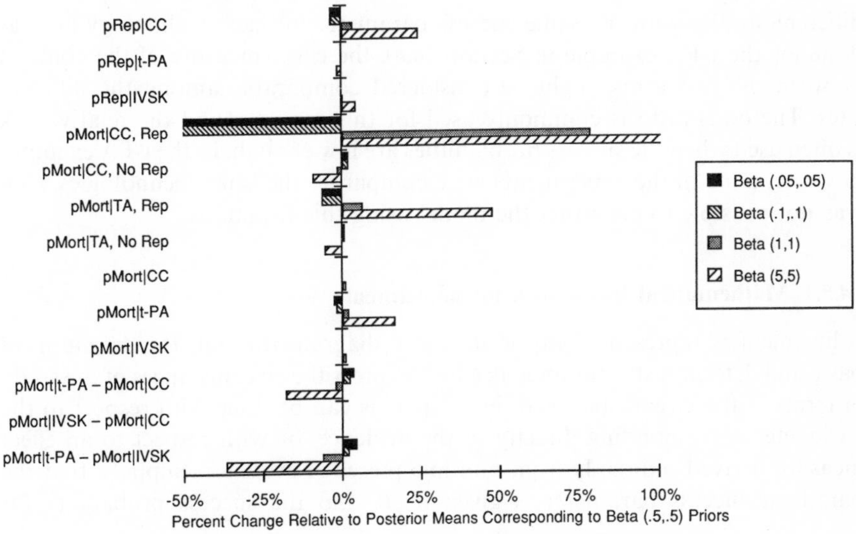

Figure 14.7 Sensitivity of posterior means to prior distributions.

choice of prior. Nonetheless, the parameter of most interest, 'pMort|t-PA – pMort|CC' is relatively insensitive to our choice of prior. As we had hoped, the results of our analysis are dominated by the data.

14.5 ADJUSTMENTS TO PARAMETERS

One of the fundamental problems in performing an assessment is that the circumstances under which the experimental evidence was gathered are rarely the paradigm circumstances of interest. In addition, the subjects in those experiments are rarely representative of the paradigm patient population and the technology is often different from the paradigm treatment or control. The highly subjective process of correcting for these differences, so that the experimental evidence can be applied is what we call *adjustment*. In this section we present adjustments to deal with a wide variety of biases to both internal and external validity of the experimental evidence.

Frequently there are features of the population, disease, technology, provider, or setting that can affect the results of an experiment. We call this combination of features the *circumstances of the study*. These differ from the paradigm circumstances of interest even under the most carefully designed experimental conditions. For example, it is common to distinguish between the efficacy of a technology, its performance under ideal medical conditions, and its effectiveness, the performance which a typical patient is likely to receive.

When multiple RCTs compare the same two technologies, it is often possible to combine their results without having to adjust for the differing circumstances of each study. Although it is rarely proper to associate the evidence from the different studies with the same success parameter for each technology (as was done for the t-PA example in Section 14.4), the effect measure of the contrast between the two arms might be considered comparable among the different sites. The odds-ratio is commonly used for this purpose, and the relative risk is often used when the success probabilities are low enough. In the t-PA example, however, none of the experiments were comparing the same technologies, so it was not possible to construct the model using this technique.

14.5.1 Mathematical framework for adjustment

Adjustment is represented in the influence diagram through the definition of basic and deterministic parameters which express the circumstances of the study in terms of the circumstances of interest. This can be done with respect to the parameter corresponding directly to the evidence, or with respect to an effect measure derived indirectly from multiple pieces of evidence. Suppose that the paradigm success parameter is given by θ_t and the success probability for evidence x is believed to be θ_S. If we express θ_S in terms of θ_t we might have the influence diagram drawn in Figure 14.8.

Almost all of the adjustments can be modeled using the framework which defines θ_S in terms of θ_t. However, there is one type of adjustment, the 'unreliable observation', which has a different mathematical form. Consider the diagram shown in Figure 14.9(a). We have evidence $X?$ about a parameter θ. Unfortunately, for one reason or another, we are uncertain about the exact experimental outcome, χ. One approach to this problem is to treat our point estimate for $X?, X$, as evidence about χ, yielding the diagram shown in Figure 14.9(b). This also requires a conditional probability distribution for χ given θ, rather than just a likelihood. Now we can take conditional expectation to obtain a likelihood relating the evidence X directly to the parameter θ,

$$L(\theta|X) \propto E_{\chi}[L(\chi|X)|\theta].$$

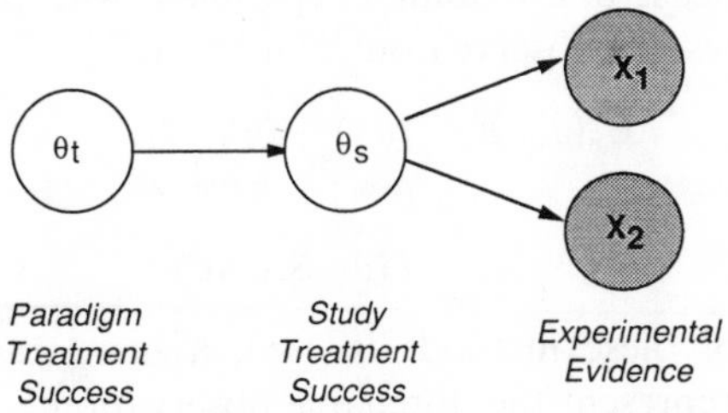

Figure 14.8 Formulation of parameter adjustment.

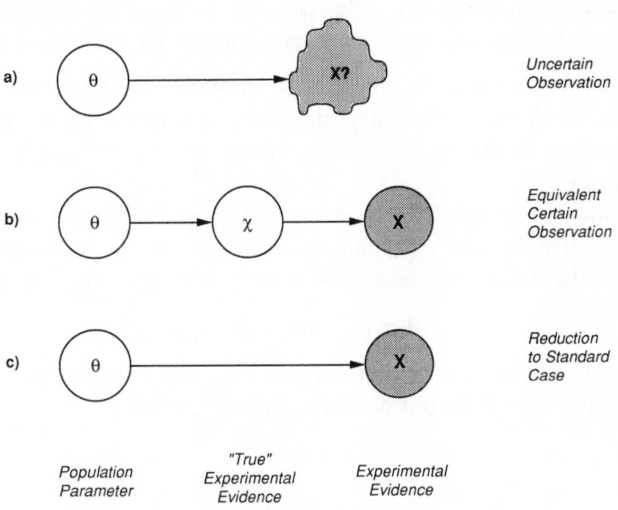

Figure 14.9 Unreliable observation of experimental evidence.

Not surprisingly, this technique is theoretically equivalent to the methods proposed by Pearl (1985) and Spiegelhalter (1986). There are important differences, however, in the questions that are asked and the manner in which the uncertainty about the observation is presented. The approach here emphasizes to the modeler that he is not specifying ‘a 10 percent chance that severe headaches occurred’ (Spiegelhalter, 1986, p. 55), but rather the relative likelihood that he would have observed what he did, given whether (or not) severe headaches occurred.

A simplified version of this method expresses uncertainty about experimental results by discounting the attention paid to the evidence using a ‘strength of evidence’ $\omega, 0 < \omega \leqslant 1$. Define the discounted likelihood function for parameter θ given evidence x to be

$$L_\omega(\theta | X) \propto [L(\theta | X)]^\omega.$$

For example, in the case of a binomial experiment with probability of success θ in which s successes are observed out of n trials,

$$\begin{aligned} L_\omega(\theta | s, n) &\propto [\theta^s(1-\theta)^{n-s}]^\omega \\ &\propto \theta^{\omega s}(1-\theta)^{\omega(n-s)} \\ &\propto L(\theta | \omega s, \omega n). \end{aligned}$$

No matter which of these methods is used, a new, transformed likelihood function is used to represent the unreliable observation.

14.5.2 Specific adjustments

There are several types of bias to internal validity. These are factors that prevent the evidence variable from accurately reflecting on its associated parameter:

1. *Measurement error* is a systematic error in the recording of the experimental outcomes. In the case of binary outcomes, this might be the tabulation of false positives; in the case of continuous outcomes, this might be due to a calibration error.

   For binary outcomes, let parameter α be the probability that a true success will be incorrectly labeled a failure, and β be the probability that a true failure will be incorrectly labeled a success. The adjustment is then

   $$\theta_s = (1-\alpha)\theta_t + \beta(1-\theta_t).$$

   For continuous outcomes, let parameters α and β instead be arbitrary real numbers (perhaps β should be nonnegative), so that

   $$\theta_s = \alpha + \beta\theta_t$$

   or

   $$\theta_s = \alpha\theta_t^\beta.$$

2. *Protocol departure* arises when subjects do not receive the technology assigned to them under the experimental design. This might arise when patients who are assigned for treatment do not receive it, such as by not showing up for their scheduled examinations or by not taking their prescribed medication. It can also arise when patients who are assigned to one treatment obtain another outside the auspices of the experiment. In the case of retrospective experiments, this definition is meant to apply to the misclassification of patients according to the technologies they received.

   Letting φ be the probability that a subject selected for treatment receives the control instead,

$$\theta_s = (1 - \varphi)\theta_t + \varphi\theta_c.$$

   A similar form can be used for departures from the control group. Of course, if adjusted parameters have been defined for those who follow or depart from the protocol, they can be used instead of the paradigm success probabilities θ_t and θ_c.

3. *Unreliable observations* arise when there is uncertainty about the true outcome of an experiment. This can be due to reporting errors, aggregation of data sets, estimates from sampling a database, or anecdotal evidence. A special case of this, *strength of evidence*, is a method for discounting the impact of experimental evidence when the experimental design is suspect.

   The adjustment for unreliable observations is in the form of a revised likelihood function rather than a parameter to describe the circumstances of interest. This adjustment was described earlier in this section.

There are also a number of biases to external validity. In the t-PA example in the preceding section, multiple evidence was associated with the same parameter. The factors below might prevent us from believing that different evidence variables should be associated with the same parameter:

1. *Selection bias* refers to a systematic difference in the types of subjects assigned to different technologies or studies. This is a common problem in retrospective studies where patients self-selected the type of treatment they received. It also arises in comparisons among studies conducted at different sites with different types of patients.

   Letting parameters α and β be arbitrary real numbers (perhaps β should be nonnegative),

$$\theta_s = \alpha + \beta\theta_t.$$

2. *Treatment intensity* is the difference in the amount of the technology made available to the subjects in different experiments. It can be the difference in treatment dosage (such as in a multiple-dose trial), number of followup examinations, or the quality of care (efficacy vs effectiveness). It can even arise from a selection bias, in which a patient who self-selected for the

treatment might be expected to receive a higher standard of care than someone not given the choice.

Letting parameter τ be the treatment intensity (one is the paradigm treatment intensity),

$$\theta_s = \theta_c + \tau(\theta_t - \theta_c).$$

Of course the other effects, relative-risk and odds-ratio, could be used instead of difference to obtain

$$\theta_s = \theta_t^{\tau}\theta_c^{1-\tau}$$

or

$$\theta_s = \left[1 + \left(\frac{\theta_t}{1-\theta_t}\right)^{\tau}\left(\frac{\theta_c}{1-\theta_c}\right)^{1-\tau}\right]^{-1}.$$

These might be used in adjusting among groups in a multiple-dose RCT, for which the treatment effect is believed to be nonlinear.

3. *Length of followup* refers to the amount of time elapsed in different studies before the outcomes are recorded. Those studies might be recording counts of repeatable events, such as episodes of asthma, or nonrepeatable events, such as death.

   Assuming a stationary arrival process and letting parameter τ be the number of paradigm periods that a subject is followed in this study, if θ_t is the paradigm arrival rate for repeated events, then the rate of arrival in the study would be

$$\theta_s = \tau\theta_t.$$

   On the other hand, if θ_t is the probability that a nonrepeatable event (the first arrival) has not yet occurred, then the probability that an event did not occur in the study would be

$$\theta_s = \theta_t^{\tau}.$$

14.5.3 Summary

This section has presented a number of internal and external biases and shown how they can be addressed by simple models within a common mathematical form. The key is expressing the circumstances of a particular study in terms of the circumstances of interest. All of the adjustments are modular, so they are easy to nest and combine. For example, subjects who depart from the protocol often demonstrate a treatment intensity or selection bias, or perhaps both. The adjustments cannot only be performed on absolute parameters, such as probabilities of success or rates of arrival, but also on effect measures, such as odds-ratios or ratios of arrival rates. The simple structure provides a rich framework in which to describe fundamental differences among experiments subjectively, so that the evidence from them can be sensibly combined.

14.6 SOLUTION METHODS

The emphasis in this chapter is on the use of influence diagrams to formulate complex medical technology assessment problems. Of course, once the model has been formulated it must be evaluated. Earlier work on our project developed numerical integration software and techniques which have been limited to single-dimension problems with little dependency (Hasselblad, 1987). On the other hand, there are several prominent techniques for solving an influence diagram model of the structure formulated here. In the unlikely case that the diagram is singly connected, for example, Pearl's (1986) algorithm would seem appropriate or, in general, the algorithm developed by Lauritzen and Spiegelhalter (1988) could be applied. However, both of those algorithms are designed for discrete evaluation and they do not seem well suited to the continuous variables in this model. Fortunately, the same continuous nature of the problem that hampers the techniques listed above suggests several alternative methods: iterative linear approximations, Monte Carlo integration, and posterior made analysis.

The iterative linear approximation method (Shachter, 1988b) is based on the Gaussian influence diagram (Kenley, 1986, Shachter and Kenley, 1989) and applies to models satisfying assumptions A1 and A2, that is, for which each piece of evidence only bears (directly) on a single parameter, and the basic parameters are independent, *a priori*. Each of the variables is transformed and then iteratively approximated by a multivariate-normal model. The linear arc (regression) coefficients for each deterministic variable are approximated by the derivatives of the deterministic functions. The Gaussian influence diagram permits simple computation of the posterior distribution for the entire model, given the evidence. This posterior distribution can then be used to improve the linear approximation, by taking a first-order approximation of the prior model about the posterior means. This method works quickly (in polynomial time) and has converged in just a few iterations for a wide variety of technology assessments. As the comparisons in Table 14.3 show, it does not underestimate the posterior variance, as one might fear from a pure linear approximation. An extra bonus is that the correlation matrix can be easily computed, showing the sensitivity of the parameters of interest to each of the parameters in the model.

The Monte Carlo integration method (Stewart, 1987; Geweke, 1986) applied to this problem appears to be more accurate but significantly slower than the iterative linear approximation method. The key to the method is the separation in the log-posterior derived in Section 14.3:

$$LP\{\beta|X\} = k + LP\{\beta\} + \sum_{j=1}^{n} LL_j\{\theta(\beta)|X_j\}.$$

The importance sampling distribution can be obtained from the mode of the posterior distribution, as explained below. This does not require any of the additional assumptions listed in Section 14.3.

Table 14.3 Evaluation of the t-PA model

	Iterative linear approximation		Monte Carlo integration		Posterior mode	
Parameter	Mean	SD	Mean	SD	Mean	SD
pRep|CC	0.115	0.027	0.115	0.027	0.115	0.027
pRep|t-PA	0.680	0.038	0.681	0.037	0.681	0.038
pRep|IVSK	0.357	0.043	0.358	0.042	0.357	0.043
pMort|CC, Rep	0.068	0.083	0.042	0.051	0.042	0.057
pMort|CC, No Rep	0.206	0.019	0.209	0.012	0.209	0.012
pMort|TA, Rep	0.063	0.025	0.064	0.025	0.063	0.025
pMort|TA, No Rep	0.203	0.020	0.203	0.019	0.203	0.019
pMort|CC	0.190	0.008	0.190	0.007	0.190	0.008
pMort|t-PA	0.109	0.015	0.108	0.015	0.108	0.015
pMort|IVSK	0.153	0.007	0.153	0.007	0.153	0.007
pMort|t-PA − pMort|CC	−0.081	0.017	−0.082	0.016	−0.082	0.017
pMort|IVSK − pMort|CC	−0.037	0.010	−0.037	0.010	−0.037	0.010
pMort|t-PA − pMort|IVSK	−0.044	0.015	−0.045	0.014	−0.046	0.015

A third approximation is to simply use the mode of the posterior distribution as the estimate. This is the fastest of the three methods even though it also requires none of the additional assumptions A1, A2, or A3. A point estimate for the basic parameters, $\hat{\beta}$, is obtained by optimization:

$$\hat{\beta} = \arg\max_{\beta} [LP\{\beta|X\}],$$

and this provides estimates for the posterior means and variances for all of the parameters:

$$E\{\theta(\beta)|X\} \approx \theta(\hat{\beta})$$

and

$$\mathrm{Var}\{\theta(\beta)|X\} \approx -\nabla\theta(\hat{\beta})\mathbf{H}^{-1}(\hat{\beta})\nabla\theta^{T}(\hat{\beta}),$$

where $\mathbf{H}$ is the Hessian of $LP\{\beta|X\}$.

The posterior mode analysis is further simplified by an asymptotic approximation to the Hessian similar to one used by Hartley (1961), based on Cramér (1946, p. 502):

$$\mathbf{H}(\beta) \approx -\nabla\theta^{T}(\beta)\mathbf{K}(\beta)\nabla\theta(\beta),$$

where

$$\mathbf{K}(\beta) = \begin{bmatrix} -\nabla^2 LP\{\beta\} & 0 \\ 0 & 0 \end{bmatrix} + \sum_{j=1}^{n} \nabla_{\theta}LL_j\{\theta(\beta)|X_j\}^{T}\nabla_{\theta}LL_j\{\theta(\beta)|X_j\}.$$

(For this approximation, it is best to break down experimental evidence into the smallest possible units. Binomial experiments, for example, should be summed as individual trials.) This approximate Hessian is not only good for approximating the variance, but it has superior convergence properties to the true Hessian, when used in Newton or modified Newton search (Luenberger, 1984), being careful to stay within the feasible region. At the same time, the errors introduced by the approximation are insignificant, at least in the t-PA example. The information requirements for the approximate Hessian are reduced as well: we need a gradient and Hessian for the log-prior $LP\{\beta\}$, deterministic function $\theta(\beta)$ and its gradient, and a gradient $\nabla_\theta LL_j\{\theta(\beta)|X_j\}$ for the individual log-likelihoods. Each of these is relatively easy to provide, even for a dynamically constructed model.

Although none of the additional assumptions for Section 14.3 are needed to perform the posterior mode analysis, we can gain insight about those assumptions from the equations above. The assumption A1, single parameter evidence, means that each outer product in the sum above would have only one nonzero entry. Assumption A2, basic parameter prior independence, results in a diagonal $\nabla^2 LP\{\beta\}$. Therefore both A1 and A2 together result in $\mathbf{K}$ diagonal. Assumption 3, $\theta(\beta) = \beta$, implies that $\nabla\theta(\beta) = I$ and $\mathbf{K} \approx -\mathbf{H}$, so $\mathbf{K}$ would be an approximation to the precision of θ. Clearly, if all three assumptions were satisfied then the (approximate) precision would be diagonal. Of course, this is the case considered in Section 14.3, in which the basic parameters are independent, *a posteriori*.

The results from using the three methods to solve the t-PA example are given in Table 14.3. The posterior mode analysis takes fifteen seconds or so (depending upon the starting point), the iterative linear approximation took under two minutes while the Monte Carlo achieved a numerical standard error (Geweke, 1986) of 0.00017 for 'pMort|t-PA – pMort|CC' after over ten hours and 300 000 iterations. The largest disagreement between the methods is for the parameter 'pMort|CC, Rep', a parameter with evidence of no deaths in only fourteen cases. (Naturally, this is difficult evidence to approximate in the Gaussian model and also proves unstable in the Monte Carlo method.) The methods essentially agree, however, with respect to all of the parameters of real interest.

The three methods provide a useful balance between speed and accuracy. The posterior mode and iterative linear approximations are fast enough to use for sensitivity analysis and model debugging, while the Monte Carlo method provides substantial accuracy for the baseline analysis.

14.7 CONCLUSIONS

This chapter has presented the confidence profile method in a representation framework based on influence diagrams. This approach can be used to quantify the effects of different health technologies on health outcomes incorporating whatever forms of experimental evidence might be available. The method allows the explicit adjustment of the evidence to account for biases to internal and external validity. Finally, techniques are developed to evaluate the models that have been formulated. These steps have been illustrated by the assessment to t-PA as a treatment for acute heart attacks.

There are a number of benefits of the confidence profile method presented in this chapter. First, it provides a general framework which helps to organize and document the model for an assessment, fostering communication among decision makers and experts. The subjective judgments and expert opinions in the model are explicitly stated. The relationships among population parameters, particularly assumptions about conditional independence, are the focus of attention. The model also makes it clear for which parameters there is little or no experimental evidence.

Second, the influence diagram precisely represents complex dependencies among the population parameters. We can decompose complex problems into manageable submodels, exploiting the natural conditional independence present in the model structure. The processes of chaining and adjustment, the basic building blocks of medical technology assessments, are modular in this framework. We can therefore represent subtle dependence relationships correctly by building models in their 'natural' direction, from population parameters to the empirical evidence (Shachter and Heckerman, 1987).

Third, the representation leads to practical methods for analysis. The model can be evaluated automatically in the influence diagram representation in which it has been formulated. The influence diagram structure can be exploited by several of the solution methods, iterative linear approximation, Monte Carlo integration, and posterior mode analysis. Because these methods evaluate the model in the same formulation, the quicker methods can be used for debugging and sensitivity analysis while the more accurate method can be used for baseline analysis.

There are two distinct areas of expertise needed to perform a medical technology assessment. Not only must the analyst be able to examine and evaluate the medical literature, but a solid statistical background is needed to integrate the evidence into a coherent analysis. The eventual goal of this research is to reduce the statistical burden on the analyst while providing him with a general modeling tool. This does not obviate the need for him to understand the underlying statistics, but allows him to focus his attention on the model formulation. There are many research challenges to be overcome in supporting the process of model construction given the basic structural assumptions of the model and the 'relevant' evidence. We believe that the methodology developed in this research provides an influence diagram representation for the target model along with techniques for evaluating it.

Many issues are common to technology assessment in general. There is a need to integrate and organize expert opinions and experimental evidence. The empirical evidence available frequently cannot be applied directly to the technology under study, and thus adjustment is required. Finally, it is most important to have a model structure which facilitates communication, so that the fundamental assumptions are explicit and their impact can be evaluated. We think that the approach presented here for addressing these issues in medical technology assessments can be successfully applied in other problem areas.

ACKNOWLEDGEMENT

This research was supported by the John A. Hartford Foundation and the National Center for Health Services Research under Grants 1R01-HS-05531-01 and 5R01-HS-05531-02. We thank Robert Wolpert for his collaboration, and John Geweke and Dale Poirer for their helpful suggestions.

14.8 DISCUSSION

14.8.1 Discussion by David J. Spiegelhalter

This is a very nice exposition of an attractive procedure. At one level, it is just a piece of applied Bayesian statistics, but the term 'just' obscures the undoubted additional value of the graphical representation which allows parameters and data to have symmetric roles and be manipulated in a flexiable manner.

At a general level, I would like to ask about the outside acceptability of such an analysis, particularly in view of the increasing importance of meta-analyses, of which this approach is the natural extension. I would, however, take some issue with the impression given in the chapter that such an approach allows nonexperts to implement analyses of similar complexity—such diagrams, although attractively simple in their final form, are not straightforward to develop unless one is skilled in structuring conditional independence arguments, which I feel will not always be the case. This approach is therefore just one more that can be abused, either intentionally or not, but it does have the great advantage that the errors may be qualitatively apparent instead of being buried in numerical form.

This particular example immediately illustrates the way in which different practitioners may come up with very different diagrams. From an expert systems perspective, one is used to thinking in terms of updating unknown parameters *case-by-case*, and so one's diagram will model an individual rather than the range of studies. Thus I would adopt an *a priori* diagram as in Figure 14.10, in which $\boldsymbol{\theta}$ expresses all conditional probabilities of reperfusion given treatment, and $\boldsymbol{\phi}$ models probabilities of mortality given treatment and reperfusion. The crucial problem is that if 'reperfusion' is not observed on a patient, this creates a posterior dependence between $\boldsymbol{\theta}$ and $\boldsymbol{\phi}$ and make updating of the distributions of these parameters complex. Various approximations could be used if this only happened on occasions, but in this case, the bulk of the data have this pattern of missing observations.

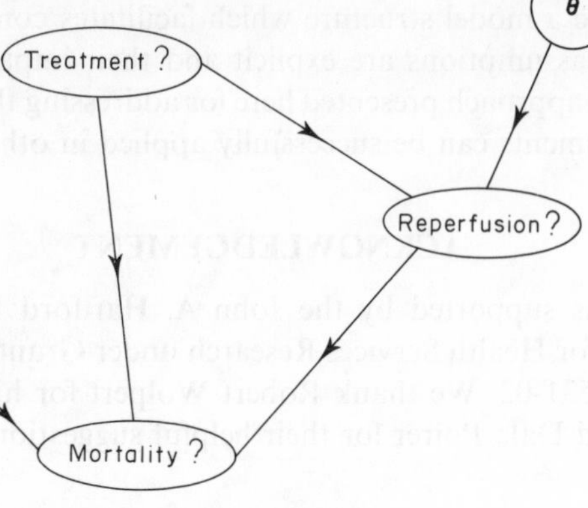

Figure 14.10

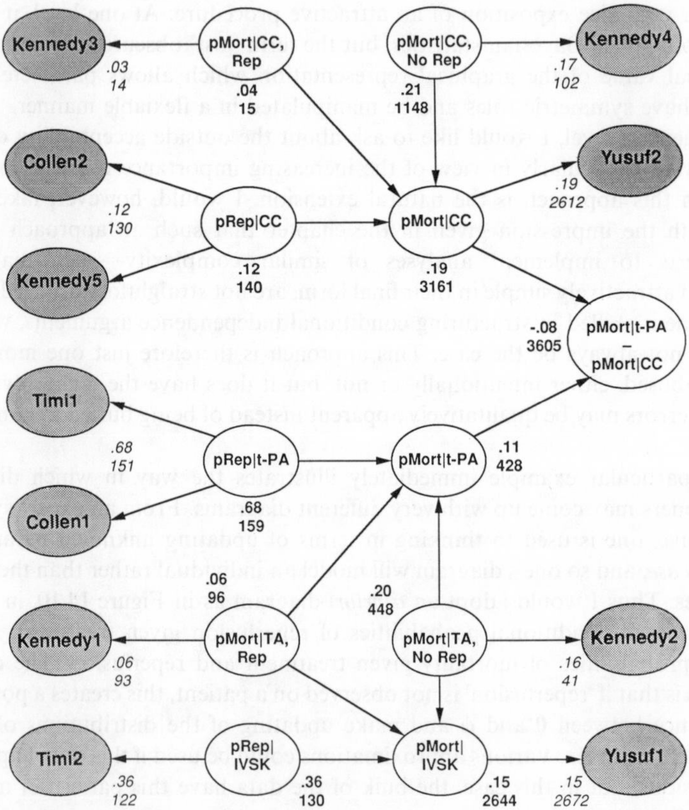

Figure 14.11

Some approximation is necessary, and it remains a question of which technique is both computationally and conceptually most appealing. First, it is worth noting that Dickey et al. (1987) have considered classes of prior distributions that remain conjugate after such sampling schemes, but it is unclear whether this provides any computational advantage. Second, it may be useful to review a computed result in terms of 'data-equivalents', to try to get an intuitive feel of what full data would have given rise to the same result. Specifically, suppose a probability is finally estimated to be $\hat{p}$ with standard error s. Then this is 'equivalent' to having observed $\tilde{n} = \hat{p}(1-\hat{p})/s^2$ patients, of which a proportion $\hat{p}$ had the condition in question. When this is done to the conclusions of the Monte Carlo study, Figure 14.11 is obtained. We can then contrast the observed proportion and sample size in the appropriate clinical trial (shown in bold) with the eventual estimate and equivalent sample size from the whole study.

We can immediately say that p(Mort|CC, Rep), p(Rep|CC), p(Mort|CC), p(Rep|t-PA), p(Mort|TA, Rep), p(Rep|IVSK), and p(Mort|IVSK) are essentially only influenced by the directly relevant evidence and nothing filters through from elsewhere in the diagram. However, both p(Mort|CC, No Rep) and p(Mort|TA, No Rep) are changed in estimate and made substantially more precise. I can understand the 'mixing' parameters p(Rep|Treatment) not being influenced, since little will be learnt by observing a mixed sample. But the pattern of 'learning' concerning the more specific probabilities seems odd. It is not related to initial precision, but is related to the size of the initial estimate, in that the smaller probabilities remain unchanged. Is this possibly an artefact of the prior assumption? (Here logit (p) has variance proportional to $\hat{p}^{-1} + (1-\hat{p})^{-1}$, and assuming constant variance on a logit scale may make low probabilities too precise?) My intuition gives up.

In conclusion, this is a good example of a carefully structured analysis, but I am not so confident that it is as straightforward as suggested.

14.8.2 Reply

I thank Dr Spiegelhalter for his kind remarks and careful review and I hope that my answers to the questions he raises might provide some additional insights.

First, I believe that many papers in the conference have illuminated an important distinction: influence diagrams provide considerable assistance in presenting, communicating, and explaining a model once it has been formulated, but they provide little assistance to the naive modeler to help him synthesize his model. I do not propose that this technique will allow proper technology assessments to be done without any statistical expertise, any more than a tool to help physicians with their role in the assessment process would render their expertise unnecessary. Rather, I have observed that the devices presented in the chapter have helped me to explain statistical models to nonquantitative experts such as physicians, and to get invaluable feedback on the appropriateness of the assumptions, where otherwise they might not have understood the assumptions well enough to actively participate.

I do appreciate how difficult it is to construct a good model of this sort, and that we are indeed creating one more tool which can be abused. Nonetheless, my understanding of the *ad hoc* manner in which many medical technology assessments are currently performed leads me to believe that there is tremendous opportunity for its proper use as well.

I appreciate the problem illustrated in Figure 14.10, and my initial tendency also was to think of the problem in a *case-by-case* fashion. I agree with you that if there are data missing only occasionally it is much simpler to use approximate deviations from the pure case-by-case method. However, when the missing data are more common, or the sources of data less compatible, the structure from the influence diagram is invaluable in helping us relate the parameters for which we have evidence. By defining an unobservable parameter and offering a prior distribution for it, we can often build a simple and defensible model which clearly explains the dependence between θ and ϕ.

A common example of this in medical technology assessment arises when the definition of a condition (or the ability to detect it) has changed since a large study was performed. We might wish to include the study in our analysis, but we must adjust the results of the study for the systematic measurement error. We cannot, of course, detect which cases were misclassified, but we can recognize and model the relationships among the parameters. In terms of your example, suppose that in an earlier study, 'Reperfusion' was observed, but by a technique we now know to be incorrect. The influence diagram approach will let us use that information, even though we do not know for any particular case whether 'Reperfusion' actually occurred.

The data-equivalents, shown in Figure 14.11, raise some interesting questions. There is the one anomalous data-equivalent, 1148 cases for 'pMort|CC, No rep'. I wish I could blame it on the logit transformation, but there is no such approximation in our Monte Carlo analysis. In fact, the iterative linear approximation, which uses the logit, obtains a more agreeable data-equivalent of 452 cases for the parameter. I cannot explain why the data-equivalent is so high for the Monte Carlo (or for the posterior mode which is 1170 cases). Perhaps it comes from the severe lack of data (and no 'successes') for 'pMort|CC, Rep'. In other respect, the data-equivalents exercise is reassuring, since the model seems to be using the data well to answer the question. In particular, it supplies a reasonable amount of data-equivalents from the large Yusuf studies to 'pMort|t-PA' and to 'pMort|t-PA − pMort|CC', the parameters about which we are most interested.

REFERENCES

Collen, D., Topol, E. J., Teifenbrunn, A. J. *et al.* (1984) Coronary thrombolysis with recombinant human tissue-type plasminogen activator: a prospective, randomized, placebo-controlled trial, *Circulation*, **70**, 1012–17.

Cramér, Harald (1946) *Mathematical Methods of Statistics*, Princeton University Press, Princeton.

Critchfield, G. C., and Eddy, D. M. (1987) A confidence profile analysis of the effectiveness of disulfiram in the treatment of chronic alcoholism, *Medical Care*, **25**(12), Supplement s66–s75.

Datta, K., and Oliver, R. M. (1989) Predicting mid-air and near mid-air collisions, to appear in *Management Science*.

Dickey, J. M., Jian, J.-M. and Kadane, J. B. (1987) Bayesian methods for censored categorical data, *J. Amer. Statist. Assoc.*, **82**, 773–81.

Eddy, D. M. (1986) The use of confidence profiles to assess tissue-type plasminogen activator. In G. S. Wagner, and R. Califfs (eds) *Acute Coronary Care* 1987, Martinus Nijhoff, Boston.

Eddy, D. M. (1989) The confidence profile method: a Bayesian method for assessing health technologies, *Operations Research*, **37**, 210–228.

Eddy, D. M., Hasselblad, V., McGivney, W. and Hendee, W. (1988) The value of mammography screening in women under age 50, *Journal of the American Medical Association*, **259**, 1512–19.

Geweke, J. (1986) *Bayesian Inference in Econometric Models Using Monte Carlo Integration*, Department of Economics, Duke University, Durham, N.C., July.

Hartley, H. O. (1961) Modified Gauss–Newton method for the fitting of non-linear regression functions by least squares, *Technometrics*, **3**, 269–80.

Hasselblad, V. (1987) *CP: Confidence Profile Computer System*, Center for Health Policy Research and Education, Duke University, Durham, N.C.

Hasselblad, V. and Critchfield, G. C. (1987) *An Analysis of Neonatal Screening for Maple Syrup Urine Disease*, Center for Health Policy Research and Education, Duke University, Durham, N.C.

Howard, R. A. and Matheson, J. E. (1981) Influence diagrams. In R. A. Howard and J. E. Matheson (eds) *Readings on The Principles and Applications of Decision Analysis*, vol. II, 1984, Strategic Decisions Group, Menlo Park, Calif., pp. 721–762.

Kenley, C. R. (1986) Influence diagram models with continuous variables, Ph.D. dissertation, EES Department, Stanford University, Stanford, Calif., June.

Kennedy, J. W., Ritchie, J. L., Davis, K. B. *et al.* (1985) The western Washington Randomized trial of intracoronary streptokinase in acute myocardial infarction. A 12-month follow-up report, *New England Journal of Medicine*, **312**, 1073–78.

Kiiveri, H., Speed, T. P. and Carlin, J. B. (1984) Recursive causal models, *Journal Australian Mathematical Society*, **A36**, 30–52.

Lauritzen, S. L. and Spiegelhalter, D. J. (1988) Local computations with probabilities on graphical structures and their applications to expert systems (with discussion), *Journal Royal Statistical Society*, **B50**, 157–224.

Luenberger, D. G. (1984) *Linear and Nonlinear Programming*, Addison-Wesley, Reading, Mass.

Oliver, R. M. and Yang, H. J. (1989) Bayesian updating of event-tree parameters to predict high risk incidents. In R. M. Oliver and J. Q. Smith (eds) (1990) *Influence Diagrams, Belief Nets and Decision Analysis*, Wiley, Chichester, pp. 277–296.

Pearl, J. (1985) *How to Do with Probabilities What People Say You Can't*, Technical Report CSD-R-49, UCLA.

Pearl, J. (1986) Fusion, propagation, and structuring in belief networks, *Artificial Intelligence*, **29**, 241–88.

Shachter, R. (1988a) Probabilistic inference and influence diagrams, *Operations Research*, **36**, 589–604.

Shachter, R. (1988b) A linear approximation method for probabilistic inference, *Proceedings of the Fourth Workshop of Uncertainty in Artificial Intelligence*, 19–21 Aug., pp. 229–307.

Shachter, R. and Heckerman, D. (1987) Thinking backwards for knowledge acquisition, *AI Magazine*, **8**, Fall, 55–61.

Shachter, R. and Kenley, C. R. (1989) Gaussian influence diagrams, *Management Science*, **35**, 527–550.

Spiegelhalter, D. J. (1986) Probabilistic reasoning in predictive expert systems. In L. N. Kanal and J. F. Lemmer (eds) *Uncertainty in Artificial Intelligence*, North-Holland, Amsterdam, pp. 47–67.

Spiegelhalter, D. J. (1987) Coherent evidence propagation in expert systems, *The Statistician*, **36**, 201–10.

Stewart, L. T. (1981) Hierarchical Bayesian analysis using Monte Carlo integration, *The Statistician*, **56**, 211–19.

TIMI Study Group (1985) The thromobolysis in myocardial infarction (TIMI) trial. Phase I findings, *New England Journal of Medicine*, **312**, 932–6.

Yusuf, S., Collins, R., Peto, *et al.* (1985) Intravenous and intracoronary fibrinolytic therapy in acute myocardial infarction: overview of results on mortality, reinfarction and side-effects from 33 randomized controlled trials, *European Heart Journal*, **6**, 556–85.

CHAPTER 15

Influence Diagrams and Medical Diagnosis

Carlos A. de B. Pereira, *Universidade de São Paulo, Brazil*

ABSTRACT

Influence diagrams operations are used to solve the following problem:

A patient consults with a specialist who is going to start a search to discover whether the patient has a disease, D, or its absence, D'. Before collecting any further information, a prior probability, $d = \Pr\{D\}$, for the presence of the disease is assessed. Looking for more information, the physician observes an indicant (E = positive response or E' = negative response), which is a new evidence associated with the patient. The experience of the physician is in part represented by the data (x, y), where $x(y)$ is the number of positive (negative), respondents among all former patients having $D(D')$. The objective is to evaluate, for the new patient, the conditional probability of D (D') given that the patient responded positively (negatively) and also that the data (x, y) have been observed. Note that the likelihood depends upon the sensitivity, $\pi = \Pr\{E|D\}$, and the specificity, $\theta = \Pr\{E'|D'\}$. However the parameters of interest, the diagnostic probability, are $p = \Pr\{D|E\}$ and $q = \Pr\{D'|E'\}$. In another context the same problem is discussed by Pereira and Pericchi (1990).

15.1 THE PROBLEM

In the search for a new indicant of a disease D, doctors in a certain clinic selected 150 patients known to have the disease and 150 patients known not to have the disease. Here **D** is the event that a patient has the disease D, while **D**$'$ is the event that a patient does not have the disease D. To each patient they apply a test obtaining a response E^+ for positive evidence or E^- for negative evidence. The results of the experiment are presented in Table 15.1.

A new patient comes to the clinic and is judged, by the doctors, to have the disease with probability 0.1. The doctors apply the same test to this patient

Influence Diagrams, Belief Nets and Decision Analysis
Edited by R. M. Oliver and J. Q. Smith

Table 15.1 Reslts of the clinical experiment

Patient's state	Patient's response E_+	E_-	Sample size
D	60	90	150
D′	9	141	150

and obtain response E^+. How does this evidence change their probability that the patient has the disease? What would be this change if the response is E^-?

15.2 DIAGNOSTIC MODEL

To present a solution for this problem we define the following quantities which we think are the elements of the model:

1. The sensitivity of the test is $\pi = \Pr\{E^+|\mathbf{D}\}$ and the specificity of the test is $\theta = \Pr\{E^-|\mathbf{D}'\}$.
2. The sampling quantities are $x|\pi \sim \text{bi}(150, \pi)$ and $y|\theta \sim \text{bi}(150, \theta)$. That is, x and y are binomial random quantities with parameters $(150, \pi)$ and $(150, \theta)$, respectively. Here x is the number of positive responses among the 150 patients having D and y is number of negative responses among the 150 patients not having D. We have observed $x = 60$ and $y = 141$.
3. The state of the new patient is

$$\delta = \begin{cases} 1 & \text{if the patient has the disease, } D \\ 0 & \text{otherwise.} \end{cases}$$

The prior diagnostic probability is $\Pr\{\delta = 1\} = 0.1$. $\delta \sim \text{ber}(0.1)$ indicates that δ is a Bernoulli random quantity.
4. The result of the test for the new patient is

$$t = \begin{cases} 1 & \text{if a positive response obtains, i.e. } E^+ \\ 0 & \text{if a negative response obtains, i.e. } E^-. \end{cases}$$

Note that $\Pr\{t = 1|\delta = 1\} = \pi$ or that $\Pr\{t = 0|\delta = 0\} = \theta$ if we judge, respectively, the new patient as we have judged the sample patients having D or the sample patients not having D. See Lindley and Novick (1981) for a complete discussion on exchangeability.
5. The posterior diagnostic probabilities, the object of the analysis, are $\Pr\{\delta = 0|t = 0, x = 60, y = 141\}$ and $\Pr\{\delta = 1|t = 1, x = 60, y = 141\}$. Note that the quantities on the right of the bar are observable and the ones on the left of the bar are the quantities of interest which at this stage are not

observable. The other quantities, π and θ, that are neither observable nor of interest, are eliminated during the analysis.

To construct the probabilistic influence diagram relating the nodes representing the above quantities we need to state the conditional independence relationships that we have judged to be relevant. The first and most important is $(x, \pi) \perp\!\!\!\perp (y, \theta)$; i.e. (x, π) and (y, θ) are independent. This is because (x, π) and (y, θ) are quantities related to two distinct and independent populations, **D** and **D**$'$. (We could think of two different urns having balls of two colors.) Since our interest is directed to a new patient (a different individual), his/her state δ is independent of the other patients in the sample. However, the response to the clinical test, t, given to a new patient depends on the value of π or θ and his/her state δ. With these restrictions in mind, Figure 15.1 presents our probabilistic influence diagram for the problem. To stress the fact that the clinical test is being given for the first time, we judge π and θ to have independent uniform distributions in the unit interval. Recall that the uniform density in the interval $(0, 1)$ is the beta density with parameters $a = b = 1$. The beta distribution with parameters a and b is denoted by $\text{Be}(a, b)$.

The diagram of Figure 15.1 (in our particular case $m = n = 150$) has four distinguished nodes. Three represent random quantities that have been observed and one represents the unknown quantity of interest, δ. The remaining nodes, π and θ, are the modeling parameters (i.e. are neither observable nor of interest) and must be eliminated. The directions of the arcs are also determined by the problem. The sample results, x and y, depend, respectively, on the chances of positive and negative responses, namely π and θ, in their respective populations **D** and **D**$'$. Analogously, the response of the new patient, t, depends on δ, the state of the patient, and on the accuracy of the test measured by π and θ.

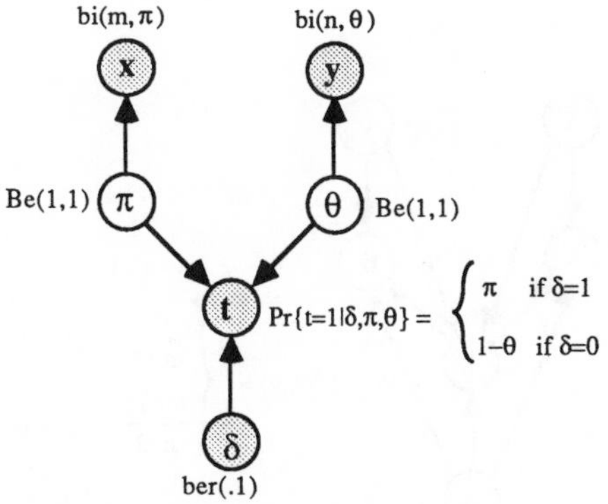

Figure 15.1 Probabilistic influence diagram for the diagnosis example.

15.3 THE 'IDEAL' SOLUTION

Figure 15.2 shows the probabilistic influence diagram after reversing arcs $[\pi, x]$ and $[\theta, y]$. After reversing these arcs, we use $\mathrm{Bb}(m, 1, 1)$ and $\mathrm{Bb}(n, 1, 1)$ to indicate that x and y are distributed as beta-binomial random quantities with parameters $(m, 1, 1)$ and $(n, 1, 1)$, respectively. (See Basu and Pereira, 1981, 1982 for a complete discussion on these distributions.) Arc reversal and node elimination, the diagram operations used here, are discussed by Barlow and Pereira (1987). After reversing arcs $[\pi, t]$ and $[\theta, t]$ we obtain the diagram of Figure 15.3. For simplicity, we give only the probability function of t since the distributions of x, y, and δ are given in Figure 15.2. Clearly the distributions of π and θ changed. Since π and θ are going to be eliminated (they are barren nodes in Figure 15.3), their probability functions do not appear in Figure 15.3. In fact we obtain:

1. $\pi|(\delta, t, x, y) \sim \pi|(\delta, t, x) \sim \mathrm{Be}(1 + x + \delta t, 1 + m + \delta - x - \delta t)$; and

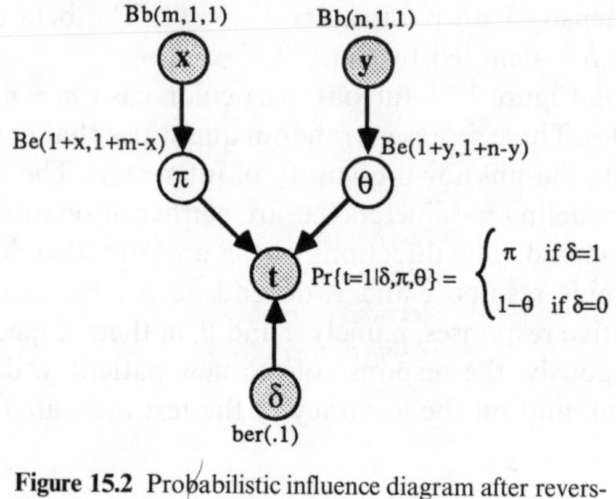

Figure 15.2 Probabilistic influence diagram after reversing arcs $[\pi, x]$ and $[\theta, y]$.

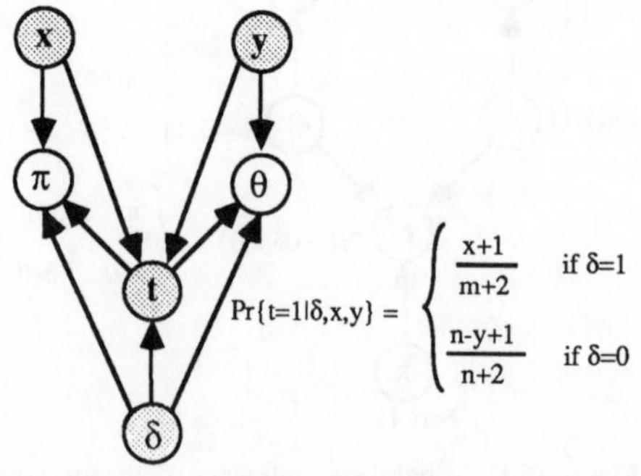

Figure 15.3 Probabilistic influence diagram.

2. $\theta|(\delta, t, x, y) \sim \theta|(\delta, t, y) \sim \text{Be}(1 + y + (1 - \delta)(1 - t),$
$1 + n + (1 - \delta) - y - (1 - \delta)(1 - t))$.

Althoughh these expressions look complicated they only reflect the fact that the new patient has to be added to the sample of **D**(**D**′) if $\delta = 1$ ($\delta = 0$) and, in this new sample, either $x(y)$ increases to $x + 1$ ($y + 1$) if his/her response was positive (negative) or $m - x$ increases to $m - x + 1$ in the case of a negative (positive) response. In addition, since the first (second) expression does not involve $y(x)$, we do not have to consider either arc $[x, \theta]$ or arc $[y, \pi]$ or arc $[\theta, \pi]$.

Figure 15.4 is our diagram after eliminating nodes π and θ. Since all the nodes are distinguished we did not shade them. The probabilistic influence diagram that permits us to evaluate the diagnostic probabilities for all possible values of the observable quantities, t, x, and y, is presented in Figure 15.5. The answer to our problem is given by the probability functions attached to node δ.

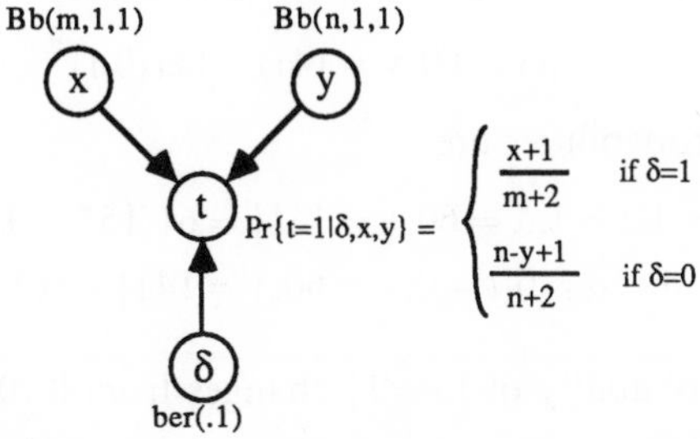

Figure 15.4 Probabilistic influence diagram after eliminating nodes π and θ.

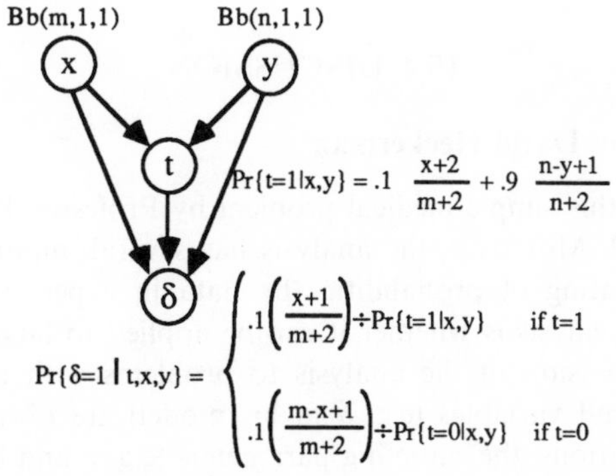

Figure 15.5 Probabilistic influence diagram after reversing node $[\delta, t]$.

Using now the experimental data displayed in Table 15.1 we obtain the following results:

1. The posterior distributions of π and θ are

$$\pi|(\delta, t, x = 60) \sim \begin{cases} \text{Be}(62, 91) & \text{if } \delta = 1, t = 1 \\ \text{Be}(61, 92) & \text{if } \delta = 1, t = 0 \\ \text{Be}(61, 91) & \text{if } \delta = 0. \end{cases} \quad \text{and}$$

$$\theta|(\delta, t, y = 141) \sim \begin{cases} \text{Be}(142, 10) & \text{if } \delta = 1 \\ \text{Be}(143, 10) & \text{if } \delta = 0, t = 0 \\ \text{Be}(142, 11) & \text{if } \delta = 0, t = 1. \end{cases}$$

2. The predictive distributions of x, y, and t are
   (a) $\Pr\{x = i\} = \Pr\{y = i\} = 1/151$, where $i = 0, 1, \ldots, 150$. That is, $x \sim y \sim$ Bb(150, 1, 1); and
   (b) $\Pr\{t = 1 | x = 60, y = 141\} = 0.1(62/152) + 0.9(10/152) = 15.2/152 = 0.1$, i.e.

$$t|(x = 60, y = 141) \sim \text{ber}(0.1).$$

3. The diagnostic probabilities are

$$\Pr\{\delta = 1 | t = 1, x = 60, y = 141\} = 61/152 = 0.40 \quad \text{and}$$
$$\Pr\{\delta = 0 | t = 0, x = 60, y = 141\} = 0.93.$$

   Hence,
   (a) if $t = 1$ the probability of $\{\delta = 1\}$ changes from 0.10, *a priori*, to 0.40, *a posteriori*, and
   (b) if $t = 0$ the probability of $\{\delta = 0\}$ changes from 0.90, *a priori*, to 0.93, *a posteriori*.

The fact that the change observed in (a) is bigger than that observed in (b) suggests that the diagnostic test in the study is more sensitive than it is specific.

15.4 DISCUSSION

15.4.1 Discussion by David Heckerman

The analysis of this simple medical problem by Professor Pereira is accurate and well presented. Moreover, the analysis has several important applications including the updating of probabilities by data in expert systems. My only concern with the analysis is whether it can be applied to larger, more realistic, problems. The extension of the analysis to problems with many variables is straightforward if all variables (e.g. δ in his model) are observed in repeated trials. In such situations, the sampling parameters (e.g. π and θ) can be updated independently. However, if one or more variables remain unobserved in certain trials, the parameters become dependent and updating becomes difficult.

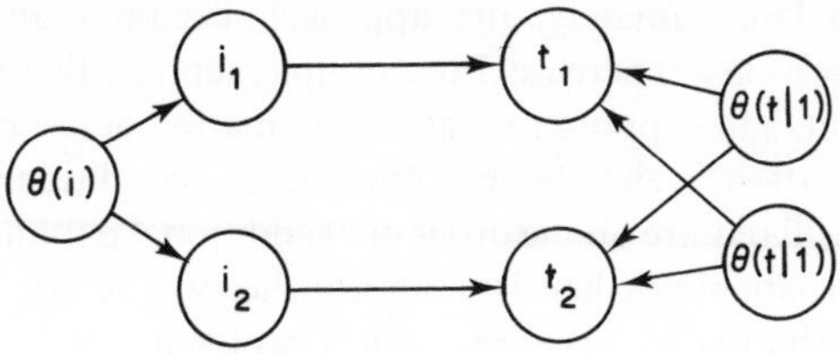

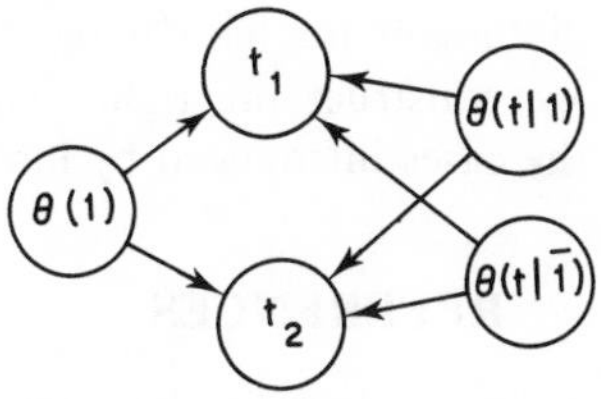

To understand this point, consider a situation that is slightly more complicated than the one presented in the chapter. Suppose the outcome of the test depends on some intermediate variable, called i, where i can be absent ($i = 0$) or present ($i = 1$). For simplicity, consider only those patients with the disease D. An influence diagram for two patients is shown in Figure 15.6. The nodes subscripted with 1 and 2 represent the observable events for the first and second patient respectively. The nodes labeled $\theta(i)$, $\theta(t\,|\,i)$, and $\theta(t\,|\,\bar{i})$ represent the sampling parameters. If the intermediate variable for each of the two patients is observed, the sampling parameters can be updated independently. However, if the intermediate states remain unobserved, then the influence diagram reduces to the one shown in Figure 15.7. The parametes are no longer independent and updating becomes difficult when many patients are considered.

An approach to circumvent this difficulty is suggested by Ross Shachter in Chapter 14 of the volume. Suppose an additional parameter $\theta(t)$ is introduced, where θ is a deterministic function of the original three parameters:

$$\theta(t) = \theta(t\,|\,i)\theta(i) + \theta(t\,|\,\bar{i})(1 - \theta(i)).$$

Conditioned on this new parameter, observations for t, where i remainsunobserved, are independent. With the introduction of $\theta(t)$, inference becomes straightforward. When i is observed, either $\theta(t\,|\,i)$ or $\theta(t\,|\,\bar{i})$ is updated. When i is unobserved, $\theta(t)$ is updated. The probabilities of interest for a new patient can be computed from the parameters using a Monte Carlo approach, discretization, or other approximate approaches.

This method works well when there are a small number of observationclasses. (In the example above, there are three classes corresponding to i absent, i present, and i unobserved.) Unfortunately, the approach becomes intractable as the number of observation classes ncrease. For example, suppose there are tests $t_1 \cdots t_n$, for which i is relevant, and suppose that any combination of test outcomes can be observed. In this situation, even if the test outcomes are conditionally independent given i, the method will require on the order of 2^n additional parameters. Therefore, it appears that approximate or heuristic approaches will be necessary to extend the analysis in this chapter to large, real-world problems.

15.4.2 Reply

I would like to thank the discussant for his comments. Also, I would like to say in a reply that, when we construct the right diagram for censored data problems, the solutions for the cases introduced by him could be obtained.

REFERENCES

Barlow, R. E. and Pereira, C. A. de B. (1987) *The Bayesian Operation and Probabilistic Influence Diagrams.* Berkeley, Department of Industrial Engineering and Operations Research, University of California, 50pp. (TR-ESRC 87-7)

Basu, D. and Pereira, C. A. de B. (1981) On Bayesian analysis of categorical survey data, *Bulletin of the International Statistical Institute*, Contributed Papers, **49**(2), 187–90.

Basu, D. and Pereira, C. A. de B. (1982) On the Bayesian analysis of categorical data: the problem of nonresponse, *Journal of Statistical Planning and Inference*, **6**(4), 345–62.

Lindley, D. V. and Novick, M. R. (1981) The role of exchangeability in inference, *Ann. Statist.*, **9**, 45–58.

Pereira, C. A. de B. and Pericchi, L. R. (1990) Analysis of diagnostability, *Applied Statistics*, **39** (1) (to appear).

Shachter, R. and Heckerman, D. (1987) Thinking backwards for knowledge acquisition, *AI Magazine*, **8** (Fall), 55–61.

PART V

Efficiency and Computational Issues

CHAPTER 16

Fast Algorithms for Probabilistic Reasoning in Influence Diagrams, with Applications in Genetics and Expert Systems

David J. Spiegelhalter, *MRC Biostatistics Unit, UK*

ABSTRACT

The manipulation of probabilities on influence diagrams has received attention from a variety of disciplines, but the relevant work in genetics appears to have gone unnoticed outside that community. We briefly consider the kind of problems considered in pedigree analysis, and use a simple example in genetic counselling to illustrate the relationship to issues in expert systems and other applications of influence diagrams. This example is analysed using the approach of Lauritzen and Spiegelhalter (1988), which provides a straightforward mechanism for carrying out many calculations of interest. Extensions to handling imprecise probabilities are briefly discussed.

16.1 INTRODUCTION

There has been a remarkable growth of interest in probabilistic inference on variables that have a dependency structure that can be represented by an acyclic directed graph. We shall term these 'probabilistic influence diagrams' (Shachter, 1986), although they have also been studied under the name of 'belief networks' (Pearl, 1986) and 'causal networks' (Lauritzen and Spiegelhalter, 1988). Research programmes in decision analysis, expert systems and statistical modelling are beginning to cross-fertilize in an invigorating manner, and the computing tools now becoming available promise attractive and efficient programming environments.

This chapter has two purposes. The first is to use a simple but challenging example to demonstrate the proposals of Lauritzen and Spiegelhalter (1988) (subsequently abbreviated to L–S). The second objective is to note the important relationship to problems of pedigree analysis, and to bring past work in that

Influence Diagrams, Belief Nets and Decision Analysis
Edited by R. M. Oliver and J. Q. Smith

area to the attention of current researchers. In Section 16.2, we briefly describe relevant issues of interest in heredity studies and, by means of a simple example in genetic counselling, we translate the specialized terminology and graphics into more generally familiar expressions.

In Section 16.3 we consider how different approaches to influence diagrams might tackle the problem, and emphasize the crucial work of Cannings, *et al.* (1978). Section 16.4 works through a solution using the L–S methodology, which—while not claimed to be the most efficient means for each specific calculation—provides a straightforward and general structure for computing many items of interest, and emphasizes the computation of a complete joint distribution conditional on observed evidence, rather than concentration on specific variables. The relationship of our work to that of Cannings *et al.* (1978) is shown.

The solution to the example is given in algebraic form, to illustrate the possibility of both likelihood and Bayesian analysis of imprecisely specified parameters. An extension in which probabilities are considered random variables is briefly discussed, both in the context of genetics and in general. Finally, the issue of criticism of graphical structure is related to inference on unknown genealogical relationships.

16.2 ISSUES IN GENETICS

In trying to understand why relevant work in genetics has not been properly related to computational problems in influence diagrams, we conclude that a major reason must be because the terminology is generally obscure to other medical statisticians, let alone those working in decision analysis and artificial intelligence. Thompson (1986) acknowledges this, and provides an admirable introduction to the issues in 'family studies', particularly in relation to inheritance of traits on extended genealogies. Below we provide the briefest of summaries in order to explain the broad issues, and apologize in advance for considering only the simplest situation, and refer to Thompson (1985, 1986) for fuller introductions.

The genetic model

Suppose we are interested in a particular individual characteristic or trait, such as blood type or the possibility of a congenital immunological deficiency. The list of states the trait can take on are the phenotypes, and the observed phenotype of an individual is governed, possibly probabilistically, by the identity of the relevant pair of genes, or genotype. Each genotype is an unordered pair of alleles, where an allele is a specific type of gene; one is a copy of a randomly chosen member of the father's relevant pair of genes, and one is derived from the mother.

In the situation considered here, a trait is dichotomous so an individual has one of two phenotypes; 'affected' and 'normal'. Only two possible alleles are considered, denoted a_1 and a_2, and hence the possible genotypes are a_1a_1, a_1a_2 and a_2a_2. If individuals carrying a_2 tend to be affected, we say a_2 is the allele for the trait; if only individuals with a_2a_2 are affected, we say the trait is '*recessive*' and a_1a_2 individuals are '*carriers*'. If a_1a_2 and a_2a_2 individuals show a similar tendency to be affected, the trait is *dominant*.

The probability model

Three probabilistic components need to be considered when analysing phenotypic data on genealogies. (If genotypes are directly measurable, the situation is greatly simplified.) First, the *penetrance probabilities* relate the underlying genotype to the observable phenotype, in specifying

$$\text{probability (phenotype} \mid \text{genotype, other characteristics)},$$

where a variety of background variables may affect the risk of displaying a condition. Here we shall consider the simple situation in which we need only to specify prob ('affected' | genotype); if we assume a recessive trait carried on a_2, then this probability is logically specified in that p ('affected' $|a_1a_1) =$ p ('affected' $|a_1a_2) = 0$, p ('affected' $|a_2a_2) = 1$.

The second probability specification concerns the *transmission* of genotype from parents to children, and the most basic model is Mendelian segregation, in which each parent independently contributes a random allele chosen with probability $\frac{1}{2}$. Thus, for example, a child of two carriers, i.e. both a_1a_2, will have probability $\frac{1}{4}$ of inheriting two a_1 alleless, probability $\frac{1}{4}$ of inheriting two a_2 alleles, and probability $\frac{1}{2}$ of being a carrier. There are a variety of ways in which these transmission probabilities can be elaborated.

The final component of the probability model concerns the genotype distribution on individuals whose parents are not contained in the genealogy; these are known as *founders*. The simplest assumption is that the genotypes of founders are independent and each has the distribution expected if the alleles are randomly allocated according to their population frequency $p(a_1) = p = 1 - p(a_2)$. In this situation of Hardy–Weinberg equilibrium the genotype distribution is $p(a_1a_1) = p^2$, $p(a_1a_2) = 2pq$, $p(a_2a_2) = q^2$. Again various perturbations may be made to this assumption.

The issues

The models outlined above may be extended to involve quantitative genotypes and phenotypes, dependency between founders, interaction between inheritance of pairs of traits and so on. However, there are a number of issues of interest that are independent of the complexity of the modelling, and which all depend on efficient computation of joint and conditional probabilities based on observed phenotype data on some or all members of a genealogy.

First, it is important to calculate the overall likelihood of the data, that is, the probability of obtaining the observed data conditional on the assumed genealogical structure and parameter values that determine the probabilistic model. If the structure is open to question due to uncertainty concerning the relationship between individuals, the likelihood may be used to contrast alternative structures. On the other hand, with fixed structure, the likelihood for different parameter values may be calculated and used for parametric inference, either directly or through combination with a prior distribution.

Second, for fixed structure and parameter values, we may wish to make statements about individuals or groups, conditional on known phenotypic data. One possibility is to provide a predictive distribution over the genotype or phenotype of a future offspring of a couple; this is required in both human genetic counselling and in selective animal breeding. The 'opposite' objective is to use data on descendants in order to draw inferences on the genotypes of founders, for example to identify the individual or individuals likely to have introduced a particular allele into a genetically isolated community such as the inhabitants of Tristan da Cunha (Thompson, 1978) or the Samaritan community in Israel (Thomas, 1986b).

Finally, we may hypothesize that founders in extensive pedigrees had a specific allele, and then the calculated probability that it is present in no descendants is the *extinction* probability.

It should be clear to those working in other applications of probabilities on directed graphs that many of the issues in genetics have parallels in their own area. For example, in expert systems there is a similar objective in drawing inferences on unobserved variables in a network given data on peripheral nodes (Pearl, 1986; Lauritzen and Spiegelhalter, 1988), and these calculations are made difficult with the existence of loops (corresponding, for example, to inbreeding in a pedigree). In medical expert systems, we wish to infer the probable nature of underlying disease nodes (the identity of founders), while predicting the consequences of carrying out additional diagnostic tests (genetic counselling). We wish to hypothesize different possible disease states and view the consequences (calculation of extinction probabilities) and we wish to learn about probabilistic relationships (parameter inference) or criticize the graphical structure itself (inference on genealogical relationships). To illustrate this close relationship, we now introduce a simple example, and describe alternative means of analysis.

16.3 AN EXAMPLE OF INBREEDING

Charles and Florence are about to reproduce. However, Florence is Charles's niece, and Florence's nephew John has a disease inherited as a recessive trait.

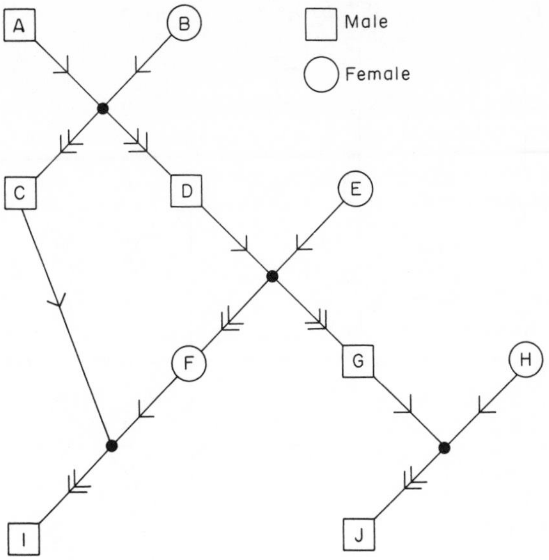

Figure 16.1 Marriage graph for genealogy example: > = marriage; ≫ = offspring.

Neither Charles nor Florence, their parents nor John's parents have the disease. What is the chance that their offspring will (a) be a carrier, (b) have the disease? (If this example appears rather contrived, remember that it is not specified whether the individuals are humans, racehorses or prize pigs.)

The structure of this example was analysed by Cannings and Thompson (1981, p. 135) as an example of their 'peeling' procedure, and Figure 16.1 shows the immediately relevant pedigree drawn as a 'marriage graph'. Here each unfilled node represents an individual, and a married pair have single arrows pointing at a marriage node. Offspring are then indicated as double arrows. If we temporarily ignore the information concerning the presence or absence of the disease, then conditional independence statements may be read off the graph representing the pedigree, in exactly the same manner as from other applications of probabilities on directed graphs. It is convenient, however, to transform the general genetic structure implicit in Figure 16.1 into a more common representation, in which only one type of arrow exists and each random quantity corresponds to a node. The nodes in Figure 16.2 marked A to J are random variables corresponding to the underlying genotypes, each of which can take on values a_1a_1, a_1a_2 or a_2a_2, while the nodes A' to J' are the phenotypes with possible values n ('normal') or a ('affected'). This now takes the form of a standard probabilistic influence diagram, and corresponds to our assumption that the joint probability distribution over the set of nodes $V = \{A, \ldots, J, A', \ldots, J'\}$ may be written

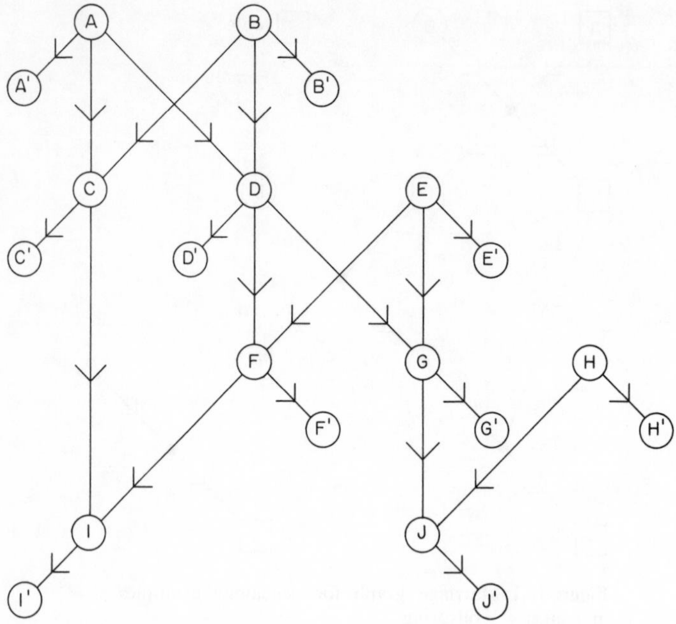

Figure 16.2 Influence diagram representation of marriage graph: (A) signifies genotype; (A′) signifies phenotype.

$$p(V) = \prod_{v \in V} p(v \mid \Pi_v), \tag{16.1}$$

where $w \in \Pi_v$ is a 'parent' of node v in the general sense of a node with a directed link $w \to v$. Assumption (16.1) and Figure 16.2 together correspond to the standard conditional independence assumptions concerning Mendelian inheritance and genotype to phenotype penetration, and means that for this particular pedigree we have

$$\begin{aligned} p(A, \ldots, J, A' \ldots, J') = {} & p(A)p(B)p(C \mid A, B)p(D \mid A, B)p(E)p(F \mid D, E) \\ & \cdot p(G \mid D, E)p(H)p(I \mid C, F)p(J \mid G, H) \\ & \cdot p(A^1 \mid A)p(B^1 \mid B)p(C^1 \mid C)p(D^1 \mid D)p(E^1 \mid E) \\ & \cdot p(F^1 \mid F)p(G^1 \mid G)p(H^1 \mid H)p(I^1 \mid I)p(J^1 \mid J). \end{aligned} \tag{16.2}$$

If we assume the founders are in Hardy–Weinberg equilibrium and the population frequency of allele a_1 is p, the genotype transmission probabilities follow simple Mendelian segregation, and the trait is simple recessive carried on allele a_2, we obtain the conditional probabilities in Table 16.1 necessary to specify the joint distribution (16.2). (We should emphasize that considerably more complex parametrizations may be more realistic and equally able to be

Table 16.1 Conditional probability tables for founder genotypes, offspring genotypes and all phenotypes, sufficient to generate joint probability distribution over all variables.

Founder nodes *A, B, E, H*

	Genotype of founder (e.g. A)		
	a_1a_1	a_1a_2	a_2a_2
$p(A) =$	p^2	$2pq$	q^2

Offspring nodes *C, D, F, G, I, J*

Genotypes of parents (e.g. D, E)		Genotype of offspring (e.g. G)		
		a_1a_1	a_1a_2	a_2a_2
$a_1a_1, \quad a_1a_1$	$p(G\|D,E) =$	1	0	0
$a_1a_1, \quad a_1a_2$		$\frac{1}{2}$	$\frac{1}{2}$	0
$a_1a_1, \quad a_2a_2$		0	1	0
$a_1a_2, \quad a_2a_2$		$\frac{1}{4}$	$\frac{1}{2}$	$\frac{1}{4}$
$a_1a_2, \quad a_2a_2$		0	$\frac{1}{2}$	$\frac{1}{2}$
$a_2a_2, \quad a_2a_2$		0	0	1

Phenotype nodes $A^1, \ldots, J^1$

Genotypes of individual (e.g. I)		Phenotype of individual (e.g. I^1)	
		n (normal)	a (affected)
a_1a_1	$p(I^1\|I) =$	1	0
a_1a_2		1	0
a_2a_2		0	1

handled by the techniques to be discussed.) Our objective in this problem is to calculate, in terms of p, the genotype and phenotype distribution of the future offspring, whom we shall call Ivan, based on knowing John is the only affected individual in the pedigree, i.e.

$$p(I \mid A' = B' = \cdots = H' = n, J' = a) \tag{16.3}$$

from which the probability that $I' = a$ is trivially obtained. For this particular example, we can immediately make some logical deductions. For example, we know $J = a_2a_2$ and $A, \ldots, H \neq a_2a_2$ since the disease is present if and only if the genotype is a_2a_2. Furthermore, since both of John's parents (George and

Hilda) must have the allele a_2, they must both be carriers, i.e. $G = H = a_1a_2$. Thus at least one of George's parents must be a carrier.

We note that expression (16.3) could be evaluated by brute force by direct enumeration and summation of relevant expressions of the full joint distribution (16.2); this is clearly chronically inefficient. The problem of finding efficient algorithms is identical to that considered by Shachter (1986), Pearl (1986), Lauritzen and Spiegelhalter (1988) and many others in the manipulation of influence diagrams, and we now consider how they might approach this example and more complex pedigrees.

Shachter's (1986) direct approach would be to marginalize out each of the unobserved nodes $A, \ldots, H, J$ in turn. This would require a series of 'arc-reversals', in which systematic use of Bayes' theorem on the joint distribution (16.2) sequentially made each of the nodes to be eliminated not a 'parent' of any other node, and hence could be trivially marginalized. For example, elimination of J requires reversing the arc $J \to J'$, which means adding new arcs $G \to J'$, $H \to J'$ and calculating the probability $p(J' \mid G, H)$. Eventually, expressions for $p(A', \ldots, H', J' \mid I)$ and $p(I)$ are obtained from which the final result may be calculated.

Alternatively, the 'conditioning' method of Pearl (1986) would disconnect the graph by conditioning on each possible value of the genotype D, which then makes the graph into a tree which can be handled using the Kim and Pearl (1983) algorithm. The resulting probability distributions on I would then be averaged with respect to the calculated conditional probability of D given the phenotypic information. We note, however, that it is not necessary to have inbreeding to stop the influence diagram being singly connected, and simply having more than one offspring makes Kim and Pearl (1983) invalid.

Before going through the L–S procedure in detail, we briefly consider the peeling method of Cannings *et al.* (1978), which is also discussed in Thompson *et al.* (1978), Cannings and Thompson (1981) and Thompson (1985). This exploits the conditional independence properties expressed in the graph in order to successively reduce ('peel') the pedigree down to the individuals of interest; at any stage in the peeling, attention focuses on a 'cutset' of individuals that would divide the marriage graph into two disjoint components: by the conditional independence property of directed graphs, conditional on the genotypes of the cutset individuals, the two parts of the pedigree are independent. It is easiest to quote from Thomas (1986a) for a summary of the technique;

> Each cutset will divide the pedigree into two parts, those marriages and individuals whose information has been fully incorporated—the 'peeled set'—and those whose information has not—the 'unpeeled set'. For each cutset we have a probability function which contains the information in the peeled set. This is called an 'R function' and is the probability of all the genetic and genealogical information in the peeled set conditional upon, or jointly with, the genotypes of the individuals in the cutset. The R function is conditional upon the genotypes of an individual if his parents' marriage is in

the unpeeled set, and joint with the genotypes of an individual if his parents' marriage is in the peeled set. The heart of the peeling method is a relationship between R functions on successive cutsets made possible by the property of conditional independence.

The references given above describe the recurrence relations in full.

In our simple example, the cutsets are $[G]$, $[D, F]$, $[C, F]$ and $[I]$ as shown in Cannings and Thompson (1981), although intermediate functions also have to be evaluated on $[C, D]$ and $[D, E]$.

16.4 A SOLUTION USING CLIQUE-TREES

Lauritzen and Spiegelhalter (1988) (L–S) describe in detail a methodology that is capable of providing a solution to many of the computational problems faced in probabilistic influence diagrams, and in particular those appropriate to pedigree analysis. However, as was emphasized in the discussion to L–S, particular diagrams may be structured in such a way as to make the L–S methodology somewhat inefficient, and the breadth of applicability remains to be explored. Nevertheless, it is informative to use our simple genetic counselling example to illustrate the essentials of the technique and draw some comparisons with the peeling method.

We first consider the general issue of manipulating the influence diagram of Figure 16.2, ignoring for the moment our specific numerical conditional probability assumptions and the observed phenotype data, and assuming there are many problems we wish to explore on this pedigree. The peeling method showed that it is vital to identify the sets of variables on which functions will have to be calculated as intermediate steps in obtaining any desired conclusion, and the L–S method emphasizes an initial restructuring of the diagram into a form that explicitly connects such groups of variables in a new undirected representation. (As we shall later emphasize, if only one calculation is required on a structure, this restructuring may be made more efficient.) Such general functions we call 'evidence potentials', and the L–S procedure is essentially concerned with systematic re-representation of the probability distribution in terms of different potential functions in order to allow efficient calculation of quantities of interest. In notation, we have that if ψ_A are non-negative functions on subsets $A \subseteq V$, then the joint distribution p may be expressed as

$$p = Z^{-1} \prod_A \psi_A, \tag{16.4}$$

where Z is a normalization constant. The ψ functions are certainly not unique, and we exploit this fact.

We see immediately from the joint distribution (16.1) that at a minimum we shall require ψ functions on each 'child' and its 'parents' in the influence diagram sense, and Figure 16.3 shows the graph arising from Figure 16.2 by joining

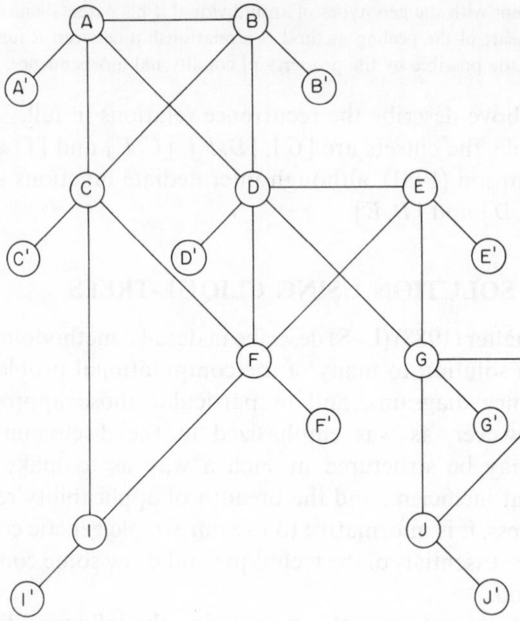

Figure 16.3 'Moral' graph formed by joining 'parents' and dropping directions.

'parents' and dropping directions, indicating that we are basing our computations on functions on unordered variables. We call this the 'moral' graph since 'parents' are joined, (which in genetics examples is literally true). Expression (16.1) is thus a potential representation (with $Z = 1$) on the moral graph. Comparison of Figure 16.3 with equation (16.2) shows that our joint distribution on V is expressed as functions of variables that form the 'cliques' (maximal sets of nodes that are all joined to each other) of the moral graph. (We therefore have a Markov field on Figure 16.3 although, since we allow zero probabilities, simple interpretation in terms of conditional independence properties requires care.)

In the peeling operation, a function on nodes C and D was required, which are not joined in Figure 16.3 and hence, under the rule relating cliques of the graph with stored functions, are not supposed to have a potential defined on them. In fact, it can be shown that for node probabilities to be calculable under all possible conditional events, using only potentials on cliques, then the undirected graph must be triangulated in the sense that no cycles of length 4 or more may exist without a chord or 'shortcut'. In Figure 16.3, the nodes A, C, F, D and B, C, F, D form two such cycles, both of which are 'short-circuited' by the edge CD.

This triangulation could have been carried out automatically using an algorithm such as 'maximum cardinality search' (Tarjan and Yannakakis, 1984): any node is selected (say A) and given a label 1; the nodes are then labelled in sequence, the next node to be chosen being the one joined to the maximum number of already labelled nodes, breaking ties at random. If lower-labelled neighbours of a node are ever unjoined, they are joined to form a 'fill-in'. In this instance, the full moral graph shown in Figure 16.4 could result, with CD joined when F is labelled as 5, since at that point C and D were F's lower-labelled neighbours and were unjoined. There is no guarantee, however, that such automatic schemes form cliques that have the minimum total number of states, and it is this factor that is crucial for computational efficiency.

The cliques of Figure 16.4 may be considered as the basic units for all future computations on the pedigree, and one of the consequences of a triangulated structure is that the joint probability (16.2) may be specified as a simple function of the marginal distributions on these cliques; we are essentially embedding our influence diagram within a decomposable model (Darroch *et al.*, 1980). Before showing how these may be derived, we first note a crucial property concerning the cliques of a triangulated graph; this *running intersection property* states that there exists an ordering of the cliques $C_1, C_2 \ldots$, such that for all $i > 1$, $C_i \cap (C_1 \cup \cdots \cup C_{i-1}) \subseteq C_k$ for some $k < i$. If we carry out maximum cardinality

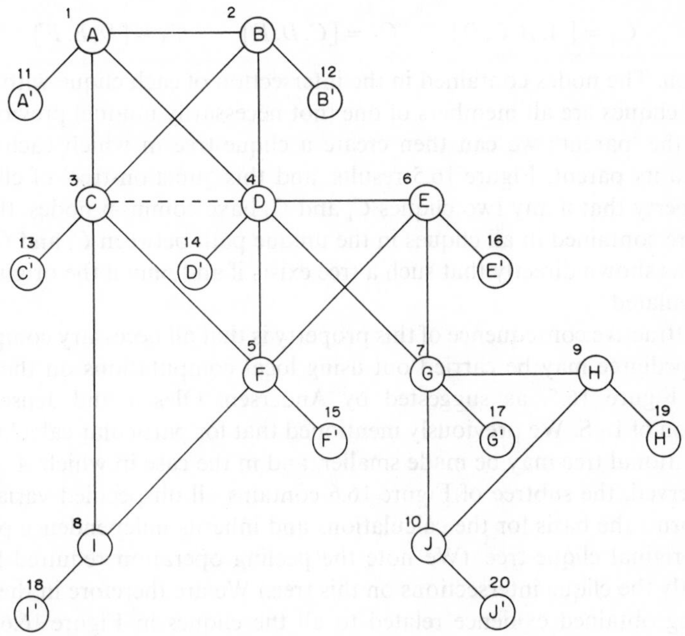

Figure 16.4 'Full' (triangulated) moral graph showing node ordering obtained by maximum cardinality search.

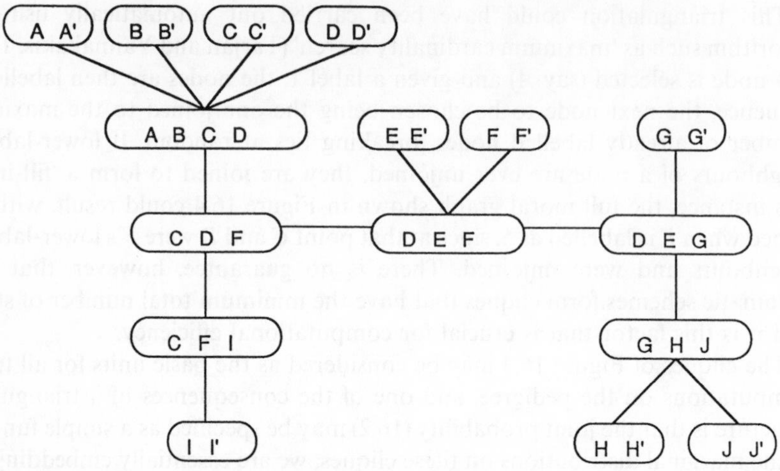

Figure 16.5 Junction tree formed by cliques of full moral graph: nodes common to any two cliques X and Y are contained in all cliques on unique path between X and Y.

search on the triangulated graph, we may consider the ordering created by the maximum label of the constituent nodes in each clique:

$$C_1 = [A, B, C, D] \qquad C_2 = [C, D, F] \qquad C_3 = [D, E, F]$$

and so on. The nodes contained in the intersection of each clique with all lower labelled cliques are all members of one (not necessarily unique) previous clique termed the 'parent'; we can then create a clique-tree in which each clique is joined to its parent. Figure 16.5 results, and this 'junction-tree' of cliques has the property that if any two cliques C_i and C_j have common nodes, then those nodes are contained in all cliques in the unique path between C_i and C_j. Jensen (1988) has shown directly that such a tree exists if and only if the original graph is triangulated.

The attractive consequence of this property is that all necessary computations on the pedigree may be carried out using local computations on the junction tree in Figure 16.5, as suggested by Andersen, Olesen and Jensen in the discussion of L–S. We previously mentioned that for particular calculations the computational tree may be made smaller, and in the case in which $A', \ldots, H', J'$ are observed, the subtree of Figure 16.6 contains all unspecified variables and hence forms the basis for the calculations and inherits independence properties of the original clique-tree. (We note the peeling operation required functions on exactly the clique intersections on this tree.) We are therefore in the position of having obtained evidence related to all the cliques in Figure 16.6, and we wish to collect this evidence onto $[I, I']$ in order to find the probability distribution on this clique. In addition to essentially repeating the peeling operation, we shall show how we may trivially find the marginal distributions (conditional on the observed phenotypes) on all the cliques and hence all the unknown genotypes.

We first order the cliques of Figure 16.6, starting at the one of interest, and arbitrarily labelling when branches occur in the tree. The clique ordering is shown in Figure 16.6 and Table 16.2, in which for each C_i the 'parent' clique C_k, $k < i$, in the tree is identified as the unique lower-labelled neighbour, and the separator defined as $S_i = C_i \cap C_k$, while the residual is $R_i = C_i \backslash S_i$. (We note that any separator set S_i is a cutset of the graph, and conditional on S_i the two components are independent; we refer to this as the 'Markov property of the clique-tree'.) The objective of the first part of the procedure is to work in reverse order down the cliques in order to find a new representation ψ^* for the distribution p^* conditional on the phenotype data; we call this a 'set-chain'

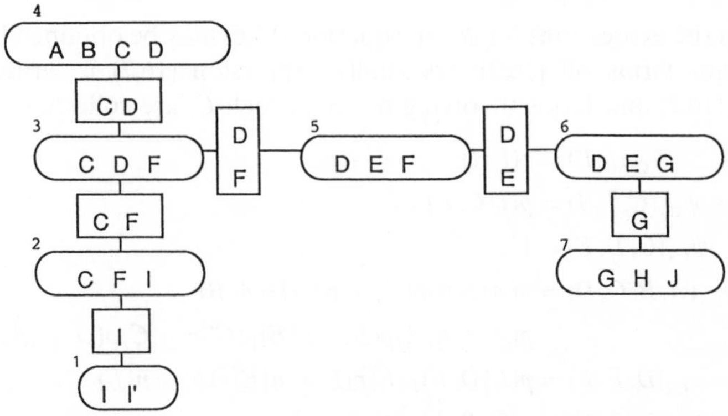

Figure 16.6 Junction tree after having observed phenotypes A′ to H′, J′, showing separator sets through which evidence is passed.

Table 16.2 Clique ordering from Figure 16.6, showing nodes S_i intersecting with parent, and residual nodes $R_i = C_i \backslash S_i$.

	Clique members	Residual R_i	Separator S_i	Parent clique label
C_1	$I\ I^1$	$I\ I^1$	$\varnothing$	—
C_2	$C\ F\ I$	$C\ F$	I	1
C_3	$C\ D\ F$	D	$C\ F$	2
C_4	$A\ B\ C\ D$	$A\ B$	$C\ D$	3
C_5	$D\ E\ F$	E	$D\ F$	3
C_6	$D\ E\ G$	G	$D\ E$	5
C_7	$G\ H\ J$	$H\ J$	G	6

representation:

$$p^* = \prod_i \psi^*_{C_i} = \prod_i p^*(R_i|S_i), \tag{16.5}$$

where $p^*(R_1|S_1) = p^*(I, I')$ is our objective (this is a correct representation because of the Markov property on the clique-tree shown in Figure 16.6).

To initialize the procedure, we need to find a potential representation

$$p^* = Z^{-1} \prod_i \psi_{C_i} \tag{16.6}$$

based on the conditional probability tables given in Table 16.1. To get this we note that

$$\begin{aligned} p^* &= p(A, \ldots, J, I' | A' = \cdots = H' = n, J' = a) \\ &\propto p(A, \ldots, J,\ A' = \cdots = H' = n,\ I',\ J' = a) \end{aligned} \tag{16.7}$$

and hence the expressions for ψ_{C_i} in equation (16.6) may be obtained by reading the relevant terms off (16.2); essentially expression (16.7) is written out as equation (16.2) and terms involving nodes in each C_i are collected:

$$\begin{aligned} \psi_{C_1}(I, I') &= p(I'|I) \\ \psi_{C_2}(C, F, I) &= p(I|C, F) \\ \psi_{C_3}(C, D, F) &= 1 \\ \psi_{C_4}(A, B, C, D) &= p(A)p(B)p(C|A, B)p(D|A, B) \\ &\quad\ p(A' = n|A)p(B' = n|B)p(C' = n|C)p(D' = n|D) \\ \psi_{C_5}(D, E, F) &= p(F|D, E)p(E)p(E' = n|E)p(F' = n|F) \\ \psi_{C_6}(D, E, G) &= p(G|D, E)p(G' = n|G) \\ \psi_{C_7}(G, H, J) &= p(J|G, H)p(H)p(J' = a|J)p(H' = n|H). \end{aligned}$$

We note there is a degree of arbitrariness to which clique some terms are assigned, for example $p(C' = n|C)$, and we have used the relevant parent in Figure 16.5. This leaves C_3 with no relevant terms and the potentials are all set, arbitrarily, to 1.

Table 16.3 shows this initial potential representation on the seven cliques, in which only rows with non-zero entries have been listed; potentials may be zero if the particular phenotype–genotype combination is ruled out, for example $\psi_{C_1}(I = a_2a_2,\ I' = n) = p(I' = n|I = a_2a_2) = 0$, or because the parent–offspring genotype combination is impossible, for example

$$\psi_{C_2}(C = a_1a_1, F = a_1a_2, I = a_1a_2) = p(I = a_1a_2|C = a_1a_1, F = a_1a_1) = 0.$$

An individual entry such as $\psi_{C_4}(a_1a_1, a_1a_2, a_1a_2, a_1a_1)$ is calculated as

$$\begin{aligned} &p(A = a_1a_1) \quad p(B = a_1a_2) \quad p(C = a_1a_2|A = a_1a_1, B = a_1a_2) \\ &p(D = a_1a_1|A = a_1a_1, B = a_1a_2) \end{aligned}$$

$$p(A'=n|A=a_1a_1) \quad p(B'=n|B=a_1a_2) \quad p(C'=n|C=a_1a_2)$$
$$p(D'=n|D=a_1a_1)=p^2\cdot 2pq\cdot\tfrac{1}{2}\cdot\tfrac{1}{2}\cdot 1\cdot 1\cdot 1\cdot 1=p^3q/2.$$

The steps in obtaining the representation (16.5) involve removing cliques one at a time, analagous to the peeling method. As the first step, we have that

$$\begin{aligned} p^*(H,J|G) &= p^*(R_7|S_7) \\ &= p^*(R_7|C_1\cdots C_6) \quad \text{by the Markov property of the clique-tree} \\ &= p^*(C_1\cdots C_7)\Big/\sum_{R_7} p^*(C_1\cdots C_7) \quad \text{by conditional probability} \\ &= \prod_{i=1}^{7}\psi_{C_i}\Big/\sum_{R_7}\prod_{i=1}^{7}\psi_{C_i} \quad \text{equation by (16.6)} \\ &= \psi_{C_7}/\phi_{S_7} \quad \text{by cancelling terms,} \end{aligned}$$

where $\phi_{S_7}=\sum_{R_7}\psi_{C_7}$. Since we therefore have that

$$p^*(C_1\cdots C_6)=p^*(C_1\cdots C_7)\phi_{S_7}/\psi_{C_7}=\prod_{i=1}^{5}\psi_{C_i}\cdot\psi_{C_6}\cdot\phi_{S_7}$$

we only need to multiply the relevant terms of C_7's parent (C_6) by ϕ_{S_7} to obtain a potential representation for $C_1\cdots C_6$. In our particular example the above calculations are rather trivial, since $\phi_{S_7}=\phi_G=(0,pq/2,0)$, and hence $p^*(R_7|S_7)=p^*(H,J|G)=1$ if $H=a_1a_2$, $J=a_2a_2$, $G=a_1a_2$; 0 otherwise. The potentials ψ_{C_6} containing $G=a_1a_2$ are thus multiplied by $pq/2$, and all others set to zero. Thus the information that G must be a carrier is propagated back up the tree.

The third column of Table 16.3 shows the conditional probability tables $p^*(R_i|S_i)$ thus obtained. We note that the functions ϕ_{S_4} and ϕ_{S_5} are multiplied together and passed to C_3. Eventually the function $\phi_{S_2}=\phi_I$ is passed to C_1 and we can think of a function $\phi_{S_1}=\Sigma\phi_{C_1}$ being passed to the empty set $\varnothing$; this function has value $p^4(1-p)^2(3p+5)(3-p)/16$.

We have now obtained our immediate objective: the probability that Ivan will be affected is

$$p^*(I'=a)=p^*(I=a_2a_2)=(-p^2-p+3)/\{2(3p+5)(3-p)\}.$$

But in getting this far we have derived a number of useful functions, and in particular we note that the conditional probability tables $p^*(R_i|S_i)$ make it trivial to chain back through the tree to obtain the clique marginals. Specifically, from the marginal on C_1 we obtain the marginal on S_2, and hence calculate $p^*(C_2)=p^*(R_2|S_2)p^*(S_2)$, and repeating this process for each clique we obtain the clique marginals in column 4 of Table 16.3.

The value for ϕ_{S_1} mentioned above is also very important. It has been obtained

Table 16.3 Functions stored in processing the example.

Clique	Configuration ($1=a_1a_1$, $2=a_1a_2$, $3=a_2a_2$)	Potentials from conditional probability tables	Potentials as set-chain probabilities	Clique marginals $\{f(p)=2(3p+5)(3-p)\}$
C_1	I I^1	$\psi_{C_1}(I, I^1)$	$p^*(I, I^1)$	$p^*(I, I^1)$
	1 1	1	$(-p^2+8p+14)/\{2(3p+5)(3-p)\}$	$(-p^2+8p+14)/f(p)$
	2 1	1	$(-4p^2+p+13)/\{2(3p+5)(3-p)\}$	$(-4p^2+p+13)/f(p)$
	3 2	1	$(-p^2-p+3)/\{2(3p+5)(3-p)\}$	$(-p^2-p+3)/f(P)$
	C F I	$\psi_{C_2}(C, F, I)$	$p^*(C, F \mid I)$	$p^*(C, F \mid I)\ p^*(I)$
	1 1 1	1	$2(1+p)(2+p)/(-p^2+8p+14)$	$2(1+p)(2+p)/f(p)$
	1 2 1	$\frac{1}{2}$	$3(1+p)/(-p^2+8p+14)$	$3(1+p)/f(p)$
	1 2 2	$\frac{1}{2}$	$3(1+p)/(-4p^2+p+13)$	$3(1+p)/f(p)$
	2 1 1	$\frac{1}{2}$	$2(2-p^2)/(-p^2+8p+14)$	$2(2-p^2)/f(p)$
	2 1 2	$\frac{1}{2}$	$2(2-p^2)/(-4p^2+p+13)$	$2(2-p^2)/f(p)$
	2 2 1	$\frac{1}{2}$	$(-p^2-p+3)/(-p^2+8p+14)$	$(-p^2-p+3)/f(p)$
	2 2 2	$\frac{1}{2}$	$2(-p^2-p+3)/(-4p^2+p+13)$	$2(-p^2-p+3)/f(p)$
	2 2 3	$\frac{1}{4}$	1	$(-p^2-p+3)/f(p)$
C_3	C D F	$\psi C_3(C, D, F)$	$p^*(D \mid C, F)$	$p^*(D \mid C, F)\ p^*(C, F)$
	1 1 1	1	$(1+p)/(2+p)$	$2(1+p)^2/f(p)$
	1 1 2	1	$(1+p)/3$	$2(1+p)^2/f(p)$
	1 2 1	1	$1/(2+p)$	$2(1+p)/f(p)$
	1 2 2	1	$(2-p)/3$	$2(1+p)(2-p)/f(p)$
	2 1 1	1	$(1-p^2)/(2-p^2)$	$4(1-p^2)/f(p)$
	2 1 2	1	$(1-p^2)/(-p^2-p+3)$	$4(1-p^2)/f(p)$
	2 2 1	1	$1/(2-p^2)$	$4/f(p)$
	2 2 2	1	$(2-p)/(-p^2-p+3)$	$4(2-p)/f(p)$
C_4	A B C D	$\psi_{C_4}(A, B, C, D)$	$p^*(A, B \mid C, D)$	$p^*(A, B \mid C, D)\, p^*(C, D)$
	1 1 1 1	p^4	$4p^2/(1+p)^2$	$16p^2/f(p)$
	1 2 1 1	$p^3q/2$	$2pq/(1+p)^2$	$8pq/f(p)$

	1	2	1	2	$p^3q/2$	$p/(1+p)$	$2p(3-p/f(p)$
	1	2	2	1	$p^3q/2$	$p/(1+p)$	$8pq/f(p)$
	1	2	2	2	$p^3q/2$	$p/2$	$2p(3-p)/f(p)$
	2	1	1	1	$p^3q/2$	$2pq/(1+p)^2$	$8pq/f(p)$
	2	1	1	2	$p^3q/2$	$p/(1+p)$	$2p(3-p)/f(p)$
	2	1	2	1	$p^3q/2$	$p/(1+p)$	$8pq/f(p)$
	2	1	2	2	$p^3q/2$	$p/2$	$2p(3-p)/f(p)$
	2	2	1	1	$p^2q^2/4$	$q^2/(1+p)^2$	$4q^2/f(p)$
	2	2	1	2	$p^2q^2/2$	$q/(1+p)$	$2p(3-p)/f(p)$
	2	2	2	1	$p^2q^2/2$	$q/(1+p)$	$8q^2/f(p)$
	2	2	2	2	p^2q^2	q	$4q(3-p)/f(p)$
C_5		D	E	F	$\psi_{C_4}(D,E,F)$	$p^*(E\|D,F)$	$p^*(E\|D,F)\ p^*(D,F)$
		1	1	1	p^2	0	0
		1	2	1	pq	1	$2(1+p)(3-p)/f(p)$
		1	2	2	pq	1	$2(1+p)(3-p)/f(p)$
		2	1	1	$p^2/2$	p	$2p(3+p)/f(p)$
		2	1	2	$p^2/2$	$p/(2-p)$	$2p(3+p)/f(p)$
		2	2	1	$pq/2$	q	$2q(3+p)/f(p)$
		2	2	2	pq	$2q/(2-p)$	$4q(3+p)/f(p)$
C_6		D	E	G	$\psi_{C_6}(D,E,G)$	$p^*(G\|D,E)$	$p^*(G\|D,E)\ p^*(D,E)$
		1	1	1	1	0	0
		1	2	1	$\frac{1}{2}$	0	0
		1	2	2	$\frac{1}{2}$	1	$4(1+p)(3-p)/f(p)$
		2	1	1	$\frac{1}{2}$	0	0
		2	1	2	$\frac{1}{2}$	1	$4p(3+p)/f(p)$
		2	2	1	$\frac{1}{4}$	0	0
		2	2	2	$\frac{1}{2}$	1	$6q(3+p)/f(p)$
C_7		G	H	J	$\psi_{C_7}(G,H,J)$	$p^*(H,J\|G)$	$p^*(H,J\|G)\ p^*(G)$
		2	2	3	$pq/2$	1	1

by systematically marginalizing over all unobserved nodes in the graph, and hence may be written as

$$\phi_{S_1} = \sum_{C_1 \cup \cdots \cup C_7} \prod \psi_{C_i} \tag{16.8}$$

and hence from equation (16.6) this means that $\phi_{S_1} = Z$, the normalization constant for p^*. But expression (16.7) and the subsequent expressions show that

$$\prod_i \psi_{C_i} = p(C_1 \cdots C_7, A' = \cdots = H' = n, J' = a)$$

and hence equation (16.8) reveals that Z is the probability of all the data observed, $p(A', \ldots, H', J')$. Since all calculations have been conditional on p, we have that $\phi_{S_1} = Z = p^4(1-p^2)(3p+5)(3-p)/16$ is the overall likelihood for the pedigree, of crucial importance for many purposes.

Even for this small example it is of interest to plot the likelihood function as in Figure 16.7. We see that the maximum likelihood estimate for p, the population frequency of the allele a_1, is 0.67. Substituting in the estimate for the marginal probabilities of each of the individuals being a carrier gives the results in Table 16.4.

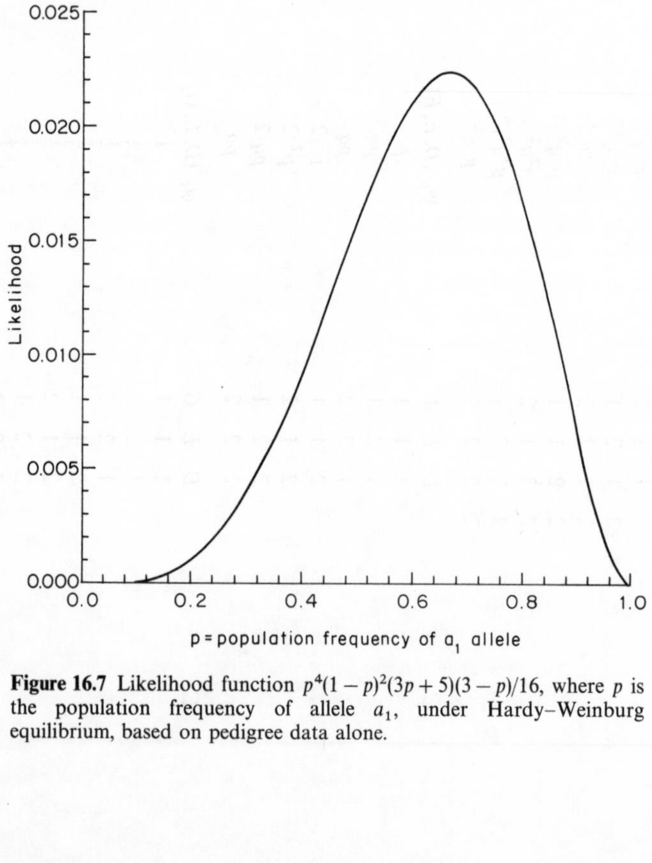

Figure 16.7 Likelihood function $p^4(1-p)^2(3p+5)(3-p)/16$, where p is the population frequency of allele a_1, under Hardy–Weinburg equilibrium, based on pedigree data alone.

Table 16.4 Probability of individuals being carriers of the allele a_2, conditional on all observed phenotype data, where $f(p) = 2(3p+5)(3-p)$, $\hat{p} = 0.67$ is maximum likelihood estimate of p based on this pedigree alone.

Individual genotype	Probability of genotype being a_1a_2 (a carrier): In terms of p	Maximum likeliood estimate
A	$2(-p^2-10p+15)/f(p)$	0.48
B	$2(-p^2-10p+15)/f(p)$	0.48
C	$4(-2p^2-p+5)/f(p)$	0.42
D	$2(-p^2+9)/f(p)$	0.52
E	$2(-5p^2-2p+15)/f(p)$	0.70
F	$2(-2p^2+p+9)/f(p)$	0.54
G	1	1
H	1	1
I	$(-p^2-p+3)/f(p)$	0.06
J	0	0

16.5 COMPARISON WITH THE PEELING APPROACH

By comparing Section 16.4 with the calculations shown by Cannings and Thompson (1981, pp. 135–7), the similarity of the approaches becomes immediately clear. First, the cutsets on which R functions are defined are precisely the clique separators through which we propagate evidence. Second, the R functions themselves are the ϕ functions that we use to pass evidence through the separators. Third, the likelihood of the pedigree is simply our normalizing constant $Z = \phi_{S_1}$.

The interpretation of the R functions given in Section 16.3 can be directly shown to apply to the ϕ functions calculated in the L–S methodology when forming a set-chain representation using original potentials which had normalization function $Z = 1$—which naturally occurs when initializing with conditional probability tables. Let S be a separator set in the junction tree, which breaks the node-set V into $V_u \cup S \cup V_p$, where V_p are the nodes from which evidence has been 'collected' (i.e. members of cliques of higher order in the appropriate set chain, which are the 'peeled' nodes in the genetics terminology), and V_u are the 'unpeeled' nodes, containing the nodes towards whom evidence is being collected. Further, let $S = S_u \cup S_p$, where $v \in S_u$ if $\Pi_v \subseteq V_u \cup S$, $v \in S_p$ otherwise; thus S_p are individuals in S who do not have any parents in V_u. (In pedigree analysis, these are nodes whose parents' marriage must have already been peeled.) Let $V_p = D_p \cup E_p$, where E_p are the nodes which have been observed to take on values E_p^*, and let $V_p \cup S$ comprise cliques $\mathbf{C}_p$.

On initialization, we have that the potentials on $\mathbf{C}_p$ are the product of conditional probability tables of nodes whose parents also lie in $V_p \cup S$, that is,

$$\prod_{C \in \mathbf{C}_p} \psi_C = \prod_{v \in V_p \cup S_p} p(v \mid \Pi_v) = p(V_p \cup S_p \mid S_u). \tag{16.9}$$

The L–S procedure for collecting evidence from E_p is equivalent to marginalizing over D_p in the potentials on $\mathbf{C}_p$, and passing the resulting function of E_p through S, i.e.

$$\begin{aligned}\phi(E_p, S) &= \sum_{D_p} \prod_{C \in \mathbf{C}_p} \psi_C \\ &= p(E_p \cup S_p \mid S_u) \quad \text{by equation (16.9).}\end{aligned}$$

When observed values E_p^* are substituted into this expression, it reveals our ϕ function to be precisely the R function defined in Section 16.3; the likelihood of the data from whom evidence has been collected and separator nodes all of whose parents are in $V_p \cup S$, conditional on the other nodes in the separator set. (It is apparent that the set-chain representation (16.5) also has $Z = 1$, but with S_p empty. Hence ϕ functions calculated on receipt of additional evidence have an even simpler interpretation as $p(E_p \mid S)$.)

One of the main difference in approaches is that we carry out an additional procedure of chaining back through the pedigree to provide marginal distributions on the cliques and hence the individual nodes. Thus the storage requirements are related to the size of the state space of the cliques rather than the clique intersections. Of course, if only probabilities of specific nodes are required, then this additional storage is not mandatory. Perhaps the main difference is that our procedure is designed for evidence that arrives in batches, in which revision of probabilities needs to repeatedly take place.

The comparison with peeling opens up consideration of using some of the approximation methods developed in genetics for more general applications such as in expert systems, both in terms of propagation through large cutsets (Thomas, 1986a) or methods for finding efficient cutsets (Thomas, 1986b).

16.6 COPING WITH INEXACT PROBABILITIES

A major issue to be faced in probabilistic influence diagrams concerns the means of dealing with conditional probabilities that are subject to some uncertainty and hence can themselves be considered as variables. Essentially, our algebraic solution to the sample problem has treated p as an unknown quantity, but in general exact algebraic methods will not be feasible.

In the discussion of L–S, we have extended the argument of Spiegelhalter

(1986) that unknown probabilities may be handled within the influence diagram structure by considering them as additional nodes directed at the relevant 'child'. We suggest that in expert systems a basic method is obtained by assuming the conditional probability tables are marginally independent random variables which form an additional quantitative level, the 'experience', standing over the qualitative 'core' of the system. Before processing a case, the current estimates (expectations) of the conditional probabilities are dropped down to the core, and all observed data processed using the L–S procedure. Finally, the distribution of each conditional probability table is revised independently by Bayes' theorem, and the result held over to the next case.

We should, however, note that this procedure is unlikely to be appropriate in genetics applications, in which a particular set of parameters will apply to many conditional probability specifications in the pedigree, but will often only hold for a single 'case', i.e. a single set of phenotype data on the pedigree. Ideally, algebraic methods may be possible in small applications; thus, in our genetic counselling example in which p is considered a parameter, the algebraic analysis provides the likelihood $p(\text{data}|p)$ and so in combination with a prior for p, a posterior distribution may be plotted: in particular, Figure 16.7 is the posterior distribution corresponding to a uniform prior for p (which would hardly be realistic). Without the algebraic solution, the likelihood could also be obtained by repeating the analysis for a range of values of p. Alternatively, if p is an object of interest, it could be modelled explicitly in the original influence diagram as a 'parent' P of the founders A, B, E and H, but the resulting triangulation produces a clique-tree with P as a member of many cliques, and hence our propagation scheme would be very inefficient.

We may also, of course, consider the transmission or penetrance probabilities as parameters. Again it would seem best to simulate from a range of possibilities, calculate the likelihood curve and combine with a prior as an external procedure—the internal method suggested in L–S is only really suitable when the conditional probabilities are independent quantities of interest (independent in both an informal and a statistical sense), and is not suitable when many conditional probability tables hold common but unknown values.

16.7 CRITICISM OF STRUCTURE

In Section 16.6 we noted that applications of influence diagrams in expert systems and genetics require different means of making inferences on inexact quantities, but in criticism of the qualitative graphical structure they may have considerably more in common. If the structure of the genealogy is itself uncertain due to some relationships being open to doubt, overall likelihoods for different postulated structures may be calculated, possibly maximized with respect to unknown parameters (Thompson, 1979). If a number of sets of phenotypic data are available on the same pedigree, these may all contribute to the likelihood comparison, analagous to a number of cases being used to criticize an expert systems structure. In expert systems, a global comparison for two competing structures H_0 and H_1 could be based on the Bayes factor $p(\text{data}|H_0)/p(\text{data}|H_1)$. The overall probability of each set of data (case) drops out of the L–S algorithm, and if conditional probabilities are completely specified, we may simply multiply the Bayes factors for each individual case to obtain the global comparison. If probabilities are themselves random quantities, then provided they are updated as each case data $\mathbf{x}_i$ is obtained, then after n cases, the overall probability under, say, hypothesis H_0, is

$$p(\mathbf{x}_1, \ldots, \mathbf{x}_n | H_0) = p(\mathbf{x}_n | H_0, \mathbf{x}_1, \ldots, \mathbf{x}_{n-1}) p(\mathbf{x}_{n-1} | H_0, \mathbf{x}_1, \ldots, \mathbf{x}_{n-2}) \cdots$$

which makes the global comparison again simply the product of the Bayes factors calculated for each case. The only difference in the contexts is that in genetics examples the likelihoods are generally maximized with respect to unknown parameters, while in the fully Bayesian expert system applications, the uncertain quantities are integrated out.

Many questions in the area of model choice need to be explored; what is a reasonable criterion to place on the Bayes factors, given the necessary trade-off between complexity and goodness-of-fit, what is a good way to break the global comparison into components to allow local model criticism, how can we best exploit the work of Edwards and Havranek (1985, 1987), Wermuth (1986) and others in graphical model choice? Cross-fertilization from different disciplines promises to provide valuable insights into these issues before too long.

16.8 CONCLUSIONS

This chapter was intended to illuminate a considerable oversight among researchers working on probabilistic influence diagrams, and to indicate the exciting possibilities in involving yet another subject area in collaborative work. The Lauritzen–Spiegelhalter procedure has been described, not as the answer to all demands, but as a straightforward computational scheme suitable for modern object-oriented programming environments. Experience will show whether the problems and structures tackled in genetics require more dedicated approaches.

ACKNOWLEDGEMENTS

I am grateful to Steffen Lauritzen and Phil Dawid for valuable discussions, and Alun Thomas for originally bringing the relevant genetics literature to our attention.

16.9 DISCUSSION

16.9.1 Discussion by Ilan Adler

This chapter describes the similarities between models and algorithms developed in mathematical genetics and influence diagrams. The chapter contains useful examples and lucidly presents the ideas in the genetic models and its application to similar models of influence diagrams.

My only comment is that a paper by Lauritzen and Spiegelhalter (1988) is an absolute required reading to really understand the one presented here. However, the Lauritzen–Spiegelhalter paper is an excellent one and the extra effort is well worth the trouble.

I highly recommend it to anybody interested in the subject, even without any prior knowledge (except for elementary probability theory).

REFERENCES

Cannings, C., and Thompson, E. A. (1981) *Genealogical and Genetic Structure*, Cambridge University Press, Cambridge.

Cannings, C., Thompson, E. A., and Skolnick, M. H. (1978) Probability functions on complex pedigrees, *Adv. Applied Prob.*, **10**, 26–61.

Darroch, J. N., Lauritzen, S. L. and Speed, T. P. (1980) Markov fields and log-linear models for contingency tables, *Annals of Statistics*, **8**, 522–39.

Edwards, D. and Havranek, T. (1985) A fast procedure for model search in multidimensional contingency tables, *Biometrika*, **72**, 339–51.

Edwards, D. and Havranek, T. (1987) A fast model selection procedure for large families of models, *J. Amer. Statist. Assoc.*, **82**, 205–11.

Jensen, F. V. (1988) *Junction Trees and Decomposable Hypergraphs*, Judex Research Report, Aalborg.

Kim, J. H. and Pearl, J. (1983) A computational model for causal and diagnostic reasoning in inference systems, *Proceedings 8th International Joint Conference on Artificial Intelligence*, Karlsruhe, West Germany, pp. 190–3.

Lauritzen, S. L. and Spiegelhalter, D. J. (1988) Local computations with probabilities on graphical structures and their application to expert systems (with discussion), *J. Royal Statist. Soc.*, **B50**, 157–224.

Pearl, J. (1986) Fusion, propagation and structuring in belief networks, *Artificial Intelligence*, **29**, 241–88.

Shachter, R. D. (1986) Evaluating influence diagrams, *Operations Research*, **34**, 871–82.

Spiegelhalter, D. J. (1986) Probabilistic reasoning in predictive expert systems. In L. N. Kanal and J. Lemmer (eds) *Uncertainty in Artificial Intelligence*, North-Holland, Amsterdam, pp. 47–68.

Tarjan, R. E. and Yannakakis, M. (1984) Simple linear-time algorithms to test chordality of graphs, test acyclicity of hypergraphs, and selectively reduce acyclic hypergraphs, *SIAM J. Comput.*, **13**, 566–79.

Thomas, A. (1986a) Approximate computation of probability functions for pedigree analysis, *IMA Journal of Mathematics Applied in Medicine and Biology*, **3**, 157–66.

Thomas, A. (1986b) Optimal computation of probability functions for pedigee analysis, *IMA Journal of Mathematics Applied in Medicine and Biology*, **3**, 167–78.

Thompson, E. A. (1978) Ancestral inference. II. The founders of Tristan da Cunha, *Annals Human Genetics*, **42**, 239–53.

Thompson, E. A. (1979) Ancestral inference. III. The ancestral structure of the population of Tristan da Cunha, *Annals Human Genetics*, **43**, 167–76.

Thompson, E. A. (1985) *Pedigree Analysis in Human Genetics* Johns Hopkins University Press, Baltimore, MD.

Thompson, E. A. (1986) Genetic epidemiology: a review of the statistical basis, *Statistics in Medicine*, **5**, 291–302,

Thompson, E. A., Cannings, C. and Skolnick, M. H. (1978) Ancestral inference. I: the problem and the method, *Annals Human Genetics*, **42**, 95–108.

Wermuth, N. (1976) Model search among multiplicative models, *Biometrics*, **32**, 153–163.

CHAPTER 17

Towards Efficient Probabilistic Diagnosis in Multiply Connected Belief Networks

Max Henrion, *Carnegie–Mellon University, USA*

ABSTRACT

Most work on large knowledge-based expert systems has eschewed the use of probabilistic representations on the grounds of representational and computational intractability. Recent developments in influence diagrams and belief networks promise ways around some of these difficulties. However, a review of available algorithms for inference in belief networks reveals that there are still serious limitations, especially for large networks which are multiply connected. A current research project is attempting to develop a coherent probabilistic interpretation of QMR, a large knowledge-based system designed to aid diagnosis in internal medicine, which currently uses a heuristic representation of uncertainty. The proposed probabilistic interpretation of QMR exemplifies the usefulness of the belief net representation, but also demonstrates the inadequacies of existing inference algorithms. Since diagnoses often involve several diseases, traditional Bayesian schemes requiring mutual exclusivity of hypotheses are inapplicable. An approach is presented to identify the *k* most probable hypotheses, based on a branch and bound search of the hypothesis space. An extension of this scheme can compute bounds on the posterior probabilities of these hypotheses. These bounds can be narrowed arbitrarily at the cost of additional search.

17.1 INTRODUCTION

Early attempts to formalize expert knowledge and to automate diagnosis and decision-making under uncertainty used schemes based on Bayesian probability and decision theory (e.g. Gorry and Barnett, 1968). Subsequent research in applying techniques from artificial intelligence (AI) to the development of 'expert

Influence Diagrams, Belief Nets and Decision Analysis
Edited by R. M. Oliver and J. Q. Smith

systems', as for example in MYCIN (Shortliffe and Buchanan, 1976), Internist-1 (Miller *et al.*, 1986), and Prospector (Duda *et al.*, 1976), also originally looked to Bayesian reasoning, if only as an unattainable ideal.

This early work encountered two kinds of difficulty in developing probabilistic knowledge representations. First, there is the *representational problem*, that is, how to structure and encode the knowledge and beliefs of human experts into a coherent probabilistic form. In Prospector, for example, it is practically impossible to assign probabilities to propositions and to express their relationships as likelihood ratios in a way consistent with the laws of probability (Duda *et al.*, 1976). Second, there is the *inferential problem*, that is, the issue of the computational tractability of algorithms for probabilistic inference and decision making. In order to achieve both representational and inferential tractability, many Bayesian diagnostic systems assumed mutual exclusivity and exhaustivity of hypotheses, for example that each patient has only one disease. This is often quite unrealistic. These difficulties led most AI researchers to turn away from exact probabilistic representations.

The recent development of the influence diagram (Howard and Matheson, 1981) and Bayes' belief network (Pearl, 1986) appears to provide a basis for resolving at least some of these problems. These schemes hold the promise of coherent representations which are practical for large knowledge-based systems. Influence diagrams and belief nets represent uncertain knowledge as directed graphs. Nodes represent variables, and arcs represent the probabilistic influence of one variable upon another. This graphical form has considerable appeal as a natural and perspicuous way to express uncertain relationships which also has a rigorously defined underlying probabilistic interpretation.

At a qualitative level, an influence arc denotes a probabilistic dependence between linked variables, and absence of a direct arc denotes various kinds of independence between variables. At a quantitative level, uncertain influences are represented by conditional probability distributions for each variable conditional on the states of its predecessors. Influence diagrams are essentially identical to belief networks, but, in addition to chance variables representing uncertain states of the world, they also contain decision variables and value variables. In this chapter our attention will be focussed on belief nets.

The use of a belief network can resolve the representational problem, found to be so acute in Prospector, of assigning probabilities to propositions and their evidential relationships in such a way that the resulting joint distribution is coherent, that is, both consistent and complete. Suppose you assign a marginal distribution for each chance variable without predecessors, and you assign a conditional distribution for each variable with predecessors to express their influence on the variable. Since a belief network (or influence diagram) has no directed cycles, it turns out that these assignments together are guaranteed to define a coherent joint distribution.

The inferential tractability of large and complex belief nets remains more of a problem, and it is the development of more efficient inference techniques for belief nets that is the focus of this chapter. I will start with an overview of the current state of the art in inference algorithms for belief nets. To provide an example to motivate the chapter, I will then discuss an existing medical expert system, Quick Medical Reference (QMR) which contains a large knowledge base for internal medicine (Miller *et al.*, 1986). A current project is proposing to develop a reinterpretation of the QMR representation into a coherent probabilistic form. The proposed belief net representation, though fairly simple in concept, is intractable for direct application of existing exact inference algorithms. The rest of the chapter outlines a new approach, involving branch-and-bound search of possible hypotheses intended to discover the most probable hypothesis, or k most probable hypotheses, without having to search the entire space. For decision making it is generally desirable to know the absolute posterior probabilities of hypotheses, not just which are most likely. Computing the exact probabilities is intractable, but a method for computing bounds on them will be described. For brevity, this chapter will emphasize the main ideas and present the key theorems on which they are based, but without providing detailed proofs or algorithms.

17.2 INFERENCE ALGORITHMS FOR BELIEF NETS

In the last few years a variety of approaches have been explored for developing more efficient algorithms for inference in belief nets*. Conceptually the simplest approach to probabilistic inference in a belief network is to explicitly compute the joint distribution over all the variables as the product of all the prior and conditional distributions. By summing over the appropriate dimensions of the joint it is straightforward to obtain arbitrary marginals. Similarly, the conditional probability $P(x|e)$ for any variable or combination of variables, x, given any evidence, e, can be computed as the ratio of the two marginals, $P(x,e)/P(e)$. The only snag with this brute force approach is that the size of the joint distribution and hence the computational effort is combinatorial in the number of variables. This is likely to be fatal if there are more than a dozen or so uncertain variables. The key to computational efficiency for inference in belief nets is to take advantage of conditional independence specified by the network topology, and so find ways of propagating the impact of new evidence locally without having to calculate the entire joint distribution explicitly.

Provided the network is a *polytree*, that is, singly connected, inference may be performed using an efficient algorithm based on constraint satisfaction for propagating the effect of new observations along the tree (Kim and Pearl, 1983). Each node in the network obtains messages from each of its parent and child nodes, representing all the evidence from the portion of the network lying

*This overview is partly based on Horvitz, Breese and Henrion (1988).

beyond them. The single-connectedness guarantees that the information in each message to a node is independent and so local updating will work. This scheme is linear in the number of variables.

Unfortunately most real networks are multiply connected, so more complex methods are required. The approach developed by Shachter (1987, 1988) focuses on the computation of the posterior distribution given the evidence of a single variable (or more generally a function of several variables), rather than all the variables in the network. It applies a sequence of operators to the network which reverse the links, using Bayes' theorem, and take expectations over nodes to eliminate them. The process continues until only the node representing the original probabilistic query remains. The scheme employs knowledge of the network topology to guide the process.

Lauritzen and Spiegelhalter (1988) describe an approach based on a reformulation of the belief network. First they 'moralize' the graph by adding arcs between all pairs of nodes that have a common successor (i.e. parents with a common child). They then triangulate it, adding arcs so that there are no undirected cycles of more than three nodes without an internal chord. They then identify *cliques*, that is, all maximal sets of nodes that are completely interconnected. They prove that by this process any network can be transformed into a singly connected 'hypernetwork' of cliques. They provide an algorithm for propagation of evidence within this polytree of cliques, which is somewhat analogous to the Kim and Pearl (1983) algorithm for a polytree of variables.

The computational complexity for these algorithms has not been completely analyzed in terms of the network topology. But all are liable to combinatorial problems if there are many intersecting cycles. More generally, Cooper (1987) has shown that the problem of inference to obtain conditional probabilities in an arbitrary belief network is NP-hard. This suggests it will be more profitable to look for approximate or bounding methods rather than exact algorithms.

One line of research has used stochastic simulation, representing a probabilistic model as a sample of deterministic cases. The accuracy of the representation depends on the size of the sample (number of simulation runs). Standard statistical techniques can be used to estimate the error in the approximation, and this can be reduced to arbitrary degree by increasing the sample size. Bundy (1985) suggested one such Monte Carlo approach for computing the probabilities of Boolean combinations of correlated logical variables, which he called the 'incidence' calculus. Henrion (1988) developed an extension to this for inference in belief nets, termed 'probabilistic logic sampling'. For each case in the sample, each source node and influence is represented as a truth value or truth table generated at random using the specified probabilities. Diagnostic inference is performed by estimating the probability of a hypothesis as the fraction of simulations that give rise to the observed set of evidence. The method is linear in the size of the network, irrespective of its degree of connectedness, but it is exponential in the number of pieces of evidence observed.

Chin and Cooper (1987) have applied the logic sampling approach to generate samples of medical cases. They avoid the exponentiality problem by rearranging the network using Shachter's algorithm, so all observed variables become source variables. Unfortunately, this is not a general solution since the rearrangement is itself liable to combinatorial problems.

Pearl (1987) has developed a stochastic sampling approach which involves propagation in both direction across each link. This method first computes the conditional distribution for each variable given all the neighbours in its Markov blanket (i.e. parents, children, and children's parents). First, all nodes are initialized at random with truth values. During simulation, each node may be updated using a truth value generated with the probability conditional on the current state of its neighbours. The probability of each node is estimated as the fraction of simulation cycles for which it is true. An advantage of this scheme is that it could be implemented as a network of parallel-distributed processors, each exchanging messages with its neighbours. Unfortunately, it is liable to convergence problems when the network contains links that are close to 0 or 1 (Chin and Cooper, 1987). Unlike in logic sampling, successive cycles are not independent.

Researchers involved in attempts to apply probabilistic methods in expert systems generally seem much more concerned about the tractability of inference methods than those in decision analysis. What is the reason for this difference? Expert system knowledge bases often contain hundreds or thousands of variables and numerous intersecting cycles. We shall discuss an example shortly. On the other hand, the models of decision analysts are generally restricted to only a handful of uncertain variables and decision variables. A model with a dozen such variables would be of unusual size. Since decision analysts generally use decision tree roll-back for inference, which is similar to a brute force approach involving calculation of the entire joint distribution, model size is drastically limited. Decision analysts try to keep the model this small by conducting extensive deterministic sensitivity analyses to discover which quantities contribute most and least to differences between decision outcomes. Those variables which turn out to be unimportant are then treated as deterministic. Common experience is that only a very few variables contribute materially to the uncertainty in the outcome, and consequently only these few need have their uncertainty modelled probabilistically.

Since an expert system may be used to support a wide variety of different decisions in a wide variety of circumstances, this kind of sensitivity analysis is much less effective in pruning the knowledge-base ahead of time to a manageable size. When it is being used to help with a particular problem or decision, it might indeed be so pruned. But assuming that the average user is unable or unwilling to do this, it requires formalization and mechanization of the sensitivity analysis and pruning process. This suggests an alternative approach to inference than those described above. Rather than develop an algorithm efficient enough to handle the hundreds or thousands of uncertain variables in a large knowledge base, we might develop a search scheme that identifies only those few that are important for the current problem. Of course, the simplification from the pruning of unimportant variables will add a degree of error. But given a formal scheme, this error may be controllable and its magnitude estimated. This insight underlies the development of the approach to be presented below.

17.3 A PROBABILISTIC INTERPRETATION OF QMR

An an example to illustrate some of the inferential problems and to motivate the following development let us briefly examine QMR (Quick Medical Reference), which is an expert system for supporting diagnosis in internal medicine (Miller *et al.*, 1986). It is a successor to the Internist-1 system (Miller *et al.*, 1982). The QMR knowledge base contains information about almost 600 diseases (of the estimated 750 diseases comprising internal medicine) and over 4000 manifestations, that is, patient characteristics, medical history, symptoms, signs, laboratory results, and other available case-specific information that a physician may use to help diagnose a patient. In this chapter I will refer to all these kinds of information generically as *findings*. QMR represents about 25 person-years of effort in knowledge engineering, and is perhaps the most comprehensive structured medical knowledge base currently existing.

The QMR knowledge base consists of a profile for each disease, that is, a list of the findings associated with it. Each such association between disease *d* and finding *f* is quantified by two numbers. The *evoking strength* is a number between 0 and 5 which answers the question 'Given a patient with finding *f*, how strongly should I consider disease *d* to be its explanation?' The *frequency* is a number between 1 and 5 answering the question 'How often does a patient with disease *d* having finding *f*?'. In addition, associated with each finding *f* is an *import*, being a number between 1 and 5 answering, 'To what degree is one compelled to explain the presence of finding *f* in any patient?' QMR also contains various relationships between diseases, such as 'disease *a* predisposes towards disease *b*'. QMR uses a heuristic scheme for diagnosis (or abductive inference) which identifies one or more hypotheses, each consisting of one or more diseases, and which are intended to explain the observed findings. This algorithm employs the evoking strengths, frequencies and imports, to identify the most plausible hypotheses and rate them in terms of the degree to which they explain the findings.

A current project is seeking to develop a coherent probabilistic interpretation of QMR*. The first goal of this work is to improve the consistency and clarity of the knowledge base, and to explicate the independence assumptions it

*This project is being conducted in collaboration with Randolph Miller of the University of Pittsburgh, and Greg Cooper, David Heckerman, and Eric Horvitz of Stanford University.

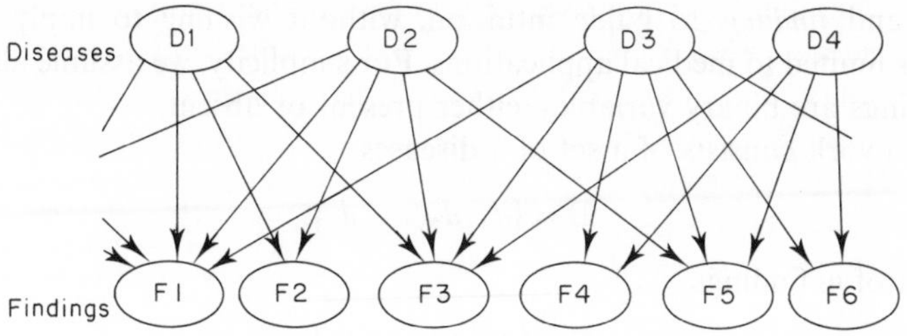

Figure 17.1 Part of a BN2 belief network.

incorporates. The second goal is to provide an example to explore the feasibility of probabilistic reasoning in a substantial real world system. Heckerman and Miller (1976) have demonstrated a fairly reliable monotonic correspondence between the frequency numbers and $P(f|d)$, the conditional probability of a finding f given only disease d. It also appears that imports can be interpreted as inversely related to $P(f|H_0)$, that is, the probability of the finding f given no explicit disease is present. Additional assumptions include conditional independence of findings given any hypothesis, and that the mechanism whereby any disease can cause a finding to appear is independent of the presence of other diseases and their causal mechanism. If we add in assessments of the prevalence rates of the diseases, it seems possible to establish a complete and coherent interpretation of the knowledge base, which formalizes the quantitative relations and assumptions in a way that seems most compatible with the informal assumptions of the current version of QMR. (These assumptions will be specified more precisely below.)

Figure 17.1 shows a schematic partial belief network corresponding to these assumptions, with the additional assumption of marginal independence of diseases. In the actual knowledge base, each disease has an average of about 80 findings, and each finding has an average of about 10 diseases that can cause it. This belief network includes a vast number of intersecting undirected cycles, for example where two or more diseases have two or more findings in common. For this reason it appears that the exact algorithms discussed above are either inapplicable or intractable.

17.4 BN2: TWO-LEVEL BELIEF NETS

We start with a more precise specification of some classes of belief network. We start with the general two-level network, BN2, to be followed by two specializations, of which the second, BN2O, corresponds to the assumptions made for the probabilistic interpretation of QMR. By 'two level', I mean the variables are divided into two classes, and elements of the second may only be influenced by elements of the first, but not by each other. I will call these classes *diseases* and *findings* to guide intuition, without wishing to imply that this scheme is limited to medical applications. For simplicity, we assume all diseases and findings are binary variables, either present or absent.

The network consists of a set of n diseases,

$$\mathbb{D} = \{d_1, d_2, \ldots, d_n\},$$

and a set of m findings,

$$\mathbb{F} = \{f_1, f_2, \ldots, f_m\}.$$

A *hypothesis* (or a diagnosis), H, consists of a set of diseases (none, one, several or all). Thus H may be any subset of $\mathbb{D}$. Note that a hypothesis is taken to specifically exclude as absent all diseases not in the hypothesis set. The set of hypotheses is the power set of the set of diseases, and so there are 2^n possible hypotheses. This includes the null hypothesis that none of the diseases are present:

$$H_0 = \varnothing.$$

Let F be a set of observed findings for a particular case. It consists of those findings observed to be present, F_1, and those findings observed to be absent, F_0. The other findings have not been observed.

Suppose we are given, or can compute, the prior probability $P(H)$ for every hypothesis, H, and the likelihood $P(F|H)$ of findings F conditional on each possible hypothesis H. We can then apply Bayes' rule in the usual way to calculate the posterior probability for the hypothesis after observing the findings:

$$P(H|F) = \frac{P(F|H)P(H)}{\sum_{\forall H \subseteq \mathbb{D}} P(F|H)P(H)}. \tag{17.1}$$

The representational problem here is that we must assess the prior $P(H)$, and conditional $P(F|H)$ for each of the 2^n hypotheses H, unless we make some independence assumptions. To compute any posterior, we need to compute the denominator of equation (17.1),

$$P(F) = \sum_{\forall H \subseteq \mathbb{D}} P(F|H)P(H),$$

which requires summing over every possible hypothesis. Hence the inferential complexity is $O(2^n)$. This means the scheme is intractable if there are more than a few possible diseases, unless we can simplify somehow.

17.5 BN2E: 'IDIOT'S BAYES'

The most common simplification of the BN2 scheme is to make the following two assumptions.

Assumption 1

All diseases are mutually exclusive and collectively exhaustive.

In this case, we need consider only singleton hypotheses, containing a single disease. It is usually helpful to add the null or residual hypothesis as an additional disease, d_0, to ensure exhaustivity. In this case there are only $n + 1$ terms in the summation. It also simplifies the assessment of the probabilities, both priors on the diseases and conditionals for the findings.

Assumption 2a

All findings are conditionally independent of each other given any disease. The conditional probability of the findings F given a disease d can be decomposed into the product of the probabilities of each finding f among those present F_1 and those absent F_0:

$$P(F \mid d) = \prod_{\forall f \in F_1} P(f \mid d) \times \prod_{\forall f \in F_0} (1 - P(f \mid d)).$$

The representational complexity of this scheme (number of probabilities that need to be assessed) is less than $(m + 1)n$, assuming not all diseases can given rise to every finding. Its computational complexity is also $O(mn)$, because the denominator of Bayes' rule,

$$P(F) = \sum_{\forall d \in \mathbb{D}} P(F \mid d) P(d),$$

need only be summed over $n + 1$ terms.

We shall call this scheme BN2E ('E' for exclusive and exhaustive). Due to its great simplicity, it is sometimes known as 'idiot's Bayes'. It has been so widely used that sometimes the mistaken impression has arisen that these assumptions are essential to any coherent Bayesian scheme.

In some cases where diseases are all independent and very rare, the assumption that there cannot be more than one is reasonable, and this provides a good approximation to the situation being modelled. But clearly in many cases it does not. For example, in internal medicine, a patient seen by an internist typically has three or more diseases. Any assumption of mutual exclusivity of diseases is out of the question.

17.6 BN2O: THE TWO-LEVEL BELIEF NETWORK WITH NOISY OR

We term the specialization of the BN2 network implied by the assumptions used in our initial probabilistic interpretation of QMR as BN2O ('O' for OR). The following are its assumptions.

Assumption 2b

All findings are conditionally independent of each other given any hypothesis. The probability of the findings F given a hypothesis H can be decomposed into the product of the probabilities of each finding given H, i.e.

$$P(F|H) = P(F_1|H)P(F_0|H) = \prod_{\forall f \in F_1} P(f|H) \times \prod_{\forall f \in F_0} (1 - P(f|H)). \qquad (17.2)$$

The only difference between the earlier independence assumption 2a and this assumption 2b is that since we no longer require mutual exclusivity of diseases, the conditioning hypothesis H may include any number of diseases.

Assumption 3

The effects of multiple diseases on a common finding are combined using a leaky noisy OR gate (LNOG). This assumes that the mechanisms whereby different diseases can cause (or are otherwise associated with) a given finding f operate independently. Suppose D_f is the set of diseases, each of which can cause finding f, and S_{df} is the event that the presence of a disease d in D_f is sufficient for f to occur. The noisy OR assumption is that each event S_{df} is independent of any other events in the model, including the presence of diseases other than d, of findings other than f, and of other sufficiency events. With a *leaky* noisy OR gate there is a further possibility of a leak event S_{0f} that finding f will occur even when none of the diseases in D_f occur. Thus, finding f will occur if there is a leak or if any disease d in D_f occurs and it is sufficient, S_{df}:

$$f \Leftrightarrow S_{0f} \vee \bigcup_{\forall d \in D_f} d \wedge S_{df}. \qquad (17.3)$$

We define $p_{df} = P(S_{df})$ as the *link probability*, i.e. the probability that d is sufficient to cause f. We define $p_{0f} = P(S_{0f}) = P(f|H_0)$ as the *leak probability*, i.e. the probability that f occurs 'spontaneously' in the absence of any diseases. For brevity we will also define the complementary link and leak probabilities

$$q_{df} = 1 - p_{df}, \qquad q_{0f} = 1 - p_{0f}.$$

From equation (17.3) we can derive the probability that f will occur given any combination of diseases in D_f as

$$P(f|H) = 1 - (1 - p_{0f}) \times \prod_{\forall d \in H} (1 - p_{df})$$

$$P(f|H) = 1 - q_{0f} \prod_{\forall d \in H} q_{df}. \qquad (17.4)$$

Assumption 4

Diseases are marginally independent. This means we can decompose the prior probability of a hypothesis H into the product of the prior probabilities of each disease in H times the product of the complements of the probabilities of the diseases not in H:

$$P(H) = \prod_{\forall d \in H} P(d) \times \prod_{\forall d \notin H} [1 - P(d)]. \tag{17.5}$$

It is convenient to rewrite this as the product of the prior odds for the diseases in H times the product of complements for all diseases, that is, the prior probability for H_0:

$$P(H) = \prod_{\forall d \in H} \frac{P(d)}{1 - P(d)} \times \prod_{\forall d \in \mathbb{D}} [1 - P(d)] = P(H_0) \prod_{\forall d \in H} \frac{P(d)}{1 - P(d)}. \tag{17.6}$$

The representational complexity of this BN2O scheme is roughly the same as the BN2E scheme. It requires a prior probability for each disease and at most a link probability for each of the $m \times n$ disease-finding pairs, plus the m leak probabilities. We can use the expressions (17.2), (17.4) and (17.6) to calculate $P(F|H)$ and $P(H)$ for use in diagnostic inference using Bayes' rule (17.1), if we wish to compute the exact posterior for any hypothesis H. But unfortunately the inferential complexity remains exponential, since the denominator of equation (17.1) still requires summing over 2^n hypotheses.

17.7 COMPUTING RELATIVE PROBABILITIES

If one is interested only in the relative posterior probabilities of alternative hypotheses, then there is a well-known strategy which avoids the combinatorial problem inherent in computing the absolute posterior. Suppose we are interested in the ratio of the posterior probability for hypothesis H_i to the null hypothesis H_0. We can then use the odds-likelihood form of Bayes' rule:

$$\frac{P(H_i|F)}{P(H_0|F)} = \frac{P(F|H_i)P(H_i)}{P(F|H_0)P(H_0)} = \frac{P(H_i, F)}{P(H_0, F)}. \tag{17.7}$$

The joint probabilities $P(H, F)$ are tractable to compute for the BN2O case. Applying equations (17.2), (17.4) and (17.6) we get:

$$\begin{aligned} P(H, F) &= P(H)P(F|H) = P(H)P(F_0|H)P(F_1|H) \\ &= P(H_0) \times \prod_{\forall d \in H} \frac{p(d)}{1 - p(d)} \times \prod_{\forall f \in F_0} q_{0f} \times \prod_{\forall f \in F_0} \prod_{\forall d \in H} q_{df} \\ &\quad \times \prod_{\forall f \in F_1} \left(1 - q_{0f} \prod_{\forall d \in H} d_{df}\right). \end{aligned} \tag{17.8}$$

Of the above five terms in $P(H, F)$ the first and third are independent of H. Hence they will cancel out when we use a ratio of two joint probabilities to compute the ratio of the two posterior probabilities as in equation (17.7). It is therefore useful to define the relative probability $R(H,F)$, so that,

$$\frac{P(H_i|F)}{P(H_0|F)}=\frac{R(H_i,F)}{R(H_0,F)}. \tag{17.9}$$

The relative probability is defined as the product of the second, fourth and fifth terms of equation (17.8), which we label R_1, R_2, and R_3:

$$R(H,F)=R_1(H)\times R_2(H)\times R_3(H), \tag{17.10}$$

where

$$R_1(H)=\prod_{\forall d\in H}\frac{p(d)}{1-p(d)},$$

$$R_2(H)=\prod_{\forall f\in F_0}\prod_{\forall d\in H} q_{df},$$

$$R_3(H)=\prod_{\forall f\in F_1}\left(1-q_{0f}\prod_{\forall d\in H} q_{df}\right),$$

$R_1(H)$ is the prior odds on hypothesis H. $R_2(H)$ represents the evidence of findings observed to be absent which might have been expected to be associated with diseases in H. If there is some disease d in H which is always sufficient to cause some finding f in F_0 (i.e. observed absent), then for some d, $p_{df}=1$, so $q_{df}=0$, and hence $R_2=0$. In other words, H would be ruled out as impossible. $R_3(H)$ represents the evidential impact of the findings observed to be present, which can be explained by one or more diseases in H. If there is any finding f in F_1 for which $p_{df}=0$, so $q_{df}=1$ for all diseases d in H, and which also has zero leak probability, $p_{0f}=0$, so $q_{0f}=1$, then $R_3=0$. In other words H can be ruled out by any finding which is not explained by any disease in H, nor can it occur 'spontaneously', that is, without any explicitly modelled disease.

17.8 FINDING THE MOST PROBABLE HYPOTHESES

While the relative score provides a tractable way to compute ratios of posterior probabilities, to find the most probable hypothesis (or the k most probable hypotheses), we still need to search in the space of 2^n possible hypotheses. This may be looked at as a problem of heuristic search (Cooper, 1984; Peng and Reggia, 1987a, b). Suppose we start from the null hypothesis H_0, that no disease is present. We then add one disease at a time to generate extension hypotheses, each of which is examined in terms of its relative score. By an extension of a hypothesis H, we mean any hypothesis H^+ that is, a proper superset of H. Figure 17.2 shows part of a search tree for a domain with seven diseases.

The key requirement is a bounding heuristic which allows us to rule out all extensions of a hypothesis as inadmissible, that is, provable not to contain more probable hypotheses. With such a heuristic we perform a branch and bound

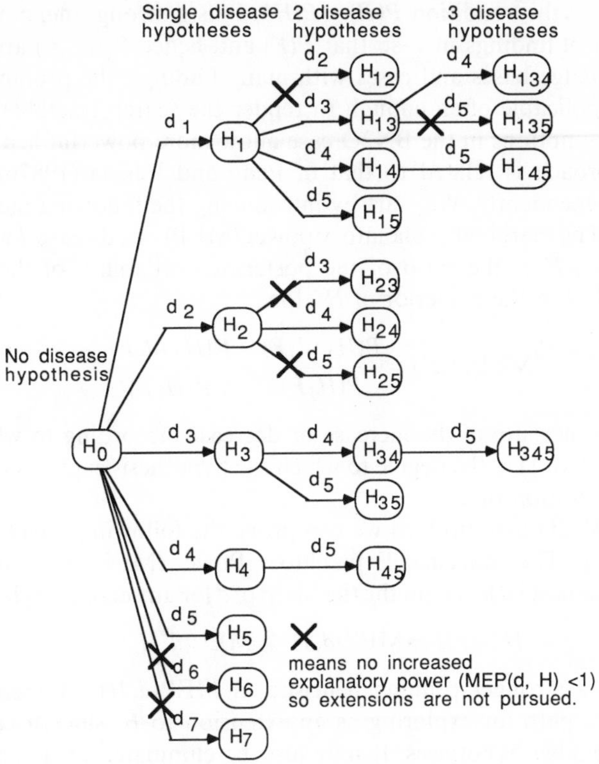

Figure 17.2 Example hypothesis search tree.

search, pruning large parts of the search tree which can be shown to be inadmissible without having to examine them.

Cooper (1984) suggests one such pruning heuristic. For a general belief network, he shows that if the prior probability for a hypothesis H is less than the joint probability of some other hypothesis H^* and the findings F, then the posterior probability of any extension H^+ of H must be less than the posterior for H^*, i.e.

$$P(H) < P(H^*, F) \Rightarrow P(H^+|F) < P(H^*|F), \qquad \forall H^+ \supset H.$$

(Cooper assumes that the prior for any hypothesis H is greater than the prior on any of its extensions, $P(H) > P(H^+)$, which certainly is true given marginal independence of hypotheses, and will usually be true in other cases.)

If H^* is the best (or kth best) hypothesis found so far, then we can test each hypothesis H. If $P(H) < P(H^*, F)$, then we know that none of its extensions can be the best (or among the kth best), and so we can prune the search tree at that point.

Unfortunately the condition $P(H) < P(H^*, F)$ is a strong one, particularly if there are a lot of findings in F so that $P(F)$ and hence $P(H^*, F)$ are small. So in large knowledge bases and cases with many findings, the pruning heuristic may not be applicable often enough to render the search tractable. However, the special assumptions in the BN2O case allow more powerful heuristics. The following approach is related to that of Peng and Reggia (1987b), although developed independently. We start by introducing the following measure:

Definition. The marginal explanatory power (MEP) of a disease d with respect to a hypothesis H is the ratio of the posterior probability of the extension hypothesis $H \cup d$ to the posterior of H alone:

$$\mathrm{MEP}(d, H) = \frac{P(H \cup d \mid F)}{P(H \mid F)} = \frac{R(H \cup d, F)}{R(H, F)}. \tag{17.11}$$

The MEP is a measure of the increase or decrease (according to whether it is greater or less than 1) in the degree to which the hypothesis explains the findings F due to the addition of d. □

Given the BN2O assumptions we can prove the following useful result:

Theorem 17.1. The Marginal Explanatory Power (MEP) of a disease d for hypothesis H cannot be less than the the MEP of d for any extension H^+ of H, i.e.

$$H^+ \supset H \Rightarrow \mathrm{MEP}(d, H) \geqslant \mathrm{MEP}(d, H^+)$$ □

This provides a stronger pruning heuristic. If $\mathrm{MEP}(d, H) \leqslant 1$ then d can be eliminated as a path for exploring as an extension to H, since it cannot lead to a more probable hypothesis. It may also be eliminated as a candidate for extending other extensions of H. Thus the only diseases which need to be considered as extensions of H are those for which $\mathrm{MEP}(d, H) > 1$.

This may be used as a basis for an algorithm to find the most probable hypothesis as follows. Let **Hbest** be the most probable hypothesis examined so far, that having relative probability $R(H, F)$ greater or equal to that of any other yet examined. It is initialized to the null hypothesis $\mathbf{H_0}$. **Elist[H]** is a list of diseases that are admissible extensions to hypothesis **H**. We initialize **Elist[$\mathbf{H_0}$]** to the set of all diseases that can explain at least one finding that has been observed present in F_1, and do not cause with certainty any findings observed to be absent, in F_0. **Hlist** is the list of hypotheses that are candidates for extension, that is, those for which we have computed the relative probability, but whose extensions we have not yet examined. **Hlist** is initialized to contain just $\mathbf{H_0}$. **Hlist** is maintained in order by decreasing $R(H, F)$. The following procedure **Extend(H)** is applied repeatedly to the first (most probable) hypothesis in **Hlist** until it is empty, at which point **Hbest** will indeed be the most probable hypothesis.

To **Extend(H)**, we first remove **H** from **Hlist.** We examine each disease **d** in **Elist[H]** and compute **MEP(d, H).** If this is less than 1, we remove **d** from **Elist[H].** If **Elist[H]** is empty we return. If it contains a single element **d**, then extension $\mathbf{H^+ = H \cup d}$ is a candidate for most probable hypothesis (none of its extensions can be more probable). We compute **R(H$^+$,F)** as **MEP(d, H)** × **R(H, F)**, and if this is greater than **R(Hbest, F)**, we substitute **Hbest:= H$^+$**, and return. Otherwise **Elist[H]** contains more than one element. We order it by decreasing **MEP(d, H).** We now go through each **d** in **Elist[H]** in sequence, and for each extension **H$^+$ = H ∪ d,** and compute **R(H$^+$, F),** and insert it in **Hlist** as a candidate. **Elist[H$^+$]** is set to the tail of **Elist[H],** that is, those remaining admissible extensions, not including the current **d.**

This describes a breadth-first/best-first search. This is likely to be the most efficient strategy in terms of computational effort, but may have large space demands as **Hlist** grows together with the **Elist[H]** for each **H** in **Hlist.** If this is a problem, an alternative is a depth-first/best-first strategy, in which **Extend** immediately calls itself recursively on its admissible extensions instead of storing them in **Hlist** for later examination.

17.9 TO FIND THE *k* MOST PROBABLE HYPOTHESES

This algorithm may be easily extended to find the *k* most probable hypotheses. **HBest** is then a set of up to *k* hypotheses. In this case, each hypothesis to which **Extend** is applied is a candidate for putting into **Hbest.** At first search continues without pruning until *k* hypotheses have been added to **Hbest.** Search is only truncated at a node **H** whose **R(H, F)** is less than the relative score of the *k*th best hypothesis so far, and for which **MEP(d, H) < 1** for all its admissible extensions **d**.

17.10 BOUNDING THE ABSOLUTE PROBABILITIES

If the results of probabilistic diagnosis are to be used to guide treatment decisions, then it is desirable to know not just the relative probabilities of the top *k* hypotheses, but also their absolute posterior probabilities. To do this exactly is intractable, as we saw, but it is possible to develop bounds on the posterior probabilities. Suppose *terminal* hypotheses are those at which we have truncated search. The basic idea is to compute an upper bound on the sum of the probabilities of all possible extensions of each terminal hypothesis. In this way it is possible to obtain an upper bound on the probability of all hypotheses which have not been considered explicitly. This approach relies on the following result:

Theorem 17.2. If E^* is the maximum MEP for any disease d with respect to hypothesis H, i.e.

$$E^* = \max_d \mathrm{MEP}(d, H),$$

then the following is an upper bound on the sum of all extensions H^+ of H:

$$\sum_{\forall H^+ \supset H} P(H^+ | F) \leqslant P(H | F)[(1 + E^*)^n - 1]. \qquad \square$$

To compute bounds we set a truncation criterion C_t with corresponding maximum MEP criterion E_t. We truncate search of extensions d of a hypothesis H whenever

$$E^* = \max_d \text{MEP}(d, H) < E_t, \qquad \text{where } C_t = [(1 + E_t)^n - 1].$$

From Theorem 17.2, for any hypothesis H_t at which search is truncated since it meets this criterion, the sum of the probabilities of all its extensions is bounded by the probability of H_t times the truncation criterion.

$$\sum_{\forall H^+ \supset H_t} P(H^+|F) \leqslant P(H|F)C_t.$$

Using equation (17.9) we can rewrite this in terms of relative probabilities:

$$\sum_{\forall H^+ \supset H_t} R(H^+, F) \leqslant R(H, F)C_t. \tag{17.12}$$

The smaller E_t and C_t, the more nodes need to be searched, but the narrower the bounds can be.

During the search we can cumulate the sum of the relative probabilities of the set $\mathbb{H}_c$ of all hypotheses examined, that is, all those for which we have computed $R(H, F)$:

$$r_c = \sum_{\forall H \in \mathbb{H}_c} R(H, F). \tag{17.13}$$

Similarly, we can cumulate the sum of the relative probabilities of the set $\mathbb{H}_t$ of all the terminal hypotheses, that is, those at which search has been truncated:

$$r_t = \sum_{\forall H \in \mathbb{H}_t} R(H, F). \tag{17.14}$$

From equation (17.12), we can compute an upper bound on the relative probability of $\mathbb{H}_e$, all the extensions of terminal hypotheses $\mathbb{H}_t$:

$$r_e = \sum_{\forall H \in \mathbb{H}_e} R(H, F) \leqslant r_t C_t. \tag{17.15}$$

Similarly, we define the sums of the posterior probabilities of all the hypotheses considered, $\mathbb{H}_c$, and of those not considered, $\mathbb{H}_e$, the extensions of the terminal hypotheses:

$$p_c = \sum_{\forall H \in \mathbb{H}_c} P(H|F), \qquad p_e = \sum_{\forall H \in \mathbb{H}_e} P(H|F).$$

Since $\mathbb{H}_e = \mathbb{H} - \mathbb{H}_c$, where $\mathbb{H} = 2^{\mathbb{D}}$, the set of all hypotheses, therefore,

$$p_c + p_e = 1$$

from which we can obtain

$$p_c = \frac{1}{1 + p_e/p_c}. \tag{17.16}$$

The ratio at the bottom can be expressed in terms of relative probabilities, from equation (17.9):

$$\frac{p_e}{p_c} = \frac{r_e}{r_c}.$$

From equation (17.15), summing over all hypotheses we can place an upper bound on the relative probability of all unexamined extensions:

$$r_e \leqslant r_t C_t.$$

Substituting these into equation (17.16) we get the following lower bound for the total posterior probability of all hypotheses considered:

$$p_c \geqslant \frac{r_c}{r_c + C_t r_t}. \tag{17.17}$$

This gives us a lower bound on the total fraction of posterior probability we have accounted for in the hypotheses we have looked at, or complementarily an upper bound on the probability of all those we have not looked at. It is therefore a useful measure of the completeness of the search. The criterion C_t controls the extent of the search and the size of this bound. By decreasing it we can increase the fraction of probability accounted for at the cost of a more expensive search. It allows us to make explicit tradeoffs between effort and accuracy.

It is also straightforward to obtain bounds on the posterior probability for any single hypothesis H. We reexpress the ratio of the posterior probabilities of H and of $\mathbb{H}_c$ in terms of their relative probabilities

$$\frac{P(H|F)}{p_c} = \frac{R(H,F)}{r_c}.$$

Hence

$$P(H|F) = \frac{p_c}{r_c} R(H,F).$$

From equation (17.17), and the fact that as a probability $p_c \leqslant 1$, we obtain:

$$\frac{1}{r_c} \geqslant \frac{p_c}{r_c} \geqslant \frac{1}{r_c + C_t r_t}$$

and hence we can compute the following bounds for the posterior probability in terms of the relative probability:

$$\frac{R(H,F)}{r_c} \geqslant P(H|F) \geqslant \frac{R(H,F)}{r_c + C_t r_t}. \tag{17.18}$$

17.11 CONCLUSIONS

Influence diagrams and belief networks appear to offer the possibility of a coherent probabilistic representation for the kind of large knowledge bases used in expert systems. The project to develop a probabilistic interpretation of QMR is intended to explore this possibility and test its practicality for a very large knowledge base. As a first step, the BN2O class of belief networks has been proposed as a reasonable, if simple, probabilistic explication of the current heuristic QMR representation.

Since multiple simultaneous diseases are common in internal medicine, we cannot assume mutual exclusivity of diseases as in the standard BN2E, or 'idiot's Bayes' approach. Therefore existing algorithms are inadequate to the task of inference in the resulting network, and indeed no exact method of computing posterior disease probabilities appears tractable. As a result we have been led to develop an alternative approximate method. This approach involves a branch and bound search of the hypothesis space, guided by a new heuristic to terminate lines of search. This heuristic is based on the concept of the MEP (Marginal Explanatory Power) which expresses the change in probability of a hypothesis resulting from extending it by an additional disease. The MEP of an additional disease decreases monotonically with extensions to the background hypothesis. This search allows identification of the k most probable multiple-disease hypotheses. An extension of the technique also allows computing an upper bound on the total probability of the hypotheses not explicitly examined, and hence bounds on the absolute probability of the most probable hypotheses. This algorithm degrades gracefully, in the sense that it can provide useful results (probable hypotheses and bounds on the probabilities) even with limited computational time. With additional time, more hypotheses can be examined and the bounds can be incrementally narrowed. This allows explicit trade-offs to be made between the precision of the results and the cost of computation.

While the BN2O class of belief networks is an important step beyond the BN2E class, and seems a reasonable interpretation of QMR, it is still fairly restrictive. Possible generalizations include extensions from binary, logical variables to multivalued variables, for example to model several levels of severity of a disease or a finding; the addition of further levels between disease and findings, such as intermediate pathophysiologic states, would allow a natural relaxation of the conditional independence assumptions between findings; and causal influences on disease prevalence, such as patient history, exposure and so on, would better modelled as additional levels of node with links into diseases rather than as standard findings as at present. Also desirable are methods for combining the probabilistic effects of multiple diseases on one finding which are richer than the leaky noisy OR gate. Each of these generalizations will provide a challenge for the design of more powerful inference algorithms.

17.12 DISCUSSION

17.12.1 Discussion by Jim Smith

While I applaud Mr Henrion's effort to bring a probabilistic interpretation and inference mechanism to Internist-1/QMR, I have reservations about the two-level belief network representation and inference mechanism presented in this chapter. Furthermore, I have deeper reservations about the use of QMR in medical diagnosis.

The chapter does, however, provide a nice review of inference algorithms for belief nets. Drawing on another recent paper by Horvitz, Breese, and Henrion (1988), the chapter does an excellent job of placing the two-level belief network in the context of other recent work. Like Henrion, I also believe that a probabilistic inference technique that allows explicit control of the accuracy/efficiency trade-off would be of great value—provided that one has a clear understanding of the nature of its errors. Most of the criticisms that follow amplify points made by Henrion himself in the chapter.

Boolean propositions are not an appropriate means of representing many findings and diseases. Both findings and diseases come in degrees, rather than being merely true or false. The degree of finding may have considerable impact on the probability of various hypotheses. Similarly, the degree of disease may have considerable impact on the treatment. For instance, one finding might be the presence of a fever. Consider the two different representations in Figure 17.3.

Errors caused by crude true/false classification could easily dominate any potential benefit from including of findings, and may in fact be multiplied by having thousands of potential findings. One might be tempted to split up variables into a set of Boolean propositions. For instance one might divide 'fever' into a more refined set of findings: 'moderate fever' (meaning a body temperature between 99.5 and 101° F) and 'fever' (meaning a temperature greater than 101° F). However, this would violate assumption 2, common to both the BN2E and BN2O representations: 'All findings are conditionally independent of each other given

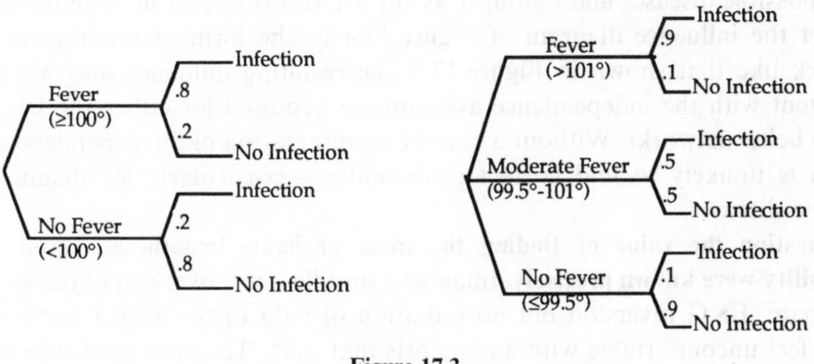

Figure 17.3

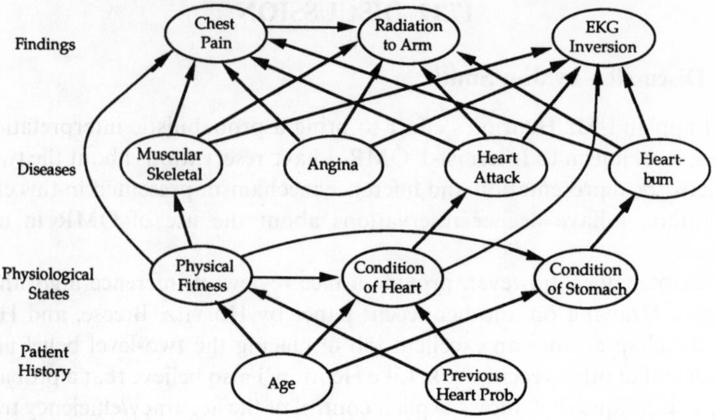

Figure 17.4

any hypothesis.' Given any hypothesis, the events 'moderate fever' and 'fever' are mutually exclusive and clearly not independent of each other.

A two-level belief network does not capture the dependencies among diseases and findings of a system as complicated as the human body. Intermediate physiological states and patient histories will almost certainly play a key role in determining the probabilities of the various hypotheses. For instance, one might draw the influence diagram of Figure 17.4 to relate four different possible causes of chest pain: muscular skeletal damage, angina, heart attack, and heartburn. Also shown are two other common and related findings: EKG inversion, a common finding sometimes related to heart problems and sometimes not; and radiation of pain to arms, symptomatic of both heart attacks and angina. Note the arc from chest pain to radiation of pain to arm—in the absence of chest pain, radiation of pain to arm does not make sense. Working backward from the diseases, we identify several relevant physiological states that would be likely to have some bearing on the probabilities of some of these diseases and findings (as well as on many other possible diseases and findings). While arc reversals and node removals can convert the influence diagram of Figure 17.4 to the form of a two-level belief network like that shown in Figure 17.5, the resulting influence diagram is not consistent with the independence assumptions required for either the BN2E or BN2O belief networks. Without a correct representation of the dependencies, the system is unlikely to find correct probabilities—particularly for diagnoses of multiple diseases.

I question the value of finding the most probable hypotheses—even if the probability were known precisely. Imagine a middle-aged, overweight patient with chest pain, EKG inversion but no radiation of pain to the arm. I know that I would feel uncomfortable with a diagnosis that said, 'The most probable disease is heartburn, with a probability of 0.42', if there were a 0.05 probability of a heart attack. In many probabilistic inference problems, particularly medical problems,

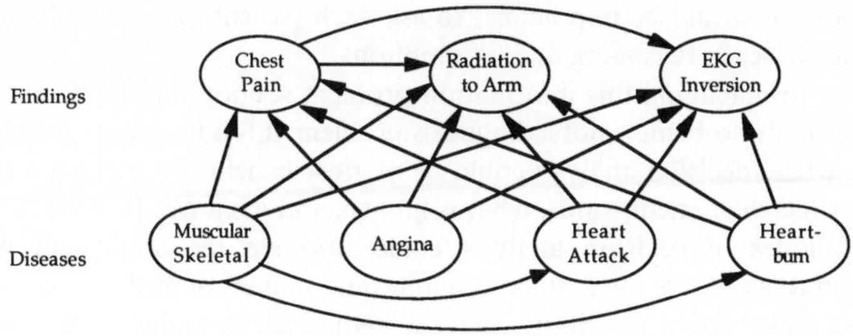

Figure 17.5

the low probability diagnoses are often most important. The possibility of a more serious condition may suggest more tests or alternative diagnostic procedures. Even if QMR were to generate the nine or ten most likely diagnoses, it might still miss a possible more serious condition. Perhaps the algorithm could be modified to find all hypotheses with probabilities above some small threshold and to find probabilities of all diseases of particular gravity or interest.

Could this branch-and-bound technique could be applied to the problem of finding probabilities for all hypotheses? Unfortunately, I believe the answer is no. Ferreting out small-probability hypotheses would seem to require examining the space of all possible events. However, given a decision problem rather than a probabilistic inference problem, branch-and-bound techniques may offer an alternative to the 'brute force' decision tree roll back. In a 1967 Stanford Research Institute report (1967), Chen and Patton present 'branch-and-bound roll back' and 'branch-and-bound probe forth' algorithms that prune nonoptimal paths to reduce the search for optimal decisions.

In the language of Winograd and Flores (1986), QMR is blind. In QMR, the belief network is constructed without a clear definition of the possible diseases or findings. In particular, the patient is not identified. Thus, the background of the patient—medical history, any peculiar symptoms, preferences—is necessarily ignored. This neglect of background renders the system and its users blind to its limitations. I do not believe that expanding QMR from a two-level belief network to an *n*-level belief network, or expanding it from 600 diseases and 4000 findings to include 10 000 diseases and 100 000 findings, will ever capture the appropriate background and certainly will not provide a mechanism for checking the assumptions made in a particular case. The fact that the system is billed as including so many diseases and findings lends QMR an air of credibility that is probably unwarranted, and potentially dangerous.

On the other hand, the QMR system, true to its name, provides quick and inexpensive access to a large body of medical knowledge. The question is how to turn the QMR system into a quality medical reference system. As I see it, the real problem with QMR is that it issues its diagnoses with minimal involvement of both physician and patient. Yet with a system containing 4000 findings and

600 diseases, it would be impractical to ask each patient and each physician to review the system's reasoning and assumptions.

The key to managing this information problem is adopting a different frame. The physician's problem is not a diagnosis problem, it is a treatment problem—a problem where decision analysis could be of great benefit. By striking a balance between what the patient wants, what is known, and what can be done, the tools and techniques of decision analysis could improve the quality of medical decision-making. While the cost, both in time and money, of professional decision assistance is likely to be prohibitive in most medical cases, sophisticated computer systems, blending the techniques of decision analysis and the technology of artificial intelligence, could provide high-quality decision assistance within the constraints of most medical decision problems. Sam Holtzman's recent book *Intelligent Decision Systems* (Holtzman, 1989) explores this idea at length and describes a prototype application of this technology to infertility treatment decisions.

17.12.2 Reply

The initial purpose of developing a probabilistic interpretation of QMR was not to defend its assumptions, but rather to elucidate what is implicit in the existing heuristic version. This was intended as a necessary preliminary to developing more sophisticated representations. The extension of variables from Boolean to multivalued and the extension of the belief network to more than two levels (e.g. with physiological states and patient history), as discussed by Jim Smith, are only two of the several possible enrichments mentioned in the paper and on our current research agenda.

While one can imagine many possible ways to enrich a model, it is not easy to decide which are or are not appropriate. No matter how large, any model is necessarily a simplification. The interesting question is not 'Can the model be made more complete?' (the answer to that is always 'yes'), but rather 'Will it be worth the effort?'. This question is usually hard to answer on a purely theoretical basis. Adequate answers will require experimental comparisons of various methods of enriching the representation in terms of the encoding effort, computational costs, and improvements in performance.

Hitherto belief networks of the size of QMR, even in simple BN2O form, have been far beyond the computational capabilities of existing algorithms. Without computationally practical inference methods the development of richer representations would be pointless. Hence the primary purpose of this paper was to present a more efficient algorithm.

Smith's suggestion is well-taken that a diagnostic system should take into account not only the probability of a hypothesis, but also the importance of a disease. Smith's criticism of QMR as 'blind' seems to be based on some misapprehensions. QMR's authors do use clear definitions of the terms used, diseases and findings. As mentioned in my paper and detailed in Miller *et al.* (1986), QMR's 'findings' (or 'manifestations') include patient demographics (age, sex, race, etc.) and medical history, as well as symptoms, signs and lab findings; indeed all kinds of potentially available information about the patient believed relevant to diagnosis. Certainly there is room for more sophisticated representations of the influence of patient background data, but it is incorrect to state that QMR currently ignores this information.

The complaint of lack of user involvement seems wide of the mark. QMR, as suggested by its name, Quick Medical Reference, was specifically designed as a rapid access electronic medical reference and textbook, to be used by the physician as an interactive tool for diagnostic problem-solving. In this role it is seeing increasingly widespread use. It may also be used in a more traditional 'expert system' role to suggest possible diagnosis. It can easily list for inspection the specific findings underlying any particular diagnostic conclusion ordered by their evidential importance (typically under a hundred). The necessity for total review of the knowledge base by each user is unclear.

Empirical evaluations of systems are often more informative than armchair speculation about what size of knowledge base and complexity of representation are necessary. Evaluations of QMR and its predecessor, Internist-1, have compared diagnostic accuracy with specialist physicians (Miller *et al.*, 1982, 1986). In these trials, the systems performed at a level comparable to top academic physicians in cases where relevant diseases were contained in the knowledge base. QMR was rated 'helpful' or 'very helpful' as an aid to the physicians' judgment in 72% of cases. Cases in which it performed poorly were much more often due to the relevant diseases being missing from the knowledge base than to faulty inference or inadequate representation. It is for this reason that the painstaking work of extending the knowledge base is continuing.

Decision analysis can provide several important ideas to medical expert systems and AI research in general, but we should recognize some important distinctions. Decision-oriented sensitivity analysis, which is so effective in pruning a decision model down to a manageable size given a specific decision problem, is unfortunately much less useful in limiting the size of a knowledge base intended to cover a large class of potential decision situations. However, such techniques may be used to prune a vast knowledge base down to a small, computationally manageable, subset for a particular case. A human decision analyst and also Holtzman's system use various heuristics for creating this subset. The scheme presented here performs an analogous task in selecting the most probable hypotheses from the vast number ($\sim 2^{600}$) of possible ones. One important difference is that the formal basis of this approach allows the possibility of proving that the identified hypotheses are the most probable (given the knowledge base and case-specific observations) and of computing an upper bound on the error introduced by the simplification.

REFERENCES

Bundy, A. (1985) Incidence calculus: a mechanism for probabilistic reasoning, *J. of Automated Reasoning*, **1**, 263–83.

Chen, K. and Patton, G. T. (1967) Branch-and-bound-approach for decision tree analysis, Stanford Research Institute, Project 188531-172.

Chin, H. L. and Cooper, G. F. (1987) Stochastic simulation of Bayesian belief networks, in *Proceedings of Third Workshop of Uncertainty in AI*, AAAI, Seattle, Wash., pp. 106–13.

Cooper, G. F. (1984) NESTOR: a computer-based medical diagnostic aid that integrates causal and probabilistic knowledge, STAN-CS-84-1031 (Ph.D. dissertation), Dept of Computer Science, Stanford University.

Cooper, G. F. (1987) *Probabilistic Inference Using Belief Networks is NP-hard*, Technical Report KSL-87-27, Knowledge Systems Lab., Stanford University.

Duda, R. O., Hart, P. E. and Nilsson, N. J. (1976) Subjective Bayesian methods for rule-based inference systems, *Proc. Nat. Computer. Conf.* (AFIPS), **45**, 1075–82.

Gorry, G. A. and Barnett, G. O. (1968) Experience with a model of sequential diagnosis, *Computers and Biomedical Research*, **1**, 490–507.

Heckerman, D. and Miller, R. A. (1976) Towards a better understanding of Internist-1 knowledge bases. In R. Salamon, B. Blum, and M. Jorgenson (eds) *MEDINFO 86*, IFIP-IMIA, Elsevier Science, North-Holland, Amsterdam, pp. 22–6.

Henrion, M. (1987) Uncertainty in artificial intelligence: Is probability epistemologically and heuristically adequate? In J. L. Mumpower (ed.) *Expert Judgment and Expert Systems*, Springer-Verlag, Berlin, pp. 105–30.

Henrion, M. (1988) Propagation of uncertainty by probabilistic logic sampling in Bayes' networks. In J. Lemmer and L. N. Kanal (eds) *Uncertainty in Artificial Intelligence*, vol. 2, North-Holland, Amsterdam, pp. 149–64.

Horvitz, E. J., Breese, J. S. and Henrion, M. (1988) Decision theory in expert system and artificial intelligence, *International J. of Approximate Reasoning*, **2**, 247–302.

Howard, R. A. and Matheson, J. E. (1981) Influence diagrams. In R. A. Howard and J. E. Matheson (eds) *Readings in Decision Analysis*, Strategic Decisions Group, Menlo Park, Calif., pp. 763–71 (Chapter 38).

Holtzman, S. (1989) *Intelligent Decision Systems*, Addison-Wesley.

Kim, J. H. and Pearl, J. (1983) A computational model for causal and diagnostic reasoning in inference engines. In *Proc of 8th IJCAI*, Int. Joint Conferences on AI, Karlsruhe, West Germany, pp. 190–3.

Lauritzen, S. L. and Spiegelhalter, D. J. (1988) Local computations with probabilities on graphical structures and their applications to expert systems, *J. Royal Statistical Society*, **B50**(2), 157–224.

Miller, R. A., Pople, E. P. and Myers, J. D. (1982) Internist-1, an experimental computer-based diagnostic consultant for general internal medicine, *New England J. of Medicine*, No. 307, 19 Aug., 468–76.

Miller, R. A., McNeil, M. A., Challinor, S. M., Masarie, F. E. and Myers, J. D. (1986) The Internist1/quick medical reference project—status report, *The Western J. of Medicine*, No. 145, 6 Dec., 816–22.

Pearl, J. (1986) Fusion, propagation, and structuring in belief networks, *Artificial Intelligence*, **29**, 241–88.

Pearl, J. (1987) Evidential reasoning using stochastic simulation of causal models, *Artificial Intelligence*, **32**, 247–57.

Peng, Y. and Reggia, J. A. (1987a) A probabilistic causal model for diagnostic problem solving. Part 1: integrating symbolic causal inference with numeric probabilistic inference, *IEEE Trans. on Systems, Man, and Cybernetics*, SMC **17**(2), Mar./Apr., 146–62.

Peng, Y. and Reggia, J. A. (1987b) A probabilistic causal model for diagnostic problem solving. Part 2: diagnostic strategy, *IEEE Trans. on Systems, Man, and Cybernetics: Special Issue for Diagnosis*,SMC **17**(3), May, 395–406.

Shachter, R. D. (1986) Evaluating influence diagrams, *Operations Research*, **34**(6), 871–82.
Shachter, R. D. (1988) Probabilistic inference and influence diagrams, *Operations Research*, **36**, (July/August), 589–604.
Shortliffe, E. H. and Buchanan, B. G. (1976), A model of inexact reasoning in medicine, *Mathematical Biosciences*, **23**, 351–79.
Winograd, T. and Flores, T. (1986) *Understanding Computers and Cognition*, Ablex.

CHAPTER 18

Optimization in Constraint Networks

Rina Dechter*, *Technion–Israel Institute of Technology, Israel*
Avi Dechter, *California State University, Northridge, CA, USA*
Judea Pearl, *University of California, Los Angeles, USA*

ABSTRACT

This chapter deals with the task of finding an optimal solution to a set of variables constrained by a network of compatibility relationships. The chapter shows that the optimization task does not require an exhaustive search among all consistent solutions but rather can be incorporated naturally into the process of finding consistent solutions. In many problem instances the interaction between the objective function and the constraints does not add any computational complexity to the task of finding a consistent solution, and when it does, the added complexity can be estimated before a solution is attempted, to decide between an exact or a heuristic approach to optimization.

18.1 INTRODUCTION

Traditionally, decision problems have been formulated as optimization takes on probabilistic knowledge bases such as statistical records or influence diagrams. In many applications it is useful to express knowledge in terms of categorical constraints among facts, events and decisions, rather than conditional probabilities. For example, resource limitations, temporal and spatial relationships, class hierarchies and object definitions are stated as black-and-white constraints which allow a limited set of feasible solutions. This categorical abstraction enjoys the advantage of descriptive simplicity because it requires fewer parameters; a model can be specified naturally and qualitatively, and the representation often leads to somewhat simpler inference procedures. For example, such a categorical abstraction (called compatibility relation) is the backbone of the

*This work was performed while the author was at UCLA.

Influence Diagrams, Belief Nets and Decision Analysis
Edited by R. M. Oliver and J. Q. Smith

Dempster–Shafer formalism, which aims at processing evidence when a complete probability model is unavailable (Pearl, 1988, Chapter 9).

Conventional influence diagrams do not offer an explicit representation for categorical constraints among decision variables. For example, to preclude the simultaneous assignment of two teachers to the same classroom in a curriculum scheduling task, we must restrict the set of options available in one decision variable as a function of choices made in others. This is normally done either by testing constraint violations during the development of the diagram into a decision tree, or by assigning each forbidden assignment a utility of $-\infty$. However, such indirect ways of representing constraints tend to mask the identity of variables which constrain one another, thus making it hard to discover a solution consistent with all the given constraints.

An alternative way of representing constraints is to organize the decision variables in a structure called a *constraint network* (Montanari, 1974) which is orthogonal to the probabilistic networks used in conventional influence diagrams. Constraint networks facilitate the control of solution strategies by the topological properties of the constraints (Dechter and Pearl, 1987). By recognizing substructures for which effective algorithms are available, the network representation enables one to determine how problems should be decomposed and which options should be tried first.

Constraint networks have so far been applied to the task of finding feasible solutions (Mackworth and Freuder, 1985; Haralick and Elliot, 1980; Dechter and Pearl, 1987), because in many artificial intelligence (AI) applications preferences play only a minor role and finding one feasible solution suffices. However, there are applications where preferences cannot be ignored. In vision we seek the most likely interpretation of a scene, given visual clues and semantic constraints on objects and their relationships. In diagnostic tasks we seek to replace the smallest set of components that is responsible for the observed behavior. In planning we normally face options leading to conflicting goals, and a strategy is sought that maximizes the overall utility of the consequences, subject to resource and feasibility constraints. It is, therefore, essential to develop efficient algorithms that find an optimal solution, not merely a satisfying one.

This study extends the applications of constraint networks to include optimization tasks, especially distributed optimization. The chapter treats the optimization of linearly decomposed criteria functions over a set of constrained variables. We provide an efficient algorithm that exploits the structure of each problem instance and give simple criteria for assessing its worst case complexity.

A constraint network involves a set of n variables, $X = \{X_1, \ldots, X_n\}$, and a set of constraints. A constraint $C_i(X_{i_1}, \ldots, X_{i_j})$ is a subset of the Cartesian product $R_{i_1} \times \cdots \times R_{i_j}$, where $X_{i_1}, \ldots, X_{i_j}$ is an arbitrary subset of variables and R_k represents the domain of variable X_k. The tuples of a constraint specify all the simultaneous assignments of values to the variables of the constraint which are, as far as this constraint is concerned, legal. An assignment of values to all the variables of the network such that all the constraints are satisfied is called a solution. The term constraint satisfaction problem (CSP) describes the computational task of finding either one solution or the set of all solutions of a given constraint network.

When a network possesses more than one solution, a natural extension of the constraint satisfaction problem is to identify the solution, or solutions, which are 'best' according to some criteria. In this chapter we provide a procedure for finding a consistent solution which maximizes the value of a criterion function of the form:

$$f(x) = \sum_{i \in T} f_i(x^i) \tag{18.1}$$

where $T = \{1, 2, \ldots, t\}$ is a set of indices, denoting subsets of variables $X^1, X^2, \ldots, X^i, \ldots, X^t, x$ is an instantiation of all considered variables, while x^i is the instantiation x restricted to variables in a subset X^i of X. The functions $f_i(x^i)$ are the components of the criterion function, also called an objective function, and are specified, in general, by means of stored tables. We focus first on the special case when the domain of values are real numbers and the objective function is a simple utility function of the form:

$$U(x_1, \ldots, x_n) = \sum_{i=1}^{n} w_i x_i. \tag{18.2}$$

The algorithm, to be presented in Section 18.3, is an adaptation of nonserial dynamic programming methods (Bertelé and Brioschi, 1972) to the case of categorical constraints, and relies on the theory of acyclic databases (Beeri *et al.*, 1983). It is guided by the network structure of each problem instance, and its complexity, for the case of utility function (18.2), does not exceed the complexity of finding an arbitrary solution.

Section 18.2 discusses the structural properties of constraint networks and defines the class of acyclic CSPs. In Section 18.3 we present an efficient scheme for finding a maximum utility solution given that the problem is an acyclic CSP. In Section 18.4 we extend the approach to objective functions whose components are contained within the CSP's constraints. Section 18.5 generalizes the scheme to any objective function and any CSP, Section 18.6 provides a complexity analysis and Section 18.7 provides concluding remarks.

18.2 STRUCTURE OF CONSTRAINT NETWORKS

The structure of a constraint network, represented by a graph, has two forms, called *primal* and *dual*. The *primal-constraint graph* represents variables by nodes and **associates** an arc with any two nodes residing in the same constraint. The *dual-constraint graph* represents each constraint by a node and associates a labeled arc with any two nodes that share at least one variable. The dual-constraint graph can be viewed as the primal graph of an equivalent constraint network, where each of the constraints of the original network is a variable (called a *c*-variable) and the constraints call for equality among the values assigned to the variables shared by any two *c*-variables.

Consider, for example, a CSP with variables A, B, C, D, E, F and constraints on the subsets (ABC), (AEF), (CDE) and (ACE), and assume that the domain of each variable is the set $\{0, 1, 2\}$. The constraint for the variable subset (ABC) is given in Table 18.1; it imposes an order on the 3-tuples and can be expressed concisely as $C_{ABC} = \{(a, b, c) | a \leqslant b \leqslant c\}$. The same constraints apply also to the subsets (AEF) and (CDE). The constraint for the variable subset (ACE) is $C_{ACE} = \{(a, c, e) | a \leqslant c < e\}$, and is given explicitly in Table 18.2. Figures 18.1(a) and (b) depict the primal- and dual-constraint graph of this problem respectively.

A constraint is considered redundant if its elimination from the network (and the possible removal of the corresponding redundant arc from the constraint graph) does not change the set of all solutions. In the dual-constraint graph it is easy to identify redundant arcs and remove them from the graph, since all constraints are equalities. Any cycle for which all the arcs share a common variable contains redundancy, and any arc having all of the variables in its label common in some cycle, can be removed. The graph resulting from such removal of arcs is called a *join-graph*, and the corresponding network is an equivalent representation of the original network.

Table 18.1

A	*B*	*C*
0	0	1
0	1	2
0	0	2
1	1	2
0	0	0
0	1	1
0	2	2
1	1	1
1	2	2
2	2	2

Table 18.2

A	*C*	*E*
0	0	1
0	1	2
0	0	2
1	1	2

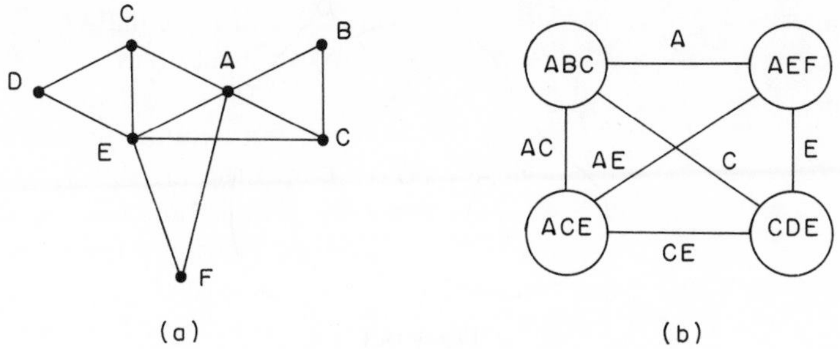

Figure 18.1 A primal and dual constraint graphs of a CSP.

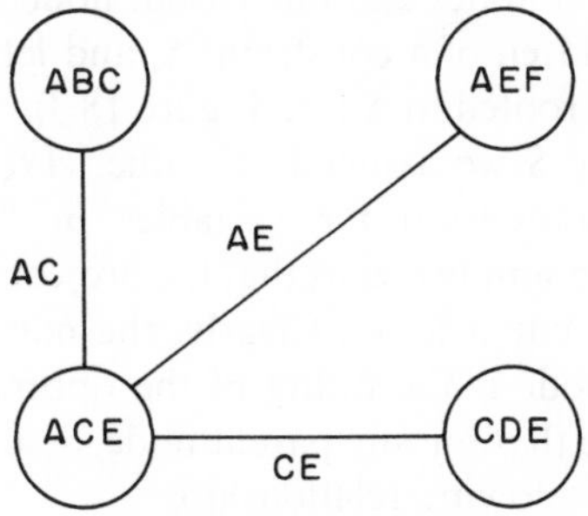

Figure 18.2 A join-tree.

For example, in Figure 18.1(b), the arc between (*AEF*) and (*ABC*) can be eliminated because the variable *A* is common along the cycle (*AEF*)–*A*–(*ABC*)–*AC*–(*ACE*)–*AE*–(*AEF*), so the consistency of variable *A* is maintained by the remaining arcs. By a similar argument we can remove the arcs labeled *C* and *E*, thus turning the join-graph into a *join-tree* (Figure 18.2).

A CSP which possesses a join-tree is said to be acyclic. Solutions for acyclic CSPs can be derived efficiently. If there are *p* constraints in the join-tree, each with at most *l* subtuples, the CSP can be solved in $O(pl \cdot \log l)$ (Dechter and Pearl, 1987).

18.3 FINDING A MAXIMUM UTILITY SOLUTION FOR JOIN-TREES

Let x_S denote a tuple in the constraint $S(S, Q, R$, etc., will stand for variable subsets as well as for the constraints provided for them). If Q is a variable subset of S then let $(x_S|Q)$ denote the projection of x_S on Q. For example, the projection of $x_{AEF} = (0, 1, 0)$ on EF is the subtuple $(1, 0)$. The utility of a subtuple is the restriction of the utility function (18.2) to the tuples' variables.

The join-tree of an acyclic CSP can be made into a directed tree by designating

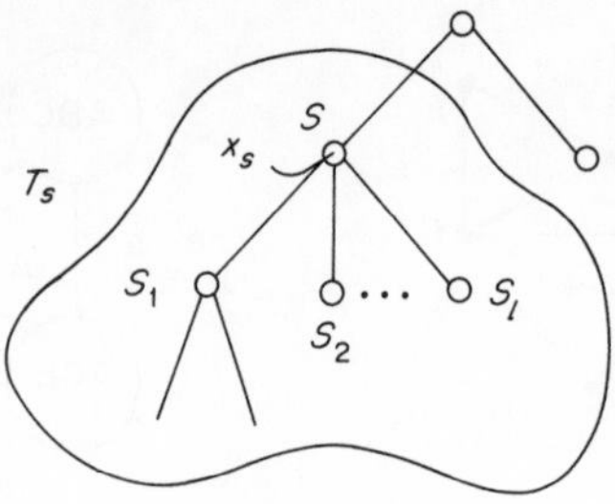

Figure 18.3

one of the nodes as the root, orienting all edges from the root outward and identifying the set of child nodes and one parent node for each of the nonroots. Let $S_1, \ldots, S_l$ be the children of a constraint S, and let T_S denote the subset of variables in the subtree rooted at S (see Figure 18.3).

With each tuple x_S of S we associate a value, $v(x_S)$, which is equal to the utility of the best instantiation for variables in T_S compatible with x_S. Compatibility in this case simply means that the projection of the partial solution on the variables in S is equal to x_S. Clearly, the max v-value associated with the tuples of the root node is the utility of the optimal solution of the entire network. It is easy to see that for any parent node, S, and its children, $S_1, \ldots, S_l$, the v-values satisfy the following relationship:

$$v(x_S) = \{u(x_{T_S}) | (x_{T_S} | S) = x_S \text{ and for every } i, v(x_{T_S} | S_i) = \max v(x_{S_i})\}. \quad (18.3)$$

Namely, the partial solution with the highest utility in T_S compatible with a tuple of S is composed of those subtuples of its child nodes having the highest value among those that are consistent with it. The reason that we can take the best in each child to find a globally optimal solution is that the join-tree property guarantees that all the variables which are shared by the children appear in this parent.

This recurrence suggests that the computation of the v-values can be performed recursively, from the leaves to the root. Values of leaf-nodes tuples are their utility values. A node will compute its v-values only after all its child nodes have computed their own values. For each tuple x_S the parent node chooses from each child node a consistent maximum-value tuple, finds a partial solution in its subtree which is composed of these subtuples, computes its utility and associates this value with x_S. The details of this computation follow from the general recurrance (18.5) to be given in section 18.4. During this recursive process the tree can also be made directional-arc consistent (Dechter and Pearl, 1987), namely, tuples of a constraint which cannot be extended to a solution can be eliminated.

As an example consider the join-tree of Figure 18.2. Here ACE is chosen to be the root of the tree, while ABC, AEF and CDE are its child nodes. Suppose

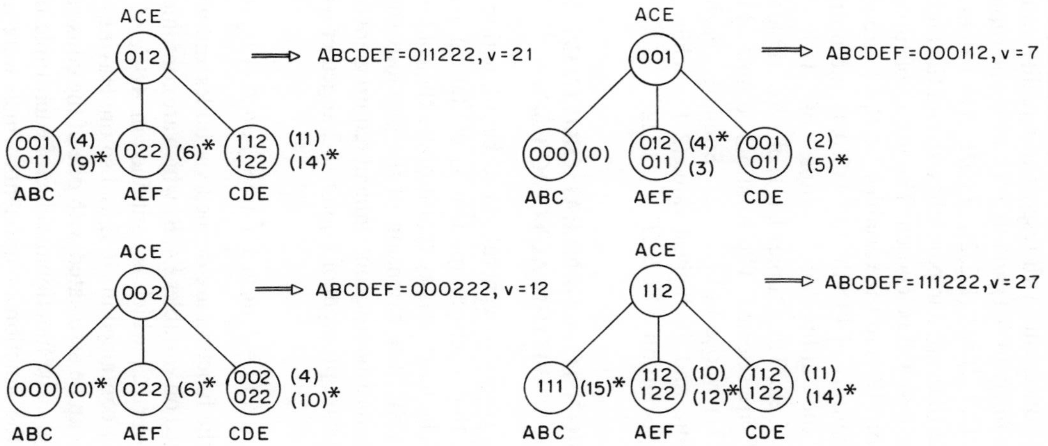
ACE
012
001
011
(4)
(9)*
022
(6)*
112
122
(11)
(14)*
ABC
AEF
CDE
ABCDEF=011222, v=21
ACE
001
000
(0)
012
011
(4)*
(3)
001
011
(2)
(5)*
ABC
AEF
CDE
ABCDEF=000112, v=7
ACE
002
000
(0)*
022
(6)*
002
022
(4)
(10)*
ABC
AEF
CDE
ABCDEF=000222, v=12
ACE
112
111
(15)*
112
122
(10)
(12)*
112
122
(11)
(14)*
ABC
AEF
CDE
ABCDEF=111222,v=27

Figure 18.4

that the utility function is given by

$$u(a,b,c,d,e,f) = 6a + 5b + 4c + 3d + 2e + f. \tag{18.4}$$

Figure 18.4 displays the computation of the values associated with each tuple of the root node ACE (given in Table 18.2). Consider for instance the tuple $(0,1,2)$. This tuple is consistent (with respect to equality constraints) with only two tuples of ABC, namely $(0,0,1)$ and $(0,1,1)$, with one tuple $((0,2,2))$ of AEF, and with two tuples $((1,1,2)$ and $(1,2,2))$ of CDE. The values of these leaf tuples are their utility values (indicated in parentheses), and the highest value (denoted by an asterisk Figure 8.4) will be chosen. The utility of tuples of ABC, derived by restricting equation (18.4) to the variables $\{A,B,C\}$, are given by $u(a,b,c) = 6a + 5b + 4c$ yielding utility of 4 to the tuple $(0,0,1)$ and utility of 9 to the tuple $(0,1,1)$. Similarly, the utility of a tuple of AEF is computed by $u(a,e,f) = 6a + 2e + f$, etc. Thus, tuple $(0,1,1)$ of ABC, tuple $(0,2,2)$ of AEF and tuple $(1,2,2)$ of CDE are selected. The solution, composed of these subtuples, is $(ABCDEF = 011222)$ whose utility is 21. Similarly, each tuple of ACE is associated with a value, and the highest value (27) is achieved for tuple $(1,1,2)$ with a corresponding solution $ABCDEF = 111222$.

18.4 EXTENSION TO A GENERAL OBJECTIVE FUNCTION WITHIN ACYCLIC CSPS

When the objective function is generalized to be a function of the form (18.1) with the restriction that each component is contained within at least one constraint in the acyclic CSP, we say that it obeys the *containment requirement* with respect to the CSP. The extension of the previous scheme to a general objective function that satisfies the containment requirement is straightforward.

Consider the constraint problem of Figure 18.2, together with a new objective function given by

$$f(a,b,c,d,e,f) = f_1(a,b) + f_2(c) + f_3(a,c,e) + f_4(a,f).$$

The components of the function are defined via tables and will not be specified here. Here, f_1 is defined on variables $\{A,B\}$ which are contained in the constraint ABC, f_2 is defined over $\{C\}$ which is contained in ABC, ACE and CDE, f_3 is defined over $\{A,C,E\}$ contained in ACE and so on. Having this property, each function component can be associated with one of the constraints that contains it and, correspondingly, the functional value of each tuple of a constraint can be computed using the function components which were assigned to it. For instance, in our example, the components f_1 and f_2 can be assigned to constraint ABC and therefore each of its tuples will attain the value $f(a,b,c) = f_1(a,b) + f_2(c)$. Similarly, tuples from ACE and AEF will attain $f(a,c,e) = f_3(a,c,e)$, $f(a,e,f) = f_4(a,f)$ respectively. Since tuples from CDE were not assigned any function component they will be assigned a constant function value of '0'.

Let us denote by $f_{/S}$ the objective function components which are assigned to constraint S (e.g. $f_{/(ABC)} = f_1(a,b) + f_2(c)$). As before, for each tuple x_S of a constraint $S, v(x_S)$ stands for the optimal value of $f_{/T_S}$ of a subtuple in T_S which is compatible with x_s (see Figure 18.3). The computation of $v(x_s)$, given the values of all subtuples in the child constraints and given that x_s is consistent with at least one tuple in each of the child nodes, is given by

$$v(x_S) = f_{/S}(x_S) + \sum_{S_i} \max_{x_S|S \cap S_i = x_{S_i}|S \cap S_i} v(x_{S_i}) \tag{18.5}$$

If x_s is not compatible with any tuple of one of its neighboring constraints it should either be eliminated or assigned a negative value which will not enable its inclusion in any solution.

The computation of the v-values will be performed from leaves to root recursively as in the case of the linear utility function. Values of leaf constraints are their assigned restricted objective function. To make the computation of a parent's values more efficient, each child can project its tuples onto the variables common to itself and its parent (i.e. the variables that label the arc between them) and associate each projected tuple with the maximum values among the corresponding child tuples. Namely, a child node will go over its tuples, project each onto the labeled variables and associate with the projection the highest value seen so far. If the projected relation is sorted, searching for a tuple can be accomplished in logarithmic rather than linear time. Thus, if the number of tuples associated with each constraint is bounded by t, projecting the constraint of a child constraint on its outgoing arc and associating the projected tuples with their maximum values is bounded by $O(t \cdot \log t)$. The parent constraint needs to look for the projected tuples that match each of its own tuples. Since the projected constraint is sorted this search takes $O(\log t)$ and, since it is done for each tuple, the overall complexity is $O(t \cdot \log t)$ as well. Thus the whole communication between a child and its parent is bounded by $O(t \cdot \log t)$. Since in acyclic CSPs there are at most n constraints (n being the number of variables), the overall computation is bounded by $O(n \cdot t \cdot \log t)$.

18.5 THE GENERAL CASE

In general, the CSP is not necessarily acyclic and the objective function's components do not obey the containment requirement. Consider the following constraint problem. The variables are denoted by A, B, C, D, E and the constraints are $(AB), (BC), (DE), (BD), (AD)$ and (CE). The objective function is: $f(a,b,c,d,e) = f_1(a,b) + f_2(b,c) + f_3(b,d,e)$. The primal- and the dual-constraint graphs are depicted in Figures 5.18(a) and (b) respectively.

It can be easily verified that the problem is not acyclic (i.e. the graph of Figure 18.5(b) does not possess a join-tree). Also, the function component f_3 is not contained in any given constraint. To deal with such cases we use the

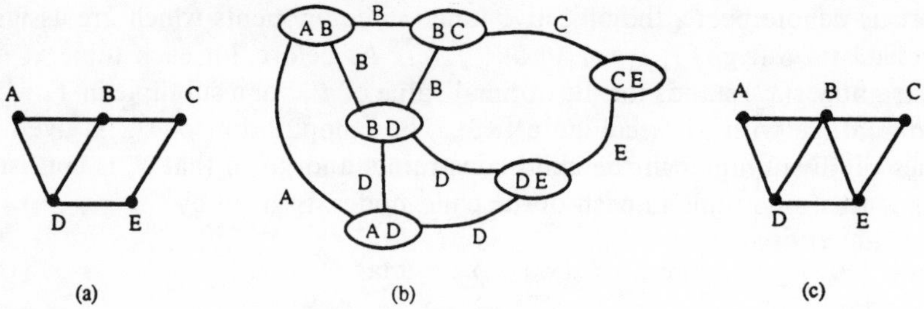

Figure 18.5

tree-clustering scheme presented in Dechter and Pearl (1989), modified to account for the structure imposed by the objective function (in addition to that imposed by the constraints themselves). Wo do that by augmenting the primal graph with arcs which are imposed by the components of the objective function, namely, connecting any two variables residing in the same function component. The resulting graph will be called the *augmented primal graph*. In our example Figure 18.5(c) depicts the augmented primal graph in which the arc (B, E) is added to account for component f_3 in the objective function.

The tree-clustering scheme will transform a general constraint optimization problem into an equivalent acyclic CSP with an objective function that satisfies the containment requirement. Once this transformation is applied, the rest of the problem can be solved using the scheme presented in Sections 18.3 and 18.4. A summary of the tree-clustering scheme is given next.

A CSP is acyclic if and only if its primal graph is both chordal and conformal (Beeri *et al.*, 1983). A graph G is chordal if every cycle of length at least four has a chord, i.e. an edge joining two nonconsecutive vertices along the cycle. A primal graph is conformal if each of its maximal cliques corresponds to a constraint in the original CSP.

The clustering scheme is based on an efficient triangulation algorithm (Tarjan and Yannakakis, 1984) which transforms any graph into a chordal graph by adding edges to it. The maximal cliques of the resulting chordal graph are the clusters necessary for forming an acyclic CSP.

The triangulation algorithm consists of two steps:

1. Compute an ordering for the nodes, using a maximum cardinality search.
2. Fill in edges between any two nonadjacent nodes that are connected via nodes higher up in the ordering.

The maximum cardinality search numbers vertices from 1 to n, in the increasing order*, always assigning the next number to the vertex having the largest set

*The order here is the reverse of that used in Tarjan and Yannakakis (1984) and was changed to simplify the presentation.

of previously numbered neighbors, (breaking ties arbitrarily). Such ordering will be called *m-ordering*.

If no edges are added in step 2, the original graph is chordal, otherwise the new filled graph is chordal. Tarjan and Yannakakis (1984) give a maximum cardinality search algorithm that can be implemented in $O(n + deg)$ where n is the number of variables and deg is the maximum degree. The fill-in-step of the algorithm runs in $O(n + m')$ *where* m' is the number of arcs in the resultant graph. There is no guarantee that the number of edges added by this process is minimal, however, since for chordal graphs the m-ordering requires no fill-in, the fill-in required for nonchordal graphs, is usually small.

The above theory suggests the following clustering procedure for CSPs:

1. Given a CSP and its primal graph, use the triangulation algorithm to generate a chordal primal graph (if the primal graph is chordal no arc will be added).
2. Identify all the maximal cliques in the primal–chordal graph. Let $C_1, \ldots, C_t$ be all such cliques indexed by the rank of its highest nodes.
3. Form the dual graph corresponding to the new clusters and identify one of its join-trees by connecting each C_i to an ancestor $C_j (j < i)$ that contains all variables that C_i shares with its ancestors (Maier, 1983).
4. Solve the constraint–satisfaction subproblems defined by the clusters $C_1, \ldots, C_t$ (this amounts to generating higher-order constraints from the lower-order constraints internal to each cluster, i.e. listing the consistent subtuples for the variables in each cluster).

In order to transform a constraint optimization problem into an acyclic CSP with an objective function that satisfies the containment requirement we will use the augmented primal graph rather than the primal graph as an input to the clustering scheme. Considering our example of Figure 18.5, the augmented primal graph (see Figure 18.5c) is already chordal; however, the maximum cliques do not correspond to the original constraints. The maximal cliques (ABC), (BDE) and (BEC) constitute an acyclic scheme whose join-tree is given in Figure 18.6. Each cluster should be solved as a separate CSP and the

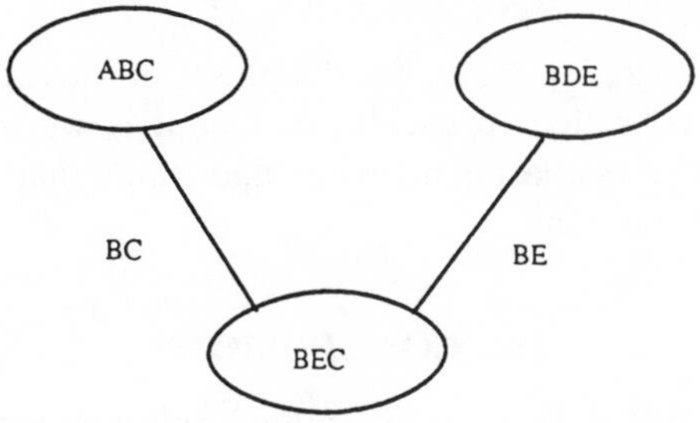

Figure 18.6

components of the objective function, now satisfying the containment requirement, can be assigned to these new constraints.

The complexity of the clustering scheme is dominated by step 4 which requires the solution of the clustered CSPs. If r is the size of the maximum cluster then its solution is bounded by $O(k^r)$ when k bounds the number of values in the domain of each variable. The space complexity is also bounded by $O(k^r)$ which is the size of the new constraint associated with the new cluster. The maximum size of clusters generated can be easily determined by computing the width of the ordered augmented primal graph resulting from the triangulation algorithm. The width of an ordered graph is the maximum width of each of its nodes, and the width of a node is the number of nodes connected to it which precedes it in the ordering. Let d denote any ordering of the variables, and $W^*(d)$ denote the width of the graph resulting from the fill-in procedure. It can be shown that the size of the maximum clique, which is also the size of the maximum cluster, is $W^*(d)$. Thus, the overall complexity of the clustering scheme is bounded by $O(k^{W^*(d)})$. Finding the ordering d for which $W^*(d)$ is minimal is an NP-complete task (Arnborg *et al.*, 1987) and the m-ordering suggested in this chapter is one possible heuristic which provides good results.

The overall computation of our algorithm (i.e. the clustering scheme and the optimal solution of acyclic CSPs satisfying the containment requirement) is given by

$$O(k^{W^*(d)}) + O(t' \cdot \log t')$$

where t' is the maximum number of tuples in each constraint of the clustered CSP. Since $t' \leqslant k^{W^*(d)}$, we conclude that the complexity is bounded by $O(k^{W^*(d)})$.

The nice feature of associating the complexity analysis with the structure of the problem is that we can isolate the additional computation imposed by the optimization task. Comparing the width resulting from processing the primal graph and the augmented primal graph gives us an upper bound on this additional complexity. Let W_p^* be the width of the processed primal graph, and W_a^* the width of the processed augmented primal graph (note that different orderings may be used in the two graphs). Since

$$k^{W_a^*} = k^{W_p^*}\{k^{W_a^* - W_p^*}\}$$

we see that the optimization multiplies the power of the exponent by W_a^*/W_p^*. This ratio can be consulted to decide whether it is worth searching for an optimal solution or to settle for heuristic algorithms that find good but not necessarily optimal solutions.

18.6 CONCLUSIONS

This chapter shows that finding an optimal solution for a CSP need not necessarily involve an exhaustive search among all consistent solutions but can be incorporated naturally into the process of finding a consistent solution. In many problem instances the interaction between the objective function and the CSP does not burden the task of finding a consistent solution, and when it does, the added computation complexity can be estimated ahead of time and be used to decide between an exact or a heuristic approach to optimization.

ACKNOWLEDGEMENT

This work was supported in part by NSF Grant IRI-851234.

18.7 DISCUSSION

18.7.1 Discussion by Ilan Adler

This chapter presents an algorithm to find an optimal solution to a constraint network problem. As such, the ideas presented are clearer and the algorithm and its complexity analysis are clearly presented. However, I have the following comments:

1. The constraint network model was presented in the context of the influence diagram framework. It is not clear by reading the chapter how it relates and applies to problems in this area. Through the comments of some of the participants it was apparent that such relationships exist. It would be useful to explore these relationships explicitly.
2. Technically, this model seems to be related to database models. It would be interesting to see how this work fits in this area (especially how it compares to algorithms developed to similar or identical mathematical models).
3. The claim that one can estimate a priori the extra effort involved in computing solution to problems that do not satisfy the simplifying assumptions (a cyclic network, and restricted objective functions) is somewhat misleading. One can compute a priori only a *bound* on the *worst case* of the extra effort. This information can be useful (due to lack to other criteria) but it should be used carefully.
4. I have the impression that the main drawback of the model is a possible large input (which partially explains how an essentially clearer enumeration leads to good complexity results). It should be very useful to report on some computational results that will help the reader to understand better its applicability under different assumptions.

18.7.2 Reply

1. The relation between constraint networks and influence diagrams is elaborated in Section 18.1 of the chapter. Succinctly, constraint networks provide a condensed graphical representation of the information that flows through the

so-called 'no forgetting' arrows in influence diagrams. Whereas the traditional representation of this information in influence diagrams requires that some temporal ordering be imposed on decisions, constraint networks permit one to articulate this information in a declarative, nontemporal and therefore, more natural manner. For example, they allow us to specify declaratively that only one course can be taught in any classroom at any given time, regardless of whether we choose to first decide on the time of a course, or first decide on its location. Moreover, whereas 'no forgetting' arcs are drawn into every decision node from all preceding decision nodes, the arcs in constraint networks are drawn only between variables that directly affect each other. Consequently, searching for a feasible solution by conventional methods amounts to running naive, chronological backtracking on the decision variables in the influence diagram, while constraint networks, by exposing relevant variables and decomposable substructures, permit a more informed search (Dechter and Pearl, 1987).

It should be noted, though, that the chapter deals only with deterministic problems, so the probabilistic portion of influence diagrams plays no role. A hybrid representation, combining categorical constraints among decision nodes and probabilistic links between decisional and chance nodes, would be a useful enrichment of current influence-diagram technology. In addition to finding decisions that maximize the expected utility, such hybrid representation should also enable us to answer Dempster–Shafer-type queries, e.g. what is the probability that a given decision $X = x$ would be necessary (denoted $Bel(x)$), or feasible (denoted $Pl(x)$), as interpreted in (Pearl, 1988, Chapter 9).

2. To the best of our knowledge, the database literature pays very little attention to problems of optimization. In this application it is normally required to retrieve all answers that match a given query, not one preferred answer. However, if we treat the additive components of the utility function as constrained hyper-edges, database and optimization techniques are often interchangeable.
3. We agree that worst case bounds offer only crude heuristics for the complexity anticipated in actual problems.
4. Computational results for several schemes of solving CSPs are reported in Dechter (1988). Performance measurements on tree-clustering techniques are available in Dechter and Meiri (1989).

REFERENCES

Arnborg, S., Corneil, D. G. and Proskurowski, A. (1987) Complexity of finding embeddings in a k-tree, *SIAM Journal of Algorithm and Discrete Math.*, **8**(2), 277–84.

Beeri, Catriel, Fagin, Ronald, Maier, David and Yannakakis, Nihalis (1983) On the desirability of acyclic database schemes, *JACM* **30**(3), 479–513.

Bertelé, U. and Brioschi, F. (1972) *Nonserial Dynamic Programming*, Academic Press, New York.

Dechter, R. (1988) Constraint processing incorporating backjumping, learning and cutset-decomposition, *Proceedings, IEEE Conference on Artificial Intelligence Applications*, San Diego, Calif., March, pp. 312–19 (to appear in *Artificial Intelligence*, Feb., 1990.

Dechter, R. and Pearl, J. (1987) Network-based heuristics for constraint satisfaction problems, *Artificial Intelligence*, **34**(1), 1–38.

Dechter, R. and Pearl, J. (1989) Tree-clustering schemes for constraint networks. *Artificial Intelligence*, **38**(3) 353–366.

Dechter, R. and Meiri, I. (1989) Experimental evaluation of preprocessing techniques in constraint satisfaction problems, In: *Proceedings Ijcai-89, Detroit, Michegan*, pp. 271–277.

Haralick, R. M. and Elliot, G. L. (1980) Increasing tree search efficiency for constraint satisfaction problems, *AI Journal*, **14**, 263–313.

Mackworth, A. K. and Freuder, E. C. (1985) The complexity of some polynomial network consistency algorithms for constraint satisfaction problems, *Artificial Intelligence*, **25**(1), 65–73.

Maier, David (1983) *The Theory of Relational Databases*, Computer Science Press, Rockville, Md.

Montanari, U. (1974) Networks of constraints: fundamental properties and applications to picture processing, *Information Science*, **7**, 95–132.

Pearl, J. (1988) *Probabilistic Reasoning in Intelligent Systems: Networks of Plausible Inference*, Morgan & Kaufmann, San Mateo, Calif.

Tarjan, Robert E. and Yannakakis, Mihalis (1984) Simple linear-time algorithms to test chordality of graphs, test acyclicity of hypergraphs and seletively reduce acyclic hypergraphs, *SIAM Journal of Computing*, **13**(3), 566–79.

CHAPTER 19

Towards Better Assessment and Sensitivity Procedures

Robert James Korsan, *Decisions, Decisions!*, *Belmont, USA*

ABSTRACT

The development of computer-assisted influence diagram systems and expert system tools to assist decision analysts is developing at a rapid pace. Implementers should take care to use procedures which make full use of the power presented by these tools. The procedures which were appropriate to the computational capabilities of a human (possibly using a hand-held calculator) must be suspect. The current generation of computers provides tremendous power that must be exploited. Our aim is to develop a more complete understanding of the problems being solved than previously available.

I use a simple example to show some pitfalls in current procedures. The information provided by the influence diagram about the existence of dependence is traditionally ignored during sensitivity analysis. The possible consequences are demonstrated in my example. A new procedure is used which avoids some of the previous traps. The new procedure takes less assessment work than traditional methods.

A change in method does call into question the ability of experts to provide other forms of information about their uncertainty. The psychological research undertaken to date has focused on direct questions about the likelihood of a particular event. Additional research is needed on the ability of experts to coherently provide moments or other indirect measures of their probability distribution.

19.1 INTRODUCTION

The procedures used by decision analysts to gain insight into decision problems have slowly evolved over more than a generation. These procedures are very

Influence Diagrams, Belief Nets and Decision Analysis
Edited by R. M. Oliver and J. Q. Smith

effective methods for decoupling uncertainty and complexity in the course of an analysis. However, these procedures have not changed greatly since their creation. The computational tools available to the decision analyst have gained tremendous power. It is my premise that the assessment and sensitivity procedures should take better advantage of the computational power which is now routinely available.

This premise is not new. The first use of some of these ideas is in Howard's paper on proximal decision analysis (Howard, 1971). The ideas presented here have been sharpened by discussions with Tom Parkinson while he was preparing his thesis (Parkinson, 1982) and Peter McNamee during the design of computer-aided software for the analysis of decision trees (McNamee, 1979, 1984). Finally, Dan Owen (1979) and Scott Olmsted's (1983) work on influence diagrams is extremely important.

In most business problems, the description of uncertainty cannot be given in an analytic form. Discrete points must be individually assessed from an expert. The amount of work to get a joint distribution of state variables is staggering. The effort for such assessments rises exponentially with the number of dependent state variables. I will suggest some changes in the method of assessment which, although computationally expensive, can lead to distinctly better insights during the early problem structuring phase of a decision analysis. This chapter will only show the process applied to a simple example. A formal definition of the process will be left to future work.

19.2 A SIMPLE EXAMPLE

The example we present is not meant to represent any real problem. It has been constructed to show some conditions under which our usual procedures would fail. During the structuring phase, the decision analyst has developed a simple influence diagram. Figure 19.1 shows a single investment decision i, and two state variables s_1, s_2, all of which influence the value function $v(s_1, s_2, i)$. The state variables also influence each other. I shall consider only the case of an expected value decision maker for simplicity.

The decision maker faces a binary choice. If we do not invest, we receive nothing. If we invest, the value function is slightly nonlinear and depends upon the state variables as shown in equation (19.1).

$$v(s_1, s_2, i) = \sum_{j=0}^{2} \sum_{k=0}^{2} v_{jk} s_1^j s_2^k \tag{19.1}$$

$$[v_{jk}] = \begin{bmatrix} -i & 52 & -42 \\ -84 & -9 & 0 \\ 84 & 0 & 0 \end{bmatrix}.$$

The value is expressed in millions of dollars. The magnitude of the investment

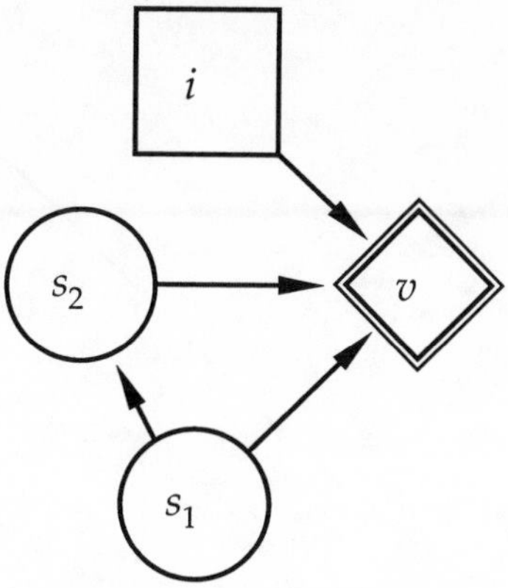

Figure 19.1 The original influence diagram.

i will be approximately $0.25 million. The decision analyst must find out if it will be profitable to invest. Following the usual procedures (Staël von Holstein, 1983) the analyst does a sensitivity analysis to determine which state variables are critical to the decision. First she/he assesses the marginal density of s_1 which is shown in Figure 19.2.

This is one of the most time-consuming aspects of decision analysis. I will use the assessment of a single-valued function such as a cumulative probability distribution as a unit of work. Usually, we ignore conditioning during sensitivity analysis. So our analyst will assess the marginal density of s_2 shown in Figure 19.3.

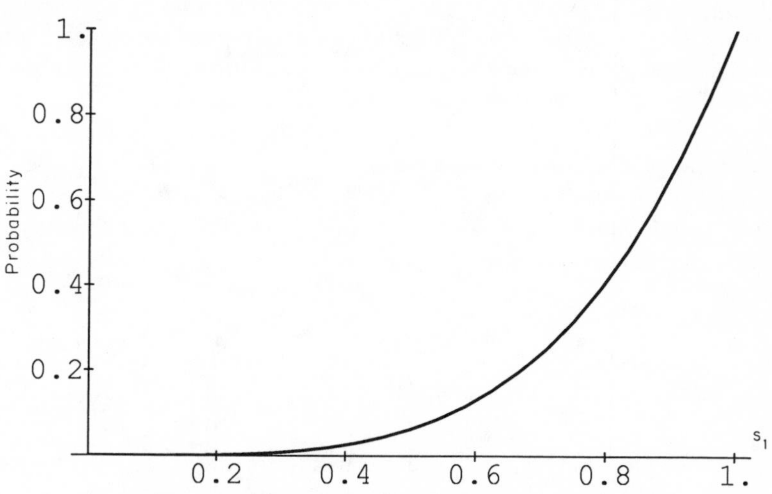

Figure 19.2 Expert marginal assessment of first state variable.

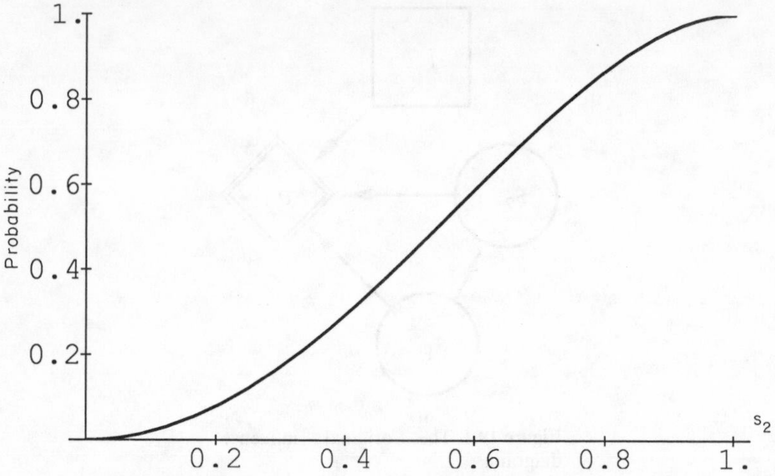

Figure 19.3 Expert marginal assessment of second state variable.

These assessments provide the information used to define a sampling region for sensitivity analysis. The changes in magnitude of the value function over the range of uncertainty allows the decision analyst to estimate changes as if they were produced by a tangent plane located at the medians. She/he uses the changes to create a ranking and then include or exclude state variables from the probabilistic analysis. She/he evaluates the value function at the 10,

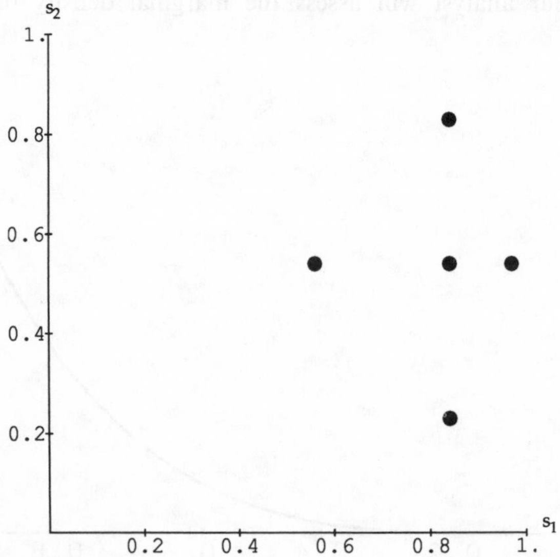

Figure 19.4 Sensitivity samples for traditional assessments.

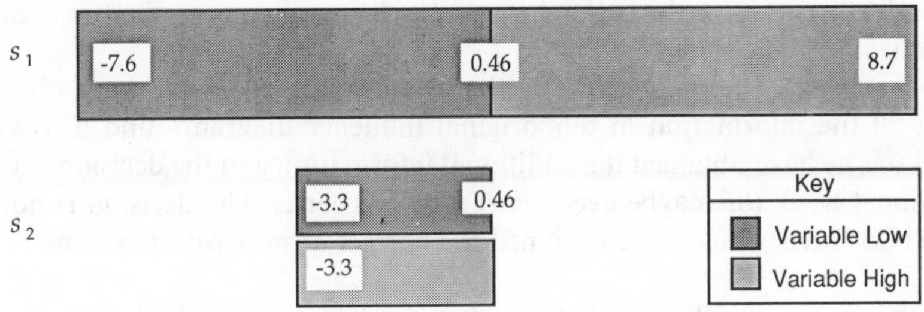

Figure 19.5 Sensitivity chart of the original sampling.

50 and 90 percent fractiles of each cumulative marginal, i.e. at the locations shown in Figure 19.4.

Once she/he has sampled the value function a 'tornado diagram' or sensitivity chart can be created to visually demonstrate the relative impacts, i.e. estimate the effect of each state variable in a planar approximation to the value function about the medians. The results of a sensitivity analysis using the value function of equation (19.1) and the previously assessed marginals are shown in Figure 19.5. We, and our analyst, immediately see that the impact of s_1 is dramatically larger than s_2* (more than a factor of four) and, according to tradition, it is eliminated from further consideration. We note that the effect of s_2 is to decrease the value of the project regardless of direction we move away from the median. This is definitely not planar behavior. However, the effect is small relative to s_1 and we proceed with our analysis under the assumption that s_2 is unimportant.

If the distribution is roughly bell shaped, a three-point approximation to the assessed marginal is usually appropriate. Using this simple approximation, we determine that the expected value, given an investment of \$0.25 million, is \$0.25 million. It is profitable. (What I call the 'rough and ready' three-point discretization is simply treating any distribution as if it were exactly normal. The approximation consists of associating 0.25 probability with the 0.10 fractile location, 0.50 probability with the 0.50 fractile location and 0.25 probability with the 0.90 fractile location.)

If our analyst is not sure this is an accurate result, she/he can do a more accurate computation. The analyst does an exact discretization using the conditional mean method† at the cost of an additional three function evaluations. She/he now computes an expected profit, given investment, of \$0.06 million. The decision analyst still recommends an investment. (It is somewhat

*This is the likely assumption if an automated aid is simply following our traditional rules of thumb.

†The conditional mean method is described in Appendix III. A one-dimensional error analysis is also presented.

discomforting to have the value drop so much, simply due to a more accurate discretization.)

So we invest. The question is: 'Is that the decision we would have obtained using all the information in our original influence diagram?' and if it is not, could she/he have obtained the additional information and the decision without additional work (or maybe even less?). The answer is 'The decision is not the same and she/he could have obtained the correct answer with the same or less work.'

In the next two sections I shall develop an alternative method of assessment which requires no more work than the assessment of marginals performed above. The method develops an approximate joint distribution directly from the assessments and thereby avoids the potential loss of information present when assuming independence. I will then use this alternative assessment, along with a continuous approximation to the value function to reanalyze our sample problem. Finally, the results obtained by the new methods are compared to our traditional methods and some directions for future research are indicated.

(Appendix II discusses our traditional sensitivity analysis and shows the incomplete use of information in our procedures. Combined with poor discretization accuracy, our techniques can lead to serious error. With a diligent analyst, this error is not likely to be overlooked. This example should, however, caution the developers of automated decision aids and influence diagram tools to implement better procedures for the diagnostic phases of a decision analysis.)

19.3 A DIFFERENT ASSESSMENT PROCEDURE

The influence diagram of Figure 19.1 shows that we can decompose the joint distribution into the following: $\{s_1, s_2 | \mathfrak{I}\} = \{s_2 | s_1, \mathfrak{I}\}\ \{s_1 | \mathfrak{I}\}$. Initially, I proceed as before. The analyst has already assessed ${}^c\{s_1 | \mathfrak{I}\}$ as shown in Figure 19.2. Suppose, however, I assess the conditional mean $\langle s_2 | s_1, \mathfrak{I}\rangle$ and the region of support (domain) of the s_2 marginal, instead of the usual marginal. (The support of a density is the region where the density is non-zero.) Figure 19.6 shows the result of the assessment. The assessment effort should be the same in either case.

Using additional analytic work, I can use this information to better advantage than our traditional procedures allow. I express the conditional density $\{s_2 | s_1, \mathfrak{I}\}$ as the Taylor series given in equation (19.2).

$$\{s_2 | s_1, \mathfrak{I}\} = \sum_{\mu=0}^{\infty} c_\mu(s_1) s_2^\mu. \tag{19.2}$$

The coefficients of this series may be formally determined by multiplying both sides of equation (19.2) by appropriate powers of s_2 and integrating over $\Omega(s_2)$, the support of s_2 given the first state variable. In this example and in most business examples, the support of the distribution is over a finite interval. Thus, the moments serve to uniquely define the series representation of the distribution.

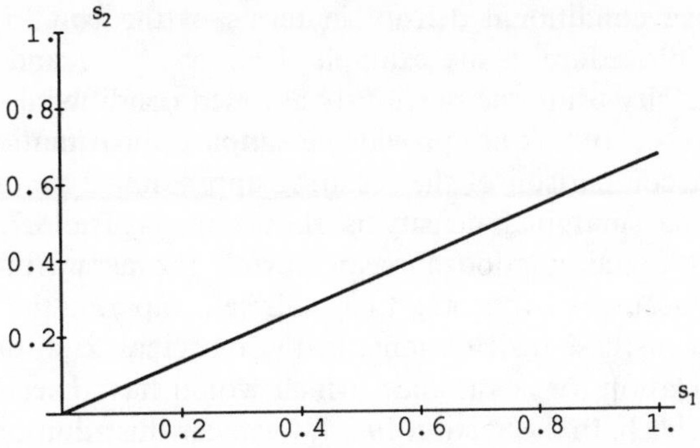

Figure 19.6 Expert assessment of the conditional mean $\langle s_2 | s_1, \mathfrak{I} \rangle$.

The result of such formal manipulation is given in equations (19.3a) and (19.3b):

$$\langle s_2^{\beta} | s_1, \mathfrak{I} \rangle = \sum_{\mu=0}^{\infty} \alpha_{\beta\mu} c_{\mu}(s_1) \quad \text{for } \beta = 0, 1, 2, \ldots \tag{19.3a}$$

where the coefficients (α) are given by

$$\alpha_{\beta\mu} = \int_{\Omega} s_2^{\beta+\mu} \, ds_2 \qquad \beta, \mu = 0, 1, 2, \ldots. \tag{19.3b}$$

If each series is truncated at fixed $\beta = \mu = n$, we have a system of $n + 1$ linear equations. The solution to this system gives the coefficients of an approximate

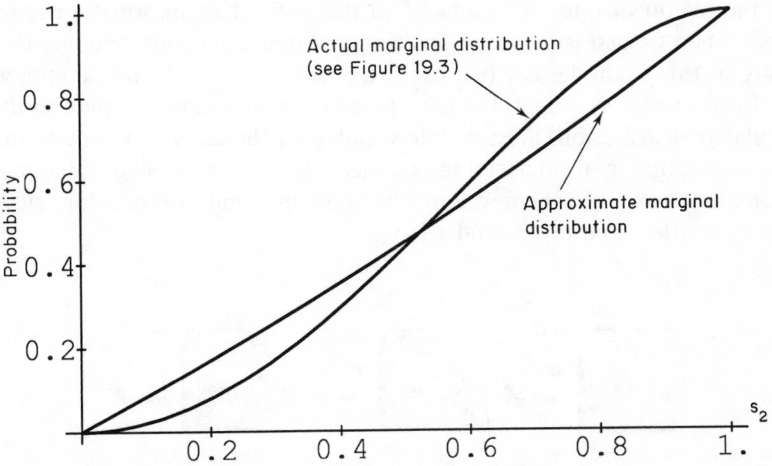

Figure 19.7 Comparing the actual and approximate cumulative distribution using the new assessment technique.

expansion of the conditional density in terms of the conditional moments. Following this procedure in my example, I choose $n = 1$ and determine this approximate density using the previously assessed conditional mean shown in Figure 19.6. This procedure has provided a simple approximation to the actual joint density. A comparison of the resulting approximate marginal density of s_2 and the actual marginal density is shown in Figure 19.7. The resulting marginal seems to be in good agreement with the actual marginal. I have sacrificed some accuracy (Appendix I has a table comparing the mean, variance and correlation of these distributions) in the marginal to obtain information about the correlation (or covariance) which would have been ignored in our traditional approach. In the Section 19.3.1, I will use this additional information to create a pattern of samples for the value function which, hopefully, provides more information for sensitivity analysis and approximations which are useful for computing expected values and solving decision problems.

19.3.1 The region of support for the density

Covariance in a distribution arises from two sources: first, the conditional structure of the decomposition of the joint density, and second, the shape of the region of support of the joint density. As part of my procedure, I implicitly estimate the region of support. (The algorithm used to estimate the support is outlined in Appendix I.) This is necessary to define the approximation and I want the information in order to perform a more accurate sensitivity analysis and to estimate expected values. A sequence of estimates is produced by the procedure which converges to a region cutting off the upper left-hand corner of $[0, 1] \times [0, 1]$. Figure 19.8 shows the first few iterates for this example. Notice that the region of interest is skewed to the right. This information is totally missed when we assess only marginals and assume the support is over the unit square in this example. On this basis, I choose a set of sample points which are also skewed to the right. I obtain these points by examining the conditional cumulative distributions in each state variable for the assessed joint distribution and determining their intersections. (A more detailed discussion of the purpose of sampling as well as the development of a consistent criterion for sensitivity analysis is attempted in Appendix II.)

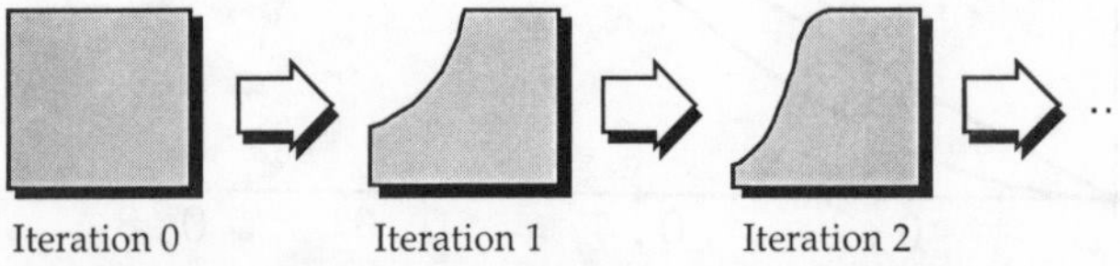

Figure 19.8 The sequence of iterates for the contractive map defining the region of support of the approximate density.

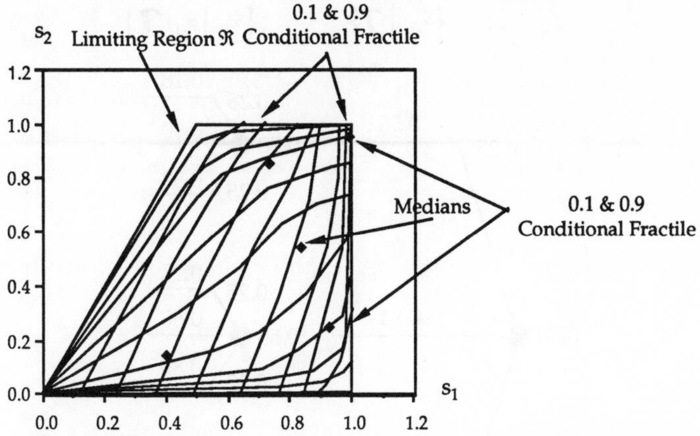

Figure 19.9 Matching the sensitivity points to the assessed distribution. Note how the conditional fractiles define a 'natural' curvilinear coordinate system over the region of support of the density. This coordinate system should give us insight into the optimal sampling structure necessary for sensitivity analysis and for estimation of expected values.

Figure 19.9 shows the resulting sampling points used for sensitivity analysis (and, later, expected value computations). This pattern attempts to place the sampling at points which are surrounding the region of probabilistic interest, denoted $\Re$ in Figure 19.9. I do not have a criterion which characterizes the best sampling points for our purposes. I have used the intersections of the 0.1 fractiles and the 0.9 fractiles since points within this region are likely to be of interest at least 80 per cent of the time by this construction. This seems to be a reasonable first guess. The characterization of 'best' and determination of an algorithm for choosing 'best points' is another area worthy of future research.

Once sampling has been performed, we will use first and second-order terms to perform a sensitivity analysis. This option is available even if we choose to use traditional methods. In each case, examination of these terms calls for the inclusion of both state variables. Then and only then will we have enough information about the value function to properly estimate expected values and make decisions. In essence, our traditional procedures have only relied on the first-order effects to make this decision. When second-order effects are taken into account, we will include the variation of both state variables.

The analysis of Appendix II could have been performed during our traditional procedures. We would have concluded that both state variables are necessary and we should assess conditional distributions. Usually, we discretize the marginal of s_1 to determine where to assess conditional distributions for s_2. We discretize again, using the conditional mean method to set up and roll back a tree. Such a tree is shown in Figure 19.10.

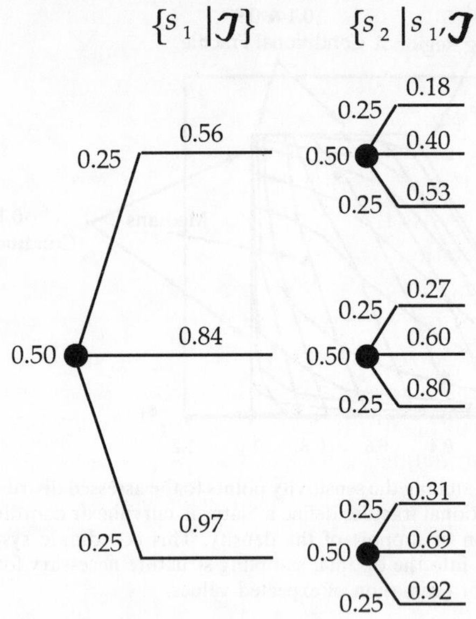

Figure 19.10 Discretized conditional assessments represented as a tree.

The result of using this information in my sample problem is to arrive at an expected value of −\$1.72 million. Hmm, the recommended decision has changed! (And drastically so.) So far, following traditional procedures has resulted in various expected values and different decisions. It is problematical that the decision has changed with differing levels of analysis. I would not be very sure about the results and would take care to explore further. However, a quick diagnostic, going no further than traditional sensitivity analysis would not have determined that care should be taken. In particular, automated tools usually engender a false sense of security in those who use them. The tools must be carefully designed to be as robust as possible.

In all of our analysis we have been using discrete methods to determine expected values and choose among alternatives. In Section 19.4, we shall explore some alternatives which can provide us with better accuracy for comparable work.

19.4 COMPUTING THE EXPECTED VALUE

There are usually two sources of computational effort in a decision analysis. The first is the assessment effort and the second is the evaluation of a complex profit model or value function. The method we have devised in Section 19.3 seeks to decrease the assessment effort necessary while providing more information than is available from just the marginals. A second objective is to provide information about where a good approximation to the value function is needed for accurate computation of expected values. I will now turn to the second area of computational effort, the accurate computation of expected values (for purposes of choosing among alternatives) with the same or less effort than using discretization-based methods.

Please notice that I computed five values at the time we performed our original sensitivity analysis. If we do a 'rough and ready' computation of the expected value along either state variable, we can reuse our original value samples. We will assume that a three-branch discrete approximation to a distribution is sufficient for our purposes. Once we discretize the distribution exactly, I must find an additional three samples of the value function to compute the expected value. This is an increase of 60 percent in work (8 points versus 5). Moreover, because of poor sensitivity analysis we excluded a necessary state variable and the recommended decision was still wrong.

Had I done better sensitivity analysis, I would still have to assess three conditional distributions and then discretize those and sample at the combinations. The additional assessments represent a 150 percent increase in assessment work (two marginals plus three conditionals versus one marginal plus one conditional mean). The additional nine samples represent a 180 percent increase in evaluation work (14 points versus 5), to compute the expected value. This seems like a waste of effort, particularly as we encounter more complicated joint distributions and value functions.

There are other ways of approximating the value function using our original sampling. If we choose a method which represents the value function in a continuous manner, we can integrate analytically for the expected value. In particular, I have found methods of approximation, borrowed from the finite

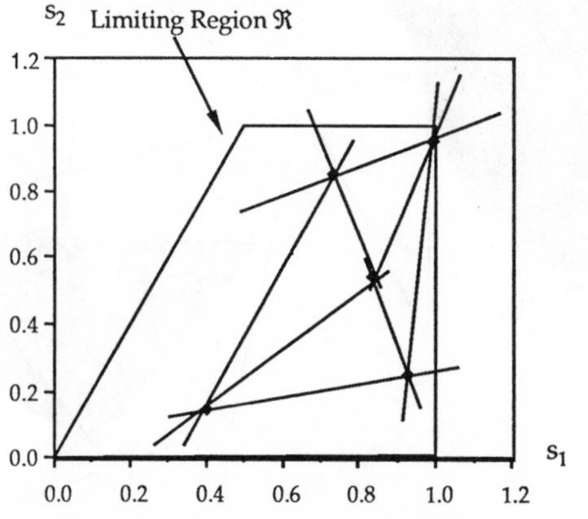

Figure 19.11 A particular triangularization of the sampled area of the support of the approximate density.

element method (Strang and Fix, 1973), that can be extremely effective. (An overview of the application of these techniques to decision analysis is given in Appendix III.)

The process of approximating the value function is as follows. First, I use only the points already sampled in Figure 19.9. I can represent the value function as a C^0 function (a C^0 function is one which may not have a continuous first derivative) by piecing together planar interpolations over each triangular region. If I use the points shown in Figure 19.11, I can extrapolate to the entire region. This process can be developed adaptively to any degree of accuracy desired. (A simplified error analysis is presented in Appendix III.)

A rough comparison of the resulting value functions is shown in Figures 19.12(a) and (b). We are looking at the value surface from slightly above and to the upper right of the sensitivity region. We see that the approximate surface lacks the high curvature of the actual surface, but the overall behavior is generally what we would like. The approximation shown here is for a nine-point sampling.

Since our example problem is polynomial in nature, we can compute the exact value lottery and it is shown in Figure 19.13(a). The exact expected value is $-\$1.72$ million. The value lottery corresponding to the tree developed in Figure 19.10 is displayed in Figure 19.13(b). We are also in a position to compute

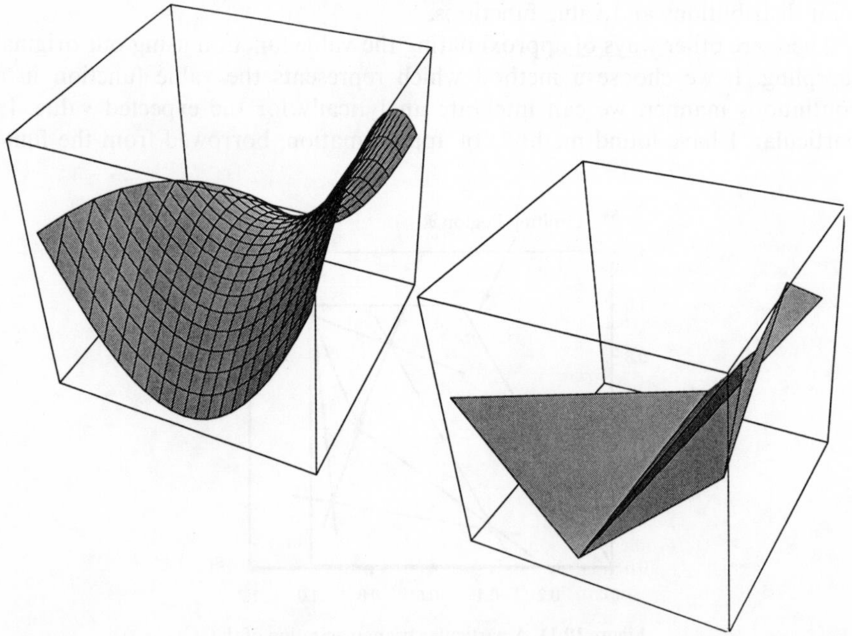

Figure 19.12 (a) The actual surface over the implied region of support; (b) a nine-point approximation to the value function over the implied region of support.

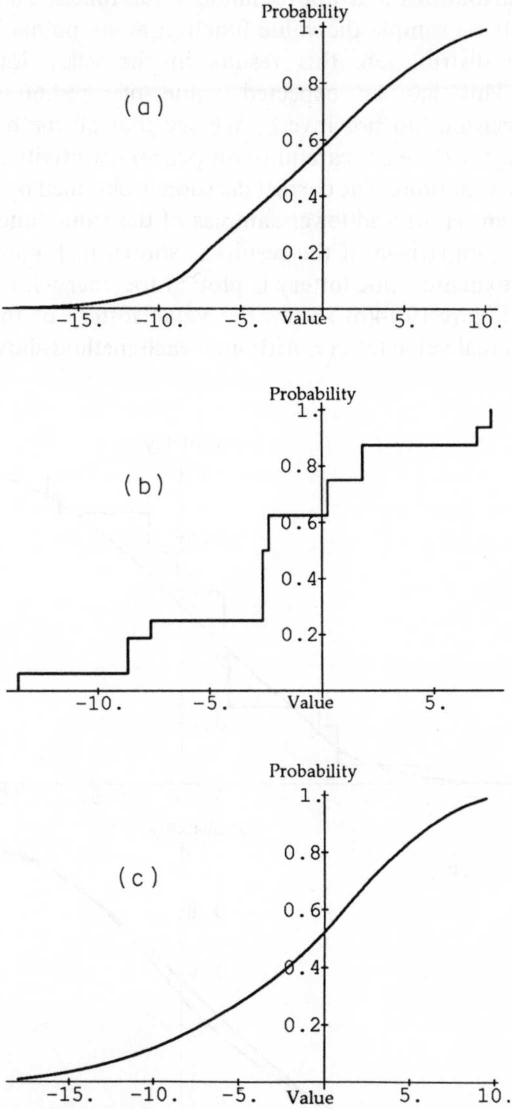

Figure 19.13 (a) The value lottery when our decision problem is solved exactly; (b) a nine-point discrete approximation to the value lottery, given exact assessments; (c) the value lottery using a six-point approximation to the value function and the new method of assessment to approximate the joint density.

an analytic solution with our new approximation method since all quantities (approximate distributions and approximate value functions) are represented by polynomials. If we sample the value function at six points in the region of the approximate distribution, this results in the value lottery shown in Figure 19.13(c). This has an expected value of −\$1.66 million and a corresponding decision 'do not invest'. We see that all methods provide the same negative answer if we are careful to do proper sensitivity analysis and use high degree approximations. The correct decision is obtained by the new method with less assessment effort and fewer samples of the value function.

A more direct comparison of the results is shown in Figure 19.14(a) where the discrete approximate value lottery is plotted together with the actual value lottery. Finally, Figure 19.14(b) shows the value lottery of the new approximation and the actual value lottery. Although each method shows discrepancies

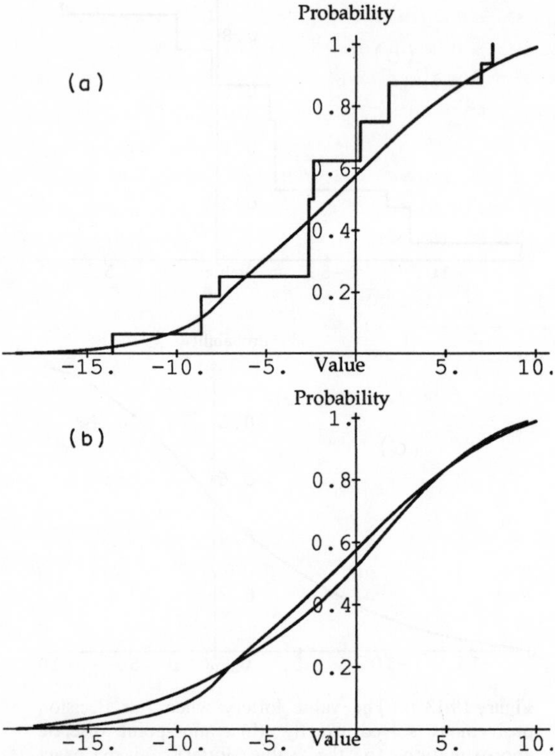

Figure 19.14 (a) A comparison of the actual distribution and the 9 point discrete approximation resulting from the tree shown in Figure 19.10. (b) A comparison of the actual value lottery and the result of using the new approximation and assessment techniques.

from the correct answer, we note that the new approximation 'hugs' the actual value lottery, crossing and recrossing the actual distribution.

The results seem to indicate that the new techniques are considerably better than traditional techniques, from a sensitivity point of view, from a decison making point of view and from an effort expended point of view. During sensitivity analysis, the joint distribution can be directly assessed with no more work than two marginals. We can use this information for a more complete sensitivity analysis. Lastly, the amount of work associated with these techniques (function evaluations and elicitation sessions) seems like a useful measure of success for each. A comparison is given in Section 19.5.

19.5 A COMPARISON OF THE DIFFERENT METHODS

Table 19.1 shows how each method detailed above compares with the other methods. The proposed method not only gives the correct decision, but requires the same or less work (in evaluations of the value function and assessment effort) than traditional methods. Of course, without the availability of a computer assistant for the decision analyst these methods are impractical. The computation necessary to arrive at meaningful results, which can be shown to experts in real time, is significantly more than an analyst can perform with a hand-held calculator. However, the current crop of influence diagram tools and the availability of prodigious computer power would seem to negate this limitation. Implementers should take this opportunity to incorporate more

Table 19.1

Decision (do not invest)	Expected value ($m)	Work in assessments	Work in function evaluations	Method
Correct	– 1.72	NA	NA	Exact
Correct	– 1.66	2 units	1 unit	Proposed procedures
Correct	– 1.72	4 units	2.8 units	Traditional 2 variables, conditional mean discretization
Incorrect	0.25	2 units	1 unit	Traditional 1 variable, rough and ready discretization
Incorrect	0.06	2 units	1.6 units	Traditional 1 variable, conditional mean discretization

accurate methods, rather than blindly use traditional methods developed for other environments and circumstances.

19.6 NEEDED RESEARCH

There are several research issues which must be addressed:

- The ability of experts to meaningfully answer questions about moments.
- The appropriate mathematical approximations for the joint distribution.
- The appropriate mathematical approximations for the value function.
- Numerical/analytic techniques necessary for implementing these new methods.
- Appropriate graphical representations for assessing information.
- Appropriate graphical representations for providing feedback.
- Methods of verifying the quality of approximations inferred.

I believe this is an opportunity to simplify the procedures and increase the power of the practical decision analysis paradigm. This research can be as innovative as the original foundations of our practice developed over two decades ago.

19.7 DISCUSSION

19.7.1 Discussion by Allen C. Miller

These comments refer to the version of the chapter that the author presented at the conference on 9–11 May 1988, plus about 30 pages of calculations that he prepared after the conference. Although the author agreed at the conference that he needed to correct problems with the chapter, he has not distributed a revised version.

The chapter and the later set of calculations present the same decision problem, but they obtain different results (including the optimum decision policy) for the exact solution to the problem. They also have different results for the various approximation methods, including some traditional approximations used in many decision analyses. The chapter should be revised to reflect the new calculations.

The chapter seeks to motivate the need for better approximation methods in decision analysis by constructing an example in which traditional methods give the wrong decision (i.e. a decision that is not optimal according to the exact solution). However, the example is sensitive to small errors in the calculations, such as rounding intermediate results to a few digits. The optimum decision associated with most of the approximation methods changed when the calculations were redone, apparently because they were recomputed with greater accuracy. It is difficult to see how we can generalize from the results of this example.

The author needs to show that the proposed approximation method is superior to traditional ones for a variety of decision problems that somehow span the range of problems we normally encounter in practice. Defining this range of decision problems will be difficult, but we need to do so before we can say with confidence that one approximation method is better than another. This does not mean that the proposed method is inferior to traditional ones, but we cannot judge them using only one (somewhat artificial) decision problem.

It would help others apply the proposed method to a variety of decision problems if the chapter explained each step in enough detail to allow readers to re-create the calculations without referring to other documents. In particular, the reference to the finite element method (on p. 452) does not show how the method is used.

In spite of these criticisms, the author should be commended for exploring an area in which we have few practical results to guide a decision analysis. Influence diagrams show us how to exploit assumptions about probabilistic independence to simplify decision analyses. However, they do not help when we are unable to make enough independence assumptions. We need a way to assess and process dependent probabilities efficiently. If we can find generally applicable methods for doing so, we could make significant advances in our ability to analyze and model complex decision problems. This chapter does not appear to present such a method, but it is a step in the right direction.

[Editor comments—the author accepted these criticisms and modified the final version of the chapter in the light of these comments.]

19.7.2 Reply

I would like to thank Samuel Holtzman, Peter McNamee, Allen C. Miller, and James Smith, as well as the other reviewers for their comments. Several of their suggestions led to much stronger conclusions and a better chapter.

REFERENCES

Howard, Ronald A. (1971) Proximal decision analysis, *Management Science*, **17**(9), May, 507–541 (1972).

McNamee, Peter (1979, 1984) Decision Analysis Group, SRI International, private communication (1979), and Strategic Decisions Group, Mento Park, Calif., private communications (1984).

Olmsted, Scott M. (1983) On representing and solving decision problems, Ph.D. dissertation, Stanford University, Stanford, Calif., December.

Owen, Daniel (1979) The use of influence diagrams in structuring complex decision problems, Ph.D. dissertation, Stanford University, Stanford, Calif., June.

Parkinson, Thomas Worthington (1982) Using complex profit models in decision analysis, Ph.D. dissertation, Stanford University, Stanford, Calif., July.

Staël von Holstein, Carl-Axel S. (1983) A tutorial in decision analysis. In *Readings on the Principles and Applications of Decision Analysis*, vol. 1, Strategic Decisions Group, Menlo Park, Calif., pp. 129–58.

Strang, Gilbert and Fix, George (1973) *An Analysis of the Finite Element Method*, Prentice-Hall, Englewood Cliffs, NJ.

APPENDIX I

An Algorithm for Approximating the Joint Density in this Example

The distribution used in our example is given by $\{s_1, s_2\} = 8s_1s_2$ over the simplex (triangle) defined by the points (0,0), (1,0) and (1,1). The density is shown in Figure AI.1(a). As you can see, it heavily weights the region to the upper right. Figure AI.1(b) shows this density with the vertical axis compressed to clarify its behavior.

The combination of nonlinearity and the fact that the region of support of our density is not a square, is the geometric reason for the failure of traditional sensitivity analysis techniques which usually assume such a square region. The assessment of the conditional mean gives the first indications that the region is not rectangular. An approximate joint distribution can be developed using the procedure outlined below.

A critical step in this algorithm is determining the region of support of the approximate density. The first approximation is to choose the entire square just as is traditional. Once this has been done, the conditional cumulative can be calculated. The surface in Figure AI.2 is the result of that calculation. As you can see, the upper left-hand corner is not part of the region of support. The next iteration uses the region pictured for its region of integration and the process is repeated. The process continues to iterate as a contractive map.

Formally, we can state the algorithm in vector/matrix terms. For $n = 1$, we rewrite equations (19.3a) and (19.3b) as

$$\mathbf{C}^{(i)} = \mathbf{M}^{-1}(\varphi_i)\mathbf{B}$$

$$\mathbf{B} = \begin{bmatrix} 1 \\ \langle s_2 | s_1, \mathfrak{I} \rangle \end{bmatrix}$$

$$\mathbf{C} = \begin{bmatrix} c_0(s_1) \\ c_1(s_1) \end{bmatrix}$$

$$\mathbf{M}(\phi_i) = [\alpha_{\beta\mu}(\phi_i)]$$

where the next region of integration is defined by the condition

$$\varphi_{i+1}(s_1) \ni \int_0^{\phi_{i+1}(s_1)} (c_0^{(i)}(s_1) + s_2 c_1^{(i)}(s_1))\mathrm{d}s_2 \equiv 1 \,\forall s_1.$$

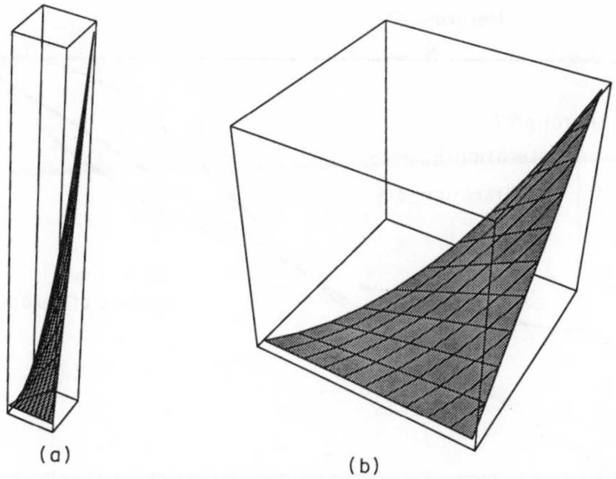

Figure AI.1

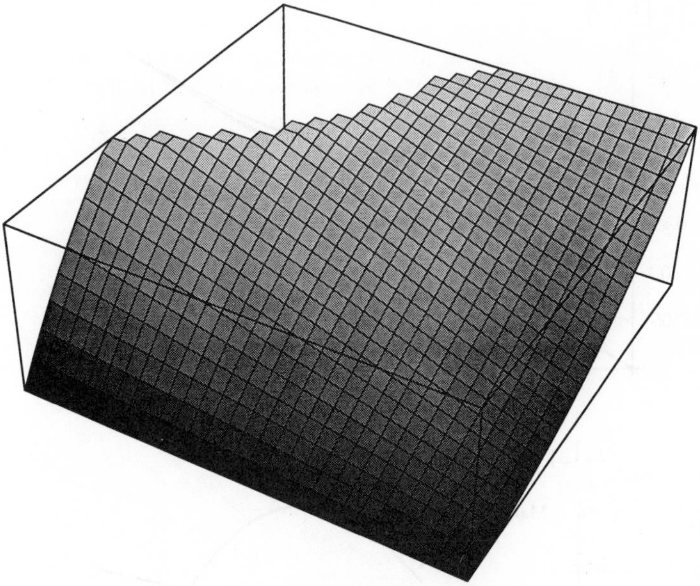

Figure AI.2

The sequence of iterates shows that the approximation converges to a limiting region. The first few iterates of the first half of the region are shown in Figure AI.3.

The conditional mean which was assessed provides a verification of our procedure. Each iteration of the algorithm results in a conditional mean which

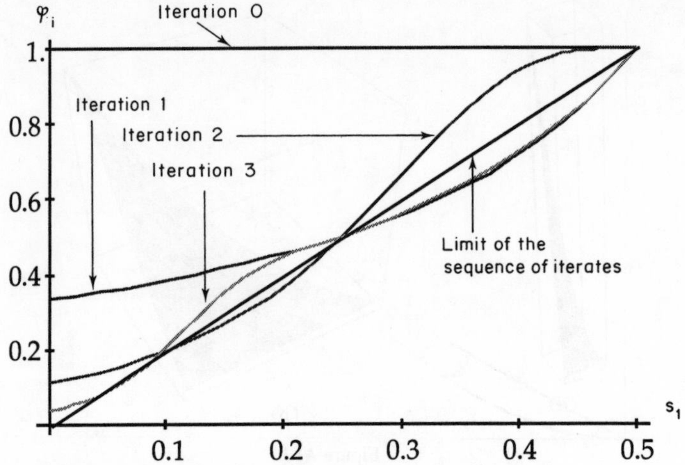

Figure AI.3

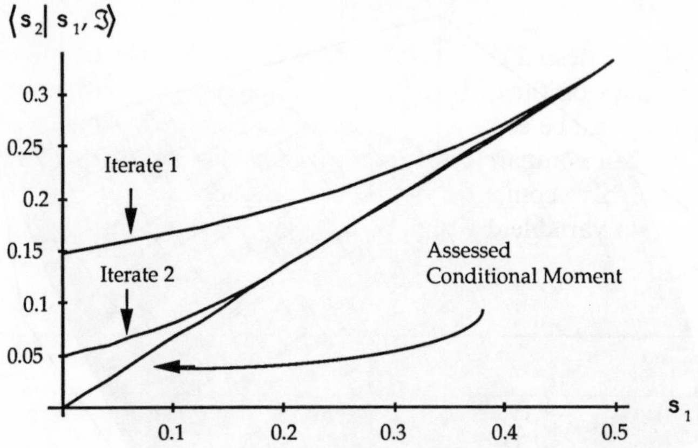

Figure AI.4

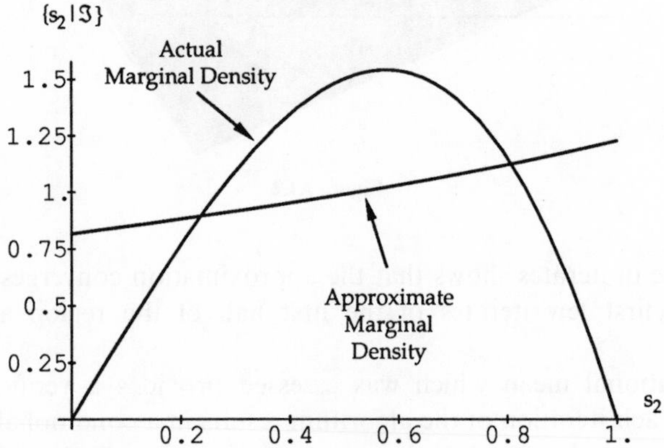

Figure AI.5

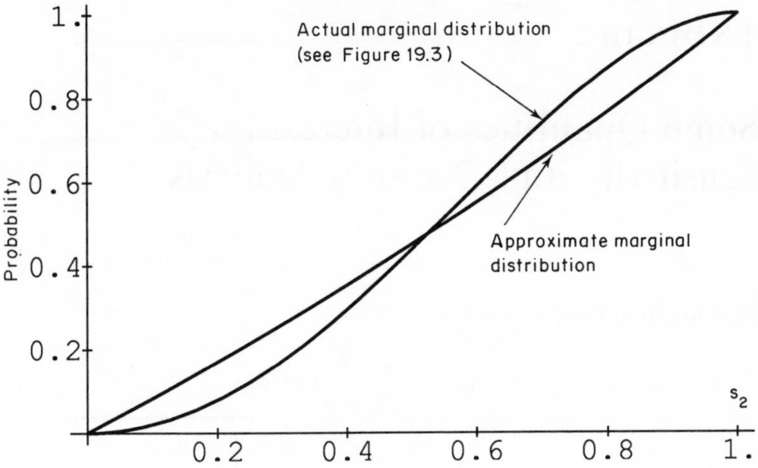

Figure AI.6

converges to the assessed conditional mean, as can be seen in Figure AI.4. As a very rough check on the quality of our resulting approximation, the resulting marginal density can be compared to the actual marginal in Figure AI.5. When integrated, we can compare the cumulatives of the marginals in Figure AI.6. Finally, Table AI.1 compares several characteristics of the exact and approximate two variable distributions.

Table AI.1

Exact density				Approximate density			
$\langle s_1 \mid \mathfrak{I} \rangle$	0.80	$\langle s_2 \mid \mathfrak{I} \rangle$	0.53	$\langle s_1 \mid \mathfrak{I} \rangle$	0.80	$\langle s_2 \mid \mathfrak{I} \rangle$	0.53
${}^{\sigma}\langle s_1 \mid \mathfrak{I} \rangle$	0.16	${}^{\sigma}\langle s_2 \mid \mathfrak{I} \rangle$	0.22	${}^{\sigma}\langle s_1 \mid \mathfrak{I} \rangle$	0.16	${}^{\sigma}\langle s_2 \mid \mathfrak{I} \rangle$	0.29
Correlation	0.51			Correlation	0.44		

APPENDIX II

Some Quantities of Interest for Sensitivity and Decision Analysis

The value function used is given by

$$v(s_1, s_2, i) = \sum_{j=0}^{2} \sum_{k=0}^{2} v_{jk} s_1^j s_2^k$$

$$[v_{jk}] = \begin{bmatrix} -i & 52 & -42 \\ -84 & -9 & 0 \\ 84 & 0 & 0 \end{bmatrix}.$$

Figure AII.1 shows this function over the unit square. The function is somewhat nonlinear, but overall rather unremarkable.

Our traditional procedure for sensitivity analysis is based on the behavior of the tangent plane to the value function about the medians of the marginal distributions as assessed from appropriate experts. A variable is accepted or rejected for inclusion in full blown probabilistic analysis based on a comparison of the components of the total differential of first order

$$\mathrm{d}v = \left| \frac{\partial v}{\partial s_1} \Delta s_1 + \frac{\partial v}{\partial s_2} \Delta s_2 \right|.$$

Geometrically, we measure the increase in the value that results from varying a particular state variable over the range of uncertainty estimated from its marginal distribution. If we have an influence diagram which declares the possible existence of dependence, we typically ignore this insight when performing sensitivity analysis.

This can lead to error arising from several sources. First, the range of uncertainty may not be as large as determined by the marginal. Depending upon the behavior of the value function this may lead to incorrect conclusions. Second, we exclude variables if the increase in value is 'small' by comparison to the increase which results from including all variables. The key observation is that this analysis describes the behavior of the tangent plane, not necessarily the value function itself.

Another question should be (and is not) asked, namely: 'How good is the tangent plane as an approximation to the value function over our joint range of uncertainty?' To answer this question, we must compute the total differential

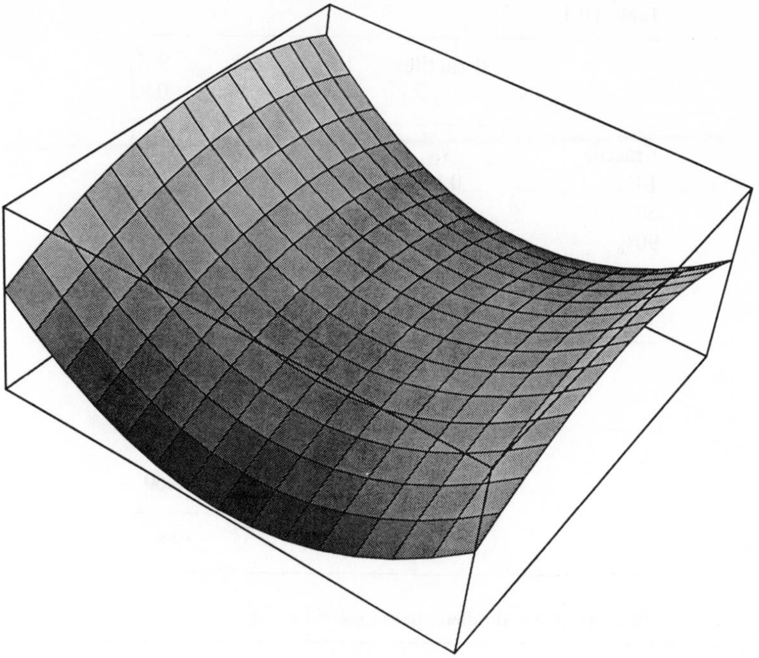

Figure AII.1

of second order

$$d^2v = \left| \frac{\partial^2 v}{\partial s_1^2}(\Delta s_1)^2 + 2\frac{\partial^2 v}{\partial s_1 \partial s_2}\Delta s_1 \Delta s_2 + \frac{\partial^2 v}{\partial s_2^2}(\Delta s_2)^2 \right|.$$

If this quantity is 'small', then our analysis in the tangent plane is appropriate and we should continue as we normally do. If it is not, then the components of this term can be analyzed further to determine the effects of inclusion or rejection. In particular, it is the cross product terms of this quantity which are sensitive to the effects of correlation and dependence.

Tables AII.1–3 use the information from our original sensitivity analysis given in Figure 19.5. They show why the behavior of the tangent plane is not a good approximation for the value function of this example.

Table AII.1 contains the information originally used for our sensitivity analysis. In any automated program the derivatives will have to be estimated by numerical techniques, but for the purposes of this example we cheat and use the exact derivatives. Table AII.2 confirms our original insight that state variable one dominates the behavior of the tangent plane about the medians. Although the total differential of second order is small (Table AII.3), each term is comparable to the terms of the total differential of first order. Tables AII.2 and

Table AII.1

	Fractile	10%	50%	90%
	s_2	0.23	0.54	0.83
Fractile	s_1			
10%	0.56		−7.60	
50%	0.84	−3.3	0.46	−3.3
90%	0.97		8.70	

Table AII.2 Total differential of first order

	Derivative	Range of uncertainty	Contribution
s_1	52.3	0.41	21.4
s_2	−0.92	0.60	−0.60
		Total	19.8

Table AII.3 Total differential of second order

	Derivative	Range of uncertainty	Contribution
s_1	168	$0.17 = 0.42^2$	28.6
$s_1 s_2$	−9	$0.25 = 0.6 \times 0.41$	−2.3
s_2	−84	$0.36 = 0.60^2$	−30.2
		Total	−3.9

3 show that second-order effects are as important as first-order ones. An analysis which neglects these effects will be in error. Finally, the effect of the nonsquare region of support of the distribution only strengthens this conclusion.

If we assume a continuous value function in two state variables and one decision variable, we can approximate the value function by a Taylor series about the medians. Further, we can compute the optimal decision up to and including covariance. If we do so, we obtain:

$$d^* \approx - \left. \frac{\dfrac{\partial}{\partial d} v + \langle s_2 | \mathfrak{I} \rangle \dfrac{\partial^2}{\partial s_2 \partial d} v + \langle s_1 | \mathfrak{I} \rangle \dfrac{\partial^2}{\partial s_1 \partial d} v + \langle s_1 s_2 | \mathfrak{I} \rangle \dfrac{\partial^3}{\partial s_1 \partial s_2 \partial d} v}{\dfrac{\partial^2}{\partial d^2} v + \langle s_2 | \mathfrak{I} \rangle \dfrac{\partial^3}{\partial s_2 \partial d^2} v + \langle s_1 | \mathfrak{I} \rangle \dfrac{\partial^3}{\partial s_1 \partial d^2} v + \langle s_1 s_2 | \mathfrak{I} \rangle \dfrac{\partial^4}{\partial s_1 \partial s_2 \partial d^2} v} \right|_{\substack{\mathrm{Frac}(s_1, 0.5), \\ \mathrm{Frac}(s_2, 0.5), d_0}}$$

It should be clear that second-order terms in both variables and the decision variable are important quantities. Thus, sensitivity analysis should be structured to capture these effects.

Finally, we can estimate the value function beyond the computed values obtained by using approximation techniques. Appendix III starts to explore some of the opportunities available. The decision analysis community should develop these techniques so that the overall procedure can be much more cost effective than is currently possible. This area is usually ignored except when decision analysts are faced with exceedingly large trees. The procedures used are always *ad hoc* and the research opportunities are great in this area as well.

An improvement in sensitivity analysis can lead to better understanding of representation, modeling, assessment, biases and solution techniques. All of these issues will have significant design implications for the influence diagram implementer.

APPENDIX III

Sensitivity Analysis and Value Function Approximation

The purpose of decision analysis is to allow a decision maker to perform a deliberate, reasoned, and logical analysis of his problems. This should lead to insights suggesting the best course of experimentation and action in an uncertain environment. To perform decision analysis without, for lack of a better term, engineering guidance about the relevance, scope and value of the analysis itself opens the way to paralysis by analysis. So we take heed of sage advice:

> A model should be as simple as possible, but no simpler.
>
> Albert Einstein

The decision analysis paradigm finds decisions which maximize the expected value of future outcomes. The algorithms used for this purpose have been around for many years and subject to considerable analysis and improvement. However, most automated tools which are in use today, or even under development, use tree-based algorithms. Typically, an analyst has worked with clients and after obtaining probability distributions, she/he discretizes them following the conditional mean algorithm. The expected value calculations used by today's tools are illustrated in Figure AIII.1.

This procedure is analogous to the midpoint rule for integration. The error of such a method is $O(h)$, where h is the maximum interval over which portions of the distribution are being approximated. Now it is easy to see that by simply using one additional point, and interpolating between values at the endpoints of each subinterval, we create a method which has an accuracy $O(h^2)$ (see Figure AIII.2). Since this cost is trivial as the number of subintervals grows, we must ask ourselves why this is not used. I believe it is because the resulting methods for multivariate distributions do not easily generalize to tensor products of one-dimensional rules. In particular, the shape of the region of support comes into explicit question and the rules must be based on triangulations of the plane and higher order spaces of approximation functions. If we are careful, we can maintain the higher degree of accuracy noted above.

This problem was faced and solved in the development of the finite element method. This has been of great utility in solving many problems of mechanical engineering and other fields. I believe we should borrow the techniques and use them in decision analysis.

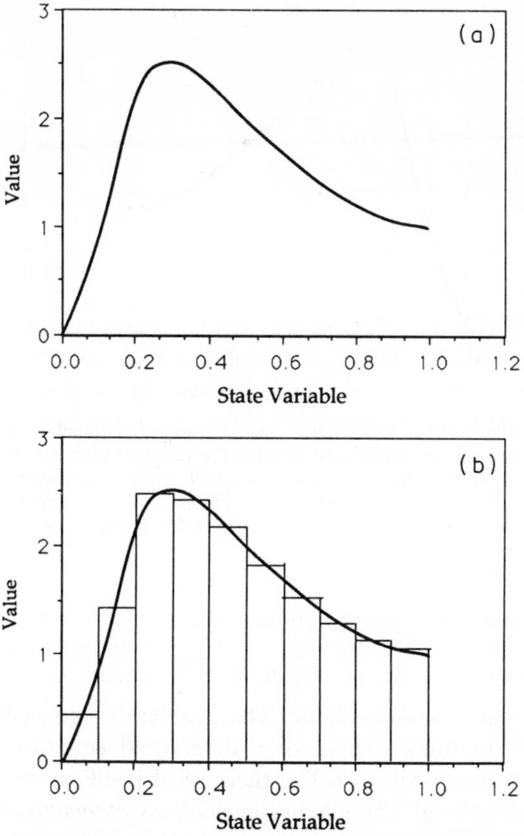

Figure AIII.1 For any given value function, the computation of expected values is the criteria used to discriminate among alternatives. Most methods use discretization to approximate these computations. The conditional mean method of discretization essentially approximates the value function by a constant. The value chosen is the value at the conditional mean in each subinterval. To compute the expected value requires N evaluations of the value function (the number of sub-intervals.)

The finite element method can be described in a few words. Suppose that the problem to be solved is in variational form—it is required to find the function u which maximizes a given expression of some potential. This maximizing property leads to a differential equation for the function u (the Euler equation), but normally an exact solution is impossible. The Rayleigh–Ritz–Galerkin idea is to choose a finite set of trial functions

$$(\phi_1, \phi_2, \ldots, \phi_N)$$

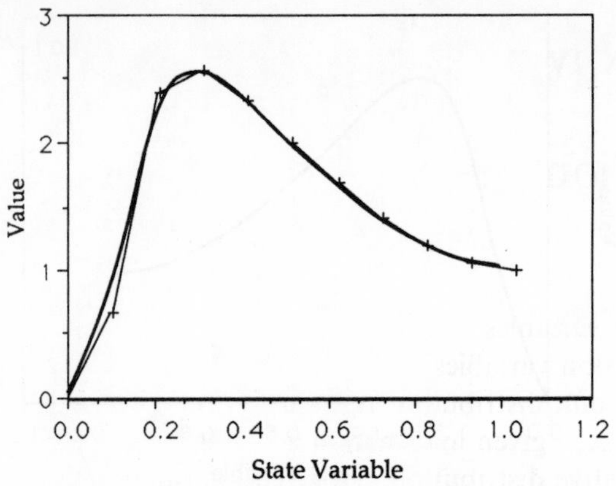

Figure AIII.2 If the value function is approximated by a straight line segment connecting the values at each end of the interval, the order of accuracy of the computed expected value can be increased by one. The added cost is simply one more evaluation of the function for a total of $N+1$ evaluations.

and among all their linear combinations

$$\sum q_j \phi_j$$

to find the one which is maximizing. The problem is reduced to a system of linear algebraic equations instead of a differential equation (in our case a dynamic programming problem). The theoretical justification for this method is simple and compelling! *The maximizing process automatically seeks out the combination which is closest to the actual solution, u.* The real difficulty is not theoretical, but one of convenience and computability. We should use the tools which the pioneers of finite element analysis have developed over the last 30 years and adapt them to our problems. In particular, I believe they will allow us to deal with time-dependent stochastic problems with much greater facility than has been heretofore possible.

In the analysis used in this chapter, I have chosen planar surfaces over triangular regions in the plane for all calculations. These elements provide simple analytic forms and can approximate connected regions of arbitrary complexity with high accuracy. This is extremely useful for problems such as the example (Section 19.2), where the region of support of the joint density is not a tensor product of intervals.

APPENDIX IV

Notation

$s_1, s_2, \ldots$ state variables
$d_1, d_2, \ldots$ decision variables
$\{s_1, s_2, \ldots | \mathfrak{I}\}$ joint distribution of state
variables $s_1, s_2, \ldots$ given information $\mathfrak{I}$
${}^c\{s|\mathfrak{I}\}$ cumulative distribution of s given information $\mathfrak{I}$
$\text{Frac}(s, f) \equiv \text{Image}({}^c\{s|\mathfrak{I}\} = f)$ the f-fractile location of the distribution of s
$\langle s|\mathfrak{I}\rangle$ expected value of s given information $\mathfrak{I}$
$\mathbf{A}, \mathbf{B}, \ldots$ matrices or vectors

CHAPTER 20

Summary observations

Ronald A. Howard, *Stanford University, USA*

This conference represents a significant opportunity to present developments in influence diagrams to a wide audience. The influence diagram was developed in the Decision Analysis Group of SRI International in the mid 1970s to meet the practical needs of professional decision analysts. The difficulty with decision tree representations was that they did not make evident the assertions that were being made about probabilistic independence of the quantities under consideration. The only way to discover these assertions was to examine the numbers actually recorded in the decision tree. The influence diagram was developed as a communication tool to bring these probabilistic dependencies into relief. However, it was soon discovered that it had a logic of its own that permitted manipulations of the diagram to yield additional insight into the problem. The influence diagram then assumed its dual role as a communication device with decision makers and experts on the one hand and as a technical and computational mechanism on the other. The early research was summarized in a report by Howard and Matheson (1983).

Since its beginnings, the use of the influence diagram has grown considerably in the decision analysis program at Stanford. A pioneering thesis by Olmsted (1983) provided the fundamental insights that allowed efficient computation. The consulting companies that grew out of the Stanford program extended the range of application. However, even a few years ago, it was rare to meet someone outside this group who knew of influence diagrams, much less used them.

Some people around the world began to see their value. When Bob Oliver, a long-time colleague, suggested holding a conference on this subject, I and my professional associates saw it as a way to disseminate the ideas of the 'Stanford school' and also to learn about related research that might be useful in our activities. Consequently, in the design of the conference there was no attempt to assure that the papers presented on influence diagrams conformed to the way the Stanford school has defined them. This policy has allowed a wide range

Influence Diagrams, Belief Nets and Decision Analysis
Edited by R. M. Oliver and J. Q. Smith

of papers to appear in the conference; however, it also could lead to confusion about just what an influence diagram is and is not.

As one of the developers of the ideas, I feel comfortable in pointing out that certain of the papers in the conference are written very much from the viewpoint of the Stanford school. They are the paper by Matheson (Chapter 2) and, of course, my own paper (Chapter 1). Certain other papers are consistent with our views; others seem to describe structures that we would not recognize as influence diagrams. Some of my colleagues have even been concerned that my appearance as a sponsor of this conference would lend support to some of the alternate (some would say heretical) views presented at the conference.

However, anyone who has read these comments up to this point should not be misled. Peruse these *Proceedings* not as a unified textbook, but as an open forum of often diverse views on how diagrams can be helpful in inference and decision. Having listened to all the presentations, I have concluded that there is nothing to be gained and much to be lost by departing from our original conception of influence diagrams. While this conception continues to evolve as the papers I have mentioned show, there is nothing in our original paper that has not stood the test of time. I believe that the practical use of influence diagrams will be most advanced by pursuing this original vision rather than any variant I have seen.

I invite you to join with us in spreading the use of a powerful tool for the formulation and evaluation of inference and decision problems.

REFERENCES

Howard, R. A. and Matheson, J. E. (1983) Influence diagrams. In R. A. Howard and J. E. Matheson (eds) *Readings on The Principles and Applications of Decision Analysis*, vol. II, Strategic Decisions Group, Mento Park, Calif., pp. 719–762.

Olmsted, S. M. (1983) On representing and solving decision problems, Ph.D. dissertation, Department of Engineering-Economic Systems, Stanford University, Stanford, Calif.

Glossary

Arc: An element of an influence diagram that joins two nodes and denotes possible statistical dependence between them. An arc may be directed or undirected. A directed arc indicates the direction of conditional dependence. If there is no directed arc from one node to another, then the corresponding quantities are conditionally independent given their immediate predecessors.

Arc reversal: Reversing the direction of a directed arc between two nodes. Arc reversal corresponds to the use of the Bayes' rule.

Arrow: used to indicate the direction of a directed arc.

Assessment: Determining values of distributions of random quantities and model parameters.

Barren node: A node with no successors.

Bayesian statistical method: Statistical procedures based on the notion that a distribution can be assigned to parameters in a statistical decision problem and that the Bayes' rule is used to make inferences.

Calibration: The process whereby the scale of a measuring instrument or model is determined or adjusted on the basis of an informative experiment.

Chance node: A node that represents a random quantity.

Conditional independence: A random quantity X is conditionally independent of Y given Z if $p(X \mid Y,Z) = p(X \mid Z)$. In an influence diagram conditional independence is denoted by the absence of *directed* arcs between nodes.

Conjugate distributions: A family of distributions with the property that, if the prior distribution of parameters belongs to some family, then for any sample size and any values of the observations in the sample, the posterior distribution also belongs to the family.

Decision node: A node that represents a decision variable with possible alternatives.

Deterministic node: A node whose value is determined exactly given the values of predecessor nodes, which may be chance, decision or deterministic nodes.

Direct predecessor: A node **P** is a direct predecessor of node **N** if there is a directed arc from **P** to **N**.

Direct successor: A node **S** is a direct successor of node **N** if there is a directed arc from **N** to **S**.

Event tree: An undirected graph that has branches extending from a root node to indicate sequences of events.

Fault tree: One of two primary Probabilistic Risk Analysis (PRA) tools that has logically organized branches leading to the top of the tree to estimate the probability of a particular undesired event.

Influence diagram: A graph using chance nodes, deterministic nodes, decision nodes, value nodes and directed arcs to represent the conditional independence structure among the random quantities and the timing of information and decisions.

Node: An element of an influence diagram used to represent random quantities, utility functions or decisions. There are four types of nodes: chance nodes, decision nodes, deterministic nodes and value nodes. Nodes may be connected by directed or undirected arcs.

Node removal (absorption): Elimination of a node and its incident arcs in an influence diagram. Elimination of a node corresponds to taking an expectation with respect to that node conditional on its predecessor nodes.

Posterior distribution: Conditional distribution of a random quantity given the observation of the outcome of an informative experiment.

Predictive distribution: Distribution of a (future) random quantity conditional on observed data or the outcome of an informative experiment.

Prior distribution: Assessed distribution of random quantity before observing the outcomes of informative experiments.

Value node: A deterministic node that represents a decision maker's preference or utility as a function of its predecessor nodes.

Index

HALD • Statistical Tables and Formulas
HALD • Statistical Theory with Engineering Applications
HAND • Discrimination and Classification
HEIBERGER • Computation for the Analysis of Designed Experiments
HOAGLIN, MOSTELLER and TUKEY • Exploring Data Tables, Trends and Shapes
HOAGLIN, MOSTELLER, and TUKEY • Understanding Robust and Exploratory Data Analysis
HOCHBERG and TAMHANE • Multiple Comparison Procedures
HOEL • Elementary Statistics, *Fourth Edition*
HOEL and JESSEN • Basic Statistics for Business and Economics, *Third Edition*
HOGG and KLUGMAN • Loss Distributions
HOLLANDER and WOLFE • Nonparametric Statistical Methods
HOSMER and LEMESHOW • Applied Logistic Regression
IMAN and CONOVER • Modern Business Statistics
JESSEN • Statistical Survey Techniques
JOHNSON • Multivariate Statistical Simulation
JOHNSON and KOTZ • Distributions in Statistics
Discrete Distributions
Continuous Univariate Distributions—1
Continuous Univariate Distributions—2
Continuous Multivariate Distributions
JUDGE, HILL, GRIFFITHS, LÜTKEPOHL and LEE • Introduction to the Theory and Practice of Econometrics, *Second Edition*
JUDGE, GRIFFITHS, HILL, LÜTKEPOHL and LEE • The Theory and Practice of Econometrics, *Second Edition*
KALBFLEISCH and PRENTICE • The Statistical Analysis of Failure Time Data
KAUFMAN and ROUSSEEUW • Finding Groups in Data: An Introduction to Cluster Analysis
KISH • Statistical Design for Research
KISH • Survey Sampling
KUH, NEESE, and HOLLINGER • Structural Sensitivity in Econometric Models
KEENEY and RAIFFA • Decisions with Multiple Objectives
LAWLESS • Statistical Models and Methods for Lifetime Data
LEAMER • Specification Searches: Ad Hoc Inference with Nonexperimental Data
LEBART, MORINEAU, and WARWICK • Multivariate Descriptive Statistical Analysis: Correspondence Analysis and Related Techniques for Large Matrices
LINHART and ZUCCHINI • Model Selection
LITTLE and RUBIN • Statistical Analysis with Missing Data
McNEIL • Interactive Data Analysis
MAGNUS and NEUDECKER • Matrix Differential Calculus with Applications in Statistics and Econometrics
MAINDONALD • Statistical Computation
MALLOWS • Design, Data, and Analysis by Some Friends of Cuthbert Daniel
MANN, SCHAFER and SINGPURWALLA • Methods for Statistical Analysis of Reliability and Life Data
MARTZ and WALLER • Bayesian Reliability Analysis
MASON, GUNST, and HESS • Statistical Design and Analysis of Experiments with Applications to Engineering and Science
MIKÉ and STANLEY • Statistics in Medical Research: Methods and Issues with Applications in Cancer Research
MILLER • Beyond ANOVA, Basics of Applied Statistics
MILLER • Survival Analysis
MILLER, EFRON, BROWN, and MOSES • Biostatistics Casebook
MONTGOMERY and PECK • Introduction to Linear Regression Analysis
NELSON • Applied Life Data Analysis
OLIVER and SMITH • Influence Diagrams, Belief Nets and Decision Analysis